GAMMA-RAY BURSTS

GAMMA-RAY BURSTS

4th Huntsville Symposium

Huntsville, AL September 1997

PART TWO

EDITORS
Charles A. Meegan
NASA/Marshall Space Flight Center

Robert D. Preece
University of Alabama, Huntsville

Thomas M. Koshut
USRA/Marshall Space Flight Center

American Institute of Physics

AIP CONFERENCE
PROCEEDINGS 428

Woodbury, New York

Editor:

Charles A. Meegan
NASA/Marshall Space Flight Center
ES84
Huntsville, AL 35812

Email: Charles.Meegan@msfc.nasa.gov

Authorization to photocopy items for internal or personal use, beyond the free copying permitted under the 1978 U.S. Copyright Law (see statement below), is granted by the American Institute of Physics for users registered with the Copyright Clearance Center (CCC) Transactional Reporting Service, provided that the base fee of $15.00 per copy is paid directly to CCC, 222 Rosewood Drive, Danvers, MA 01923. For those organizations that have been granted a photocopy license by CCC, a separate system of payment has been arranged. The fee code for users of the Transactional Reporting Service is: 1-56396-766-9/ 98 /$15.00.

© 1998 American Institute of Physics

Individual readers of this volume and nonprofit libraries, acting for them, are permitted to make fair use of the material in it, such as copying an article for use in teaching or research. Permission is granted to quote from this volume in scientific work with the customary acknowledgment of the source. To reprint a figure, table, or other excerpt requires the consent of one of the original authors and notification to AIP. Republication or systematic or multiple reproduction of any material in this volume is permitted only under license from AIP. Address inquiries to Office of Rights and Permissions, 500 Sunnyside Boulevard, Woodbury, NY 11797-2999; phone: 516-576-2268; fax: 516-576-2499; e-mail: rights@aip.org.

L.C. Catalog Card No. 97-70902
ISBN 1-56396-766-9 (Set)
ISBN 1-56396-812-6 (Part One)
ISBN 1-56396-813-4 (Part Two)
ISSN 0094-243X
DOE CONF- 9709153

Printed in the United States of America

CONTENTS

PART TWO

OPTICAL OBSERVATIONS

Optical Transients as a Class: Reality and Background 473
    R. Hudec
The Decay of Optical Emission from the γ-Ray Burst GRB970228 478
    T. J. Galama, P. J. Groot, J. van Paradijs, and C. Kouveliotou
**The Gamma-Ray Bursts GRB970228 and GRB970508:
What Have We Learnt?** 483
    M. Livio, K. C. Sahu, L. Petro, A. S. Fruchter, E. Pian, F. D. Macchetto,
    J. van Paradijs, C. Kouveliotou, P. J. Groot, and T. J. Galama
Optical/IR Follow-up Observations of GRBs Detected by BeppoSAX 489
    A. J. Castro-Tirado, J. Gorosabel, E. Costa, M. Feroci, L. Piro,
    F. Frontera, D. Dal Fiume, L. Nicastro, E. Palazzi, J. Greiner, K. Birkle,
    R. Fockenbrock, E. Thommes, C. Wolf, C. Bartolini, A. Guarnieri,
    N. Masetti, A. Piccioni, M. Mignoli, J. Heidt, T. Seitz, H. Pedersen,
    S. Guziy, A. Shlyapnikov, L. Metcalfe, R. Laureijs, B. Altieri, M. Kessler,
    L. Hanlon, B. McBreen, N. Smith, J. Studt, N. Benítez, E. Martínez-González,
    H. Kristen, A. Broeils, M. Wold, M. Lacy, and M. V. Alonso
Is the Fuzziness of GRB970228 Constant? 494
    P. A. Caraveo, R. Mignani, and G. F. Bignami
Optical Follow-up of GRB970508 499
    P. J. Groot, T. J. Galama, J. van Paradijs, C. Kouveliotou,
    R. A. M. J. Wijers, M. Centurion, P. Smith, C. Mackay, S. Smartt,
    and C. Benn
Hubble Space Telescope Imaging of the Field of GRB970508 504
    E. Pian, A. S. Fruchter, L. E. Bergeron, S. E. Thorsett, F. Frontera,
    M. Tavani, E. Costa, M. Feroci, J. Halpern, R. A. Lucas, L. Nicastro,
    E. Palazzi, L. Piro, W. Sparks, A. J. Castro-Tirado, T. Gull, K. C. Hurley,
    and H. Pedersen
HST/STIS Observations of the Optical Counterpart to GRB970228 509
    A. S. Fruchter, E. Pian, S. E. Thorsett, R. Gonzalez, K. C. Sahu,
    M. Mutchler, F. Frontera, T. J. Galama, P. J. Groot, R. Hook,
    C. Kouveliotou, M. Livio, F. D. Macchetto, J. van Paradijs, E. Palazzi,
    L. Petro, and M. Tavani
Konus-Wind and Konus-A Observations of GRB970228 516
    R. L. Aptekar, P. S. Butterworth, T. L. Cline, D. D. Frederiks,
    S. V. Golenetskii, V. N. Il'inskii, E. P. Mazets, D. E. Stilwell,
    and M. M. Terekhov
**The Galactic Extinction Toward the GRB970228 Field
and Its Implications** 520
    F. J. Castander and D. Q. Lamb

Multicolor Photometry of the GRB970508 Optical Remnant 525
    V. V. Sokolov, A. I. Kopylov, S. V. Zharikov, M. Feroci, E. Costa,
    L. Nicastro, and E. Palazzi

On the Light-Curve of OT/GRB970508 530
    H. Pedersen, A. O. Jaunsen, R. Østensen, T. Grav, M. Wold,
    M. Lacy, H. Kristen, A. Broeils, A. J. Castro-Tirado, J. Gorosabel,
    J. M. Rodriguez Espinosa, A. M. Perez, M. Näslund, C. Fransson,
    M. I. Andersen, C. Wolf, R. Fockenbrock, L. Piro, M. Feroci,
    E. Costa, L. Nicastro, E. Palazzi, F. Frontera, L. Monaldi, J. Heise,
    and J. Hjorth

The Redshift of GRB970508 535
    D. E. Reichart

Did GRB970228 and GRB970508 Present Similar Optical Properties? 540
    C. Bartolini, G. M. Beskin, A. Guarnieri, N. Masetti, and A. Piccioni

Source Confusion in Optical Searches for GRB Counterparts 545
    J. J. Brainerd

Optical Gamma-Ray Burst Followup at Kitt Peak National Observatory 548
    J. E. Rhoads, T. von Hippel, B. Carney, C. F. Claver, A. D. Code,
    A. Cole, C. Conselice, A. Dey, C. Howk, G. Jacoby, B. Jannuzi,
    J. Lee, J. A. Orosz, D. J. Pisano, D. Sawyer, N. Sharp, and P. Smith

**Archival Searches for Optical Emission from GRB910122
and GRB970228** 552
    J. Gorosabel and A. J. Castro-Tirado

A Search for Optical Afterglow from GRB970828 557
    P. J. Groot, T. J. Galama, J. van Paradijs, C. Kouveliotou,
    R. K. Vanderspek, A. Castro-Tirado, J. E. Rhoads, M. Lehnert,
    H. Hoekstra, and N. Metcalfe

RADIO OBSERVATIONS

Radio Counterparts of Gamma-Ray Bursts 563
    D. A. Frail

VLBA Observations of the Radio Counterparts to γ-Ray Bursters 571
    G. B. Taylor, A. J. Beasley, D. A. Frail, and S. R. Kulkarni

**The BIMA Search for Millimeter-Wavelength Counterparts
to Gamma-Ray Bursts** 576
    R. A. Gruendl, I. A. Smith, R. Forester, K. Y. Lo, E. P. Liang,
    and M. Leventhal

The STARE Project: A Progress Report 581
    C. A. Katz, J. N. Hewitt, C. B. Moore, and B. E. Corey

FLIRT Update: Rapid Radio Observations of GRBs 585
    R. J. Balsano, S. E. Thorsett, W. A. Coles, B. J. Rickett, P. S. Ray,
    S. D. Barthelmy, P. S. Butterworth, T. L. Cline, and N. Gehrels

**SCUBA Sub-millimeter Observations of Gamma-Ray
Burst Counterparts** 590
    I. A. Smith, J. van Paradijs, T. J. Galama, P. J. Groot,
    L. B. F. M. Waters, and C. Kouveliotou

HOST GALAXIES

Severe Constraints on Possible Cosmological Models Caused by Limits on Host Galaxies. 595
    B. E. Schaefer

Results of an Extragalactic Survey of Gamma-Ray Burst Localizations. 600
    S. B. Larson and I. S. McLean

A Statistical Treatment of the Gamma-Ray Burst "No Host Object" Issue 605
    D. L. Band and D. H. Hartmann

Study of the Possible Connection between Gamma-Ray Bursts and Active Galactic Nuclei 610
    J. Gorosabel and A. J. Castro-Tirado

Properties of GRB Host Galaxies. 615
    D. H. Hartmann and D. L. Band

The Complex No-Host Problem: Real or Imaginary? 620
    S. B. Larson

The USNO Deep Optical Survey of Small IPN^3 Localizations: Where are the Hosts? 625
    F. J. Vrba, C. B. Luginbuhl, M. C. Jennings, D. H. Hartmann, K. C. Hurley, C. Kouveliotou, C. A. Meegan, G. J. Fishman, T. L. Cline, and M. Boër

Deep Imaging of the IPN^1 Localization of GRB790325 (GBS 1810+31) 630
    F. J. Vrba, C. B. Luginbuhl, D. H. Hartmann, and M. C. Jennings

NIR Imaging of Gamma-Ray Burst Error Boxes 635
    S. Klose, B. Stecklum, J. Eislöffel, J. L. Hora, and R. Tuffs

On the Possible GRB/QSO and GRB/Abell Cluster Correlations. 640
    K. C. Hurley, D. H. Hartmann, C. Kouveliotou, C. A. Meegan, G. J. Fishman, and T. L. Cline

SHOCKS

Theoretical Models of Gamma-Ray Bursts. 647
    P. Mészáros

Kinematic Arguments Against Single Relativistic Shell Models for GRBs 657
    E. E. Fenimore, E. Ramirez, and M. C. Sumner

The Internal-External GRB-Afterglow Model 662
    T. Piran and R. Sari

Internal Shock Models for Gamma-Ray Bursts: Temporal and Spectral Properties. 667
    R. Mochkovitch and F. Daigne

Can Internal Shocks Produce the Variability in GRBs? 672
    S. Kobayashi, T. Piran, and R. Sari

Hydrodynamical Study of Internal Shocks in a Relativistic Wind 677
    F. Daigne and R. Mochkovitch

Relativistic Shocks and Gamma-Ray Burst Afterglow Lightcurves 682
    T. Bulik and M. Sikora

AFTERGLOWS

What Have We Learned From GRB Afterglows? 689
    J. I. Katz and T. Piran
Afterglows as Diagnostics of Gamma-Ray Burst Beaming 699
    J. E. Rhoads
Measuring the GRB Parameters from Afterglow Observations 704
    R. Sari and T. Piran
Theory of GRB Afterglows 709
    M. Tavani
Limit on the Distance to Gamma-Ray Bursts from Parallax 714
    T. Bulik and B. Czerny
Afterglow Hydrodynamics 720
    R. Sari

THEORY

Radio Constraints on Shell Afterglow Theories 727
    J. J. Brainerd
Spectral Properties of Gamma-Ray Bursts in the 1 GeV—1 TeV Range 732
    M. G. Baring and A. K. Harding
Spectral Variability in Relativistic MHD Winds 737
    C. Thompson
Cosmology with GRBs 742
    R. J. Nemiroff, G. F. Marani, J. P. Norris, and J. T. Bonnell
The Implications of Direct Red-Shift Measurement of γ-Ray Bursts 747
    E. Cohen and T. Piran
Cooling Synchrotron Spectra and GRB Theory 752
    J. J. Brainerd
Emission and Cooling Processes in a Hybrid Thermal-Nonthermal Plasma 756
    D. Lin and E. P. Liang
Constraining the Intergalactic Magnetic Field with Cascading
TeV Emission from Cosmological GRBs 760
    B. Roscherr and P. S. Coppi
Variability in Shell Models of GRBs 765
    M. C. Sumner and E. E. Fenimore
Radiative Efficiency in Gamma-Ray Bursts and Afterglows 771
    A. Panaitescu and P. Mészáros
Cosmic Rays and Neutrinos from Gamma-Ray Bursts 776
    J. P. Rachen and P. Mészáros

MODELS

Gamma-Ray Bursts as Hypernovae 783
    B. Paczyński
A Binary Neutron Star GRB Model 788
    J. R. Wilson, J. D. Salmonson, and G. J. Mathews
Coalescing Neutron Stars as Possible Gamma-Ray Burst Sources 793
    M. Ruffert and H.-T. Janka
Black Hole-Neutron Star Coalescence as a Source of Gamma-Ray Bursts 798
    W. H. Lee and W. Kluźniak
Gamma-Ray Bursts Near the Horizon 803
    R. A. M. J. Wijers
Gamma-Ray Bursts from Closed and Filled Neutron Star Magnetospheres 808
    M. Böttcher, B. Eastlund, and B. Miller
Saturated Compton Cooling Model of Cosmological Gamma-Ray Bursts 813
    E. P. Liang
Single Black Holes as Parent Bodies of Galactic and Extragalactic Gamma-Ray Bursts 815
    G. M. Beskin, A. Shearer, M. Redfern, A. Golden, R. C. Butler,
    C. Bartolini, A. Guarnieri, N. Masetti, and A. Piccioni
The Galactic Model of GRBs 820
    S. A. Colgate and H. Li
Gamma-Ray Bursts from Electrical Discharges 825
    T. Li
The Physical Source of Gamma-Ray Bursts 830
    W. Kluźniak and M. Ruderman

INSTRUMENTATION AND TECHNIQUES

First Year Results from LOTIS 837
    G. G. Williams, H. S. Park, E. Ables, D. L. Band, S. D. Barthelmy,
    R. M. Bionta, P. S. Butterworth, T. L. Cline, D. H. Ferguson,
    G. J. Fishman, N. Gehrels, D. H. Hartmann, K. C. Hurley, C. Kouveliotou,
    C. A. Meegan, L. Ott, E. Parker, and R. Wurtz
Super-LOTIS: A High-Sensitive Optical Counterpart Search Experiment 842
    H. S. Park, E. Ables, D. L. Band, S. D. Barthelmy, R. M. Bionta,
    P. S. Butterworth, T. L. Cline, D. H. Ferguson, G. J. Fishman, N. Gehrels,
    D. H. Hartmann, K. C. Hurley, C. Kouveliotou, C. A. Meegan, L. Ott,
    E. Parker, and G. G. Williams
TAROT: A Status Report 846
    M. Boër, J.-L. Atteia, M. Bringer, A. Klotz, C. Peignot,
    R. Malina, P. Sanchez, H. Pedersen, G. Calvet, J. Eysseric,
    A. Leroy, M. Meissonier, C. Pollas, and J. de Freitas Pacheco

EN: Real-Time Optical Data for GRBs.. 851
    R. Hudec, Z. Ceplecha, P. Spurný, J. Florián, A. Kovář, J. Boček,
    and J. Borovička

BART: Burst Alert Robotic Telescope 855
    J. Soldán, R. Hudec, M. Němček, and T. Rezek

OMC Camera Experiment for INTEGRAL and Search
for Compton GRO BATSE LOCBURST Optical Transients 859
    T. Rezek, R. Hudec, F. Hroch, J. Soldán, M. Mas-Hesse,
    and A. Giménez

Optical Transient Monitor 864
    M. Bernas, P. Páta, R. Hudec, J. Soldán, T. Rezek,
    and A. J. Castro-Tirado

Optical Imaging of Gamma-Ray Bursts with the LONEOS Telescope 869
    R. M. Wagner, E. Bowell, K. H. Cook, S. B. Howell, B. W. Koehn,
    C. R. Shrader, S. G. Starrfield, and C. W. Stubbs

The Status of the Burst Observer and Optical Transient
Exploring System (BOOTES).. 874
    A. J. Castro-Tirado, J. Gorosabel, R. Hudec, J. Soldán, M. Bernas,
    P. Páta, and T. Rezek

Observing GRBs with Integral... 879
    C. Winkler

Simulated Observations of GRBs with Glast............................... 884
    J. T. Bonnell, J. P. Norris, B. L. Dingus, and J. D. Scargle

Development of a Hard X-Ray Polarimeter for Gamma-Ray Bursts.......... 889
    M. L. McConnell, D. J. Forrest, J. Macri, J. M. Ryan,
    and W. T. Vestrand

The Konus-Wind and Konus-A Instrument Response Functions
and the Spectral Deconvolution Procedure............................... 894
    M. M. Terekhov, R. L. Aptekar, D. D. Frederiks, S. V. Golenetskii,
    V. N. Il'inskii, and E. P. Mazets

The BATSE Trigger Efficiency as a Function of Intensity
and Energy Range 899
    G. N. Pendleton, J. Hakkila, and C. A. Meegan

Wide Field X-Ray Telescopes: Detecting X-Ray Transients/Afterglows
Related to GRBs... 904
    R. Hudec, L. Pina, A. Inneman, and P. Gorenstein

A Search for Gamma-Ray Burst Optical Emission
with the Automated Patrol Telescope 909
    B. Grossan, S. Perlmutter, and M. Ashley

Search for GeV GRBs at Chacaltaya 914
    A. Castellina, P. L. Ghia, C. Morello, G. Trinchero, P. Vallania,
    S. Vernetto, G. Navarra, O. Saavedra, H. Yoshii, T. Kaneko,
    K. Kakimoto, K. Nishi, R. Cabrera, D. Urzagasti, A. Velarde,
    S. D. Barthelmy, P. S. Butterworth, T. L. Cline, N. Gehrels, G. J. Fishman,
    C. Kouveliotou, and C. A. Meegan

SOFT GAMMA-RAY REPEATERS

Observations of SGR1806-20 with the Konus-Wind and Konus-A Experiments in 1996–97 .... **921**
    D. D. Frederiks, R. L. Aptekar, S. V. Golenetskii, V. N. Il'inskii,
    E. P. Mazets, and M. M. Terekhov

HEXTE Observations of SGR 1806-20 During Outburst **926**
    D. Marsden, R. E. Rothschild, C. Kouveliotou, S. Dieters,
    and J. van Paradijs

Mid-Infrared Spectra of SGR 1806-20 and SGR 1900+14 .... **931**
    W. A. Mahoney, S. Corbel, Ph. Durouchoux, J. C. Higdon,
    M. E. Ressler, and P. Wallyn

Infrared Observations of Soft Gamma-Ray Repeaters **936**
    I. A. Smith

Lognormal Properties of SGR1806-20 and the Possibility of a Quiescent Population of Other SGR Sources **939**
    B. McBreen and K. J. Hurley

Testing Models of the Soft Gamma Repeaters **944**
    C. Thompson

X-Ray Spectroscopy of Bursts from SGR1806-20 with RXTE **947**
    T. E. Strohmayer and A. Ibrahim

Symposium Participants **953**
Author Index **961**

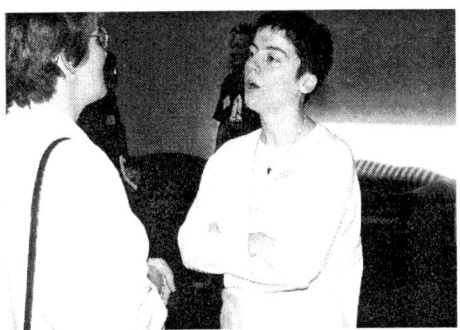

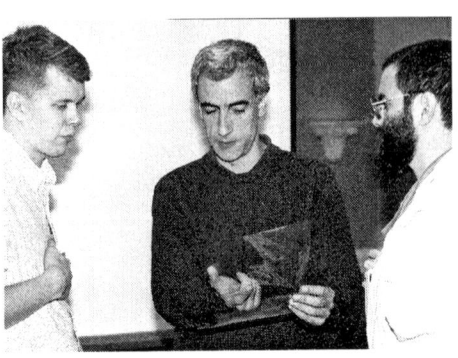

OPTICAL OBSERVATIONS

Optical Transients as a Class: Reality and Background

René Hudec

Astronomical Institute of Czech Academy of Science, Observatory Ondřejov, 251 65 Ondřejov, Czech Republic

Abstract. The recent detection of Optical Transients (OTs) related to GRBs arises question about their physical nature which seems to be not clearly understood. I review and discuss OTs from a more general point of view as a class including summary of previously and recently detected OTs from positions inside or close to GRBs error boxes and discussion whether or not they can be related to GRB phenomena. I show that, especially at faint magnitudes, the background of unrelated triggers such as variable stars and variable extragalactic objects may be rather high.

INTRODUCTION

The recent GRB discoveries related to precise BeppoSAX localizations yield exciting new results. On the other hand, there is still skepticism: for GRB970616, two X-ray fading counterparts and two optically fading objects have been found, one inside the gamma-ray box but both outside the X-ray boxes [1]. For GRB970402, an optically variable/fading source has been found outside the X-ray error box [3]. For GRB970508, several optical variable sources are present in the GRB box, only one in the X-ray box [2]. The obvious conclusion is that not all optical and probably also not all X-ray variable/fading objects found at or close to the GRB positions are related to the GRBs in question.

There are obvious questions arising: **(1)** Can we be absolutely sure that the optically variable and/or fading sources inside X-ray or GRB error boxes are indeed related to GRBs? **(2)** How to distinguish real OTs related to GRBs and random coincidences? **(3)** How large is the surface density of faint variable optical (and X-ray) sources? **(4)** What categories exist of OTs and what is their nature? **(5)** How large is the OT rate and the related background level? **(6)** Are the newly discovered OTs analogous to previously detected OTs?

OPTICAL TRANSIENTS AS A CLASS?

Although the term Optical Transients (OTs) is now widely used for the recently detected optical afterglows of GRBs, there are different categories of OTs both of astrophysical as well as of false (background) origin. Every survey project and/or attempt to identify GRBs with OTs must take this into account. Further, it should be noted that originally, the term OT has been used for short–lived optical flaring phenomena [4], while the recently studied optical afterglows decline over a time scale of order of months.

1. OTs of true astrophysical origin

    (a) True GRB counterparts (b) Variable stars (c) Variable Extragalactic Objects (AGNs) (d) Supernovae (e) Others

2. OTs of Non–Astrophysical Origin

    (a) Detection Faults
        (i) Emulsion Defects (ii) CCD Defects (iii) Others
    (b) Man–made Artefacts
        (i) Aircraft Flashes (ii) Satellite Glints (iii) Others
    (c) Natural Artefacts
        (i) Head–On Meteors (ii) Others

OTS DETECTED PRIOR SAX

Archival

Analyses of archival astronomical plates still play an important role in detections of OTs mainly due to the enormous fractions of monitoring times available. Recently, deep searches for quiet counterparts at the positions of OTs detected on archival plates represent a valuable tool for their verification [5].

AGN candidates

Three pairs of OTs-AGNs(-and perhaps GRBs) have been identified [5,6]:

- OT050510/GRB910219 7 arcmin outside the IPN#3 error box
- OT1946,1954/GRB790325b 3 arcmin outside the IPN#1 error box
- OT630724/GRB920406 2 arcmin outside the IPN#3 error box

The flare amplitude was ≥ 10 mag in all three cases (the largest QSO flare amplitude known before was more than 6.7 mag for 3C279 [7]). The relation of OTs to AGNs seems to be unambigous (the random probability is very low since the OTs error boxes are very small). However, the relation of OT/AGNs to GRBs still remains questionable [5,6].

Flare star candidates

Two pairs of OTs-flare stars have been detected in GRB error boxes [8,6] as large amplitude flares (5-9 mag) from otherwise typical dMe flare stars. The OT660127(OT660814) has been found inside the error box of GRB910522(GRB781006B) with area of 20(3200) square arcmin. The quiet star is of mag B=20.5 (18.5) and the peak flare magnitude reached 11.6 mag (13.5). Only one flare has been found in 700 (1400) monitoring hours. Although the stellar flares are probably not related to GRBs, they contribute essentially to the OT background of astrophysical origin. The rate of analogous stellar flares is however unknown so nothing can be said about the statistics.

Are Archival OTs Related to GRBs?

This is not yet clear due to some controversies listed below. CONTRA: (1) OT/AGN positions lie close but outside GRB IPN#3 error boxes. (2) OT/flare stars are inside large GRB error boxes; hence, the positional confidence is weak. PRO: (1) The IPN error boxes may be slightly in error due to instrumental effects such as spacecraft timing changes (see the recent displacement of IPN#3 boxes). (2) Analogous flaring behaviour. (3) AGNs represent the most common class of recent GRBs candidates, being found at positions of GRB920508, GRB960720 and probably also GRB970508 [10–12]. (4) OTs may be related to others (unknown) GRBs since their surface density seems to be large (BATSE: soon 2000 bursts with still little hints for repetitions) (5) Recurrences of GRBs are now possible (possible recurrent BATSE triggers GRB961027a,27b,29a,29b).

Follow–Up/Temporal Coincidences

The only known optical candidate prior to SAX to a particular GRB both in time and position is the OT Borovička (OT790930) detected 1979 Sep29/30 20:30-00:49 UT i.e. 7.1-9.6 hrs after the GRB790929 inside its error box (150,000 arcmin^2) [9]. This OT is either related to a flare on HDE 249119 (F8 V=10 mag star) with amplitude of ∼1.5 mag and duration of ∼hours or, alternatively, to a large flare of an underlying faint object.

Random Discoveries

There are real OTs detected both on emulsions and CCDs. Examples (real CCD detections): **(1) OT 970215**. Real CCD detection: guiding error (trailing), V 13 mag, not on any of other 800 CCD observations, nothing down to 20 mag on the position, amplitude more than 7 mag, [13]. **(2) OT 950806** Real object: detected on 20 CCD frames, peak magnitude I 7.5, amplitude more than 10 mag, nothing down to mag 21 48 hrs after detection [14].

THE BACKGROUND

Variable Stars Surface Density

The estimated rates for variable stars brighter than 20 mag are ~ 80 deg^{-2} for |bII| less than 20 deg, ~ 4 deg^{-2} for bII more than 40 deg [15], but the discovery probability is ~ 0.1 (blinkmicroscope use). No statistics for variable stars below 20 mag are available (no systematic surveys). Variable stars are observed more commonly in decline than in increase since the declines are typically slower, such as delta Cep stars, U Gem stars, flare stars, novae etc. (W. Wenzel, priv. comm.).

AGNs/QSOs Surface Densities

The expected surface rate of faint AGNs/QSOs is rather high: ~ 10 deg^{-2} with limiting mag B 20.5 has been found by Iovino et al. [16], ~ 30 deg^{-2} (limiting mag V 20.5) by Hartwick and Scade [17], and 111 deg^{-2} (limiting mag B 22.6) by Trevese et al. [19]. A significant fraction of AGNs/QSOs is variable. The typical QSO variability in a sample of 149 optically selected QSOs has been found to be 0.26-0.33 in B and 0.22-0.30 in R [18]. The study by Trevese et al. counts ~ 100 variable QSOs deg^{-2} (by more than 0.1 mag); i.e., 1 variable QSO in 6x6 arcmin^2 brighter than B=22.6 [19]. Ninety-seven percent of QSOs below B=22.5 have been found to be variable [20]. The QSO variability seems to increase with decreasing luminosity [18]. Smith [21] lists typical amplitudes of AGNs (QSOs, BL Lacs, Radio-Loud, Galaxies) between 0.5 and 1.5 (base level) and between 0.5 and 1.3 (flares).

For AGNs/blazars, a 2-day delay has been observed for the X-ray flare of blazar PKS2155-304 [22] with a very steep flare seen in X-rays followed by a broader, lower amplitude UV flare 2 days later. The amplitude of the flare decreased and the duration increased with increasing wavelength. The gamma/X-ray and gamma-ray/optical correlations for blazar flares are still unknown for blazars as a class, due primarily to the small number of blazar outbursts detected to date with EGRET.

SUMMARY

Even the recent high localization accuracy of GRBs cannot exclude positionally coinciding OTs as just random unrelated coincidences. The exact background rate of unrelated triggers such as variable stars and extragalactic objects is unknown but may be rather high. The expected rate of faint (limiting mag B=22.6) variable AGNs is 1 in a 5 x 5 arcmin box. The statistics of large-amplitude (≥ 0.5 mag) variable AGNs is unknown. The rate of faint variable stars is unknown. For brighter variable stars, the expected rate is 0.03 - 0.5 (depending on galactic latitude) in a 5 x 5 arcmin box (limiting mag 20). It is obvious that we need better statistics of variable faint optical objects. The future survey optical projects are important especially at fainter magnitudes.

ACKNOWLEDGEMENTS

The investigations of GRBs are supported by The Ministery for Education and Youth of the Czech Republic, Projects No. ES002/1996 (CEI/KONTAKT) and ES036/1996 (ESA/KONTAKT).

REFERENCES

1. Greiner, J. et al., *IAU Circ.* No. 6722 (1997).
2. Djorgovski, S. G. et al., *Nature* **387**, 876 (1997).
3. Pedersen, H. et al., *IAU Circ.* No. 6628 (1997).
4. Schaefer, B. E., *ApJ* **286**, L1 (1984).
5. Hudec, R. et al., in *Proceedings of the 2nd INTEGRAL Workshop*, ESA SP-382, 1997, pp. 481-484.
6. Hudec, R. et al., in *Blazar Continuum Variability*, ASP Conf. ser. Vol 110, 1996, pp.129-134
7. Eachus, L. J. and Liller, W., *ApJ* **200**, L61 (1975).
8. Greiner, J. and Motch, C., *A&A* **294**, 177 (1995).
9. Borovička J. et al., *A&A* **258**, 379 (1992).
10. Drinkwater, M. J. et al., *IAU Circ.* No. 6600 (1997).
11. Piro, L. et al., *IAU Circ* No. 6570 (1997).
12. Metzger, M. R. et al., *IAUCirc.* No. 6655 (1997).
13. Vidal-Saiz, J. et al., *IBVS Budapest* 4324, (1996)
14. Toth, I. et al., *A&A* **315**, 153 (1996).
15. Hudec, R. and Wenzel, W., *A&ASS* **120**, C707-C710 (1996).
16. Iovino, A. et al., *A&ASS* **119**, 265-269 (1996).
17. Hartwick, F. D. A. and Scade, D., *ARA&A* **28**, 437 (1990).
18. Cristiani, S. et al., *A&A* **321**, 123-128 (1997).
19. Trevese, D. et al., *AJ* **98**, 108 (1989).
20. Trevese, D. and Kron, R. G., in *Multi-Wavelength Continuum Emmission of AGNs* (Eds. T. Courvoisier and A. Blecha), 412.
21. Smith, A. G., in *Blazar Continum Variability*, ASP Conf Ser 110, 1996, pp. 3-16.
22. Urry, C. M., in *Blazar Continum Variability*, ASP Conf Ser 110, 1996, pp. 391-401.

The Decay of Optical Emission from the γ-Ray Burst GRB970228

Titus J. Galama*, Paul J. Groot*, Jan van Paradijs*† and Chryssa Kouveliotou‡∥

* *Astronomical Institute "Anton Pannekoek", University of Amsterdam, & Center for High Energy Astrophysics, Kruislaan 403, 1098 SJ Amsterdam, The Netherlands.*
† *Physics Department, University of Alabama in Huntsville, Huntsville, AL 35899, USA*
‡ *Universities Space Research Association*
∥ *NASA Marshall Space Flight Center, ES-84, Huntsville, AL 35812, USA*

Abstract. We present the R_c band light curve of the optical transient (OT) associated with GRB970228, based on re-evaluation of existing photometry. Data obtained until April 1997 suggested a slowing down of the decay of the optical brightness. However, the HST observations in September 1997 show that the light curve of the point source is well represented by a single power law, with a "dip", about a week after the burst occured. The exponent of the power law decay is $\alpha = -1.10 \pm 0.04$. As the point source weakened it also became redder.

INTRODUCTION

The γ-ray burst of February 28, 1997, detected [1] with the Gamma-Ray Burst Monitor on the BeppoSAX observatory, and located with an $\sim 3'$ radius position with the Wide Field Camera on the same satellite, was the first for which a fading X-ray [1] and optical counterpart [2,3] were found.

The optical counterpart was discovered from a comparison of V and I_c band images taken with the William Herschel Telescope (WHT) on February 28.99 UT, and the Isaac Newton Telescope (INT; V band) and the WHT (I_c band) on March 8.86 UT. After the counterpart had weakened by several magnitudes, it was found to coincide with an extended object [3–5]. In subsequent observations with the Hubble Space Telescope (HST) on March 26 and April 7, 1997, it was found that the optical counterpart consists of a point source and an extended ($\sim 1''$) object, offset from the point source by $\sim 0.5''$ [6].

We here reassess the photometric information presented by Galama et al. [7] in the light of the recent HST findings [8,9], and present the R_c band optical light curve of the GRB counterpart.

OBSERVATIONS

In Table 1 we have collected the optical photometry reported on GRB970228, obtained in the V, R_c, and I_c passbands (effective wavelengths $\sim 5500, \sim 6500$, and ~ 8000 Å, respectively, corresponding closely to the Cousins VRI system). In the interpolations to the R_c band (see Table 2) we have assumed that the spectra of both the point source and the extended emission are smooth (i.e., not dominated by emission lines). We have used the relation between the color indices V-R_c and V-I_c given by Thé et al. [10] for late-type stars; for bluer stars we have inferred this relation from the tables given by Johnson [11] for main-sequence stars and the color transformations to the Cousins VRI system given by Bessel [12]. We have tested the validity of these color-color relations from numerical integrations of power law flux distributions and of Planck functions, and conclude that if the flux distribution of the optical counterpart is smooth, the uncertainty in the interpolated R_c magnitude is unlikely to exceed 0.1 magnitude [7]. Here we discuss the differences with respect to Galama et al. [7].

The earliest image of the OT was obtained by Pedichini et al. [13]. This observation was obtained with a wide filter; we have transformed this wide filter magnitude, using the reported filter characteristics, to the R_c band. In the images of Guarnieri et al. [14], the OT is blended with the nearby late type star, due to bad seeing. We corrected for the contribution of the late type star (for which $R_c =$

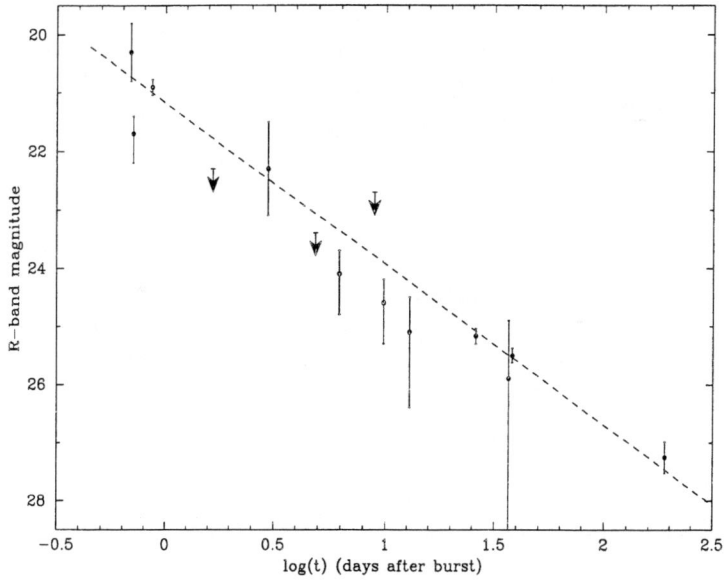

FIGURE 1. The R_c band lightcurve of GRB970228. Indicated is a power-law fit, $F_\nu \propto \nu^\alpha$, to the data with slope $\alpha = -1.10 \pm 0.04$ ($\chi_r^2 = 2.3$ with 9 degrees of freedom).

22.1 ± 0.1 [7]) and find for the OT $R_c = 21.7^{+0.5}_{-0.3}$ (Note that in [7] the time of the observation of Guarnieri et al. [14] is given slightly incorrect). We have included a lower limit to the R_c magnitude on Mar. 1.791 [14], which after correction for the contribution of the late type star, gives $R_c > 22.2$. We have not included the subsequent measurements and upper limits as given by Guarnieri et al. [14] as they are consistent with detections of the late-type star only. We noticed that the Keck calibration [5,15] differs by +0.3 magnitudes; we have corrected for this in Table 2. HST STIS observations between Sept. 4.65 and 4.76 UT [8,9] show that both the nebula and the point-source optical transient found in earlier HST WFC2 observations [6], are also detected in the STIS images at a level of $V = 28.0 \pm 0.25$ for the point source and $V = 25.7 \pm 0.15$ for the nebula. Reanalysis of the earlier HST WFPC2 F606W observations gives $V = 25.6 \pm 0.25$ for the nebula [8]. From the recent HST observations and the reanalysis we infer that the R_c band magnitude of the nebula is $R_c = 25.0 \pm 0.3$, fainter than but consistent with $R_c = 24.7 \pm 0.3$ [7]. Assuming that the colors of the OT remained constant during the decay (i.e., taking the observed $V - I = 1.85$ from the HST March 26 and April 7 observations) we infer from the Sept. 4.71 HST observations that $R_c = 27.25 \pm 0.27$ for the OT. In all ground-based photometry we corrected for the contribution of the extended object ($R_c = 25.0 \pm 0.3$) and show the results in Table 2 and in Fig. 1. We have fitted a power law, $F_R = F_0 t^\alpha$, to the detections and find a magnitude $m_0 = 21.14 \pm 0.13$ (corresponding to F_0) and a slope $\alpha = -1.10 \pm 0.04$ ($\chi_r^2 = 2.3$; the three upper limits are not included in this fit).

DISCUSSION

The lightcurve can be well represented by a power law. In the interval between $\log(t) = 0.7$ and 1.2 we have three detections and one upper limit located below the power-law fit. The three detections deviate from the power law by: 1.9 σ for the Keck observation (Mar 6.32 UT), 1.7 σ for the INT observation (Mar 9.90 UT) and 1.4 σ for the NTT observation (Mar 13.00 UT). This might indicate that the lightcurve is not smooth, but superposed on the power law behaviour we have deviations of small amplitude. A similar result has been found for the OT of GRB 970508 [16,17]. As the point source weakened it also became redder ($V - I = 0.7 \pm 0.14$ on Feb. 28.99 to $V - I = 1.90 \pm 0.14$ on March 26 and $V - I = 1.80 \pm 0.14$ on April 7; see Table 1).

REFERENCES

1. Costa, E., et al., *Nature* **387**, 783 (1997).
2. Groot, P.J., et al., *IAU Circular* No. 6584 (1997).
3. Van Paradijs, J., et al., *Nature* **386**, 686 (1997).
4. Groot, P.J., et al., *IAU Circular* No. 6588 (1997).
5. Metzger, M.R., et al., *IAU Circular* No. 6588 (1997).

TABLE 1. Summary of optical observations.

Date (UT)	Telescope[a]	Magnitude	Remarks[b]
Feb. 28.81	RAO	wide $R = 20.5 \pm 0.5$	OT
Feb. 28.83	BUT	$R = 21.1 \pm 0.2$	OT+LTS
Feb. 28.99	WHT	$V = 21.3 \pm 0.1$	OT
Feb. 28.99	WHT	$I = 20.6 \pm 0.1$	OT
Mar. 01.79	BUT	$R > 21.4$	OT+LTS
Mar. 03.10	APO	$B = 23.3 \pm 0.5$	OT
Mar. 04.86	NOT	$V > 24.2$	OT+EXT
Mar. 06.32	Keck	$R = 24.0$	OT+EXT
Mar. 08.86	INT	$V > 23.6$	OT+EXT
Mar. 08.88	WHT	$I > 22.2$	OT+EXT
Mar. 09.85	INT	$B = 25.4 \pm 0.4$	OT+EXT
Mar. 09.90	INT	$R = 24.0 \pm 0.2$	OT+EXT
Mar. 13.00	NTT	$R = 24.3 \pm 0.2$	OT+EXT
Mar. 26.38	HST	$V = 26.1 \pm 0.1$	OT
Mar. 26.47	HST	$I = 24.2 \pm 0.1$	OT
Mar. 26.38 & Apr. 07.22	HST	$V = 25.6 \pm 0.25$	EXT
Mar. 26.47	HST	$I = 24.5 \pm 0.3$	EXT
Apr. 05.76	Keck	$R = 24.9 \pm 0.3$	OT+EXT
Apr. 07.22	HST	$V = 26.4 \pm 0.1$	OT
Apr. 07.30	HST	$I = 24.6 \pm 0.1$	OT
Apr. 07.30	HST	$I = 24.3 \pm 0.35$	EXT
Sept 04	P5m	$R = 25.5 \pm 0.5$	OT+EXT
Sept 04.71	HST	$V = 28.0 \pm 0.25$	OT
Sept 04.71	HST	$V = 25.7 \pm 0.15$	EXT

[a] Abbreviations: RAO, Rome Astrophysical Observatory; BUT, Bologna University Telescope; WHT, William Hershell Telescope; APO, Apache Point Observatory; NOT, Nordic Optical Telescope, INT, Isaac Newton Telescope; HST, Hubble Space Telescope; P5m, Palomar 5-m Hale telescope.

[b] Abbreviations: OT, optical transient; EXT, extended source; LTS, late-type star.

TABLE 2. The R-band lightcurve of GRB970228.

Date(UT)	Telescope[a]	R(OT+LTS+EXT)	R(OT+EXT)	R(OT)	Reference
Feb. 28.81	RAO		20.3 ± 0.5	20.5 ± 0.5	[13]
Feb. 28.83	BUT	21.1 ± 0.2	$21.7^{+0.5}_{-0.3}$	$21.7^{+0.5}_{-0.3}$	[14]
Feb. 28.99	WHT		20.9 ± 0.14	20.9 ± 0.14	[7]
Mar. 01.79	BUT	> 21.4	> 22.2	> 22.3	[14]
Mar. 03.10	APO		22.2 ± 0.7	$22.3^{+0.8}_{-0.8}$	[7]
Mar. 04.86	NOT		>23.3	>23.4	[7]
Mar. 06.32	Keck		23.7 ± 0.2	$24.1^{+0.5}_{-0.4}$	[5]
Mar. 08.88	INT+WHT		>22.6	>22.7	[7]
Mar. 09.90	INT		24.0 ± 0.2	$24.6^{+0.7}_{-0.4}$	[7]
Mar. 13.00	NTT		24.3 ± 0.2	$25.1^{+1.3}_{-0.6}$	[7]
Mar. 26.20	HST		24.3 ± 0.2	25.17 ± 0.13	[7,6]
Apr. 05.76	Keck		24.6 ± 0.3	$25.9^{+\infty}_{-1.0}$	[15]
Apr. 07.23	HST		24.5 ± 0.15	25.50 ± 0.13	[7,6]
Sep. 04	P5m		25.5 ± 0.5		[18]
Sep 04.71	HST		24.9 ± 0.3	27.25 ± 0.27	[8,9]

[a] Abbreviations as in Table 1.

6. Sahu, K., et al., *Nature* **387**, 476 (1997).
7. Galama, T.J., et al., *Nature* **387**, 479 (1997).
8. Fruchter, A., et al., *IAU Circular* No. 6747 (1997).
9. Fruchter, A., et al., these proceedings.
10. Thé, P.S., Steenman, H., & Alcaino, G., *A&A* **132**, 385 (1984).
11. Johnson, H.L., *ARA&A* **4**, 191 (1966).
12. Bessel, M.S., *UBVRI Photometry with a Ga-As Photomultiplier*, *PASP* **88**, 557 (1976).
13. Pedichini, F., et al., *A&A* **327**, L32 (1997).
14. Guarnieri, A., et al., *A&A*, submitted; Astro-ph 9707164 (1997).
15. Metzger, M.R., et al., *IAU Circular* No. 6631 (1997).
16. Galama, T.J., et al., *ApJL*, submitted (1997).
17. Groot, P.J., et al., these proceedings.
18. Djorgovski, S.G., et al., *IAU Circular* No. 6732 (1997).

The Gamma-Ray Bursts GRB970228 and GRB970508: What Have We Learnt?

Mario Livio[1], Kailash C. Sahu[1], Larry Petro[1], Andrew Fruchter[1], Elena Pian[2], F. Duccio Macchetto[1], Jan van Paradijs[3,4], Chryssa Kouveliotou[5], Paul J. Groot[3], Titus J. Galama[3]

[1] *Space Telescope Science Institute, 3700 San Martin Drive, Baltimore, MD 21218, USA*
[2] *Instituto di Technologie e Studio delle Radiazioni Extraterrestri, C.N.R., Via Gobetti 101, I-40129 Bologna, Italy*
[3] *Astronomical Institute "Anton Pannekoek", University of Amsterdam, & Center for High Energy Astrophysics, Kruislaan 403, 1098 SJ Amsterdam, The Netherlands*
[4] *Physics Department, University of Alabama in Huntsville, Huntsville, AL 35899, USA*
[5] *Universities Space Research Association, NASA Marshall Space Flight Center, ES-84, Huntsville, AL 35812, USA*

Abstract. We examine what we regard as key observational results on GRB970228 and GRB970508 and show that the accumulated evidence strongly suggests that γ-ray bursts (GRBs) are cosmological fireballs.

We further show that the observations suggest that GRBs are not associated with the nuclear activity of active galactic nuclei, and that late-type galaxies are more prolific producers of GRBs.

We suggest that GRBs can be used to trace the cosmic history of the star-formation rate. Finally, we show that the GRB locations with respect to the star-forming regions in their host galaxies and the total burst energies can be used to distinguish between different theoretical models for GRBs.

INTRODUCTION

It is very often the case in astronomy that multiwavelength observations of a single object allow a dramatic progress in the understanding of the object, and γ-ray bursts (see e.g. Fishman and Meegan [10]) proved to be no exception. The identification of the X-ray and optical counterparts of the two γ-ray burst sources GRB970228 and GRB970508 marks a remarkable milestone in the research of these enigmatic objects (e.g. Costa et al. [6,7]; Heise et al. [14]; Piro et al. [29]; van Paradijs et al. [27]; Sahu et al. [30,31]; Bond [1]; Djorgovski et al. [8]; and references therein). In the present short note, we first present what we regard as the key findings and their potential implications, and we then examine possibilities to make further progress with regard to specific γ-ray burst models.

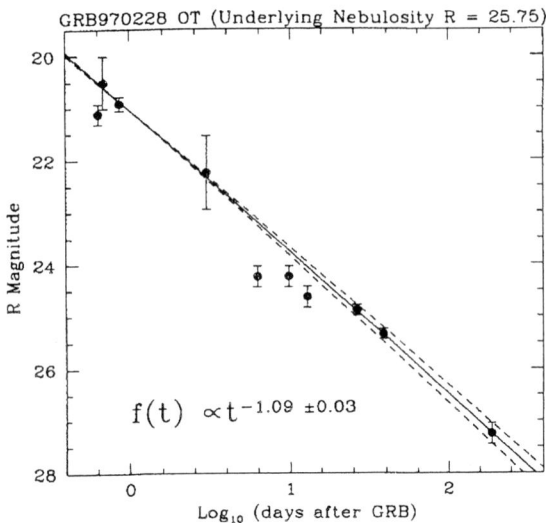

FIGURE 1. The decay behavior of the optical transient associated with GRB 970228 in R-band.

KEY OBSERVATIONAL RESULTS AND THEIR MAIN IMPLICATIONS

We regard the following observational findings (in no particular order) as providing the main clues for the understanding of the nature of γ-ray bursts:

(1) Contrary to claims made by Caraveo et al. [2,3], *no significant proper motion has been detected for GRB970228* (the limit is 36 milliarcsec per year [30,31,11]). This makes an extremely close origin for the GRB very unlikely.

(2) *An absorption- and emission-line system at z=0.835 has been detected in the optical spectrum of GRB970508* [20,21].

Since this absorption/emission system probably arises from a host (or intervening) galaxy, this provides direct evidence that this GRB is at a cosmological distance (assuming, of course, that the identification of the source is correct).

(3) The principal features of the afterglow in the optical (see Fig. 1) are well represented by a forward-radiating blast-wave model [18]. In fact, given the fact that the afterglows may be expected to depend on the properties of the environment of the GRB and on the angular anisotropy of the fireball itself (e.g. [19]), the agreement with the simplest model can be considered quite remarkable.

(4) *No break towards a more rapid decline has been seen so far in the power-law behavior of the light curve (Fig. 1).* Such a break is expected to occur after the blastwave has 'snowploughed' through a rest-mass energy of the order of the burst energy, since the remnant then becomes nonrelativistic (e.g. [37]). The timescale

for such a break to occur is given by

$$t_{break} \simeq \left(\frac{3E_\gamma}{4\pi\rho c^5}\right)^{1/3} = \begin{cases} 1 \text{ yr} \left(\frac{E_\gamma}{10^{51} erg}\right)^{1/3} \left(\frac{n}{0.1 cm^{-3}}\right)^{-1/3} \\ 2 \text{ days} \left(\frac{E_\gamma}{10^{42} erg}\right)^{1/3} \left(\frac{n}{0.001 cm^{-3}}\right)^{-1/3} \end{cases} \quad (1)$$

where ρ is the medium density (n is the number density), E_γ is the burst energy, and we have scaled these quantities with values appropriate for a cosmological or extended halo origin, respectively. Eq. (1) clearly demonstrates that the fact that a break has not yet been observed in GRB970228, more than six months after the burst, strongly favors a cosmological origin.

(5) *A potential host galaxy has been identified in the case of GRB970228* ([30,11]; see Fig. 2). The probability of a chance superposition of the source with a galaxy of that magnitude (V $\sim$ 25.7) is of the order of 2% [11].

An examination of points (1)-(5) above clearly suggests that GRBs are cosmological fireballs. While it is certainly true, that with only two optical afterglows observed so far, an alternative explanation can be found for each one of the above points (e.g. the host galaxy of GRB970228 could be a chance superposition afterall; what is thought to be the counterpart of GRB970508 may be an unrelated BL Lac object, etc.), the combined weight of all the observational facts strongly argues against a local origin for the GRBs. We will therefore from here on *assume* that GRBs are cosmological, and that GRB970228 is indeed located in what appears to be its host galaxy.

GENERAL IMPLICATIONS AND SPECIFIC MODELS FOR GRBS

A close examination of the HST images of GRB970228 [30,11] reveals two more observational facts:

(i) The GRB is not at the center of the host galaxy.

(ii) The galaxy looks like a dwarf irregular or spiral, not like an elliptical galaxy.

The first of these facts implies that *GRBs are probably not associated with the central massive black holes in their host galaxies*, or in general, with the nuclear activity of active galactic nuclei. This rules out, for example, tidal disruption of stars [4] as potential models for GRBs. The second point is somewhat less certain, because the faintness of the galaxy makes any morphological determination not entirely conclusive. Nevertheless, taken at face value, this observation suggests that the frequency of GRBs may be *higher in late-type galaxies*. This, in turn, favors models for GRBs which involve a young stellar population. Leading models in this category include: merging neutron stars (e.g. [9]) or a neutron star and a black hole (e.g. [22]), "failed type Ib supernovae" [39], "hypernovae" [25], radio pulsar glitches [17] and, to a lesser extent, collapsing white dwarfs [33]. For example, Sahu et al. [30] have shown that the ratio of the neutron-star merger rate in disk galaxies to

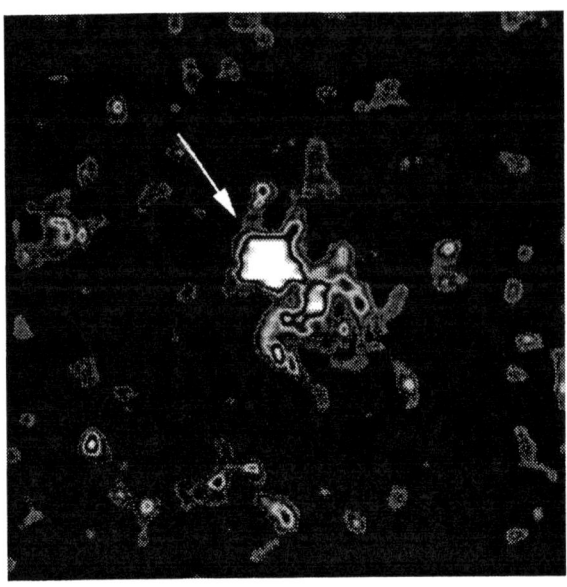

FIGURE 2. Smoothed HST image of GRB970228, taken in March/April 1997 with the WFPC2 camera [30].

that in ellipticals is about 80 (see also Phinney [28]; Narayan, Piran and Shemi [23]).

Another important thing to realize is the fact that *all* the leading models for GRBs have relatively short delays ($\leq 10^9$ yr) with respect to the star formation process. Consequently, *GRBs essentially trace the redshift distribution of the star-formation rate* [31,32,38]. This means that one can use the inferred cosmic history of the star-formation rate [16,5] to construct a synthetic log N - log P relation for GRBs. Turned around, given enough statistics, the redshift distribution of GRBs can provide *an independent test* for the cosmic history of the star formation rate.

Finally, we can ask: is the available data sufficient to point us towards a specific model for the GRBs?

There are at present two pieces of information which can, in principle, provide clues in this direction. One is related to the burst energy and the other to the burst location (see also Paczyński [26]). In Table 1, we list some of the most popular GRB models, the total energy expected from these models, and the expected location of the GRBs with respect to the birth place of their progenitors. One should note that failed supernovae [39], hypernovae [25] and white dwarf collapses [33] are all expected to be found near the birth place of their progenitors for the following reasons. Since failed supernovae and hypernovae originate from very massive stars, the lifetime of their progenitors is short, $\sim 10^6$ yrs, and consequently, they cannot travel very far. The progenitor lifetime of a white dwarf collapse can be longer, but

TABLE 1. Locations of GRBs and Total Energies in Different Models.

Model	Location of GRB	Total Energy (ergs)
Failed supernova	In star-forming region	$\sim 10^{51}$
Hypernova	In star-forming region	$\sim 10^{54}$
White-dwarf collapse	In disks or globular clusters	$\sim 10^{51}$
Merging neutron stars or neutron star+BH	Up to tens of kpc away from the star-forming region	$\sim 10^{51}$

its space velocity is typically low. On the other hand, double neutron star systems are expected to be born with high kick velocities of the order of a few hundred km s^{-1} [15,36,13], and it takes them typically a long time ($\sim 10^8$ yr) to merge. Consequently, the GRBs in this case can be expected to be found typically 30 kpc away from their birthplace. Unfortunately, the existing data do not provide yet a clear picture concerning the burst locations. For example, while the detection of the [OII] 3728Å emission line [21] in GRB970508 seems to indicate the presence of a relatively dense medium (and thus, a potential association with a star-forming region), HST failed so far to detect a host galaxy at the burst location [12], while two nearby faint galaxies (which could, in principle, be the hosts) were detected at distances of $\sim$30 kpc and 35 kpc (for z=0.835, $H_0 = 65$ km s^{-1} Mpc^{-1}). Similarly, the location of GRB970228 near the edge of its host galaxy does not give us conclusive information on whether the progenitor has moved from its birth place or not.

Concerning the total energy, again the present data are still very inconclusive. For example, the total energy inferred for GRB970508 was $\sim 10^{52}$ ergs, assuming spherical emission [34,35]. The situation is further complicated by the possibility that the γ-ray emission is beamed (e.g. Wijers et al. [38]).

In spite of the difficulties pointed out above, the burst locations and the total energies do hold the potential of providing in the future the next step from generic fireballs to specific physical models.

REFERENCES

1. Bond, H.E., IAU Circ. No. 6654 (1997).
2. Caraveo, P.A., Mignami, R., Tavani, M., and Bignami, G.F., IAU Circ. No. 6629 (1997).
3. Caraveo, P.A., Mignami, R., Tavani, M., and Bignami, G.F., *Astron. Astrophys* **326**, L13 (1997).
4. Carter, B., *ApJ.* **391**, L67 (1992).
5. Connolly, A.J., Szalay, A.S., Dickinson, M., Subbarao, M.V., Brunner, R.J., *ApJ.* **486**, L11 (1997).
6. Costa, E. *et al.*, preprint astro-ph/9706065 (1997).
7. Costa, E. *et al.*, IAU Circ. No. 6572 (1997).
8. Djorgovski, S.G. *et al.*, *Nature* **387**, 876 (1997).

9. Eichler, D., Livio, M., Piran, T., and Schramm, D.N., *Nature* **340**, 126 (1989).
10. Fishman, G.J., & Meegan, C.A., *Ann. Rev. Astron. Astrophys.* **33**, 415 (1995).
11. Fruchter, A. *et al.*, *ApJ*, submitted (1997).
12. Fruchter, A. *et al.*, IAU Circ. No. 6674 (1997).
13. Fryer, C. and Kalogera, V., *ApJ*, in press (1997).
14. Heise, J. *et al.*, IAU Circ. No. 6654 (1997).
15. Lyne, A.G. and Lorimer, D.R., *Nature* **369**, 127 (1994).
16. Madau, P. *et al.*, *MNRAS* **283**, 1388 (1996).
17. Melia, F., Fatuzzo, M., *ApJ.* **398**, L85 (1992).
18. Mészáros, P., and Rees, M.J., *ApJ* **476**, 232 (1997).
19. Mészáros, P., Rees, M.J., Wijers, R.A.M., astro-ph/9704153 (1997).
20. Metzger, M.R. *et al.*, IAU Circ. No. 6676 (1997).
21. Metzger, M.R. *et al.*, *Nature* **387**, 879 (1997).
22. Mochkovitch, R., Hernanz, M., Isern, J., Martin, X., *Nature* **361**, 236 (1993).
23. Narayan, R., Paczyński, B., and Piran, T., *ApJ* **395**, L83 (1992).
24. Narayan, R., Piran, T., Shemi, A., *ApJ* **379**, L17 (1991).
25. Paczyński, B., astro-ph/9706232 (1997).
26. Paczyński, B., astro-ph/9710086 (1997).
27. van Paradijs, J. *et al.*, *Nature* **386**, 686 (1997).
28. Phinney, E.S., *ApJ.* **380**, L17 (1991).
29. Piro, L., *et al.*, IAU Circ. No. 6656 (1997).
30. Sahu, K.C. *et al.*, *Nature* **387**, 476 (1997).
31. Sahu, K.C. *et al.*, *ApJL* **489**, L127 (1997).
32. Totani, T., *ApJ.* **486**, 71 (1997).
33. Usov, V., *Nature* **357**, 472 (1992).
34. Waxman, E., *ApJ* **489**, L33 (1997).
35. Waxman, E., *ApJ* **485**, L5 (1997).
36. White, N.E., and van Paradijs, J., *ApJ* **473**, L25 (1996).
37. Wijers, R.A.M.J., Rees, M.J., and Mészáros, P., *MNRAS* **288**, L51 (1997).
38. Wijers, R.A.M.J., Bloom, J.S., Bagla, J.S., Natarajan, P., astro-ph/9708183 (1997).
39. Woosley, S., *ApJ.* **405**, 273 (1993).

Optical/IR Follow-up Observations of GRBs Detected by BeppoSAX

A. J. Castro-Tirado[1], J. Gorosabel[1],
E. Costa[2], M. Feroci[2], L. Piro[2],
F. Frontera[3], D. Dal Fiume[3], L. Nicastro[3], E. Palazzi[3],
J. Greiner[4],
K. Birkle[5], R. Fockenbrock[5], E. Thommes[5], C. Wolf[5],
C. Bartolini[6], A. Guarnieri[6], N. Masetti[6], A. Piccioni[6],
M. Mignoli[7], J. Heidt[8], T. Seitz[8],
H. Pedersen[9], S. Guziy[10], A. Shlyapnikov[10],
L. Metcalfe[11], R. Laureijs[11], B. Altieri[11], M. Kessler[11],
L. Hanlon[12], B. McBreen[12],
N. Smith[13], J. Studt[14],
N. Benítez[15], E. Martínez-Gozález[15],
H. Kristen[16], A. Broeils[16], M. Wold[16],
M. Lacy[17], M. V. Alonso[18]

[1] *Laboratorio de Astrofísica Espacial y Física Fundamental (LAEFF-INTA) P.O. Box 50727, E-28080, Madrid, Spain*
[2] *IAS, Frascati, Italy*
[3] *ITESRE, Bologna, Italy*
[4] *AIP, Potsdam, Germany*
[5] *MPIA, Heidelberg, Italy*
[6] *Università di Bologna, Italy*
[7] *Obs. Astronomico di Bologna, Italy*
[8] *Landessternwarte Heidelberg, Germany*
[9] *Copenhagen University Observatory, Denmark*
[10] *Nikolaev Univ. Observatory, Ukraine*
[11] *ESA-Vilspa, Madrid, Spain*
[12] *Univ. College, Dublin, Ireland*
[13] *Regional Tech. College, Dublin, Ireland*
[14] *Hamburg Observatory, Germany*
[15] *IFCA, Santander, Spain*
[16] *Stockholm Observatory, Sweden*
[17] *NAL, Oxford, United Kingdom*
[18] *Observatorio de Córdoba, Argentina*

Abstract. The first two optical transients associated to GRBs were identified in 1997, thanks to the capabilities of the instruments aboard the BeppoSAX X-ray satellite. As part of a programme initiated in 1996, we performed optical/IR follow-up observations following the BeppoSAX alerts. The error boxes for GRB970111, GRB970228, GRB970402 and GRB970508 were imaged at different observatories. We briefly discuss here the observational characteristics of the first two optical counterparts: for GRB970228 and GRB970508. The latter allowed to prove the cosmological origin for a significant fraction of GRBs. No optical/IR counterparts, however, were found for GRB970111 or GRB970402, in spite of intensive searches. It is clear now that a significant fraction are beyond the reach of optical telescopes, possibly due to intrinsic absorption in the host galaxies.

GRB970111

Optical and IR images of the refined IPN/BeppoSAX error box [1] were obtained 19 hr after the onset of the event. The B and R-band frames were obtained at the 2.2-m telescope at the German-Spanish Calar Alto Observatory (CAHA) on Jan 12 and Feb 10-11. Additional B and R-band frames were taken at Loiano Observatory (Italy) on Jan 14 and 19.

The main result is that no variable optical counterpart was found within the entire IPN/BeppoSAX GRB error box. Any fading or increase was ≤ 0.2 mag for $B \leq 21.0$, $R \leq 20.8$, ≤ 0.5 mag for $B \leq 23.0$, $R \leq 22.6$ [2].

No variation was observed in the brightness of the radiogalaxy [3] coincident with one of the two X-ray sources found by BeppoSAX. A more detailed study is published elsewhere [4].

GRB970228

Optical/IR imaging of the GRB error box as given by BeppoSAX [5] was obtained 16 hr after the burst. B and R-band frames were obtained at Loiano (1.5-m telescope) on Feb 28, Mar 1,3-5, 12-13 and 18. An unfiltered frame was obtained at CAHA (2.2m) on Mar 7. Unfiltered frames were also obtained at La Silla (1.5m Danish) on Mar 10, 11, 13, 15 and 16. A K-band image was obtained on 2 Apr at Mauna Kea (3.8m Ukirt).

This is the first burst for which an optical counterpart was discovered [6,7]. The object was also detected 4 hr earlier (Feb 28.82) in the B and R-band frames at Loiano, with $R = 21.1 \pm 0.2$, $B = 22.3 \pm 0.3$. This was the first evidence for the rising light curve of an optical transient related to a GRB [8], with a maximum reached about 18 hr after the onset. Then, the optical flux declined following a power-law decay as $t^{-1.1}$. No object was seen down to $K = 20$ in the Ukirt images.

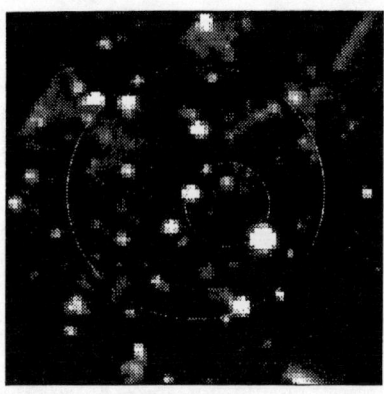

FIGURE 1. The content of the GRB970402 error box as observed by ISO. 9' x 9' FOV. North is up and east to the left. From [14].

GRB970402

The optical imaging of the BeppoSAX error box [9] began 30 hr after the burst. The B, V, R and I-band frames were obtained at La Silla on Apr 3-6, 11 (3.5m NTT) and May 8 (3.6m).

No variable optical counterpart is found within the entire IPN/BeppoSAX GRB error box. Any fading or increase was ≤ 0.3 mag for $V \leq 22.5$, ≤ 0.3 mag for $I \leq 21.7$ [10].

The star BL Cir is at the edge of the X-ray error box [11], and was found to be in a faint state [10,12]. A low dispersion spectrum at the 1.5m Danish indicates that it is a deeply reddened star. More details are given in [13].

The infrared imaging of the BeppoSAX error box with the ESA's Space Infrared Observatory (ISO) begun 55 hr after the event. CAM (12 μm) and PHT (174 μm) observations were performed on Apr 5 (see Fig. 1). Further observations with CAM (12 μm), for comparison purposes, were obtained on Apr 13. About 50 sources are detected in the GRB field, with 7 within the 1' radius X-ray error box.

No variable IR source is seen in the error box with the exception of the variable star BL Cir, the strongest source in the field. Upper limits for any new object are ≤ 0.14 mJy at 12 μm, and ≤ 350 mJy at 174 μm [14].

GRB970508

The first images of the BeppoSAX error box [15] were obtained 4 hr after the burst. R-band and unfitered frames were taken at Calar Alto (2.2m) on May 9-12. Further R-band frames obtained at Loiano (1.5m) on May 10, 13-14 and La Palma (4.2m WHT) on Aug 25. U-band frames were acquired at La Palma (4.2m WHT) on May 9-12 and 15 (2.5m NOT). Further U-band images were obtained on 25 Aug

(WHT). Additional images (40) with 1.5 min time resolution were obtained on May 12 (4.2m WHT).

We also performed low-resolution spectroscopy of the counterpart beginning only 47 hr after the event. Spectra were taken on May 10-11 at CAHA (2.2m), and May 16-17 at La Palma (2.5m NOT).

The optical counterpart was discovered by Bond [16]. The object was present in our images only 4 hr after the burst, in both the U and R-bands. It displayed a strong UV excess. The image taken on May 11 showed that the object was declining in brightness (see Fig. 2). No variation ≥ 0.2 mag was seen on a 1.5 min timescale over 1.5 hr on May 12. The first optical spectrum ever taken was obtained ~ 9 hr before the Keck observations, and did not show any strong emission lines [17]. A more extensive discussion of these results can be seen in [18].

The Keck spectrum, with a better signal-to-noise ratio, allowed a direct determination of the redshift of GRB970805 ($z \sim 0.835$) and was the first proof that at least a fraction of the GRB sources lie at cosmological distances [19].

Unlike GRB 970228, no host galaxy has been seen, but the optical spectroscopy and the flattening of the decay in late August [20] suggest that we may have detected the host galaxy, which should be very underluminous and could be a dwarf, blue rapidly forming starburst galaxy, as proposed elsewhere [21].

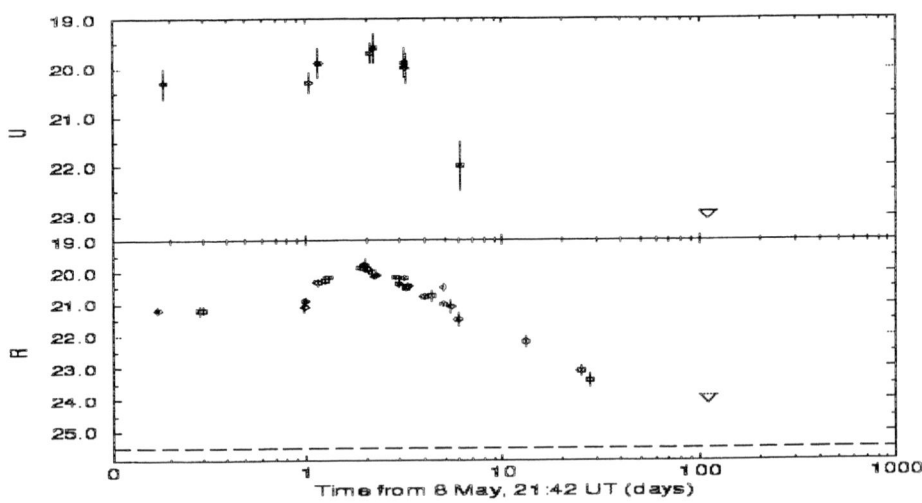

FIGURE 2. The light curve of GRB970508 in the U and R-bands. See [18] for more details.

CONCLUSIONS

As part of a programme initiated in 1996 in collaboration with the BeppoSAX team, we performed optical/IR follow-up observations of the first optical GRB counterparts, and pointed out the difference between GRB970111 and GRB970228 [2]. It is clear now that not all GRBs have similar counterparts to GRB970228 and GRB970508. In fact, only the latter two have been found in the rapid searches performed this year for seven GRBs (see also [22]). This means that there is a large fraction of sources which may be difficult to detect in the optical possibly due to intrinsic absorption in the host galaxy.

ACKNOWLEDGEMENTS

We are indebted to many persons, in particular to the staff at the observatories, for their help, when the images have to be taken at short notice. We also thank our relatives (especially wives/husbands), because they wait patiently for our return when we have to leave, while on holidays, for the nearest observatory to perform the follow-up observations for three or four days.

REFERENCES

1. in't Zand, J. ,et al., IAUC 6569 (1997).
2. Castro-Tirado, A. J., Gorosabel, J., Heidt, J., et al., IAUC 6598 (1997).
3. Frail, D., et al., IAUC 6545 (1997).
4. Gorosabel, J., Castro-Tirado, A. J., Heidt, J., et al., A&A, in press (1998).
5. Costa, E., et al., IAUC 6572 (1997).
6. Groot, P. J., et al., IAUC 6584 (1997).
7. van Paradijs J., et al., Nature **386**, 686 (1997).
8. Guarnieri, A., et al., A&A **328**, L13 (1997).
9. Feroci, M., et al., IAUC 6610 (1997).
10. Pedersen, H., et al., IAUC 6628 (1997).
11. Piro, L. et al., IAUC 6617 (1997).
12. Harrison, E., et al., IAUC 6632 (1997).
13. Pedersen, H., et al., A&A, in press (1998).
14. Castro-Tirado, A. J., et al., A&A, in press (1998).
15. Costa, E., et al., IAUC 6649 (1997).
16. Bond, H., IAUC 6655 (1997).
17. Castro-Tirado, A. J., et al., IAUC 6657 (1997).
18. Castro-Tirado, A. J., Gorosabel, J., Benítez, N., et al., Science **279**, 1011 (1998).
19. Metzger, M., et al., Nature **387**, 878 (1997).
20. Pedersen, H., et al., ApJL, in press (1998).
21. Pian, E., et al., ApJL, in press (1998).
22. Castro-Tirado, A. J., et al., these proceedings.

Is the Fuzziness of GRB970228 Constant?

P.A. Caraveo*, R.Mignani[†] and G.F.Bignami[+,*]

*Istituto di Fisica Cosmica del CNR, Milano, ITALY
[†]MPE, Garching, GERMANY
[+]Agenzia Spaziale Italiana, Roma, ITALY

Abstract. In view of the data gathered in September 1997, we review the flux values collected so far for the "fuzziness" seen in the optical counterpart of GRB970228. Comparison between the ground based data collected in March and the data of September 1997 suggests a fading of the fuzz. Given the diversity of the data in hand, the magnitude of the effect and its significance are not easy to quantify. Only new images, both from the ground and with the Space Telescope, directly comparable to the old ones could settle this problem.

INTRODUCTION

After the SAX positioning of GRB970228 [1], and the discovery of an optical transient in the refined error box [10](hereafter vP97), the optical counterpart of GRB970228 has been observed many times both with ground based instruments and with the Hubble Space Telescope. Several days after the event, an extended optical emission was detected where the Optical Transient (OT) had been seen in the discovery image, taken 21 h after the event (vP97). Since then, the magnitude of such an extended emission has been measured many times, by several observers, using different instrumental set-ups. In this paper we review and compare the measurements gathered so far to investigate if the flux values recently measured by STIS on HST [3] and by the 5m Palomar telescope [2] are consistent with the ground based ones obtained at early epochs.

THE DATA

Table 1 summarizes the data collected so far, both for the OT integrated magnitude (ground measurements) and for the contribution of the two components: point source and extended emission (HST data). The first claim for an extended object, elongated in the N-S direction, gave, on March 13th, $m_R = 23.8 \pm 0.2$ (vP97). An elongated object of $m_R = 24.0 \pm 0.2$, was also seen in the Keck data [7] taken on March 6th.

Using HST, the extended source was resolved into a point source superimposed to a "fuzz". Comparison of the March and April images showed that the point source was most probably fading, while nothing definite could be said on the diffuse emission [9]. More Keck observations [8], taken on April 5th and 6th gave, for the total emission, a flux lower than both the HST one and the Keck March 6th data. On Sept 4th, using the Palomar 5m, Djorgovski et al [2] detected the extended source at $m_R \sim 25.5$. The STIS instrument on board HST also observed the source on Sept 4th with the Clear filter. Fruchter et al [10], are barely able to detect the point source over a diffuse emission of $m_V = 25.7 \pm 0.15$, i.e. 0.8 magnitude fainter than in the HST March observation. This prompted a reanalysis of the March/April WFPC data which yielded for the fuzz $m_V = 25.6 \pm 0.25$, i.e. half of the flux of Sahu et al [9]. However, even accepting that the HST data, after re-analysis, can be rendered consistent, it seems very difficult to reconcile the September STIS/Palomar data with the NTT/Keck ones of early March.

HST VS NTT

In the following, we shall compare the NTT data (kindly provided to us by Jan van Paradijs) with the HST ones. If we assume that the September STIS/Palomar flux values for the extended source are correct, and if we further assume no fading, we have to explain the extended total emission seen both by NTT and by Keck with a combination of the STIS/Palomar fluxes plus a point source of suitable magnitude. Even considering the revised NTT mag value [4], we have to account for a total flux of $\sim 5.4 \; 10^{-30}$ erg cm^{-2} s^{-1} Hz^{-1}. Since the extended source observed in September provides $\sim 1.9 \; 10^{-30}$ erg cm^{-2} s^{-1} Hz^{-1}, the unseen point source should have been $\sim 3.5 \; 10^{-30}$ erg cm^{-2} s^{-1} Hz^{-1}, i.e. definitely brighter than the extended one. However, in order to simulate the appearance of such a combination, one should be able to locate the HST point source into the NTT nebulosity. This calls for an accurate superposition of the HST March data onto the NTT/SUSI frame. After correcting for geometric distorsions using the IRAF/STSDAS task *mosaic*, the PC image has been rebinned and rotated onto the SUSI one (see Figure 1) with an accuracy certainly of better than 1/2 pixel (actually 1/10 would be a more realistic estimate). Zooming on the OT, one sees clearly that the HST point source falls in the central part of the NTT nebulosity, where the emission is less intense and no hint of a point-like object is seen. However, the central region of the nebulosity is just where one should put a hypothetical point source of $3.5 \; 10^{-30}$ erg cm^{-2} s^{-1} Hz^{-1} ($m_R \sim 24.65$). This value is similar to the flux measured by the NTT for star #3 in Figure 1. Such a source is an interesting test case, since it is point-like in the HST/PC image but it looks extended in the NTT one. However, inspection of Figure 2, where we have compared the Right Ascension and Declination tracings of the two sources, shows unambigously their difference in shape for a comparable

Date	Telescope	Filter	mag			flux $(10^{-30} erg\, cm^{-2}\, s^{-1}\, Hz^{-1})$		Ref.
			total	point source	extended	total	point source	extended	
March 6th	Keck	R	$m_R = 24 \pm 0.2$			$7.2^{+1.4}_{-1.2}$			1.
March $13th$	NTT	R	$m_R = 23.8 \pm 0.2$			$8.6^{+1.4}_{-1.7}$			2.
			$m_R = 24.3 \pm 0.2$			$5.4^{+0.9}_{-1.1}$			3.
March $26th$	HST/PC	606W	$m_R = 24.1 \pm 0.2$	$m_R = 25.17 \pm 0.13$	$m_R = 24.7 \pm 0.3$	$6.5^{+1.1}_{-1.3}$	$2.4^{+0.3}_{-0.3}$	$3.8^{+0.9}_{-1.2}$	3.
				$m_V = 26.1 \pm 0.1$	$m_V = 24.9 \pm 0.3$		$1.3^{+0.1}_{-0.1}$	$4.03^{+0.9}_{-1.3}$	4.
					$m_V = 25.6 \pm 0.25$			$2.11^{+0.4}_{-0.5}$	5.
April $6th$	Keck	R	$m_R = 24.9 \pm 0.3$			$3.11^{+0.7}_{-0.9}$			6.
April $7th$	HST/PC	606W	$m_R = 24.2 \pm 0.15$	$m_R = 25.5 \pm 0.13$	$m_R = 24.7 \pm 0.3$	$5.94^{+0.7}_{-0.8}$	$1.79^{+0.2}_{-0.2}$	$3.74^{+0.9}_{-1.2}$	3.
				$m_V = 26.4 \pm 0.1$	$m_V = 25.2 \pm 0.35$		$1.01^{+0.08}_{-0.09}$	$3.05^{+0.8}_{-1.1}$	4.
					$m_V = 25.6 \pm 0.25$			$2.11^{+0.4}_{-0.5}$	5.
Sept $4th$	5 m Hale	R	$m_R \sim 25.5$			~ 1.8			7.
Sept $4th$	HST/STIS	clear		$m_V = 28 \pm 0.25$	$m_V = 25.7 \pm 0.15$		$0.23^{+0.05}_{-0.06}$	$1.92^{+0.2}_{-0.3}$	5.

TABLE 1. Ground-based and HST observations of the GRB970228 optical counterpart. For each observation the telescope as well as the "original" filter are indicated. Columns 4-6 list the total magnitude of the GRB counterpart as well as the magnitudes of the point and extended source (when resolved). Corresponding monochromatic fluxes are given in column 7-9. We note that the V/R magnitudes obtained from HST broad band observations are based on ad hoc color transformation. Therefore, they are not directly comparable to the ground based R measurements which provide the long term coverage of the source evolution.

1. Metzger et al, 1997a; 2. Groot et al, 1997b (see also, van Paradijs et al, 1997); 3. Galama et al, 1997; 4. Sahu et al, 1997; 5. Fruchter et al, 1997; 6. Metzger et al, 1997b; 7. Djorgovski et al, 1997;

flux. While star #3 is dominated by a clear peak superimposed to a region of higher background, the GRB nebulosity does not show any obvious point-like contribution. This is somewhat surprising, since a point source of $m_R \sim 24.6$ should have been far easier to detect than a $m_R \sim 25.5$ extended one. Moreover, we note that such a faint extended source would be hardly within reach of an 1 hour NTT exposure. Thus, the truly extended nature of the NTT source, coupled with the lack of point source at the HST location, leads to the conclusion that the nebulosity itself has faded away between March 13th and Sept 4th.

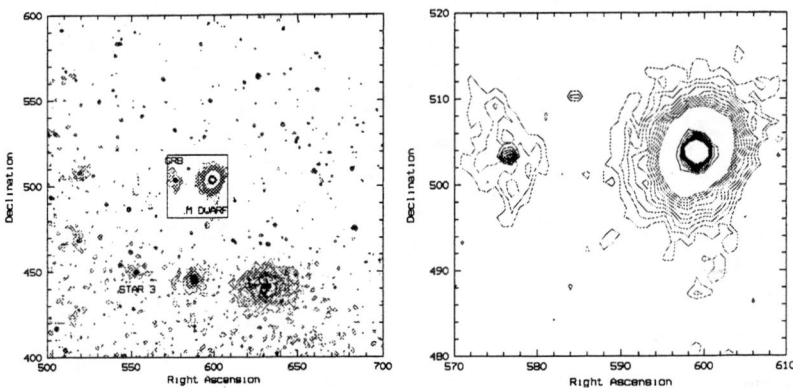

FIGURE 1. Superposition (26×26 arcsec) of the March HST/PC frame onto the NTT/SUSI one. North to the top and East to the left. Axis units are SUSI pixels (0.13 arcsec). The GRB counterpart as well as the nearby M dwarf and star #3 (see text) are labelled. A zoom of the central square is shown in the right panel. The GRB position measured by the HST/PC falls exactly at the center of the NTT nebulosity.

CONCLUSIONS

Although comparing non-homogeneous data sets is not straightforward, the compilation of the available magnitudes for the optical counterpart of GRB970228 (Table 1) points toward a fading both of the point source and of the diffuse emission. While the fading of the point source is expected in all theoretical scenarios, the fading of the diffuse emission has far reaching consequences and, as such, is in need of a dedicated observing campaign. The data available are numerous, but too diverse to provide the constraints needed to assess with certainty if, and how much, the nebulosity has faded. Indeed, for GRB970228, it looks as if every new observation results in a downward revision of the values previously published. Only more observations, directly comparable with those already in hand (i.e. obtained with identical instrumental set-ups) can provide a definite answer to this all-important point.

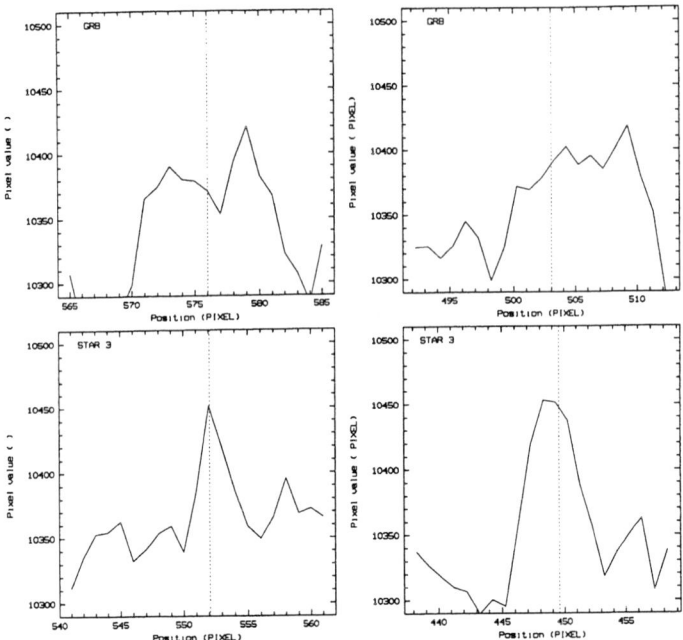

FIGURE 2. Right ascension (left) and declination (right) tracings for the GRB counterpart (upper panels) and for star #3 (lower panels) obtained from Fig. 1. In both cases, the tracings have been computed at the expected location of the point sources. Their coordinates, obtained by registering the HST/PC frame onto the NTT/SUSI one, are marked as vertical lines. For star #3, the presence of a point source is clear.

REFERENCES

1. Costa, E., et al., *Nature* **387**, 783 (1997).
2. Djorgovski, S., et al., *IAU Circ 6732* (1997).
3. Fruchter, A., et al., *IAU Circ. 6747* (1997).
4. Galama, T., et al., *Nature* **387**, 479 (1997).
5. Groot, J.P., et al., *IAU Circ. 6584* (1997a).
6. Groot, J.P., et al., *IAU Circ. 6588* (1997b).
7. Metzger, M.R., et al., *IAU Circ. 6588* (1997a).
8. Metzger, M.R., et al., *IAUCirc. 6631* (1997b).
9. Sahu K., et al., *Nature* **387**, 476 (1997).
10. van Paradijs, J., et al., *Nature* **386**, 686 (1997).

Optical Follow-up of GRB970508

Paul J. Groot[1], Titus J. Galama[1], Jan van Paradijs[1,2], Chryssa Kouveliotou[3], Ralph Wijers[4], Miriam Centurion[5,6], Paul Smith[7], Chris Mackay[7], Stephen Smartt[6], Chris Benn[6]

[1] *Astronomical Institute 'Anton Pannekoek', University of Amsterdam, Kruislaan 403, 1098 SJ Amsterdam, The Netherlands*
[2] *Physics Department, University of Alabama in Huntsville, Hunstville AL, USA*
[3] *USRA at NASA/MSFC, Code ES-84, Huntsville AL 35812, USA*
[4] *Institute of Astronomy, Madingley Road, Cambridge, UK*
[5] *Instituto Astrofisica de Canarias, La Laguna, Tenerife, Spain*
[6] *ING Telescopes, Apartado 321, Sta. Cruz de La Palma, Tenerife 38780, Spain*
[7] *Kitt Peak National Observatories*

Abstract. We report on the results of optical follow-up observations of the counterpart of GRB970508 between 7 hours and 100 days after the event. Multi-colour U, B, V, R_c and I_c band observations were obtained during the first three consecutive nights. The R_c light curve rises to a maximum 1.8 days after the burst and after this declines with a power law decay, the exponent of which is $\alpha=-1.21\pm0.01$. Superposed on this power law decay, we detect small amplitude (< 0.5 magnitude) variations on timescales as short as half a day. The optical energy distribution can be well represented by a power law, the slope of which became steeper during the rise to maximum light and stayed constant afterwards.

INTRODUCTION

On May 8.904 UT BeppoSAX recorded a moderately bright gamma-ray burst [1] with the Gamma-Ray Burst Monitor, which was also recorded with the Wide Field Cameras on-board BeppoSAX. Analysis of the WFC observations gave a 3' (3 σ radius) error box centered on RA=$06^h53^m28^s$, Dec=+79°17.'4 (J2000; [2]).

From optical observations of the WFC error box, made on May 9 and 10 with the Kitt Peak 0.9m reflector, a variable object was found [3] at RA = $06^h53^m49^s.2$, Dec = +79°16'19"(J2000), which showed an increase by ~1 mag in the V band. BeppoSAX Narrow Field Instruments observations revealed an X-ray transient [4] whose position is consistent with that of the optical variable. Based on the absence of extended emission associated with this optical counterpart [5,6] it was shown [7] that either GRB970508 originated from an intrinsically very faint dwarf galaxy or that it occured at a large distance from a host galaxy (> $25h_{70}^{-1}$ kpc).

OPTICAL OBSERVATIONS

We here report on the results of optical photometry of GRB970508, made between 0.3 and 110 days after the initial GRB. We have used the 4.2m William Herschel Telescope (WHT) at La Palma to make optical observations in U, B, V, R_c and I_c. The WHT was used with both the Prime Focus Camera (PF), with a field of view of 9.0×9.0, and with the Auxiliary Port Camera (AUX) with a circular field of view of 0.6 radius. All WHT PF exposures lasted 600 seconds, except the one made on Aug 26.9 UT which lasted 1800 seconds. The WHT AUX exposures were 3600 seconds each.

As part of the Kitt Peak National Observatory Queue observations, the error box of GRB970508 was also observed with the 3.5m WIYN telescope, using the WIYN Imager S2KB at the Nasmyth focus. Ten 300 s, 6.8×6.8 arcmin R-band images were obtained between May 9.15 and 9.23 UT at several pointings to cover the whole error box.

We have corrected for the non-linearity at low counts of the LORAL2 CCD Chip[1], which we used for the WHT observations on May 10, 11, and 12, for each pixel in the U and B band images (the effect of non-linearity is negligible in the V, R_c and I_c observations).

We made a photometric calibration using WHT observations on May 9.99 UT of the standard fields PG1047+03 and PG1530+57 [8]. These observations were made in U, V, R_c and I_c. For V, R_c and I_c we find good agreement with the results of Sokolov et al. [9]. To include a B band calibration, and avoid small calibration differences with respect to Sokolov we used their secondary photometric B, V, R_c and I_c standards (stars 1, 2 and 4 in Fig 1) (their star 3 was close to saturation and therefore not used). Stars 2, 5, 6 and 7 (see Fig 1) were used as secondary photometric standards for the U band. We have corrected for a U band atmospheric extinction of 1.3 times the nominal extinction[1] (as Carlsberg Automatic Meridian Circle V band extinction measurements indicate).

In Fig. 2 we present the R_c band differential light curve, as constructed from our own and published data. Data have been calibrated with respect to stars 1, 2 and 4 [9], or only star 4, when no information on the other was available. For star 4 we used $R_c(4) = 19.49$, and have corrected for any differences if necessary. The light curve shows that maximum light is reached about 1.9 days after the initial gamma-ray burst. When the light curve before and after this maximum is fitted with a power law, the slopes of this power law are: $\alpha = 1.67 \pm 0.06$ ($-0.1 < \log(t) < 0.27$) and $\alpha = -1.21 \pm 0.01$ ($\log(t) > 0.27$). In order to test whether a flattening off of the light curve [10] exists, we also fitted the decay part of the light curve with a power law decay plus a constant contribution ($F = F_0 t^\alpha + C$). In this case the power law exponent $\alpha = -1.31 \pm 0.02$, and the constant contribution corresponds to $R_c = 25.35 \pm 0.15$. However, the introduction of a constant does not significantly improve the fit and it can therefore not be concluded whether the flattening off is

---

[1] La Palma webpage http://www.ing.iac.es

real. When we subtract the power law fit from the decay part of the R_c light curve of GRB970508 we find erratic statistically important low amplitude (<0.5 mag) variations, on timescales as short as half a day.

Between May 9.905 UT and May 10.033 UT the R_c band magnitude increased by 0.137 ± 0.014 magnitude per hour. In estimating the spectral energy distribution we have corrected the May 9.9 UT observations for this brightening, i.e., we have assumed that in all passbands the rate of brightening is the same, and reduced the magnitudes to a single epoch (May 9.93 UT). The May 10.98 UT observations show no evidence for a source brightening or decay (these observations occur at maximum light; see Fig. 2), so we have not applied any corrections. From May 11 onwards the OT decreases in a power-law manner. Corrections on the May 12 data have been calculated using a power-law fit to the R_c band light curve after maximum light ($\log(t) > 0.27$). The corrections are minor (< 0.04 mag) and marginally (< 0.02) affect the spectral slope.

Using the calibration of Bessel [11] we have fitted the broadband photometry with a power law, $F_\nu = \nu^\beta$, (Fig. 3) after making corrections for the interstellar extinction A_V, for which we derive a value of $A_V < 0.01$ from IRAS 100μ flux measurements. However, to compare our results with those of Djorgovski et al. [12], we will use $A_V = 0.08$. If we use $A_V = 0$ instead, the slopes are affected by −0.1. The results of the power law fitting are represented in Fig. 4. When we include all other available data [9,12,13], we derive Fig. 4 for the time dependence of the optical slope. The spectrum clearly reddens on the rise to maximum light and stays constant afterwards. For a more extensive discussion of these results see Galama et al. [14].

REFERENCES

1. Costa, E., et al., *IAU Circular* No. 6649 (1997).
2. Heise, J., et al., *IAU Circular* No. 6654 (1997).
3. Bond, H., *IAU Circular* No. 6654 (1997).
4. Piro, L., et al., *IAU Circular* No. 6656 (1997).
5. Fruchter, A., et al., *IAU Circular* No. 6674 (1997).
6. Pian, E., et al., *ApJ*, submitted (1997).
7. Natarajan, P., et al., *New Astr.* **2**, 461 (1997).
8. Landolt, A., *AJ* **104**, 340 (1992).
9. Sokolov, V.V. et al., these proceedings and astro-ph 9709093.
10. Pedersen, H., et al., *ApJ*, submitted (1997).
11. Bessel, M.S., *PASP* **91**, 589 (1997).
12. Djorgovski, S.G., et al., *Nature* **387**, 876 (1997).
13. Metzger, M., et al., *Nature* **387**, 878 (1997).
14. Galama, T.J., et al., *ApJ*, submitted (1997).

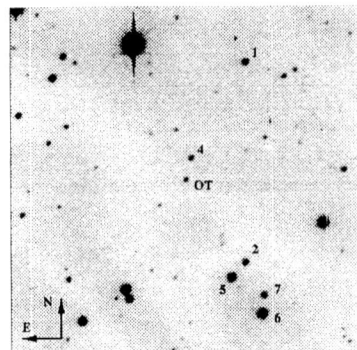

FIGURE 1. The field of GRB970508 ($3.5 \times 3.5'$), taken with the WHT on May 10, showing the optical transient (OT) and local standard stars used.

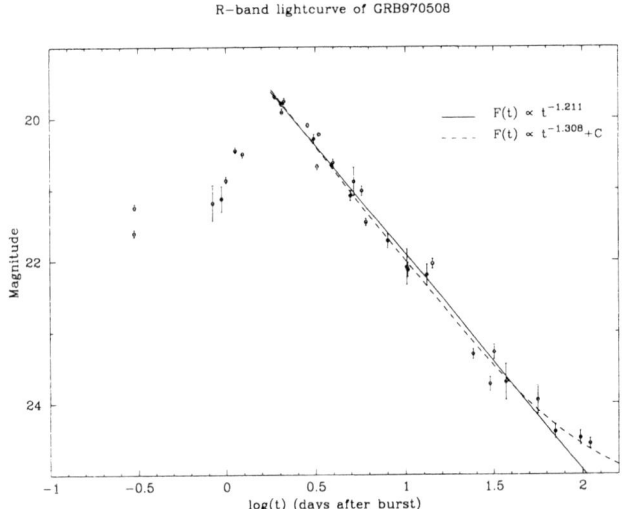

FIGURE 2. Differential R_c band light curve of GRB970508, based on our and published data. The solid line indicates the power law fit, and the dashed line the power law plus constant fit.

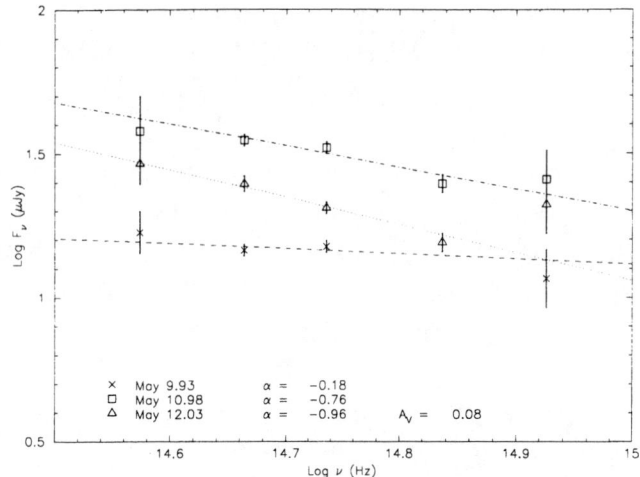

FIGURE 3. Broad band optical measurements of GRB970508 taken with the WHT at La Palma. Indicated are the power law fits.

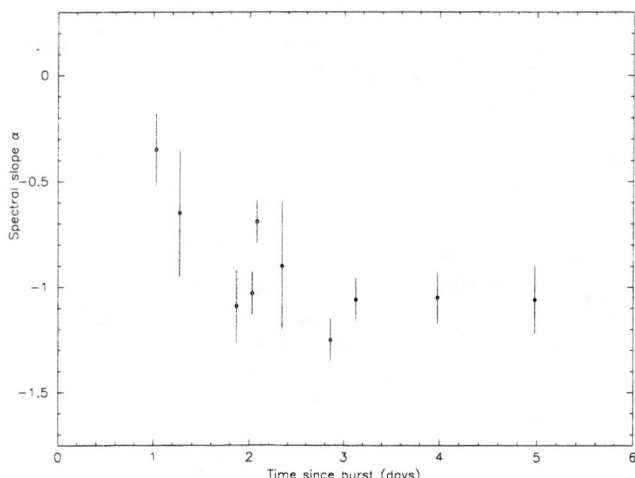

FIGURE 4. Time dependence of the power law slope to the optical broadband spectrum of GRB970508. A clear reddening occurs towards the maximum in the R_c band light curve after which it stays constant.

Hubble Space Telescope Imaging of the Field of GRB970508

E. Pian[1], A. Fruchter[2], L. E. Bergeron[2], S. E. Thorsett[3],
F. Frontera[1,4], M. Tavani[5,6], E. Costa[7], M. Feroci[7],
J. Halpern[6], R. A. Lucas[2], L. Nicastro[1], E. Palazzi[1],
L. Piro[7], W. Sparks[2], A. J. Castro-Tirado[8], T. Gull[9],
K. Hurley[10], H. Pedersen[11]

[1] *ITESRE-CNR, Via Gobetti 101, I-40129 Bologna, Italy*
[2] *STScI, 3700 San Martin Drive, Baltimore, MD 21218*
[3] *Joseph Henry Labs. and Dept. of Physics, Princeton University, Princeton, NJ 08544*
[4] *Dip. Fisica, Università di Ferrara, Via Paradiso 12, I-44100 Ferrara, Italy*
[5] *Columbia Astrophysics Laboratory, Columbia University, New York, NY 10027*
[6] *IFCTR-CNR, Via Bassini 15, I-20133 Milano, Italy*
[7] *IAS-CNR, Via E. Fermi 21, I-00044 Frascati, Italy*
[8] *Laboratorio de Astrofísica Espacial y Física Fundamental, INTA, P.O. Box 50727, 28080 Madrid, Spain*
[9] *NASA/ Goddard Space Flight Center, Greenbelt, MD 20071*
[10] *University of California Space Sciences Laboratory, Berkeley, CA 94720*
[11] *Copenhagen University Observatory, Juliane Maries Vej 30, DK 2100 Copenhagen, Denmark*

Abstract. We report on Hubble Space Telescope (HST) observations of the optical transient (OT) discovered in the error box of GRB970508. The object was imaged on 1997 June 2 with the Space Telescope Imaging Spectrograph (STIS) and Near-Infrared Camera and Multi-Object Spectrometer (NICMOS). The observations reveal a point-like source with $R = 23.1 \pm 0.2$ and $H = 20.6 \pm 0.3$, in agreement with the power-law temporal decay seen in previous ground-based monitoring. Unlike the case of GRB970228, no nebulosity is detected surrounding the OT of GRB970508, although Mg I absorption and [O II] emission seen in Keck spectra at a redshift of $z = 0.835$ suggest the presence of a dense, star-forming medium. The HST observations set very conservative upper limits of $R \sim 24.5$ and $H \sim 22.2$ on the brightness of any underlying extended source. If this subtends a substantial fraction of an arcsecond, then the R band limit is ~ 25.5. Subsequent photometry suggests a flattening of the light curve at later epochs. Assuming the OT decline follows a pure power-law and ascribing the flattening to the presence of an underlying component of constant flux, we find that this must have $R = 25.4$, consistent with the upper limits determined by HST. At $z = 0.8$, this would correspond to an absolute magnitude in the U band of ~ -18, similar to that of the Large Magellanic Cloud (LMC). We propose a scenario in which the host galaxy of the GRB is of Magellanic type, possibly being a "satellite" of one of the bright galaxies located at few arcseconds from the OT.

INTRODUCTION

Optical and infrared imaging of GRB fields has been pursued for several years with the aim of studying the environments of the gamma-ray events, and of localizing and characterizing their host sources. However, till recently, due to the insufficient angular precision in the knowledge of GRB positions, the results have been inconclusive, as reviewed by Band and Hartmann [1].

The high positional accuracy attained by the BeppoSAX Wide Field Cameras has allowed the pointing of optical telescopes at the error circles of GRB970228 and 970508, and the discovery of optical transients (OT) associated with them [2,3]. For the former GRB, HST has resolved an extended source surrounding the point-like OT [4], whose shape, colors and constant brightness led to the identification with a galaxy of irregular morphology.

R band photometry of the OT of GRB970508 [9–11,7,12–18,6,19,20] shows that, following a first increase, the flux started subsiding ~ 2 days after the GRB (see Figure 1). This decay can be modeled, until the epoch of the HST observation, with a power-law temporal dependence of index -1.17 ± 0.04.

Spectroscopy at the Keck II telescope [5] reveals absorption systems at $z = 0.767$ and 0.835 superposed on the continuum as well as [O II] line emission at $z = 0.835$ [6]. These features, besides proving unambiguously the extragalactic location of the OT, suggest a line of sight through a dense interstellar medium; however, the only potential host galaxy detected from ground-based imaging is a faint blue object lying 5".2 away from the OT [7].

The HST observations presented here were designed to search for a host galaxy and to obtain late-time photometry of the OT of GRB970508. A detailed description of the data and results has been given in Pian et al. [8], to which we refer for a complete presentation.

OBSERVATIONS, DATA ANALYSIS AND RESULTS

Four 1250-second exposures were obtained of the GRB970508 field using the STIS CCD in Clear Filter mode during 1997 June 2.52-2.66 (UT). They have been dithered to allow removal of hot pixels and to obtain the highest possible resolution. The images were bias subtracted, flat-fielded, corrected for dark current and calibrated by the newly created STIS pipeline. The final "drizzled" [21] image is available in the Web (URL http://www.stsci.edu/~fruchter/GRB/data_970508). Four exposures of 514 seconds each were also made with the NICMOS Camera 2 on 1997 June 2.67-2.74 (UT) and dithered using the NICMOS spiral dither pattern. The F160W filter (close to the standard near-infrared H band) was used. The OT point-like source is easily visible in all of the data sets.

The photometric calibration of the images was done using the synthetic photometry package SYNPHOT in IRAF/STSDAS. For STIS, given the broad-band response of the instrument, a power-law spectral shape with index $\alpha_\nu \simeq 1$ was assumed, based on Keck spectrophotometry and on the photometric colors. This yields for the OT V= 23.45 ± 0.15 (1σ) and R= 23.10 ± 0.15. For NICMOS, our calibration gives an OT magnitude of H = 20.6 ± 0.3. The faint galaxies located at North-East (G1, adopting the convention of Djorgovski et al. [7]) and North-West of the OT (hereafter G2) are found to have apparent magnitudes R = 24.8 ± 0.2, H = 22.8 ± 0.1 (G1), and R = 25.5 ± 0.2, H = 21.9 ± 0.1 (G2), respectively.

The R band STIS magnitude lies within the 1σ uncertainty of the extrapolation to June 2.5 of the power-law decay fit to earlier data. An R band measurement (R = 23.4) taken at Keck after our HST observation (5 June) confirms this trend. However, subsequent photometry suggests a slight flattening of the light curve (see Figure 1). Under the assumption that a power-law correctly reflects the behavior of the OT, the flattening in the temporal descent might be the signature of an underlying component of constant brightness, such as a galaxy. While a fit to the

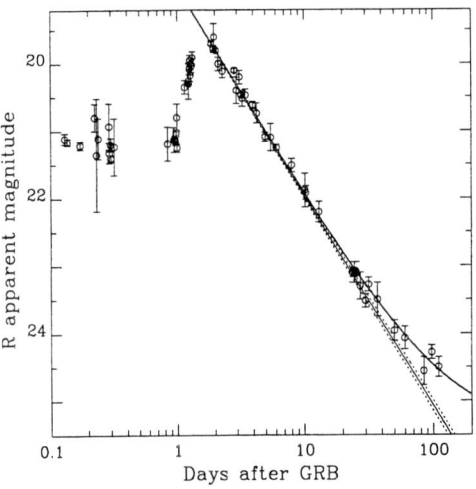

FIGURE 1. R band light curve of the OT associated to GRB970508. Photometry is from ground-based telescopes (open circles) and HST-STIS (filled circle). All magnitudes have been converted to Kron-Cousins R. Uncertainties have been rounded up to 0.1 magnitudes when smaller values were reported in the literature, to take into account possible systematic photometric offsets due to instrumental differences. The fit with a power-law plus a constant with R = 25.4 is reported (thick solid line) along with the power-law curve of index −1.24 (thin solid line) and its 1σ uncertainty range (dashed lines).

data with a simple power-law yields an index of -1.14 ± 0.02 with a $\chi^2 = 1.6$, a fit with a power-law plus a constant gives a power-law best-fit index of -1.24 ± 0.02, with $\chi^2 = 1.2$. The fitted magnitude of the constant component is R = 25.4, with a 90% lower limit of R = 24.7.

As visible from Figure 1, at the epoch of the HST observation it would have not been possible to appreciate a deviation of the measured flux from a simple power-law behavior. Consistently, direct inspection and analysis of the HST-STIS image does not reveal any significant residual flux from an extended source underlying the OT either by subtracting the STIS scaled stellar point spread function (PSF) from the image, nor by subtracting a "compact galaxy" PSF, i.e. the convolution of a normal PSF with a Gaussian of intrinsic FWHM = 0".15. This allows us to conclude that any underlying galaxy must be no brighter than R= 24.5; if it is an extended object with a scale size greater than a few tenths of an arcsecond, it must be even fainter (R ~25.5). These limits are consistent with the best fit value derived above.

Similarly, after subtraction of a scaled artificial PSF, the NICMOS image is also consistent with sky noise statistics and we estimate any underlying, extended component must have H> 22.2 within 0".4 of the point-like source.

DISCUSSION

The detection of Mg I absorption and [O II] emission in the Keck spectra of the OT of GRB970508 implies that the absorbing medium is not highly excited and that active star formation is occurring, respectively. While there are several galaxies with V> 24.5 within a few arseconds of the OT (e.g., G1 and G2), this corresponds to a projected distance of tens of kiloparsecs at $z = 0.8$. It seems unlikely that either the high density or low excitation necessary for the formation of the Mg I line could be maintained this far out in a galactic halo [22]. Therefore, we believe that the absorbing medium responsible for these lines is presently hidden by the light from the OT and is almost certainly the underlying host galaxy.

The apparent magnitude of the constant component derived from the fit to the photometric points, dereddened with $A_R = 0.07$ [7], would correspond, at $z = 0.8$, to an absolute magnitude in the U band of -17.8, with a 90% lower limit of -18.5, assuming $H_0 = 75$ km s^{-1} Mpc^{-1}, and no K-correction. This is consistent with the absolute de-extincted U magnitude of the Large Magellanic Cloud (LMC), -18 [23]. If a K-correction of half a magnitude is assumed (appropriate for the optical-UV spectrum of the LMC [23]), the absolute magnitude of the putative host galaxy would be $M_U = -17.1$, and the absolute magnitude corresponding to the 90% lower limit would be -17.8, still consistent with the luminosity of the LMC, given the uncertainties in the cosmological parameters. Therefore, it is possible that the GRB occurred in a small galaxy of brightness similar to that of the LMC and of comparable size and shape. Strong star formation takes place in Magellanic-like galaxies [24], so that the proposed scenario would be still consistent with the speculation of Pian et al. [8] about the link of GRBs to star formation.

The OT is located at ~5 arcseconds from the bright galaxies located on the North (G1 and G2). If all three objects are at $z = 0.8$, the GRB host galaxy would be at a few tens of kiloparsecs away from those bright galaxies, and probably be dynamically related to one of them. Indeed, G1 is extremely blue, as reported by Djorgovski et al. [7], and the colors are consistent with a rapidly star-forming galaxy at any reasonable redshift. G2 is somewhat redder, but has the colors of a nearby late-type spiral galaxy whose spectrum has been shifted to $z \sim 0.7 - 0.8$.

In order to confirm or disprove this scenario, further optical and near-infrared imaging of the OT of GRB970508 is necessary, which would extend the sampling of the light curve and possibly directly detect the hiding host galaxy.

REFERENCES

1. Band, D. L., and Hartmann, D. H. these proceedings; astro-ph/9711328.
2. van Paradijs, J., et al., *Nature* **386**, 686 (1997).
3. Bond, H. E., *IAU Circ.* No. 6654 (1997).
4. Sahu, K. C., et al., *Nature* **387**, 476 (1997).
5. Metzger, R. M., et al., *Nature* **387**, 879 (1997).
6. Metzger, R. M., et al., *IAU Circ.* No. 6676 (1997).
7. Djorgovski, S. G., et al., *Nature* **387**, 876 (1997).
8. Pian, E., et al., *Ap. J.* in press (1998); astro-ph/9710334.
9. Pedersen, H., et al., *Ap. J* **496**, in press (1998).
10. Castro-Tirado, A. J., et al., *IAU Circ.* No. 6657 (1997).
11. Sahu, K. C., et al., *Ap. J. Lett.* **489**, L127 (1997).
12. Galama, T. J., et al., *IAU Circ.* No. 6655 (1997).
13. Schaefer, B., et al., *IAU Circ.* No. 6658 (1997).
14. Sokolov, V. V., et al., these proceedings.
15. Mignoli, M., et al., *IAU Circ.* No. 6661 (1997).
16. Groot, P. J., et al., *IAU Circ.* No. 6660 (1997).
17. Garcia, M., et al., *IAU Circ.* No. 6661 (1997).
18. Chevalier, C., and Ilovaisky, S. A., *IAU Circ.* No. 6663 (1997).
19. van Paradijs, J., et al., these proceedings.
20. Groot, P. J., et al., these proceedings.
21. Fruchter, A. S., and Hook, R. N. *Publ. Astron. Soc. Pac.* in press (1997); astro-ph/9708242.
22. Steidel, C. C., and Sargent, W. L. W., *Ap. J. Suppl.* **80**, 1 (1992).
23. De Vaucouleurs, G., et al., *Third Reference Catalog of Bright Galaxies*, V. 3.9 (1991).
24. Hunter, D., *Publ. Astron. Soc. Pac.* **109**, 937 (1997).

HST/STIS Observations of the Optical Counterpart to GRB970228

Andrew S. Fruchter[1], Elena Pian[2], Stephen E. Thorsett[3], Rosa Gonzalez[1], Kailash C. Sahu[1], Max Mutchler[1], Filippo Frontera[2,6], Titus Galama[7], Paul Groot[7], Richard Hook[9], Chryssa Kouveliotou[8], Mario Livio[1], Duccio Macchetto[1], Jan van Paradijs[7], Eliana Palazzi[2], Larry Petro[1], Marco Tavani[4,5]

[1] *Space Telescope Science Institute, 3700 San Martin Drive, Baltimore, MD 21218, USA*
[2] *Istituto di Tecnologie e Studio delle Radiazioni Extraterrestri, C.N.R., Via Gobetti 101, I-40129 Bologna, Italy*
[3] *Joseph Henry Laboratories and Dept. of Physics, Princeton University, Princeton, NJ 08544, USA*
[4] *Columbia Astrophysics Laboratory, Columbia University, New York, NY 10027, USA*
[5] *Istituto di Fisica Cosmica e Tecnologie Relative, C.N.R., Via Bassini 15, I-20133 Milano, Italy*
[6] *Dip. Fisica, Università di Ferrara, Via Paradiso 12, I-44100 Ferrara, Italy*
[7] *Astronomical Institute "Antonton Pannekoek", University of Amsterdam, Kruislaan 403, 1098 SJ Amsterdam, The Netherlands*
[8] *NASA Marshall Space Flight Center, ES-84, Huntsville, AL 35812, USA*
[9] *Space Telescope European Coordinating Facility, D-85748 Garching, Germany*

Abstract. We report on observations of the fading optical counterpart of the gamma-ray burst GRB970228, made on 4 September 1997 using the STIS CCD on the Hubble Space Telescope. The unresolved counterpart is detected at $V = 28.0 \pm 0.25$, consistent with a continued power-law decline with exponent 1.14 ± 0.05. No proper motion is detected, in contradiction of some earlier claims. The counterpart is located within, but near the edge of, a faint extended source with diameter $\sim 0''.8$ and integrated magnitude $V = 25.7 \pm 0.25$. Comparison with WFPC2 data taken one month after the initial burst and NTT data taken on March 13 shows no evidence for fading of the extended emission.

After adjusting for the probable Galactic extinction in the direction of GRB970228 of $A_v \sim 0.7$, we find that the observed nebula is consistent with the sizes of galaxies of comparable magnitude found in the Hubble Deep Field and other deep HST images, and that only 2% of the sky is covered by galaxies of similar or greater surface brightness. We therefore argue that the extended source observed about GRB970228 is most likely a galaxy at moderate redshift, and is almost certainly the host of the gamma-ray burst.

INTRODUCTION

Identification and analysis of long wavelength counterparts of gamma-ray bursts (GRBs) has for many years been considered a promising path towards understanding the nature of the burst events. But while many attempts have been made to identify GRB counterparts [21,20,6,10] until recently the uncertainties in the position of the gamma-ray sources proved too large to allow sufficiently sensitive surveys for associated optical transients (OTs). The situation improved dramatically in early 1997, when gamma- and X-ray observations by the BeppoSAX satellite provided a sub-arcminute position of burst GRB970228, allowing the first firm optical identification of a fading GRB counterpart [23]. A second optical GRB counterpart, this time of GRB970508, was discovered two months later [1,4].

Although broadly similar in their fading behavior, phenomenological differences between the two counterparts in the weeks after their discoveries seemed to compound rather than clarify the mystery of the GRBs. HST imagery of GRB970228 suggested the presence of a nebulosity centered 0.3 arcsecond from the point-like fading transient source [18], while no extended source brighter than $R = 24.5$ has been found near GRB970508 [16]. Furthermore, a proper motion of 550 mas yr^{-1} was reported [3] for GRB970228, though the measurement was disputed [19]. Additionally, tentative evidence for the fading of the adjacent nebulosity [14] was proposed. Either result would ineluctably lead to the conclusion that GRB970228 was a Galactic event. In contrast, the measurement of absorption lines with redshift $z \geq 0.835$ in the spectrum of GRB970508 [15] demonstrates its extragalactic nature.

To help resolve the situation, we have reobserved the GRB970228 with HST six months after the initial outburst. Although the earlier HST observations of GRB970228 employed WFPC2, we have availed ourselves of the newly installed STIS CCD camera. The excellent throughput and broad bandpass of this instrument, combined with the long time baseline since the gamma-ray burst, provide us with a superb opportunity to study the nature of the source and its environment.

OBSERVATIONS, IMAGE REDUCTION AND PHOTOMETRY

The field of GRB970228 was imaged during two HST orbits on 1997 September 4 from 15:50:33 to 18:22:41 UT, using the STIS CCD in Clear Aperture (50CCD) mode. Two exposures of 575s each were taken at each of four dither positions for a total exposure time of 4600s. The exposures were dithered to allow removal of hot pixels and to obtain the highest possible resolution. The images were bias and dark subtracted, and flat-fielded using the the STIS pipeline. The final image was created and cleaned of cosmic rays and hot pixels using the Drizzle and Blot

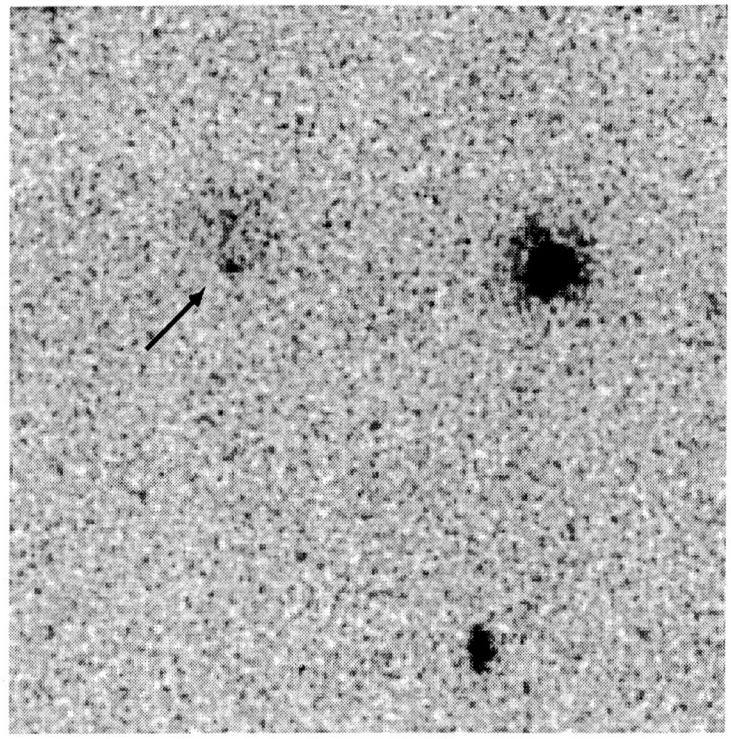

FIGURE 1. The STIS image of GRB970228. North is up; East is to the left. An arrow points to the OT. The nebula can be seen extending to the north of the OT.

algorithms developed for the Hubble Deep Field (HDF) [25,7]. An output pixel size of 0″.025 cross, or one-half the size of the input pixels was used.

The magnitude of the OT was determined from the drizzle image via aperture photometry. The flux in an aperture of radius four (0″.025) pixels was determined, and our best estimate of the surrounding nebular background was subtracted. An aperture correction of 0.50 magnitudes was derived for this aperture using the bright star visible in Figure 1 to the west of the nebula. We find a total magnitude for the point source OT of V = 28.0 ± 0.15.

In Figure 2 we plot the magnitude of the OT as a function of time since the burst last February. The STIS magnitude as been converted to the R by interpolating the WFPC2 V and I colors. A power law of the form $f(t) = a_0 * t^{-\alpha}$ has been fitted to the HST points and extrapolated back to earlier times. We find a best fit of $\alpha = 1.14 \pm 0.05$. All non-HST photometry has been adjusted under the assumption that the R magnitude of the nebula is 25.3, again obtained by interpolating the WFPC2 V and I measurements (see below).

We determined the magnitude of the nebula by summing all pixels in a region of

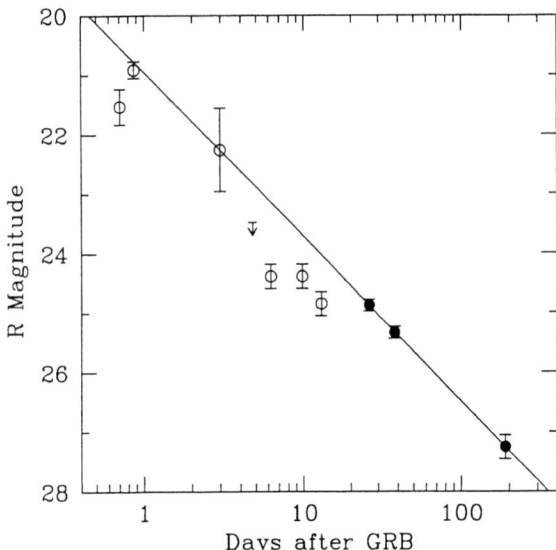

FIGURE 2. The R magnitude of the OT as a function of time. A nebular R magnitude of 25.3 has been subtracted from all non-HST magnitudes. The line shows the best fit power law through the three HST observations (shown in black). See Rees and Meszaros [17] for a possible explanation for the deviation of the observed points deviation from the power law.

approximately 1.4 sq. arcsec. surrounding the object. The flux of the point source was then subtracted from the sum. We derive a magnitude of V = 25.7 ± 0.25 for the nebula. Due to the very wide bandpass of the STIS clear observations (the only constraint on the bandpass comes from the optics and the response of the CCD), the primary source of photometric error is the correction of the STIS magnitude for the color of the object. This is particularly large for the nebula, as the only measurement of the color of this object comes from the previous, rather noisy, Planetary Camera observations of the field (these images are discussed in more detail in the next section).

In order to determine whether the optical transient displayed proper motion or the nebula faded we have compared the STIS images to the previous HST WFPC2 images obtained of GRB970228 [18]. The images taken on March 26 with WFPC2 provide a baseline of 162 days and are used here to look for proper motion of the optical transient.

The positions of the four reference stars used in Sahu et al. [19] agree with their positions in the STIS images to within the expected uncertainties of 2 to 3 milliarcsec, which shows that the transformations between the two images have been done correctly. The uncertainty in the position of the OT is about 10 mas in each of the two colors. We find that any motion of the GRB between the two epochs is less than about 16 milliarcsec. This corresponds to a motion of less than 36 milliarcsec per year. This is inconsistent with the value claimed by Caraveo et al. [3]; however, it does improve the upper limit on the proper motion reported by Sahu et al. [19] by a factor of six.

To check on the photometry of the optical transient and nebula, the point source magnitude was determined by using 1 and 3 pixel radius circular apertures and adjusting the observed fluxes according to the aperture corrections found by Holtzman et al. [9]. The nebular magnitude was redetermined in the WFPC2 images by taking the sum of all counts above sky in a box approximately $1\rlap{.}''5 \times 1''$, and subtracting the counts (estimated as above) attributable to the point source. The position of this box was determined by the position of the nebula in the STIS image. It is, however, somewhat larger than the observed nebula in all directions. Averaging together the two WFPC2 observations, we obtain an I magnitude of 24.4 ± 0.2 and a V magnitude of 25.5 ± 0.2. These magnitudes are easily consistent with the STIS magnitude.

We have also re-examined the NTT observation of March 13 [8] to further test whether the nebular magnitude may have varied with time. We have again used the stellar image $\sim 2\rlap{.}''5$ to the west of the OT as a point spread function. We find that we can subtract a point source from the position of the OT which is fainter than the extrapolation of the power-law, yet which leaves behind a "neblula" which is as faint, or fainter than, the HST nebular magnitude. Thus we find no evidence that the nebula has changed magnitude with time.

ASTROPHYSICAL IMPLICATIONS

There is little room for doubt that the fading point source is associated with the gamma-ray event. Between 28 February and 4 September 1997, the source faded by $\Delta V \sim 6.5$, or a factor of ~ 350, and as shown in Figure 2, this dramatic fall in luminosity largely followed a power law whose index, -1.14 ± 0.05, is, within the errors indistinguishable from the index of power-law decline of the optical counterpart to GRB970508 [16]. Given the lack of any other astrophysical objects with similar behavior, and the theoretical prediction of a power-law fall-off with time of the luminosity of afterglow [13], we believe there is no reasonable alternative to the conclusion that we are observing the optical afterglow of GRB970228.

Furthermore, in simple blast wave models, a break in the power-law to $F \sim t^{-1.8}$ is expected [24] when the remnant enters a Sedov-Taylor phase after sweeping up a rest mass energy equal to its initial energy E at time:

$$t \approx 1 \,\text{yr} \left(\frac{E_{52}}{n}\right)^{1/3}, \tag{1}$$

where n is the density of the surrounding medium in protons per cubic centimeter. However, were the GRB a Galactic rather than an extragalactic phenomenon, the amount of energy available would only be of order 10^{41} ergs, and for any imaginable density the break would occur on a timescale of days rather than many months. Therefore the power-law fit is, in itself, a strong argument for the extragalactic nature of the burst.

If the burst is extragalactic, then it is natural to inquire whether the apparently constant nebula seen under the OT is the host galaxy. The galactic extinction in the direction of GRB970228 has been estimated as $A_v \sim 0.7$ [2,22] – a figure that we have been able to independently verify by comparing the counts and colors of background galaxies in the WFPC2 field with the HDF. Adjusting the surface brightness limit to reflect the extra ~ 0.7 mags of extinction in the direction of GRB970228, we find that only about 2% of the sky in the HDF is covered by galaxies of comparable magnitude, and that the size of the putative host of GRB970228, while larger than the mean 25th magnitude galaxy in the HDF, is not, by any means, extraordinary.

Although we have no spectroscopic information on the redshift of this object nor do we have sufficient colors to attempt a photometric redshift (though planned NICMOS observations may rectify this problem), we can attempt to place a crude constraint on the plausible redshift simply from the luminosity of the object. Were the object closer than $z \sim 0.5$ it would be more than four magnitudes fainter than L_* [11], and this is unlikely even given the steep luminosity function at that redshift [5]. On the other hand, the apparent host is as bright as any "U dropout" in the HDF [12], and therefore would be an unusually bright galaxy were it at the typical redshift of these dropouts, $z \sim 2.5$. Thus a plausible redshift range for the host is $0.5 \lesssim z \lesssim 2.5$. However, while the luminosity function of galaxies is a rather blunt instrument for estimating the redshifts of GRB hosts, we will show in the journal paper associated with this work that GRB hosts may prove a rather better tool for determining the luminosity function of galaxies.

ACKNOWLEDGEMENTS

We thank the Director of STScI, Bob Williams, for allocating Director's Discretionary time to this program.

REFERENCES

1. Bond, H. E., *IAUC*, 6654 (1997).
2. Burstein, D., and Heiles, C., *Astron. J.* **87**, 1165–1189 (1982).

3. Caraveo, P. A., Mignani, R. P., Tavani, M., and Bignami, G. F., *Astr. Astrophys.* **326**, L13–L16 (1997).
4. Djorgovski, S. G., et al. , *Nature* **387**, 876–878 (1997).
5. Ellis, R. S., Colless, M., Broadhurst, T., Heyl, J., and Glazebrook, K., *Mon. Not. R. Astr. Soc.* **280**, 235–251 (1996).
6. Fenimore, E. E. et al. , *Nature* **366**, 40 (1993).
7. Fruchter, A. S. and Hook, R. N., in *Applications of Digital Image Processing XX, Proc. SPIE* **3164**, ed. A. Tescher, SPIE, 120–125 (1997).
8. Galama, T. et al. , *Nature* **387**, 479–481 (1997).
9. Holtzman, J. A., Burrows, C. J., Casertano, S., Hester, J. J., Trauger, J. T., Watson, A. M., and Worthey, G., *Publ. Astr. Soc. Pacific* **107**, 1065 (1995).
10. Larson, S. B., *Astrophys. J.* **491**, 86 (1997).
11. Lilly, S. J., Tresse, L., Hammer, F., Crampton, D., and Le Fevre, O., *Astrophys. J.* **455**, 108–124 (1995).
12. Madau, P., Ferguson, H. C., Dickinson, M. E., Giavalisco, M., Steidel, C. C., and Fruchter, A. S., *Mon. Not. R. Astr. Soc.* **283**, 1388–1404 (1996).
13. Meszaros, P. and Rees, M. J., *Astrophys. J.* **476**, 232 (1997).
14. Metzger, R. M., Cohen, J. G., Chaffee, F. H., and Blandford, R. D., *IAUC* 6676 (1997).
15. Metzger, R. M., Djorgovski, S. G., Kulkarni, S. R., Steidel, C. C., Adelberger, K. L., Frail, D. A., Costa, E., and Frontera, F., *Nature* **387**, 878 (1997).
16. Pian, E. et al. , *Astrophys. J.* **492**, L103 (1998).
17. Rees, M. J. and Meszaros, P., astro-ph/9712252 (1997).
18. Sahu, K. C. et al. , *Astrophys. J.* **489**, L127 (1997).
19. Sahu, K. C. et al. , *Nature* **387**, 476 (1997).
20. Schaefer, B. E., in *Gamma-Ray Bursts: Observations, Analyses and Theories*, ed. C. Ho, R.I. Epstein, and E.E. Fenimore, Cambridge University Press, 107 (1992).
21. Schaefer, B. E. et al. , *Astrophys. J.* **313**, 226–230 (1987).
22. Schlegel, D. J., Finkbeiner, D. P., and Davis, M., astro-ph/9710327 (1997).
23. van Paradijs, J. et al. , *Nature* **386**, 686–689 (1997).
24. Wijers, R. A. M. J., Rees, M. J., and Meszaros, P., *Mon. Not. R. Astr. Soc.* **288**, L51–L56 (1997).
25. Williams, R. E. et al. , *Astron. J.* **112**, 1335–1389 (1996).

Konus-Wind and Konus-A Observations of GRB970228

R.L.Aptekar[1], P.S.Butterworth[2], T.L.Cline[2], D.D.Frederiks[1],
S.V.Golenetskii[1], V.N.Il'inskii[1], E.P.Mazets[1], D.E.Stilwell[2],
and M.M.Terekhov[1]

[1] *Ioffe Physical-Technical Institute, St.Petersburg, 194021, Russia*
E-mail: aptekar@mz.ioffe.rssi.ru
[2] *Goddard Space Flight Center, Greenbelt, MD 20771, USA*
E-mail: butterworth@lheavx.gsfc.nasa.gov

Abstract. The gamma-ray burst of 970228 was observed with Konus instruments on both the US GGS-Wind spacecraft and the Russian Kosmos-2326 Earth orbiter. The time histories and incident photon spectra obtained throughout the broad energy band above 12 keV and the source localization results are presented.

INTRODUCTION

An ordinary weak GRB of 970228 was observed with gamma-ray burst detectors on many spacecraft including BeppoSAX, Ulysses, GGS-Wind, and Kosmos-2326.
Highly extraordinary results were obtained from studying this event. The BeppoSAX Wide Field Camera provided a small error box with radius of 3' for the GRB position [1], which has been reduced by a factor of 7 with the IPN triangulation data [2]. In this small area, a fading X-ray source [3], and an optical transient [4], consistent with an extended source [5,6], have been observed. Some radio sources as possible counterparts to the GRB have been reported [7]. These data, assuming a genuine association with the GRB, imply an extragalactic origin of GRB's [8]. In this report, we present the time history, the photon spectrum, and source localization data obtained for this event with the Konus-Wind [9] and Konus-A [10] experiments.

RESULTS

Figure 1 displays with a resolution of 256 ms the time histories of GRB970228 recorded in three energy windows on both the Wind and Kosmos-2326 spacecraft.

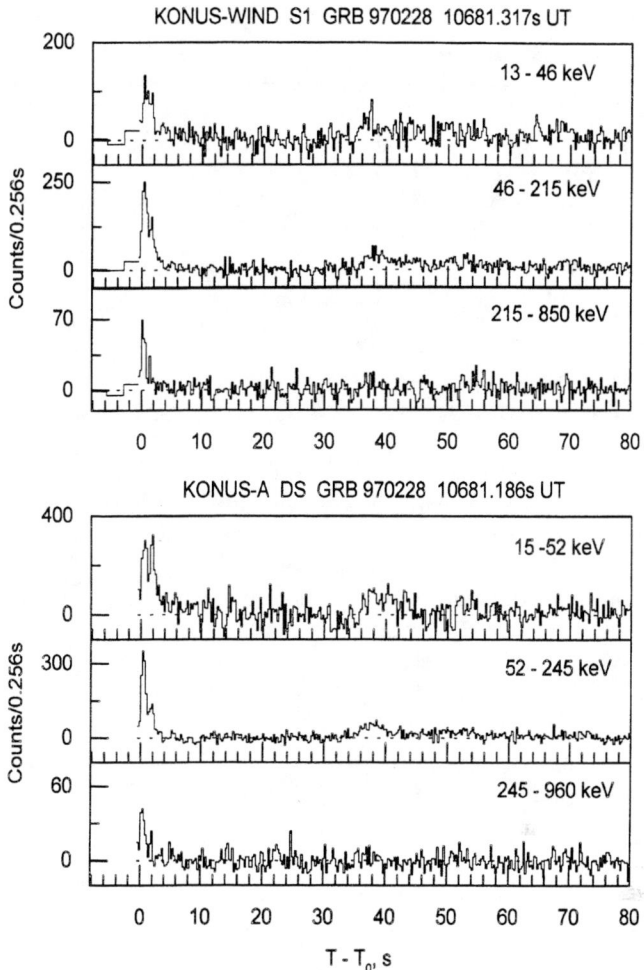

FIGURE 1. Background-subtracted time histories of GRB970228 observed by the Konus-Wind and Konus-A instruments.

The burst is weak in intensity. Its time profile consists of an initial strong pulse separated by a 25 s time interval from several weaker subsequent pulses.

Averaged photon spectra of the whole initial pulse measured by means of three different detectors are presented in Figure 2. These measurements agree very well with each other. The spectrum below 250 keV is well fitted by an approximate expression $E^{-1}\exp(-E/150\text{keV})$ exhibiting a power-law tail $\sim E^{-2.7}$ at higher energy. The total fluence above 15 keV for this burst is 8.2×10^{-6} erg cm^{-2}.

An area of possible location for this burst is shown in Figure 3. The annulus of possible arrival directions obtained by triangulation of Konus-Wind and Konus-A

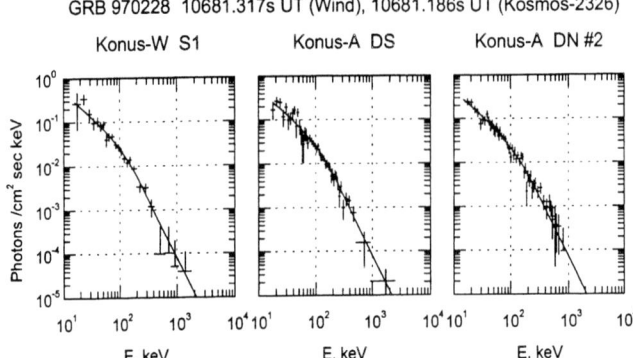

FIGURE 2. Time-averaged photon spectra of the initial pulse of GRB970228.

data, intersects a rather large error box provided by the detector array with known angular sensitivity on board Kosmos-2326. The triangulation annulus between Ulysses and GGS-Wind spacecraft [2] as well as BeppoSAX WFC position [1] are also shown. These results both add to the compilation of phenomenological data concerning this historically significant event and confirm the directional capabilities of the Konus-Wind/Konus-A network.

ACKNOWLEDGMENTS

This work was partly supported by RSA contracts, by RFBR grants N96-02-16860 and N97-02-18067, and by the CRDF grant RP1-236.

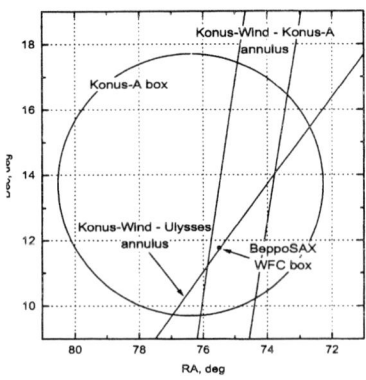

FIGURE 3. Konus-Wind – Konus-A – Ulysses localization of GRB970228

REFERENCES

1. Costa, E., et al., *IAU Circ.* *6572* (1997).
2. Hurley, K., et al., *Ap.J* **485**, L1 (1997).
3. Costa, E., et al., *IAU Circ.* *6576* (1997).
4. Groot, P., et al., *IAU Circ.* *6584* (1997).
5. Groot, P., et al., *IAU Circ.* *6588* (1997).
6. Sahu, K., et al., *IAU Circ.* *6606* (1997).
7. Frail, D.A., et al., *IAU Circ.* *6576* (1997).
8. Fruchter, A., et al., *IAU Circ.* *6747* (1997).
9. Aptekar, R.L., et al., *Space Science Rev.* **71**, 265 (1995).
10. Aptekar, R.L., et al., *Astronomy Letters* **23**, 265 (1997).

The Galactic Extinction Toward GRB970228 and Its Implications

Francisco J. Castander and D. Q. Lamb

Department of Astronomy and Astrophysics, University of Chicago

Abstract. The IRAS 100 micron image of the GRB970228 field shows that the amount of galactic dust in this direction is substantial and varies on arcminute angular scales. From an analysis of the observed surface density of galaxies in the $2' \times 2'$ HST WFPC image of the GRB970228 field, we find $A_V = 1.1 \pm 0.10$. From an analysis of the observed spectra of three stars in the GRB970228 field, we find $A_V = 1.71^{+0.20}_{-0.40}$. This value may represent the best estimate of the extinction in the direction of GRB970228, since these three stars lie only $2.7''$, $16''$, and $42''$ away from the optical transient. If instead we combine the two results, we obtain a conservative value $A_V = 1.3 \pm 0.2$. This value is significantly larger than the values $A_V = 0.4 - 0.8$ used in papers to date. The value of A_V that we find implies that, if the extended source in the burst error circle is extragalactic and therefore lies beyond the dust in our own galaxy, its optical spectrum is very blue: its observed color $(V - I_c)_{obs} \approx 0.65^{+0.60}_{-0.85}$ is consistent only with a starburst galaxy, an irregular galaxy at $z > 1.5$, or a spiral galaxy at $z > 2$. On the other hand, its observed color and surface brightness $\sigma_V \approx 24.5$ arcsec^{-2} are similar to those expected for the reflected light from a dust cloud in our own galaxy, if the cloud lies in front of most of the dust in this direction.

INTRODUCTION

The IRAS 100 micron image of the GRB970228 field shows that the amount of galactic dust in this direction is substantial and varies significantly on arcminute angular scales (see Figure 1). Here we report a determination of the visual extinction A_V toward GRB970228 using two methods: (1) the observed versus the expected surface density of galaxies in the HST WFPC 606 and 814 nm images of the GRB970228 field; and (2) the observed color versus the spectral type of three stars that lie near the position of the optical transient (see Figure 2).

ANALYSIS AND RESULTS

Galaxy number counts can be used to measure directly the relative extinction between two fields. Since absorption due to dust increases the observed apparent magnitude of a galaxy at infrared, optical, and ultraviolet wavelengths, the number

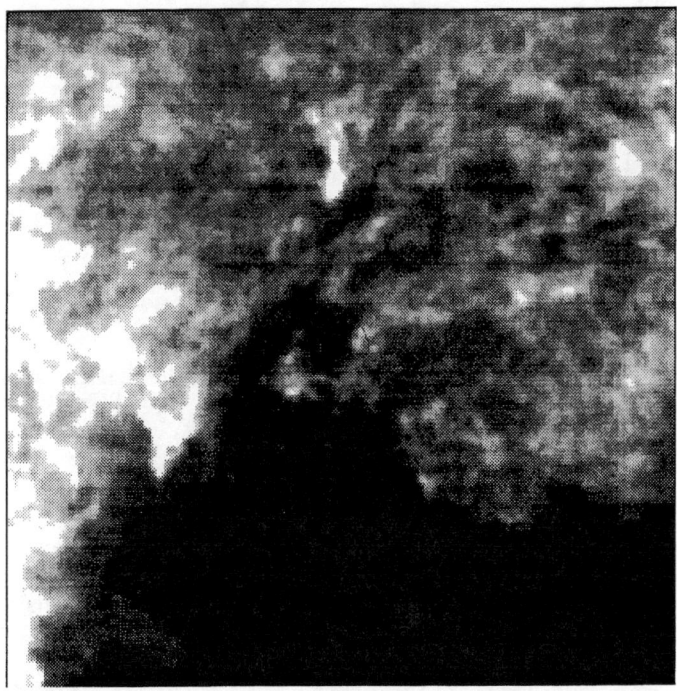

FIGURE 1. IRAS 100 µm map of the GRB970228 field, covering 8.5° × 8.5° and having a resolution of 2'. The bright regions correspond to strong dust emission, the dark regions to weak dust emission. We have superposed on the map a circle 20' in radius, centered on the position of the optical transient (R.A. = $5^h01^m46.7^s$, Decl. = +11°46'54'', J2000).

of galaxies per unit area brighter than a given apparent magnitude is reduced if extinction is present. We have compared the surface density of galaxies as a function of apparent magnitude in the HST WFPC 606 and 814 nm images of the GRB970228 field [1] with similar images of the Hubble Deep Field [2] (HDF) (Figure 3). We have used the same procedure to analyze both fields; our results for the HDF field agree with those reported earlier [2]. We restrict our comparison to galaxies in the WFPC images of the GRB970228 field brighter than $m_{606} \leq 26.5$ and $m_{814} \leq 25.9$, apparent magnitudes for which the galaxy counts in this field appear to be complete.

We determine the extinction in the HST WFPC images of the GRB970228 field using a maximum likelihood method. We construct a likelihood function that is a product of four individual likelihood functions, one for the 606 and 814 nm images of each field. We model the surface density of galaxies as a function of apparent magnitude in each image as a power-law distribution. We assume that the slopes α_{606} and α_{814} of the power-law distributions in the 606 and 814 nm images are the same. We further assume that the extinction in both fields is described by the

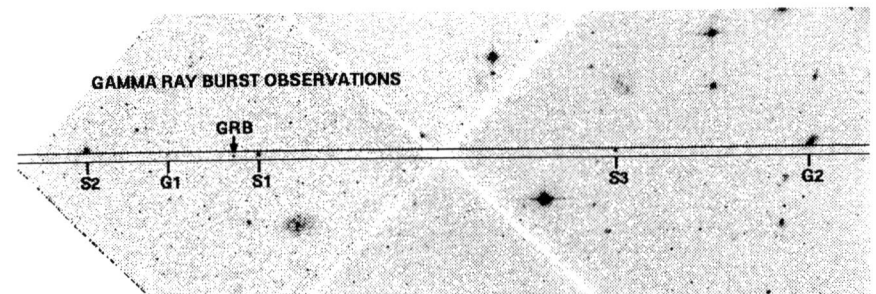

FIGURE 2. Position of the 1″ wide slit used in observations made on 1997 March 31, April 1, and April 2 UT using the LRIS spectrograph on the Keck II 10-meter telescope [3,4], superposed on the HST WFPC 606 nm images of the GRB970228 field obtained using the HST on 1997 March 26 [1]. The small truncated square to the left is the field of the PC chip; the other three truncated squares are the fields of the WF chips. The position of the optical transient is labeled "GRB;" also labeled are the positions of stars S1, S2, and S3; and galaxies G1 and G2.

typical extinction behavior of the interstellar medium, so that $A_{606} = 0.92 A_V$ and $A_{814} = 0.61 A_V$. The resulting likelihood function only depends on five parameters: the normalizations and the slopes α_{606} and α_{814} of the power-law distributions in the two filters and the difference in the extinction between the GRB970228 field and the HDF. We then marginalize this likelihood function over the normalizations in the 606 nm and 814 nm filters and use the measured value $A_V^{HDF} = 0.0$. [2] This reduces the number of parameters to three: α_{606}, α_{814}, and A_V. Maximizing the likelihood of the observed surface density of galaxies as a function of apparent magnitude in each image of each field, given the model, we obtain a best-fit value $A_V = 1.1 \pm 0.10$.

We also determine the visual extinction A_V toward GRB970228 using the spectra of three stars, denoted S1, S2, and S3, that lie near the optical transient in the GRB970228 error circle (see Figure 2). The spectra of these stars were obtained on 1997 March 31, April 1, and April 2 UT using the LRIS spectrograph on the Keck II 10-meter telescope [3,4]. Comparing the observed spectra with the stellar spectral atlases [5,6], we find that S1, S2, and S3 lie in the spectral ranges K3v-K5v, K4v-K7v, and M0v-M3v, respectively. Reddening the spectra in the stellar spectral atlases, we obtain best-fit spectra and visual extinction values for S1, S2, and S3 of K4v and $A_V = 1.4^{+0.5}_{-0.8}$, K4v and $A_V = 1.8^{+0.2}_{-0.5}$, and M2v and $A_V = 1.8^{+0.5}_{-1.0}$, respectively. Weighting the individual values of A_V by the signal-to-noise of their spectra, we obtain a combined value $A_V = 1.71^{+0.20}_{-0.40}$. This may represent the best estimate of the extinction in the direction of GRB970228, since these three stars lie only 2.7″, 16″, and 42″ away from the position of the optical transient.

Combining the results of our analysis of the surface density of galaxies in the GRB970228 field and the results of our analysis of the spectra of stars S1, S2, and S3, we obtain a conservative value $A_V = 1.3 \pm 0.2$. This value is consistent with the X-ray spectrum of the gamma-ray burst itself, which yields $n_H = 3.5^{+3.3}_{-2.3} \times 10^{21}$

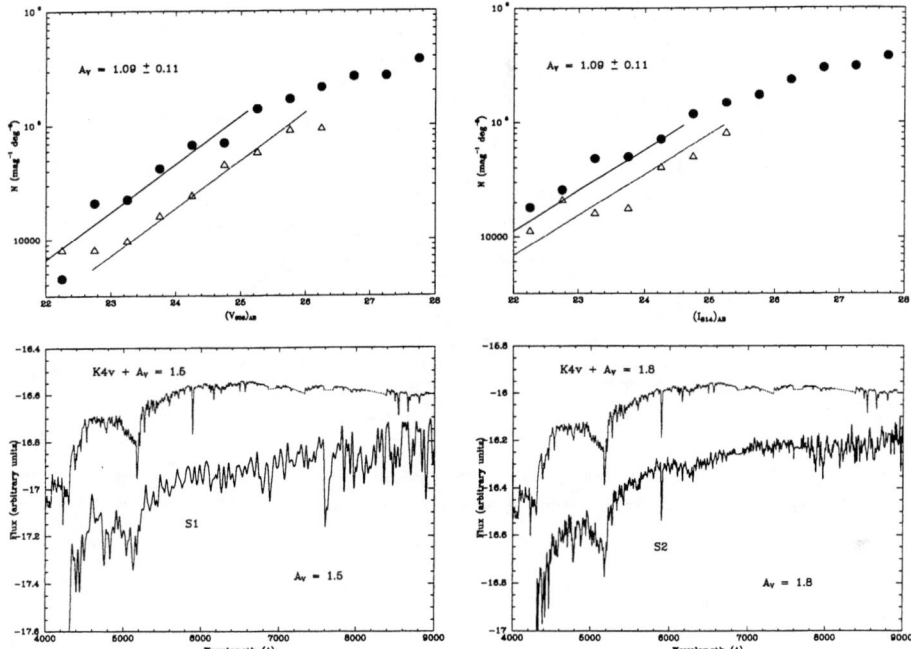

FIGURE 3. Upper panels: Surface density of galaxies as a function of apparent magnitude in the WFPC 606 nm (left panel) and 814 nm (right panel) images of the HDF (filled circles) and the GRB970228 field (open triangles). Also shown is the surface density of galaxies as a function of apparent magnitude for the HDF (solid lines) and the GRB970228 field (dotted lines), given by the best-fit model we use to describe the combined data for the HDF and GRB970228 fields. Lower panels: Observed (solid curves) and model (dotted curves) spectra of stars S1 (left panel) and S2 (right panel). The observed spectra of S1 and S2 are best fit by that of a K4v star, reddened by $A_V = 1.4^{+0.5}_{-0.8}$ and $A_V = 1.8^{+0.2}_{-0.5}$, respectively.

cm^{-2} [7], implying $A_V = 2.3^{+2.1}_{-1.5}$.

CONCLUSIONS

From an analysis of the observed surface density of galaxies in the $2' \times 2'$ HST WFPC image of the GRB970228 field, we find $A_V = 1.1 \pm 0.10$. From an analysis of the observed spectra of three stars in the GRB970228 field, we find $A_V = 1.71^{+0.20}_{-0.40}$. This value may represent the best estimate of the extinction in the direction of GRB970228, since these three stars lie only 2.7", 16", and 42" away from the optical transient. If instead we combine the two results, we obtain a conservative value $A_V = 1.3 \pm 0.2$.

This value is significantly larger than the values $A_V = 0.4 - 0.8$ used in papers to

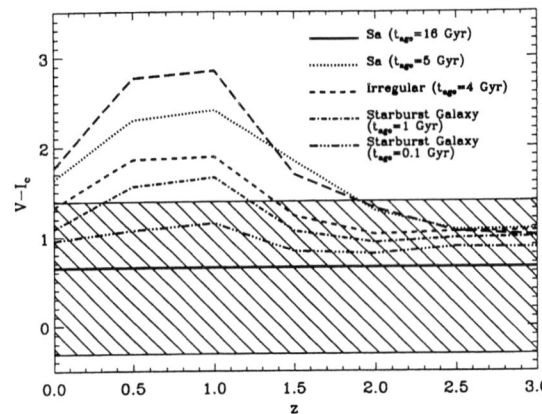

FIGURE 4. Expected colors for several kinds of galaxies at different ages as a function of redshift z. The curves include the appropriate K-correction, but assume no galaxy evolution. The thick solid line shows $(V - I_c)_{\rm obs}$ for the extended source in the burst error circle; the hatched region indicates the range of uncertainty in this color. This figure shows that the extended source is very blue: its $V - I_c$ color is consistent only with that expected for starburst galaxies, for irregular galaxies at $z > 1.5$, or for spiral galaxies at $z > 2$.

date. The value of A_V that we find implies that, if the extended source in the burst error circle is extragalactic, its optical spectrum is very blue: its observed color of $(V - I_c)_{\rm obs} \approx 0.65^{+0.60}_{-0.85}$ is consistent only with a starburst galaxy, an irregular galaxy at $z > 1.5$, or a spiral galaxy at $z > 2$. On the other hand, its observed color and surface brightness $\sigma_V \approx 24.5$ arcsec^{-2} are similar to those expected for the reflected light from a dust cloud in our own galaxy, if the cloud lies in front of most of the dust in this direction.

We thank John Tonry, Esther Hu, Len Cowie, and Richard McMahon for making available the spectra of the galaxies and stars in the GRB970228 field. We acknowledge support from NASA grants NAGW-4690 and NAG 5-1454.

REFERENCES

1. Sahu, K. C. et al., *Nature* **387**, 476 (1997).
2. Williams, R. E. et al., *AJ* **112**, 1335 (1996).
3. Tonry, J. L., Hu, E. M., Cowie, L. L. and McMahon, R. G., IAU Circular No. 6620 (1997).
4. Tonry, J. L. et al. http://www.ifa.hawaii/faculty/hu/grb.html (1997).
5. Jacoby, G. H., Hunter, D. A. & Christian, C. A. 1984, *ApJS* **56**, 257 (1984).
6. Silva, D. R. & Cornell, M. E., *ApJS* **81**, 865 (1992).
7. Costa, E. et al., *Nature* **387**, 783 (1997).

Multicolor Photometry of the GRB970508 Optical Remnant

V.V. Sokolov[*], A.I. Kopylov[*], S.V. Zharikov[*], M. Feroci[†],
E. Costa[†], L. Nicastro[‡], E. Palazzi[‡].

[*] *Special Astrophysical Observatory of RAS, Karachai-Cherkessia, Nizhnij Arkhyz, 357147 Russia; sokolov,akop,zhar@sao.ru*
[†] *Istituto di Astrofisica Spaziale CNR, 00044 Frascati, Italy*
[‡] *Istituto Tecnologie e Studio Radiazioni Extraterrestri CNR, 40129 Bologna, Italy*

Abstract. We report results of follow-up multicolor photometry of the optical variable source that is a probable remnant of the gamma-ray burst GRB970508 discovered by the BeppoSAX satellite [3]. Observations were carried out in Johnson-Kron-Cousins BVR_cI_c system with the 1-m and 6-m telescopes of SAO RAS. Between the 2nd and the 5th day after the burst a fading of the remnant is well fitted with an exponential law in all four bands. During this period the 'broadband spectrum' of the object was unchanged and can be approximated by a power-law, $F_\nu \propto \nu^{-1.1}$. After the 5th day the decline of brightness is slowed down. In the R_c band until the 84th day, the light curve can be described by a power-law relation, $F \propto t^{-1.179(\pm 0.016)}$.

THE SEARCH FOR AN OPTICAL COUNTERPART

The first coordinates of the 10′ radius error box for the GRB970508 (May, 8.904 UT) were received at SAO RAS from the BeppoSAX team on May 9.05 UT. At that time observations were not possible because of the beginning of twilight. The refined coordinates of 5′ radius error box were received on May 9.3 UT.

The search for an optical counterpart began with the 1-m telescope on May 9.74 UT. The 5′ error box for the GRB970508 localization was completely covered with the CCD mosaic of 29 images in R_c band with 300 and 600 s exposure times. The images from the 1-m telescope were compared to the corresponding fields of the Digitized Sky Survey (DSS). No new bright object was found up to the DSS limit for this field.

On the next night, May 10/11, a better position was available: $\alpha_{2000} = 06^h 53^m 28^s$; $\delta_{2000} = +79°17'.4$ with a 3′ error radius (99% confidence level). Photometric observations of GRB970508 field were then continued with the 6-m telescope with a CCD photometer installed at the Primary Focus.

The first image at the 6-m telescope was obtained on May 10.76 UT and a variable

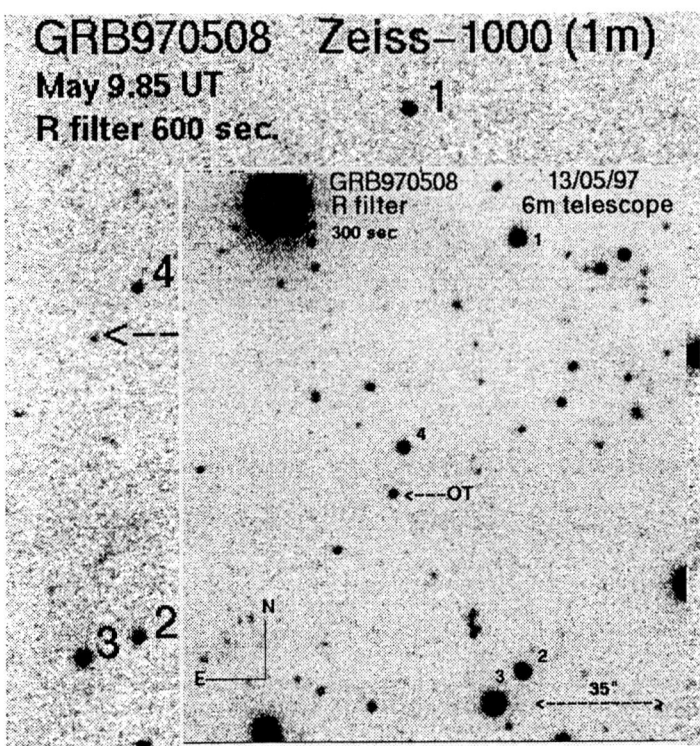

FIGURE 1. Field of GRB970508 optical counterpart from the 1-m and 6-m telescopes.

object was discovered by comparison with the image taken the night before. Its brightness from May 9.85 UT to May 10.76 UT increased by about 1.5 magnitudes. This object was first reported by H. Bond as a possible optical counterpart of GRB970508 [1] but was independently found in our data only about 0.5 day later.

PHOTOMETRY

Observations were carried out with filters closely matching the BVR_cI_c Johnson-Kron-Cousins system. The data were processed using the ESO-MIDAS software. Standard data reduction includes bias subtraction, flat-fielding and cosmic particle traces removal. Photometric conditions remained stable during the two nights of May 13/14 and May 21/22. Four bright stars (Fig. 1) in the GRB970508 field were used as secondary photometric standards. The magnitudes of these stars were determined on the night of May 13/14 with good photomeric conditions using four standard stars in the field of PG1657+078 [5]. Zero-point errors are better than 0.05^m. Coordinates and magnitudes of secondary photometric standards are given in Table 1. Our R_c magnitudes of stars 2, 3, 4 are 0.20 ± 0.01 higher than the magnitudes measured by Schaefer et al. [9].

TABLE 1. Magnitudes of secondary standard stars.

NN	$\alpha_{2000.0}$	$\delta_{2000.00}$	B	V	R_c	I_c
1	06:53:37.19	79:17:30.7	20.44	19.14	18.31	17.53
2	06:53:36.30	79:15:30.0	19.93	19.17	18.71	18.27
3	06:53:39.23	79:15:21.1	17.94	17.40	17.06	16.71
4	06:53:48.50	79:16:32.7	21.93	20.43	19.49	18.53

RESULTS AND CONCLUSIONS

Johnson-Kron-Cousins magnitudes in the BVR_cI_c bands with the associated errors for GRB970508 optical counterpart are given in Table 2. We have:
1) In the period of May ~9.13 UT [3,4] to May 9.85 UT the R_c brightness of the object seems to remain constant.
2) Object brightness in R_c band from May 9.85 UT to May 10.76 UT increased 1.5 mag. The magnitude increase rate using ours 1-m data and the data from Palomar [4] amount to 0.12 mag. per hour.
3) The brightness maximum was at $t_{max} \approx 1.5$ day after the burst. On May 10.76 UT the R_c magnitude was 19.70 and from that date a decline of brightness began.
4) Magnitude values during the beginning (2–5 days) of the fading follow an exponential law in all four bands:

$$\begin{aligned} B &= 19.689(\pm 0.036) + 0.452(\pm 0.014)(t - t_0) \\ V &= 19.264(\pm 0.053) + 0.449(\pm 0.020)(t - t_0) \\ R_c &= 18.874(\pm 0.029) + 0.443(\pm 0.011)(t - t_0) \\ I_c &= 18.355(\pm 0.050) + 0.450(\pm 0.019)(t - t_0) \end{aligned} \quad (1)$$

where $(t - t_0)$ is in days. The BVR_cI_c light curves of the source during the first 5 days after the burst are shown in Figure 2 only for the 6-m telescope data. For

TABLE 2. Results of photometry of GRB970508 OT at the SAO RAS.

UT	$t - t_0$	B	V	R_c	I_c
9.75 May	0.841			21.19 ± 0.25	
9.85	0.944			21.13 ± 0.18	
10.77	1.866	20.50 ± 0.03	20.06 ± 0.03	19.70 ± 0.03	19.1 ± 0.04
10.93	2.026	20.60 ± 0.03	20.22 ± 0.03	19.80 ± 0.03	19.3 ± 0.03
11.76	2.856	21.03 ± 0.04	20.52 ± 0.03	20.10 ± 0.03	19.5 ± 0.04
12.87	3.966	21.48 ± 0.06	21.10 ± 0.04	20.63 ± 0.05	20.1 ± 0.06
13.88	4.976	21.92 ± 0.07	21.47 ± 0.05	21.09 ± 0.07	20.5 ± 0.09
22.00	13.096			22.20 ± 0.15	
9.60 Jun	31.696			23.28 ± 0.10	
27.90 Jun	49.999			23.96 ± 0.15	
7.95 Jul	60.050			24.07 ± 0.17	
1.15 Aug	84.246			24.56 ± 0.21	

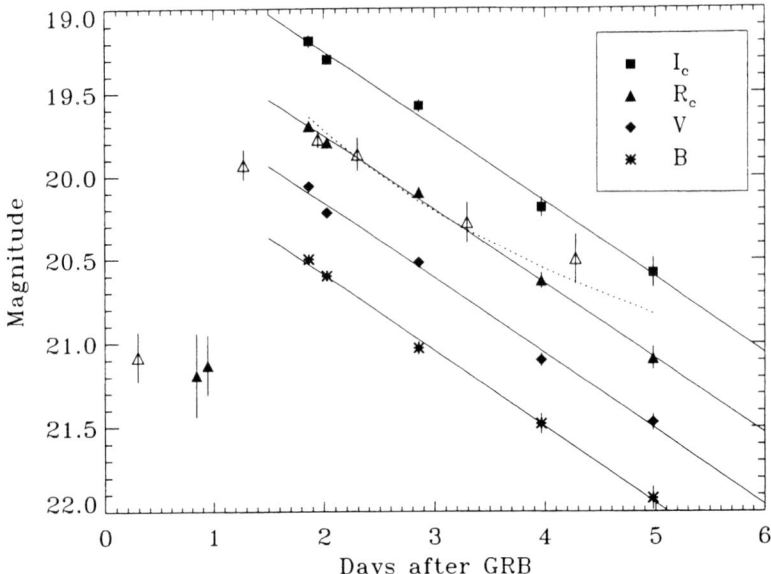

FIGURE 2. The light curves of GRB970508 optical counterpart during 5 days after the burst. SAO RAS (filled symbols), Loiano [7] ($t - t_0 = 1.95$) and Palomar [4] (transformed to $R_c = r - 0.34 + A_r$) (open triangles) magnitudes with their errors are shown. Lines correspond to equations (1) of exponential decline of brightness reported in the text. For R_c also the best fit power law is shown (dotted line).

the same data a power law did not fit the data. For example for the R_c data it is $\chi_n^2 = 0.97$ for the exponential fit while it is $\chi_n^2 = 4.5$ for the power law. All the R_c data are used: SAO, Loiano and Palomar.

The spectrum of the object was close to a power law and its slope $F_\nu \propto \nu^{-1.2}$ did not change in time. $(B-V) = 0.43$, $(V-R) = 0.39$, $(R-I) = 0.52$. The account for the galactic absorption $E(B-V) = 0.03$ gives $F_\nu \propto \nu^{-1.1}$ and the following color indices: $(B-V)_0 = 0.40$, $(V-R)_0 = 0.37$, $(R-I)_0 = 0.50$.

5) The observations of the object taken after May 13 show that, after the 5th day from the burst, the brightness fading is better described by a power law (2) with $\alpha = 1.179(\pm 0.016)$ obtained only from SAO data. Figure 3 shows the light curve in the R_c band.

$$R_c = 18.888(\pm 0.078) + 2.948(\pm 0.040) \times \log(t - t_0) \quad (2)$$

The data obtained with the 1-m and 6-m telescopes SAO RAS allow dividing the brightness change curve into three stages: 1) the increase of brightness on a time scale of about one day; 2) the exponential brightness fall during about 4 days with the conservation of broadband power-law spectrum; 3) the further slowing down of the brightness fading according to a power law.

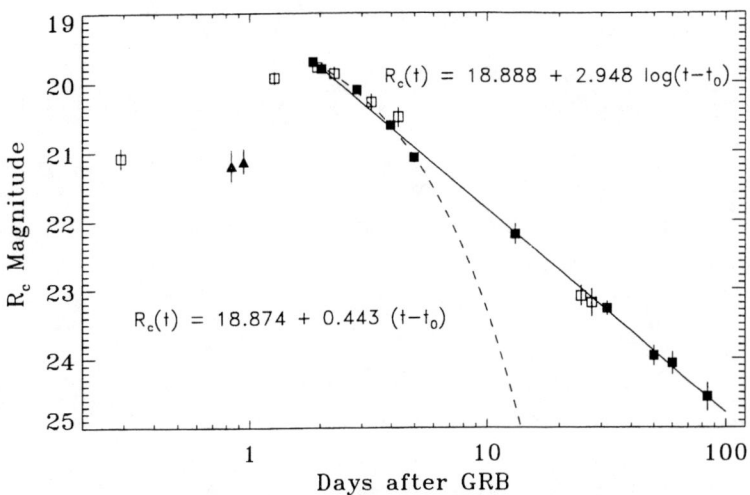

FIGURE 3. R_c light curve of the optical counterpart of GRB970508 during 84 days after the burst. SAO RAS (filled marks) and Palomar [4], Loiano [7], HST [8], Keck II [6] (open squares, transformed from Schaefer's photometric system to ours) magnitudes are shown. Lines correspond to exponential and power law for fading brightness. Errors on measurements after May 20.00 UT are presented.

REFERENCES

1. Bond, H. E., IAU Circ. No. 6654 (1997).
2. Castro-Tirado, A. J., et al., IAU Circ. No. 6657 (1997).
3. Costa, E., et al., IAU Circ. No. 6649 (1997).
4. Djorgovski, S., et al., *Nature* **387**, 876 (1997).
5. Landolt, A. U., *Astron.J.* **104**, 340 (1992).
6. Metzger, M. R., et al., IAU Circ. No. 6676 (1997).
7. Mignoli, M., et al., IAU Circ. No. 6661 (1997).
8. Fruchter, A., et al., IAU Circ. No. 6674 (1997).
9. Schaefer, B., et al., IAU Circ. No. 6658 (1997).

On the Light-curve of OT/GRB970508

H.Pedersen[1], A.O.Jaunsen[2], R.Østensen[3], T.Grav[2], M.Wold[4], M.Lacy[5], H.Kristen[4], A.Broeils[4], A.J.Castro-Tirado[6], J.Gorosabel[6], J. M. Rodriguez Espinosa[7], A.M.Perez[7], M.Näslund[4], C.Fransson[4], M.I.Andersen[3], C.Wolf [8], R.Fockenbrock[8], L.Piro[9], M.Feroci[9], E.Costa[9], L.Nicastro[10], E.Palazzi[10], F.Frontera[11], L.Monaldi[11,12], J.Heise[13], J.Hjorth[14]

(1) Copenhagen University Observatory, Denmark
(2) ITA, Oslo, Norway
(3) NOT, La Palma, Spain
(4) Stockholm Observatory, Stockholm, Sweden
(5) Nuclear and Astrophysics Laboratory, Oxford, UK
(6) LAEFF, INTA, Madrid, Spain
(7) IAC, Tenerife, Spain
(8) MPA, Heidelberg, Germany
(9) Frascati, Italy
(10) TESRE, CNR, Bologna, Italy
(11) Universita di Ferrara, Italy
(12) DSRI, Copenhagen, Denmark
(13) SRON, Utrecht, The Netherlands
(14) NORDITA, Copenhagen, Denmark

Abstract. Using the Nordic Optical Telescope, La Palma, Spain, optical studies of OT/GRB970508 were initiated 3 hours 5 minutes after the high energy event (This appears to be sooner than achieved elsewhere, for any optical transient). The OT was clearly detected in the earliest images. The last observation was done August 13-15, i.e. 95-97 days after the event. We have combined the NOT R and unfiltered data with similar photometry from elsewhere. We conclude that the observed source brightness cannot easily be modeled by theories predicting a monotonic rise to maximum. Also the post-maximum power-law decay, albeit predicted by several models, is no longer valid at the moment of our last observation. The contribution of a constant source, $m_R = 25.5$ is indicated, but this does not show up in the image profile.

INTRODUCTION

For several years, a programme has been conducted at the 2.56-m Nordic Optical Telescope, La Palma, aiming at the identification of optical counterparts to gamma-ray burst sources. The observations were proposed as searches for quiescent radiation from well-established GRB error boxes, and as searches for rapidly fading optical emission. Positional input was to be derived from individual spacecraft and from joint analysis of timing data (the 3rd Interplanetary Network).

The advent of swift Beppo-SAX X-ray localizations starting with the event of 1997 January 11, gave the first chance to put the latter part of this programme into action. Unfortunately, the weather at La Palma was poor during this period. For the event of 1997 February 28, observations started in response to IAU Circular No. 6572. J.-E.Solheim, Tromsø, kindly executed a series of exposures during the night March 3 to 4. Although transparency was good, the 'seeing' was abysmal, 3″ to 4″ so no useful information can be derived from these data. During the next night, a series of V, B, and unfiltered exposures were obtained, of which the V image was reported by van Paradijs et al. [1].

GRB970508

The third Beppo-SAX event which could be seen from Northern latitudes was GRB970508. Due to very fortunate circumstances, NOT observations started only three hours, five minutes after the high energy event. This appears to be the earliest observation from anywhere. The auxiliary instrument was ALFOSC, which covers an area of 6′ x 6′. For reason of the faintness of OT/GRB970228 [1] and the absence of optical transients from GRB970111 [2] and GRB970402 [3,4], it was decided to conduct the first two nights' observations in unfiltered light. It was the hope that intercomparison of these should serve to locate the OT, while precision photometry was relinquished.

The position of OT/GRB970508 is included in the very first exposure, which lasted 5 minutes, Figure 1. The two other exposures conducted during this night lasted 10 minutes each. One similar 10-minute exposure was done during the next night, 24 hours after the GRB. Intercomparison of these first four images failed to locate the OT. During May 10/11 and 11/12, several V and I exposures were done. Subsequent to H. E. Bond's detection of the OT [5], observations were exclusively done through a Bessel R filter, which reproduces well the Kron-Cousins system. The most recent images were obtained during the period August 13-15.

We have argued [6] that the unfiltered exposures have a central wavelength not far from that of the R band, allowing us to represent all unfiltered and R band observations in one diagram, Figure 2. In this, we have also included 'red' photometry from elsewhere (if available prior to the Symposium).

THE LIGHT CURVE

Inspection of Figure 2 reveals a light curve which is more complex than the one derived for OT/GRB970228. The early data suggests that the OT remained constant, or was slowly declining in brightness during the period from 3 until 8 hours after the GRB. The accuracy of this statement depends on the color index of the OT, which unfortunately is not known at this early moment (but see the caption to Figure 1).

The light-curve as presented here contradicts the simplest afterglow model, in which the OT rises in brightness until a single maximum is reached. From ~8 until ~20 hours after the GRB, no data are available (this being due to lack of observations from Pacific - Asian longitudes), but at +24 hours, the brightness is close to the level between +3 and +8 hours. Hereafter, the brightness increased rapidly (when inspected on the logarithmic time scale), and reached its maximum around +40 hours. The R magnitude at this moment was 19.7, which is about one magnitude brighter than the corresponding value for OT/GRB970228.

From +40 hours onwards, the brightness faded approximately as a power law. The data are in general agreement with the model by Chiang & Dermer [7], which was developed over a subset of the GRB970508 data. The very last data point, at +96 days, is, however, ~0.5 magnitude brighter than predicted by the power-law trend of the rest of the data. We see this as an indication that either the physical model for the power law decay has failed, e.g. due to inhomogeneities of

FIGURE 1. The earliest image of OT/GRB970508, a 5-minute, unfiltered exposure, which began 3 hours, 5 minutes after the GRB. North is up, East is left. The OT image (center of figure) is slightly elongated, possibly caused by atmospheric refraction (airmass = 1.9). Unfortunately, it does not appear possible to thereby characterize the spectrum at this early moment.

the interstellar medium near the OT, or that a source of constant brightness, such as a host galaxy, has been detected. A least squares analysis gives $m_R = 25.5 \pm 0.5$ for this constant light, and index 1.21 for the power law component. We have searched for the contribution of the galaxy to the image profile, finding none; the profile of the OT (at +96 days) is consistent with that of field stars, which is 0.8″ FWHM. The value $m_R = 25.5$ is consistent with the upper limit derived from HST data [8].

FROM NOW ON

Whether or not the deviation from a power law is caused by the additive signal of a constant source, the OT should remain visible to ground-based telescopes for years to come. For studies of image profile, however, the Hubble Space Telescope appears indispensable.

For future gamma-ray bursts, efforts should be taken to trace the optical light curves at shorter delays than so far achieved; this would support the modeling of

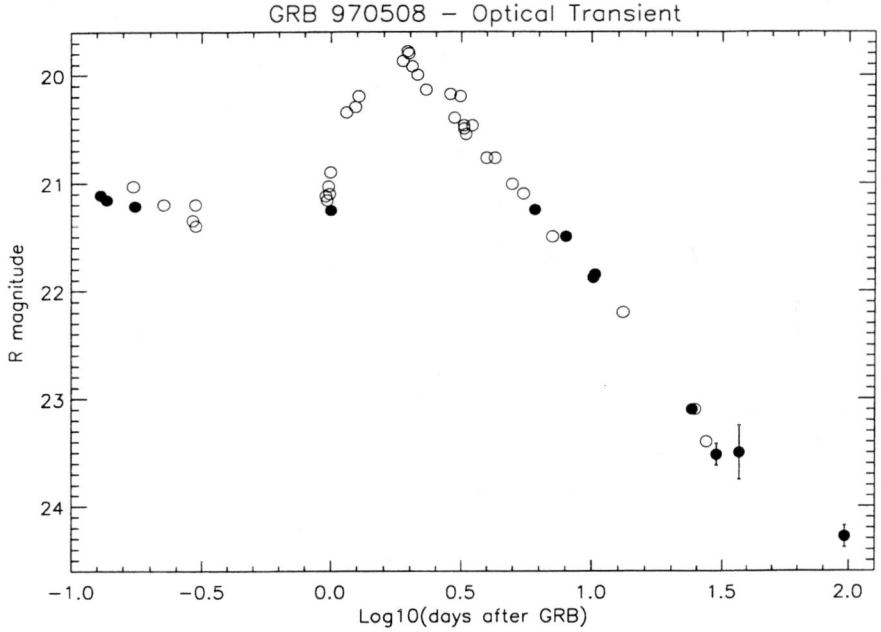

FIGURE 2. The brightness of OT/GRB970508, as function of time. Observations from NOT are denoted by filled symbols, observations from elsewhere by open symbols. The time-scale is logarithmic; the observations span the time interval from 3 hours to 96 days after the high energy event.

the physical emission process.

Although the different sets of R-band measurements, as presented here, have been easy to combine, we would recommend some standardization of observing procedures. This should include at least two widely separated color-bands, to monitor the OT colors as function of time.

REFERENCES

1. van Paradijs, J., et al., *Nature* **386**, 686 (1997).
2. Castro-Tirado, A.J., et al., *IAU Circular* No. 6598 (1997).
3. Groot, P.J., et al., *IAU Circular* No. 6616 (1997).
4. Pedersen, H., et al.. *IAU Circular* No. 6628 (1997).
5. Bond, H.E., *IAU Circular* No. 6654 (1997).
6. Pedersen, H., et al., *Ap.J.* **496**, in press (1998).
7. Chiang, J., & Dermer, C.D., *astro-ph9708035* (1997).
8. Pian, E., et al., *Ap.J.*, in press (1997).

The Redshift of GRB970508

Daniel E. Reichart

Department of Astronomy and Astrophysics, University of Chicago, Chicago, IL 60637

Abstract. GRB970508 is the second gamma-ray burst (GRB) for which an optical afterglow has been detected. It is the first GRB for which a distance scale has been determined: absorption and emission features in spectra of the optical afterglow place GRB970508 at a redshift of $z \geq 0.835$ [1,2]. The lack of a Lyman-α forest in these spectra further constrains this redshift to be less than $z \sim 2.1$. I show that the spectrum of the optical afterglow of GRB970508, once corrected for Galactic absorption, is inconsistent with the relativistic blast-wave model unless a second, redshifted source of extinction is introduced. This second source of extinction may be the yet unobserved host galaxy. I determine its redshift to be $z = 1.09^{+0.14}_{-0.41}$, which is consistent with the observed redshift of $z = 0.835$. Redshifts greater than $z = 1.40$ are ruled out at the 3 σ confidence level.

INTRODUCTION

Discovered by the BeppoSAX Gamma-Ray Burst Monitor [3], GRB970508 is the second gamma-ray burst (GRB) for which an optical afterglow has been detected (e.g., [4]). Transient X-ray, near-infrared, millimeter, and radio emission have also been detected. GRB970508 is the first GRB for which a distance scale has been determined: Metzger et al. [1,2] report the existence of absorption and emission features in spectra of the optical afterglow taken with the Keck II 10-m telescope ≈ 2 days and ≈ 26 days after the GRB event. Their identification of these features places GRB970508 at a redshift of $z \geq 0.835$. The lack of a Lyman-α forest in these spectra further constrains this redshift to be less than $z \sim 2.1$. Consequently, GRB970508 is almost certainly cosmological in origin.

A host galaxy for GRB970508 has not yet been observed. However, Metzger et al. [1] report that the line emission observed in the Keck II spectra is consistent with constancy between their May 11 and June 5 observations. Over this same period, the continuum emission has faded. This suggests that a host galaxy of relatively weak continuum emission may be present at the observed redshift of $z = 0.835$ [2].

GRB afterglows are believed to be described by the relativistic blast-wave model [5-13]. Furthermore, the afterglow of GRB970228, the only other GRB for which sufficient afterglow information is available, has been shown to be compatible with this model [14,11,15-17,13]. A basic prediction of the relativistic blast-wave model is that after an initial period of increasing optical flux, lasting hours to days, the optical flux of the afterglow will decrease as a power-law of index $-1.5 \lesssim b \lesssim -0.5$. During this period of declining optical flux, the optical spectrum will be described by a power-law of index $a = 2b/3$ (e.g., [15]). However, in the case of GRB970508, $a \approx b$ [18], which is inconsistent with the relativistic blast-wave model.

In §2, I show that the reported GRB970508 optical afterglow measurements, once corrected for Galactic absorption, are inconsistent with this prediction of the relativistic blast-wave model unless a second, redshifted source of extinction is introduced. This second source of extinction may be the yet unobserved host galaxy. I determine its redshift and I estimate its hydrogen column density along the line of sight. An observing strategy for future optical afterglow observations is recommended in §3.

DATA ANALYSIS & MODEL FIT

As of 1997 November 3, >70 optical and near-infrared measurements of the GRB970508 afterglow have been reported. These measurements span >80 days, the earliest of which was taken ≈ 4 hours after the GRB event. Although the photometry is generally quite good, with quoted errors that are often less than 0.1 mag, zero-point errors are evident between different observing groups in various bands. Consequently, I consider only the largest self-consistent subset of these data: Sokolov et al. [18] report 22 optical measurements (B, V, R_C, and I_C bands) taken with a 6m telescope between ≈ 2 days and ≈ 31 days after the GRB event. All of these measurements were taken after the optical flux had peaked (§1), also ≈ 2 days after the GRB event.

I have corrected these measurements for Galactic absorption using the IRAS 100 μm V-band absorption measure of Rowan-Robinson et al. [19] and the interstellar absorption curve [20,21]. The V-band correction is $A_V = 0.09$ mag. I use the absorption measure of Rowan-Robinson et al., which measures the dust directly, instead of measures of the hydrogen column density, i.e. [22-24], since the IRAS 100 μm flux about the location of GRB970508 varies significantly on angular scales that are smaller than the scales over which these measures of the hydrogen column density are averaged. The corrected measurements are plotted in Figure 1.

The following model is now χ^2-fitted to these corrected measurements:

$$F_\nu = F_0 \nu^{\frac{2b}{3}} t^b - F_{ext}(\nu; A_V(z), z). \tag{1}$$

The first term is the extinction-free prediction of the relativistic blast-wave model (§1). The second term is the correction that a second, redshifted source of extinction introduces. It is given by redshifting the interstellar absorption curve and by specifying the magnitude of the extinction, which I parameterize as the V-band absorption magnitude *at the redshift of the source*: $A_V(z)$. A constant error of ≈ 0.07 mag must be added in quadrature to the quoted errors of Sokolov et al. [18] for the model to fit the data ($\chi^2 \approx \nu$). This suggests that either the quoted errors are underestimated by a factor ~ 2, or that the flux is varying by $\approx 6\%$ on timescales of days. Reichart [16] noticed possibly related temporal behavior in the light-curve of GRB970228.

The best fit is: $\log F_0 = -9.03^{+0.44}_{-0.44}$ cgs, $b = -1.22^{+0.03}_{-0.03}$, $A_V(z) = 0.24^{+0.12}_{-0.08}$ mag, and $z = 1.09^{+0.14}_{-0.41}$. The quoted uncertainties are 1 σ confidence intervals for one interesting parameter. The best-fit redshift is consistent with the observed redshift of $z = 0.835$. Furthermore, redshifts greater than $z = 1.40$ are ruled out at the 3 σ confidence level. The best-fit V-band absorption magnitude corresponds to a hydrogen column density of $\approx 4.5 \times 10^{20}$ cm^{-2}. The possibility that there is no second source of absorption, i.e., $A_V(z) = 0$, is ruled out at the 3.8 σ confidence level.

DISCUSSION & CONCLUSIONS

Using Equation (1) and the best-fit temporal decline power-law index, I have scaled each of the measurements that were fitted to in §2 to its corresponding value for May 11, just shortly after the optical peak. These points define the post-peak, time-independent optical spectrum and are plotted in Figure 2. The best-fit spectrum, as well as the extinction-free, relativistic blast-wave component of this spectrum (the first term of equation (1)), are also plotted in Figure 2. A broad

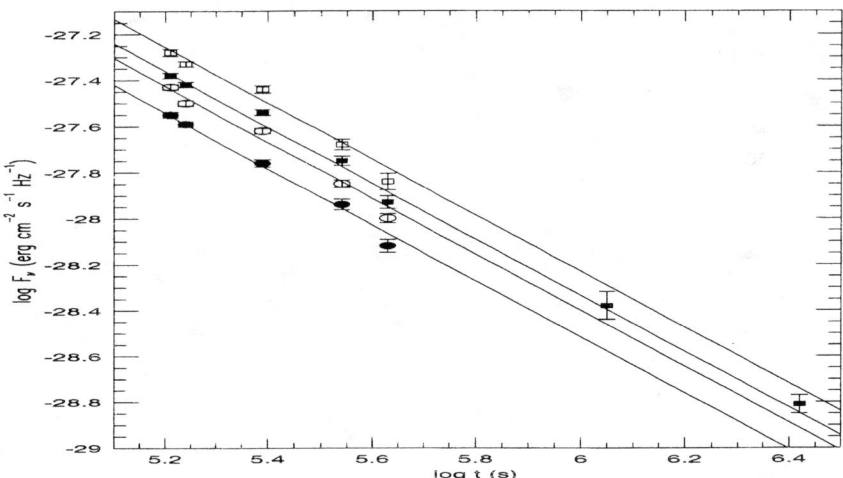

FIGURE 1. Fluxes of Sokolov et al. [18] corrected for Galactic absorption (§2) and the best fit to equation (1). Solid circles are B-band measurements, open circles are V-band measurements, solid squares are R_C-band measurements, and open squares are I_C-band measurements.

absorption feature is apparent in the best-fit spectrum. This is the ultraviolet absorption feature of the interstellar absorption curve, redshifted into the B band. Had the redshift of GRB970508 been less than $z = 0.835$, U-band measurements would also have been necessary for this, the only strong feature of the interstellar absorption curve, to have been detected. Consequently, in this type of analysis, different bands most sensitively probe different redshift ranges, depending on whether or not this absorption feature has been redshifted into the band in question. Since GRBs are generally believed to have redshifts of $z \sim 1$, self-consistent sets of U-, B-, and V-band measurements should be a goal of future optical afterglow observations. In this letter, I present a method by which redshifts can be determined for GRBs that are associated with host galaxies, even if absorption or emission lines are not observable or if spectra of sufficient quality are unattainable. This method also yields hydrogen column densities along the line of site, which provides valuable information about the distribution of GRBs within galaxies. For GRB970508, I find that the redshift of its host galaxy, or possibly that of an intermediate galaxy, is $z = 1.09^{+0.14}_{-0.41}$, which is consistent with the observed redshift of Metzger et al.: $z = 0.835$. Redshifts greater than $z = 1.40$ are ruled out at the 3 σ confidence level.

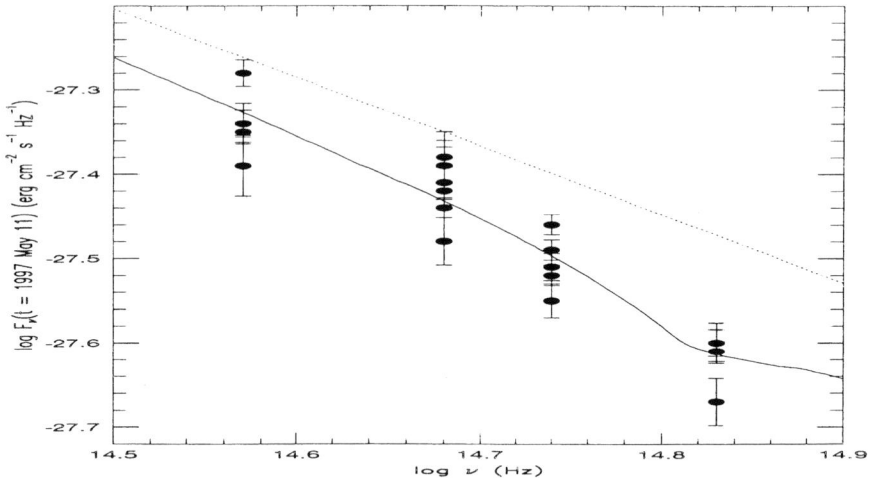

FIGURE 2. The post-peak, time-independent optical spectrum of the afterglow of GRB970508, scaled to 1997 May 11. Plotted from left to right are the I_C-, R_C-, V-, and B-band fluxes of Sokolov et al. [18] corrected for Galactic absorption. The solid line is the best-fit spectrum and the dotted line is the extinction-free, relativistic blast-wave component of this spectrum.

ACKNOWLEDGMENTS

I am grateful to D. Q. Lamb, R. C. Nichol, B. P. Holden, F. J. Castander, W. B. Burton, and D. Hartmann for valuable discussions and information.

REFERENCES

1. Metzger, M. R., et al., *IAU Circ.*, 6676 (1997).
2. Metzger, M. R., et al., *Nature* **387**, 879 (1997).
3. Costa, E., et al., *IAU Circ.*, 6649 (1997).
4. Bond, H. E., *IAU Circ.*, 6654 (1997).
5. Paczyński, B., & Rhoads, J., *ApJ* **418**, L5 (1993).
6. Katz, J., *ApJ* **422**, 248 (1994).
7. Katz, J., *ApJ* **432**, L107 (1994).
8. Mészáros, P. & Rees, M. J., *ApJ* **476**, 232 (1997).
9. Sari, R., & Piran, T., *ApJ* **485**, 270 (1997).
10. Vietri, M., *ApJ* **478**, L9 (1997).
11. Waxman, E., *ApJ* **485**, L5 (1997).
12. Sari, R., *ApJ* **489**, L37 (1997).
13. Katz, J., & Piran, T. *ApJ*, in press (1997).
14. Katz, J., Piran, T., & Sari, R., preprint (1997).
15. Wijers, R. A. M. J., Rees, M . J., & Mészáros, P., *MNRAS* **288**, L51 (1997).
16. Reichart, D. E., *ApJ* **485**, L57 (1997).
17. Sahu, K., et al., *Nature* **387**, 476 (1997).
18. Sokolov, V. V., et al., preprint (1997).
19. Rowan-Robinson, M., et al., *MNRAS* **249**, 729 (1991).
20. Johnson, H. L., *ApJ* **141**, 923 (1965).
21. Bless, R. C., & Savage, B. D., *ApJ* **171**, 293 (1972).
22. Hartmann, D., & Burton, W. B., *Atlas of Galactic Neutral Hydrogen*, Cambridge: Cambridge University Press, 1978.
23. Stark, A. A., et al., *ApJ Supp.* **79**, 77 (1992).
24. Burstein, D., & Heiles, C., *AJ* **87**, 1165 (1982).

Did GRB970228 and GRB970508 Present Similar Optical Properties?

C. Bartolini[1], G.M. Beskin[2], A. Guarnieri[1],
N. Masetti[1] & A. Piccioni[1]

[1] *Dipartimento di Astronomia, Università di Bologna, Italy*
[2] *Special Astrophysical Observatory of RAS, Nizhnij Arkhyz, Russia*

Abstract. The analysis of the light curves of the optical counterparts of GRB970228 and GRB970508 points out remarkable similarities. The spectral distribution obtained from the color indices shows that both the transients became bluer during the increasing stage, and redder after the maximum. A main difference concerns the behaviour of the optical fading which is well fitted by a single power law in the case of GRB970508 but not in the case of GRB970228.

GRB970228

The optical counterpart of GRB970228 was discovered by van Paradijs et al. [18] who observed it for the first time on February 28.99 UT, 1997. Early observations by Guarnieri et al. [11] in the B and R bands on February 28.83 allowed the detection of the presence of a rising branch in the light curve.

Fig. 1 (left) shows the R light curve obtained by using the collection of observations reported by Galama et al. [8] and the data by Guarnieri et al. [11] and Fruchter et al. [6]. The observations of Pedichini et al. [15] were not included because of the difficulty in reducing their color system to the R band. A fitting of the optical data with a single power law yields an index $\alpha_{opt} = 1.21 \pm 0.02$; however, data reported by Galama et al. [8] show a decay behaviour which seems to follow a power law with spectral index $\alpha_{opt} = 2.1$ before March 6, 1997, and with $\alpha_{opt} < 0.35$ after that day.

The evolution of the optical spectrum of GRB970228 is presented in Fig. 1 (right). The flux densities were evaluated from the photometric data taken on February 28.83 [11], February 28.99 [18], March 26.4 and April 7.2 [17] using the formulae by Fukugita et al. [7]. The optical transient became bluer during the rise and then, on its way to quiescence, significantly reddened, going from $B - V = 0.5$ and $V - R = 0.5$ on February 28.83 to $V - R = 0.4$ and $V - I = 0.7$ on February 28.99, to $V - R = 0.9$ and $V - I = 1.9$ on March 26.4.

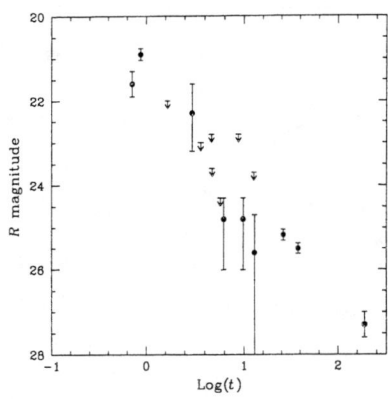 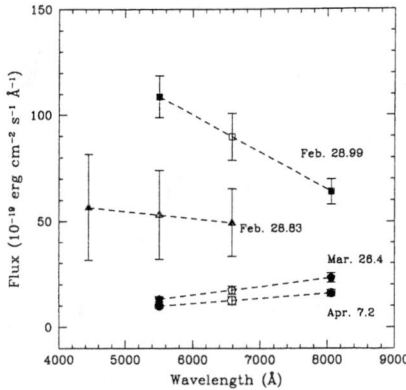

FIGURE 1. Left: R light curve of the optical transient associated with GRB970228. Data were taken from literature (see text), t is the time interval, in days, since the γ-burst. **Right**: broadband spectra of GRB970228; the flux densities were computed from the photometric data by Guarnieri et al. [11] (triangles), van Paradijs et al. [18] (squares) and Sahu et al. [17] (circles). The HST data were multiplied by a factor of 10 for sake of clarity. The data corresponding to the open symbols are evaluated from the measured values (filled symbols) using a linear interpolation of the flux densities.

GRB970508

Surprisingly — but not too much in the light of the behaviour found by Guarnieri et al. [11] for GRB970228 — the optical transient discovered by Bond [1] inside the GRB970508 error box increased in brightness during the first two days after the γ-burst, reaching the peak around May 10.8 UT. According to Castro–Tirado et al. [2], during this phase the object was very blue.

Its R light curve is shown in Fig. 2 (left). The data were taken from the IAU Circulars, from several papers published in these proceedings and from Kelemen [12]. An upper limit to the brightness was obtained with the Bologna Telescope on May 23.89 UT. A power–law decay fits the data until 50–60 days after the γ-burst. We obtain $\alpha_{opt} = 1.34 \pm 0.02$, which is slightly higher than the values found by Djorgovski et al. [3], Kopylov [13] and Fruchter et al. [5]. Then, the light curve seems to flatten about at $R \approx 25.2$, which might be the magnitude of the host galaxy [16].

The observations by Galama et al. [9], Groot et al. [10], Kopylov et al. [14] and Kopylov [13] are homogeneous and cover almost all the optical band in the same time lapse. From these data, we evaluated the flux densities and the broadband spectral evolution of GRB970508 using the same procedure applied to GRB970228 data; the results are reported in Figs. 2 (right) and 3 (left). On May 10.05 the object was red; then it became progressively bluer on May 10.77, 10.93 until it

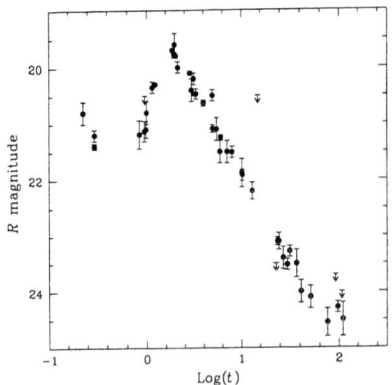
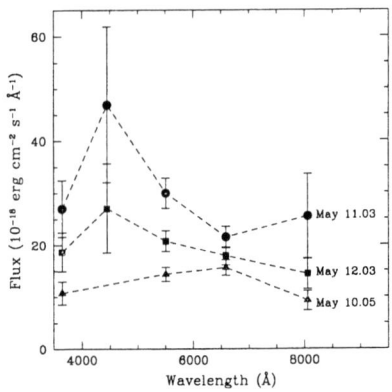

FIGURE 2. Left: R light curve of the optical transient associated with GRB970508. Data and upper limits are from various authors (see text for details). t represents the time interval, in days, since the γ-burst occurred. Right: broadband spectra of GRB970508 built with the flux densities computed from the photometric data by Galama et al. [9] on May 10.05 (triangles) and on 11.03 (squares), and by Groot et al. [10] on 12.03 (circles).

reached a maximum on May 11.03, close to the light peak. On May 11.76, the spectrum became flatter (Fig. 3, left) but on May 12.03, the slope rose again at short wavelengths (Fig. 2, right). Subsequently the spectrum started to redden, peaking in the R on May 22.00. This behaviour is confirmed by the trend of the color of the optical transient (Fig. 3, right).

It is noteworthy that the $B - V$ secondary maximum occurred around the appearance of a transient radio emission associated with GRB970508. Indeed, nearly a week after the onset of the γ-burst, Frail et al. [4] found a flaring radio source, increasing in brightness, which was coincident with the positions of the X–ray and optical counterparts of GRB970508.

COMPARISON BETWEEN THE LIGHT CURVES

Although the light curve of GRB970228 is less sampled, we can see remarkable similarities with that of GRB970508, by comparing the left panels of Figs. 1 and 2. Both light curves show a rising branch. In particular, the maximum occurred $\sim$1 day after the burst for GRB970228, and around 2 days after for GRB970508. Therefore, the presence of a delayed maximum in the optical light curve appears to be a common feature for all the GRBs observed thus far.

The two light curves show similar overall decays. But if the R light curve of GRB970508 can be fitted with a single power law until 60 days from the onset of the γ-burst, the one of GRB970228 does not; indeed, Guarnieri et al. [11] noticed

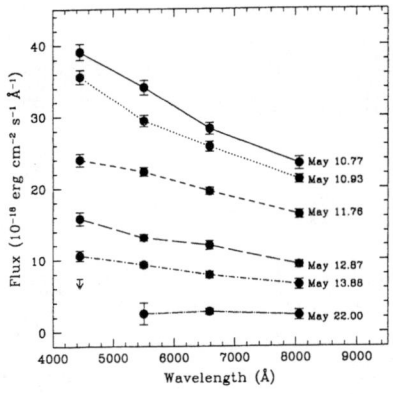 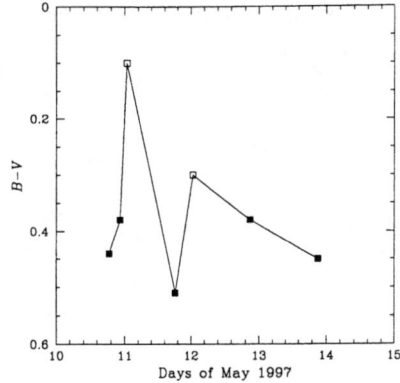

FIGURE 3. Left: broadband spectra of GRB970508 built with the flux densities computed from the photometric data by Kopylov et al. [14] and Kopylov [13]. The upper limit in the figure refers to May 22.00. Right: $B - V$ colors of GRB970508 from May 10 to May 14. Dutch data are indicated with open squares, and Russian ones with filled squares.

a more rapid decrease during the first 3 days after the γ event.

COMPARISON BETWEEN THE SPECTRA

The spectral distribution (Figs. 1 and 2, right, and Fig. 3, left) and the $B - V$ color index (Fig. 3, right), show that both the optical transients became bluer during the rising phase, and then reddened during their approach to quiescence. This could be simply understood in the light of the 'fireball' model (e.g. Wijers et al. [19]): a moving shock wave heats the surrounding medium, which then cools down. In this framework, the fast spectral and color variation of GRB970508 in the optical is therefore surprising.

Due to the paucity of the observations, we cannot say if GRB970228 showed a similar behaviour during the first days after the burst. However, if these variations are connected with the radio transient and since no radio emission was revealed for GRB970228, it is possible that no such changes in the optical spectral energy distribution took place in GRB970228.

Another explanation for these spectral changes could arise from the presence of cool absorbing material placed on the line–of–sight at a distance of about 2 or 3 light–days from the center of the burst; it could have produced a 'dip' at shorter wavelengths suddenly after the light peak, and then could have been shocked by the blast wave in the following days.

It seems that GRB970228 was redder than GRB970508 at light peak ($B - V =$ +0.5 instead of +0.1). However we are not sure that the minimum $B - V$ for

GRB970228 has been observed. Moreover, this value could be affected by interstellar (and/or intergalactic) absorption; indeed, GRB970228 lies closer to the galactic plane than GRB970508. Actually, the absorption in the V towards GRB970228 is $A_V = 0.4$ mag [18], while towards GRB970508 is only $A_V = 0.08$ mag (derived from the A_B value reported by Djorgovski et al. [3]).

ACKNOWLEDGEMENTS

This investigation is supported by the University of Bologna (Funds for selected research topics). We thank G. Valentini for several useful discussions.

REFERENCES

1. Bond, H.E., IAU Circ. 6654 (1997).
2. Castro–Tirado, A.J., Gorosabel, J., Wolf, C., et al., IAU Circ. 6657 (1997).
3. Djorgovski, S.G., Metzger, M.R., Kulkarni, S.R., et al., *Nature* **387**, 876 (1997).
4. Frail, D.A., Kulkarni, S.R., Nicastro L., et al., *Nature* **389**, 261 (1997).
5. Fruchter, A., Bergeron, L., & Pian, E., IAU Circ. 6674 (1997).
6. Fruchter, A., et al., these proceedings.
7. Fukugita, M., Shimasaku, K., & Ichikawa, T., *PASP* **107**, 945 (1995).
8. Galama, T., Groot, P.J., van Paradijs, J., et al., *Nature* **387**, 479 (1997).
9. Galama, T., Groot, P.J., van Paradijs, J., et al., IAU Circ. 6655 (1997).
10. Groot, P.J., Galama, T., van Paradijs, J., et al., IAU Circ. 6660 (1997).
11. Guarnieri, A., Bartolini, C., Masetti, N., et al., *A&A* **328**, in press (astro–ph/9707164) (1997).
12. Kelemen, J., IBVS No. 4496 (1997).
13. Kopylov, A.I., IAU Circ. 6671 (1997).
14. Kopylov, A.I., Sokolov, V.V., Zharikov, S.V., et al., these proceedings.
15. Pedichini, F., Di Paola, A., Stella, L., et al., *A&A* **327**, L36 (1997).
16. Pian, E., et al., these proceedings.
17. Sahu, K.C., Livio, M., Petro, L. et al., *Nature* **387**, 476 (1997).
18. van Paradijs, J., Groot, P.J., Galama, T., et al., *Nature* **386**, 686 (1997).
19. Wijers, R.A.M.J., Mészáros, P., & Rees, M.J., *MNRAS* **288**, L51 (1997).

Source Confusion in Optical Searches for GRB Counterparts

J. J. Brainerd

Physics Dept., University of Alabama in Huntsville, Huntsville AL 35899

Abstract. Searches for optical transients from gamma-ray bursts employ location boxes that are several to tens of arc-minutes across. With boxes of this size, the probability of finding a type 1A supernovae in a search down to V magnitude 24 is substantial. For instance, in an error box of 1 arc-minute radius, one expects the probability of finding a supernovae is approximately 0.15%.

INTRODUCTION

A recurring difficulty in gamma-ray burst studies has been identifying sources at other wavelengths. The literature over the past decade contains associations with quiescent x-ray counterparts, active galaxies, radio galaxies, and other unusual sources, all occurring at the several percent significance level. As a consequence, a certain scepticism is warranted in judging any claim of an association.

In this paper I discuss limits on type 1A supernovae occurring in deep searches of gamma-ray burst error boxes. Type 1A supernovae bear a passing resemblance to the optical transient observed for GRB970228, making the misidentification of a type 1A supernovae with a burst location a real possibility. The other optical transient, that associated with GRB970508, is clearly not a supernova, as it is much too bright for the observed redshift. The question is then how often will a type 1A supernovae fall into a gamma-ray burst error box.

SUPERNOVAE CHARACTERISTICS AND RATES

The characteristics that make type 1A supernovae a source of confusion are their occurrence rates, rise and decay timescales, and magnitudes. The absolute magnitude for a type 1A is $M \approx -19.8$, which gives a magnitude of $m \approx 22.5$ at 3 Gpc, in line with the observed magnitude of the optical transient associated with GRB970228. The type 1A supernovae persist with a peak emission of roughly two weeks, after which they decay at a rate of approximately 0.12 magnitudes per day in the B band. The decay rate is very dependent on the photon energy, with decay rates approaching 0.17 magnitudes per day in the U band. Finally, the type 1A supernovae rate is $0.28 \, h^2/10^{10} \, L_\odot/100 \, \text{year}$ [1].

DETECTION RATE ESTIMATE

An estimate of the supernovae rate expected when observing down to approximately 24 magnitude is derived under the assumption that the effects of cosmological expansion are negligible. From the supernovae rate given above, the supernovae rate for magnitudes greater than m_l for the full sky is given by

$$R_T = 4 \times 10^{6+0.6(m_l-22)} \, \text{yr}^{-1} \, h^4 \left(\frac{20}{A}\right) \left(\frac{\Omega_B}{0.025}\right)$$

In this equation, h is the Hubble constant in units of $100 \, \text{km s}^{-1} \, \text{Mpc}^{-1}$, Ω_B is the density of baryons in units of the closure density, and A is the mass to light ratio. For a circle on the sky with radius of θ, the supernovae rate is

$$R_T = 0.8 \times 10^{0.6(m_l-22)} \, \text{yr}^{-1} \left(\frac{\theta}{1 \, \text{arcmin}}\right)^2 h^4 \left(\frac{20}{A}\right) \left(\frac{\Omega_B}{0.025}\right)$$

Assuming that the supernovae is detectable at 22 magnitude for two weeks, one finds that the probability of finding a supernovae associated with a gamma-ray burst at the time of the burst is

$$P_T = 0.003 \times 10^{0.6(m_l-22)} \left(\frac{\theta}{1 \, \text{arcmin}}\right)^2 h^4 \left(\frac{20}{A}\right) \left(\frac{\Omega_B}{0.025}\right)$$

Given that GRB 970228 had an x-ray error box of 50 arcsec, the probability of finding a supernovae associated with the gamma-burst is $\approx 0.2\%$, which is small but not insignificant. A larger location box of 3 arcmin increase the probability to $\approx 3.0\%$.

DETECTION RATE FROM SURVEYS

Automated searches for supernovae to very low magnitude now exist that can be used to place a limit on the detection rate of supernovae in gamma-ray burst error boxes. One such search [2] found 3 type 1A supernovae and one additional transient event that may be a supernova in a $1.7 \, \text{deg}^2$ region that was searched only once. This survey is sensitive to supernovae at $R \approx 23$. Using the three certain detections alone, one finds a probability of a supernovae falling within a gamma-ray burst error box to be $\approx 0.0015 \, (\theta/1 \, \text{arcmin})^2$. This value is in line with the rough estimate given above. The probability of a chance association of GRB 970228 with a type 1A supernovae is 0.10%. The probability of finding a supernovae in a 3 arcmin box is a substantial 1.4%.

CONCLUSION

The probability that the optical transient associated with GRB970228 is a type 1A supernova is quite small. On the other hand, the probability of eventually finding a type 1A supernovae within a gamma-ray burst error box is quite high. For error circles with a 3 arcmin radius (28 arcmin^2), one should find at least one supernova per 35 to 75 locations. Error boxes of radius $\theta = 8$–11.5 arcmin, for an area of ≈ 200–400 arcmin^2, contain a supernova once in every 5. Since the success rate of finding an optical transient in an error box is of order 20% over the past year, one must go considerably below 10 arcmin in radius to avoid source confusion.

REFERENCES

1. Evans, R., van den Bergh, S., & McClure, R. D., *ApJ* **345**, 752 (1989).
2. Pain, R., et al. , *ApJ* **473**, 356 (1996).

Optical Gamma-Ray Burst Followup at Kitt Peak National Observatory

James E. Rhoads[1], Ted von Hippel[2], Bruce Carney[3], Charles F. Claver[1], Arthur D. Code[2], Andrew Cole[2], Christopher Conselice[2], Arjun Dey[1], Chris Howk[2], George Jacoby[1], Buell Jannuzi[1], Jaewoo Lee[3], Jerome A. Orosz[4], Daniel J. Pisano[2], Dave Sawyer[1], Nigel Sharp[1], & Paul Smith[1]

[1] *Kitt Peak National Observatory*
[2] *University of Wisconsin, Madison, WI 53706*
[3] *University of North Carolina, Chapel Hill, NC 27599*
[4] *Pennsylvania State University, University Park, PA 16802*

Abstract. The discovery of probable X-ray, optical, and radio counterparts to gamma-ray bursts (GRBs) may allow significant advances in GRB research. More data are sorely needed, however. We have a target of opportunity program this semester at the 0.9 meter telescope to search for optical counterparts to GRBs. Images from the search will be made publicly available at our web site (**http://www.noao.edu/noao/grb/**) as quickly as we can reduce them.

OBSERVING PROGRAM

The 0.9 meter telescope at Kitt Peak National Observatory is an appropriate tool for GRB afterglow searches, as demonstrated by Howard Bond's discovery of the GRB970508 afterglow [1]. Our team is organizing target of opportunity observations of gamma-ray burst locations with this telescope.

During the fall 1997 semester, we expect to observe up to 6 GRB error boxes at up to three epochs each. Ideally, we would like two epochs on the first night following the GRB trigger, and a third epoch on the following night. We will decide whether or not to pursue individual bursts based on several factors, including error box size, coordinates (reasonable airmass and Galactic latitude), and moon phase. The 0.9 meter telescope will be most valuable when the error box is between 10 and 20 arcminutes in size, as this fits within a single 0.9 m field and does not fit in a typical field for larger telescopes.

RESULTS SO FAR

During summer 1997, we attempted followup of two gamma-ray bursts: GRB970616 and GRB970828. While our data did not reveal any confirmed counterparts to either burst, we believe the upper limits we obtained may be of value.

GRB970616

For GRB970616, George Jacoby (NOAO) and Bruce Carney and Jaewoo Lee (University of North Carolina) obtained photometry with the 0.9 meter telescope. This reached limiting magnitude around V=19. In addition, Art Code (Wisconsin) and Chuck Claver (NOAO) observed the burst location with the 3.5 meter WIYN telescope. We believe our data provide the earliest upper limits on the optical flux for this burst. The total data set spans four nights (1997 June 18.4 to 21.4 UT). Note that the 970616 burst was unfavorably placed for optical observations, with airmass > 2 at morning twilight and a nearly full moon to make matters worse.

GRB970828

GRB970828 was observed with three telescopes at Kitt Peak on the night of 1997 August 29 (30 August UT). Paul Smith (NOAO) obtained four R band images (total integration time 3600 seconds) with the 0.6m Burrell Schmidt telescope. We made these available on the Web within two hours of the observations.

Jerry Orosz (Penn State) obtained three R band images (total integration time 3000 seconds) with the 0.9 meter telescope. The telescope was operating in f/13.5 mode, with field of view 13' rather than the more usual 23'.

Finally, Chris Howk (U Wisconsin) obtained two R band images with WIYN (600 and 900 sec) which he, Andrew Cole, Chris Conselice, and D. J. Pisano reduced. Groot et al. [2] have incorporated these WIYN data in a paper on upper limits for the optical afterglow flux of GRB970828.

All three data sets are now publically available.

AN APPEAL

Poor weather prevented further observations of GRB970828 from Kitt Peak. While we made our images publically available, we were unable to easily locate data from other observatories for comparison with other epochs.

We would like to encourage others to make afterglow search images rapidly available. Please let us know if you have a web site or FTP address where you might put such data in future; we would be happy to link it to our pages.

CONTACT INFORMATION

We hope to use feedback from the larger GRB community to help optimize our observing strategy and coordinate our efforts with other ongoing efforts. If you would like to discuss any of this, you can reach us by email to grb@noao.edu, which will automatically be forwarded to members of our followup group. If you prefer, you can also contact James Rhoads (rhoads@noao.edu) or Ted von Hippel (ted@noao.edu) directly.

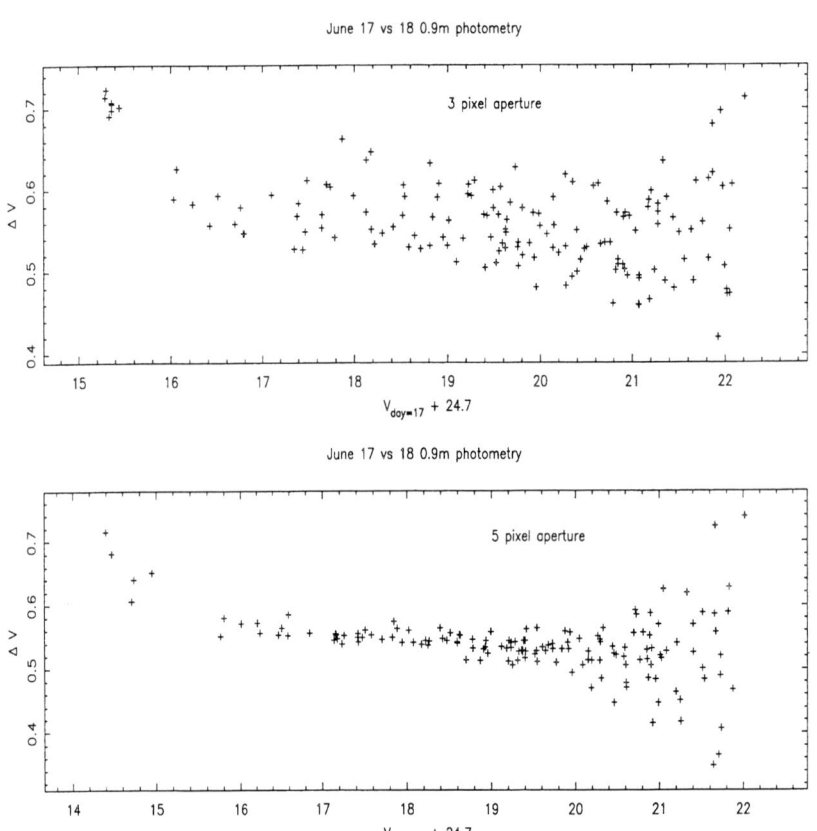

FIGURE 1. We performed differential aperture photometry on the GRB970616 error box field taken in the V filter. Here we see a comparison between instrumental magnitudes from data taken on June 17 and 18 civil time (June 18 and June 19 UT). No variable candidates appeared in this comparison. (The upturn at bright magnitudes is due to saturation effects. The absolute zero points are not calibrated, giving the nonzero mean offset between the two nights.)

REFERENCES

1. Bond, H., *IAU Circular* No. 6654 (1997).
2. Groot, P. J., Galama, T. J., van Paradijs, J., Kouveliotou, C., Wijers, R.A.M.J., Bloom, J., Tanvir, N., Vanderspek, R., Greiner, J., Castro-Tirado, A. J., Gorosabel Urkia, J., von Hippel, T., Murakami, T., Lehnert, M., Kuijken, K., Hoekstra, H., Metcalfe, N., Howk, C., Conselice, C., Telting, J., Rutten, R. G. M., Rhoads, J., Cole, A., Pisano, D.J., Naber, R., & Schwarz, R., "A Search for Optical Afterglow from GRB 970828", submitted to *The Astrophysical Journal Letters* (1997).

Archival Searches for Optical Emission from GRB910122 and GRB970228

J. Gorosabel[1] and A. J. Castro-Tirado[1]

[1] *Laboratorio de Astrofísica Espacial y Física Fundamental (LAEFF-INTA)*
P.O. Box 50727, E-28080, Madrid, Spain.

Abstract. We present here the results of a study based on the Harvard College Observatory Plate Collection. One part of our study focus on the unique archival search carried out so far for a GRB optical counterpart (GRB970228). We examined $\sim$ 12000 plates: 3995 at the position of GRB910122 and 8004 at the position of GRB970228, over a span of 90 years (from 1889 to 1979). The total exposure time is $\sim$ 0.55 and $\sim$1.1 yr respectively. Additionally, 18 deep Schmidt plates containing the GRB910122 error box were blinked at the Royal Observatory of Edinburgh. No convincing optical transient was found for either GRB970228 or GRB910122. We also discuss the reality as an OT of a spot found outside the GRB910122 error box.

INTRODUCTION

Archival searches are extremely important, because they allow detection of possible recurrent optical emission. Therefore, these studies rely upon the assumption that GRBs do repeat and show optical transient emission. This is already assumed, at least for some of them. The detection of optical emission of GRB970228 [1,2] and GRB970508 [3,4] confirms this fact. However, for many of them no optical emission was detected although very fast and deep follow up observations were carried out [5,6]. Transients have been found in archival searches, however the confirmation of optical recurrent emission associated with GRBs is still unclear [7-12]. Table 1 shows a summary of the archival searches carried out so far.

TABLE 1. The most important GRB archival searches carried out to date

Plate collection	Number of boxes	Monitoring time (yr)	Reference
Ondřejov, Sonneberg, Bamberg	21	$\sim$10	Hudec et al. [9,10,20]
Harvard	16	4.25	Schaefer et al. [8,17]
Sonneberg	14	2	Greiner et al. [11,19]
Harvard, ROE	2	1.65	Gorosabel et al. (in preparation)
Odessa	40	1.26	Moskalenko et al. [12,21]

GRB910122

GRB910122 is one of the few GRBs with error boxes ~ 20 arc min^2. The measured X-ray flux in the 6-15 keV range, as seen by WATCH [15] was 6×10^{-8} erg cm^{-2} s^{-1}. The γ-ray flux was 4×10^{-6} erg cm^{-2} s^{-1} (E $>$ 100 keV) [13]. The detection of GRB910122 by *Ulysses*, SIGMA/*Granat* and *PVO* allowed the intersection of three annuli which determined a tiny error box [14]. Furthermore, the position of the GRB was consistent with that provided by WATCH/*Granat* [15]. This fact, together with the small area and the low surface density of objects ($l = 30°$) made the search in the Harvard College Observatory Plate Collection (HCO) very suitable. The total exposure time for GRB910122 was ~ 0.55 yr. Table 2 shows the plates series examined.

TABLE 2. HCO Plates examined for GRB910122.

Plate series	Limiting magnitude	Plate scale (arc s mm^{-1})	Number of plates	Total exposure (hours)
ADH	18.0	68	0	0.0
AM	14.1	600	3001	4077.7
B	15.2	179	277	78.0
Damon	15.2	580	254	285.1
MF	16.4	167	135	81.6
RB	14.8	391	312	282.4
SB	18.0	26.3	0	0.0

Spots were found on 89 plates, although 82 were easily discarded using amplification lenses. The remaining 7 plates were checked using the reflected-transmitted light microscope. None of the 7 plates shows any spot within the GRB910122 IPN error box. On the other hand, if we assume an extended halo radius equal to 150 Kpc and a neutron star speed of 1000 km/s the apparent movement on the sky in 100 years would be only ~ 0.07". Therefore, theories based on fast neutron stars in the halo cannot place the optical source outside the IPN error box [16]. Once the reflected-transmitted light microscope was used, only plate AM 24589 was selected. It shows a spot ~ 0.6 mag over the plate limit being more compact than the stars on the plate. It shows a small coma distortion, but not so large as the surrounding stars. However the reality of the spot is under debate because plate AM 24589 was blinked and another two spots were found, thus a double exposure cannot be ruled out. Generally, the lack of precise pointing makes the flash images on large exposure plates to be steeper than those of stars, although they could be shallower [17]. This fact could explain the different compactness of the spot. Changing the focal length of the microscope, we were able to examine the spot at different depths. So, we excluded the spot to be a plate default on the surface of the emulsion. The great ecliptic latitude of the GRB ($\beta = -32°$) argues against the presence of a minor planet in the plate. Fig. 1 shows the position of the spot. Plate

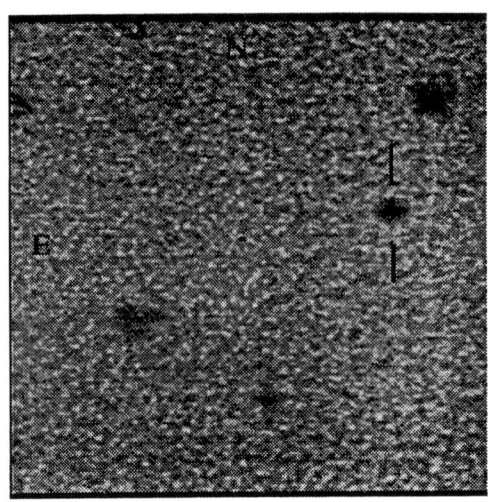

FIGURE 1. A spot found on plate AM 24589. Further details are published in [22].

AM 24589 is a 90 min blue exposure taken in 1945 May 18, so any relationship with aircraft or satellite glints is excluded. The spot appears as a $12.5^m \pm 0.5^m$ star image (or $\sim 4^m$ if we assume a 1 s flash duration) and the corresponding S_γ/S_{opt} (B band) ratio is ~ 300, and S_x/S_{opt}(B band) is ~ 1. The astrometry of the object was carried out resulting the following coordinates: $\alpha_{2000} = 19^h 47^m 23^s \pm 1.8^s$ $\delta_{2000} = 70°33'7.3'' \pm 9''$.

Additional Schmidt plates were blinked at Royal Observatory of Edinburgh (ROE). No variable object was found near or inside the IPN error box with ~ 0.2 mag above the plate limit. Table 3 shows a list of the plates examined at ROE.

TABLE 3. HCO Plates examined at ROE.

Plate number	Date	Exposure time (s)	Plate number	Date	Exposure time (s)
UB 53	730727	200	J 7019S	810511	600
J 897	740908	400	J 7024	810512	700
J 916	740916	450	J 7192	810917	100
J 1751	750807	700	UJ10444P	850911	250
J 1758	750808	700	B11967	870622	150
J 1791	750829	700	U11968	870622	600
HA 2560	760829	1300	OR14496	910831	650
HA 2585	760912	2400	OR15122	920821	600
BJ 4251	780506	150	I15763	930912	1200

GRB970228

GRB970228 is the first GRB with a convincing optical counterpart [1,2]. The accurate position made the search very easy in comparison with other GRBs with large error boxes. The possible proper motion of the GRB970228 counterpart [18] of 0.5"/yr would be undetectable even in the oldest HCO plates. The oldest plate examined was taken in 1889, so the possible proper motion would imply a maximum shift of 0.83 mm on the A plates (the smallest scale series) and 0.51 mm on the MA and MC series. This shift is smaller than the size of a spot on that plates series, so we restrict our search to a single point.

TABLE 4. Plates examined at HCO for GRB970228

Plate series	limiting magnitude	scale (arc s mm^{-1})	number of plates	total exposure (hours)
A	16.5	60	3	0.5
AC	14.1	600	2906	3834
FA,AI	12.5	1200	4602	5287
MA	16.5	97	5	5.2
MC	16.5	98	7	5.4
MF	16.4	167	35	14.2
RH	14.8	391	446	472

The total monitoring time covered for GRB970228 is ~1.1 yr. Many star-like spots were found, however none of them was consistent with the position of the optical counterpart of GRB970228. We find a lower limit of ~1.1 years for any optical transient emission activity brighter than 12.5^m (or 3.2^m if a 1 s flash is assumed).

CONCLUSIONS

We have examined ~12000 plates at HCO for GRB910122 and GRB970228, covering ~0.55 and ~1.1 yr respectively. For GRB910122 a spot was found outside the error box, however the presence of other two spots on the plate make the object unreliable. Additional plates blinked at ROE do not show any star like spot near or inside the IPN error box ~0.2 mag over the plate limit. For GRB970228 many star-like spots were found, however none of them was consistent with the position of the optical counterpart of GRB970228.

ACKNOWLEDGMENTS

The authors warmly thank the Harvard College Observatory plate curator Martha Hazen and Alison Doane for allowing us to develop this study at the HCO plate stacks and Rene Hudec for valuable comments and advice. Also, we would like to thank Prof. Ofer Bar-Yosef, (Stone Age Laboratory), Department of Anthropology, Harvard University, for giving us the opportunity of using the reflected-transmitted light microscope. The authors wish also to acknowledge Sue Tritton for the support provided by The Royal Observatory of Edinburgh.

REFERENCES

1. van Paradijs, J., et al., *Nature* **386**, 686 (1997).
2. Guarnieri, A., et al., *A&A Letters*, in press (1997).
3. Bond, H., IAU Circular No. 6654 (1997).
4. Castro-Tirado, A. J., et al., in preparation (1997).
5. Castro-Tirado, A. J., et al., IAU Circular No. 6598 (1997).
6. Castro-Tirado, A. J., et al., IAU Circular No. 6688 (1997).
7. Schaefer, B. E., *Nature* **294**, 722 (1981).
8. Schaefer, B. E., et al., *ApJL* **286**, L1 (1984).
9. Hudec, R., et al., A&A **175**, 71 (1987).
10. Hudec, R., et al., *Adv. Space. Res.* **8**, 665 (1988).
11. Greiner, J., et al., *Astroph. Space Sci.* **138**, 155 (1987).
12. Moskalenko, E.I., et al., *A&A* **223**, 141 (1989).
13. Therekov, O.V., et al., *Astronomy Letters* **20**, 265 (1994).
14. Hurley, K., private communication (1996).
15. Castro-Tirado, A. J., Ph. D. Thesis, University of Copenhagen (1994).
16. Li, H., & Dermer, C., *Nature* **259**, 514 (1992).
17. Schaefer, B. E., *ApJ* **364**, 590 (1990).
18. Caraveo, P.A., et al., IAU Circular No. 6688 (1997).
19. Greiner, J., et al., *A&A* **227**, 115 (1990).
20. Hudec, R., AIP Conference proceedings **265**, 337 (1991).
21. Moskalenko, E.I., et al., *Gamma Ray Bursts Observations, Analyses and theories*, C, Ho, R. I., Epstein and E. E. Fenimore eds., Cambridge University Press, Cambridge, p. 127 (1992).
22. Gorosabel, J. & Castro-Tirado, A. J., *A&A*, in press (1998).

A Search for Optical Afterglow from GRB970828

Paul J. Groot[1], Titus J. Galama[1], Jan van Paradijs[1,2], Chryssa Kouveliotou[3], Roland Vanderspek[4], Alberto Castro-Tirado[5], James Rhoads[6], Matt Lehnert[7], Henk Hoekstra[8] and Nigel Metcalfe[9]

[1] *Astronomical Institute 'Anton Pannekoek', University of Amsterdam, and Center for High Energy Astrophysics, Kruislaan 403, 1098 SJ Amsterdam, The Netherlands*
[2] *Physics Department, University of Alabama in Huntsville, Hunstville AL, USA*
[3] *USRA at NASA/MSFC, Code ES-84, Huntsville AL 35812, USA*
[4] *Center for Space Research, MIT, Cambridge MA 02139, USA*
[5] *Laboratorio de Astrofísica Espacial y Física Fundamental, P.O. Box 50727, E-28080, Madrid, Spain*
[6] *NOAO, Tucson AZ 85726, USA*
[7] *Sterrewacht Leiden, Postbus 9513, 2300 RA Leiden, The Netherlands*
[8] *Kapteyn Astronomical Institute, Postbus 800, 9700 AV, Groningen, The Netherlands*
[9] *Physics Department, University of Durham, South Road, Durham, UK*

Abstract. We report on the results of R band observations of the error box of the γ-ray burst of August 28, 1997, made between 4 hours and 8 days after this burst occurred. No counterpart was found varying by more than 0.2 magnitudes down to $R = 23.8$. We discuss the consequences of this non-detection for relativistic blast wave models of γ-ray bursts, and the possible effect of redshift on the relation between optical absorption and the low-energy cutoff in the X-ray afterglow spectrum.

INTRODUCTION

GRB970828 was discovered with the All-Sky Monitor (ASM) on the Rossi X-ray Timing Explorer (RXTE) on August 28, 1997, UT $17^h44^m36^s$ from an elliptical region centered at RA=$18^h08^m29^s$, Dec=$+59°18'.0$ (J2000), with a major axis of $5'.0$, and a minor axis of $2'.0$ [1,2]. Within 3.6 hours the RXTE/PCA scanned the region of the sky around the error box of the ASM burst, and detected a weak X-ray source, located in the ASM error box with a 2–10 keV flux of 0.5 mCrab [3]. The burst was also detected with the Burst And Transient Source Experiment (BATSE) and the GRB experiment on Ulysses. Its fluence and peak flux were 7×10^{-5} erg cm^{-2}, and 3×10^{-6} erg cm^{-2} s^{-1}, respectively. From the difference between burst arrival times, its position was constrained to lie within a 1.62 arcminute wide annulus, that intersected the RXTE error box [4]. In an ASCA observation made between

Aug. 29.91 and 30.85 UT, a weak X-ray source was detected at an average flux level of 4×10^{-13} erg cm^{-2} s^{-1} (2–10 keV). The ASCA error box is centered on RA=$18^h08^m32\overset{s}{.}3$, Dec =$+59°18'54''$(J2000) and has a $0\overset{'}{.}5$ radius [5].

OBSERVATIONS AND DATA ANALYSIS

We observed the GRB error box with the Prime Focus Camera of the WHT, on 9 nights between August 28, UT 21^h47^m, and September 5, UT 22^h07^m (see Table 1). The first observation was made just over 4 hours after the γ-ray burst. All observations were made with a Cousins R band filter [6]. During the first two nights and the last three nights, we used a LORAL 2048×2048 CCD chip, with 15μ pixels, giving a field of view of $8\overset{'}{.}45 \times 8\overset{'}{.}45$. During the intervening nights we used an EEV CCD chip (2048×4096), windowed at 2048×2400, with 13.5μ pixels giving a $8\overset{'}{.}1 \times 9\overset{'}{.}5$ field of view. On August 30 two R band images were made with the WIYN Telescope. The camera contained a 2048 × 2048 CCD, giving a field of view of $6\overset{'}{.}8 \times 6\overset{'}{.}8$.

We obtained a photometric calibration of the CCD images from observations of Landolt Selected Area 113, stars 281, 158, 183 and 167 [7], on Aug 31, 0^h14^m UT with the WHT.

A region of $2' \times 2'$ centered on the ASCA position in the bias-subtracted and flatfielded images was analyzed using DoPhot [8], in which astrometric and photometric information of all objects are determined from bivariate Gaussian function fits to the brightness distribution in their image; the parameters of these fits also tell us whether an object is stellar (i.e., unresolved) or a galaxy. In this region (see Fig. 1) we find a total of 63 objects, 36 of which are stellar, and 27 galaxies, down to R=23.8.

We have searched for variable objects by comparing the magnitudes of each star as determined for each of the images. Comparison of images taken on different nights showed no variation on time scales between a day and a week in excess of 0.2 mag for $R \leq 23.8$ (for the last three nights the limit on variability is 0.3 mag for $R \leq 23.8$). Comparison of three images taken on the night of August 29 to 30 showed no variations on time scales of several hours in excess of 0.2 mag for $R < 22.5$.

ABSORPTION IN REDSHIFTED MATERIAL

An explanation, pointed out to us by Dr. B. Paczyński, for the non-detection of optical afterglow could be photoelectric absorption, also visible as a low-energy cut-off in the X-ray spectrum.

In case the absorption takes place at some redshift z the relation between the optical and X-ray absorption is affected. The cross section for photoelectric absorption in the (0.2–5) keV range depends on energy roughly as $E^{-2.6}$ [9]. Then the factor by which the apparent N_H, inferred from the low-energy cut-off in the

X-ray spectrum, has to be increased is approximately $(1 + z)^{2.6}$. If we assume, for example, that the GRB occured at a redshift of $z=1$, the factor by which the apparent value of N_H has to be increased would be ~ 6. Moreover, the photons in the R band we observe would be at wavelengths near 3200 Å at the source, at which wavelength the interstellar absorption is approximately a factor 2.5 larger than in the R band [10,11]. These combined effects would lead, for a GRB at $z=1$ and an apparent, moderate, $N_H=10^{21}$ atoms cm^{-2} to an R band extinction of ~ 5 mags.

If absorption is the correct explanation, a substantial fraction of GRB sources (those with a small optical response) would be located close to where large column densities are available, i.e., in disks of galaxies. This would link GRBs to a population of massive stars. This is expected for the failed-supernova model and for the hypernova model, proposed by Woosley [12] and Paczyński [13], respectively. In view of the large kick velocities imparted on neutron stars at birth [14–16] it remains to be seen whether a merging neutron star binary model would be consistent with this consequence.

ACKNOWLEDGMENTS

We thank the RXTE ASM and PCA teams for their very fast response to and communications regarding the γ-ray burst of August 28, 1997. TG is supported by NFRA under grant no. 781.76.011.

REFERENCES

1. Remillard, R.A. et al., *IAU Circular*, 6726 (1997).
2. Smith, D. et al., *IAU Circular*, 6728 (1997).
3. Marshall, F.A. et al., *IAU Circular*, 6727 (1997).
4. Hurley, K. et al., *IAU Circular*, 6728 (1997).
5. Murakami, T. et al., *IAU Circular*, 6732 (1997).
6. Bessel, M.S., *PASP* **91**, 589 (1979).
7. Landolt, A., *Astron. Journ.* **104**, 340 (1992).
8. Schechter, P.L., Mateo., M., & Saha, A., *PASP* **105**, 1342 (1993).
9. Morrison, R., & McCammon, D., *ApJ* **270**, 119 (1983).
10. Gorenstein, P., *ApJ* **198**, 95 (1975).
11. Cardelli, J.A, Clayton, G.C, & Mathis, J.S., *ApJ* **345**, 245 (1989).
12. Woosley, S.E., *ApJ* **405**, 273 (1993).
13. Paczyński, B., *ApJ*, in press (1997).
14. Lyne, A., & Lorimer, *Nature* **269**, 127 (1994).
15. Hansen, B.M.S, & Phinney, E.S., *MNRAS*, in press (1997).
16. van den Heuvel, E.P.J., & van Paradijs, J., *ApJ* **483**, 399 (1997).

TABLE 1. Log of observations GRB970828

Date	Telescope	UT Start	Exp. time (s)	Seeing
Aug 28	WHT	21:47	900	$0\rlap{.}''86$
Aug 29	WHT	21:15	900	$0\rlap{.}''74$
Aug 30	WIYN	05:08	600	$0\rlap{.}''8$
Aug 30	WIYN	07:38	900	$1\rlap{.}''2$
Aug 30	WHT	23:22	900	$0\rlap{.}''90$
Aug 31	WHT	20:54	900	$0\rlap{.}''71$
Sep 1	WHT	21:16	600	$0\rlap{.}''80$
Sep 2	WHT	20:53	600	$0\rlap{.}''76$
Sep 3	WHT	22:44	600	$0\rlap{.}''88$
Sep 4	WHT	21:53	600	$0\rlap{.}''79$
Sep 5	WHT	22:07	600	$0\rlap{.}''86$

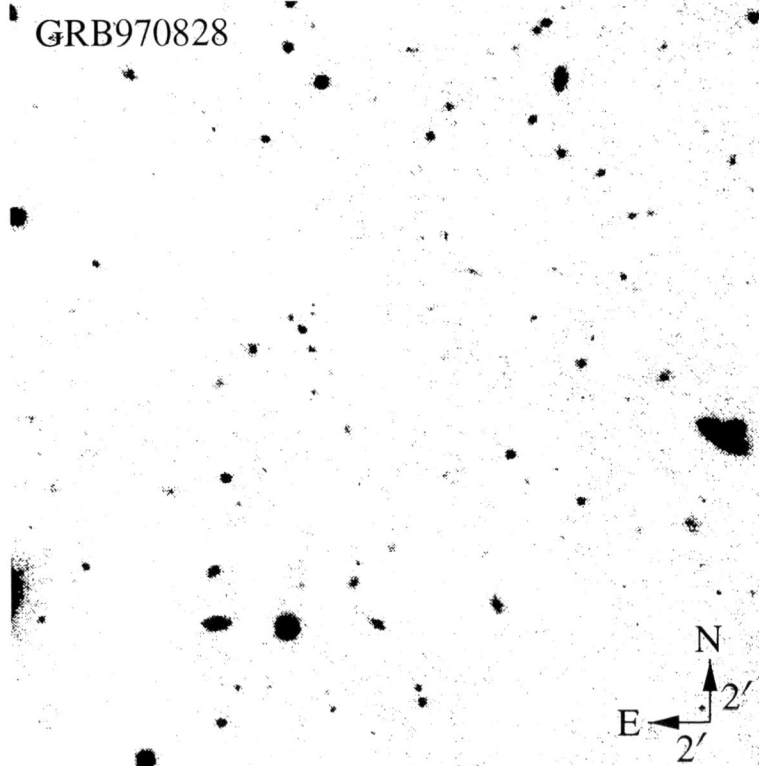

FIGURE 1. $2' \times 2'$ R-band image of the sky region centered on the $0.5'$ radius ASCA error box of GRB970828, taken at the WHT on Sept 2.

RADIO OBSERVATIONS

The Radio Counterparts of Gamma-Ray Bursts

D. A. Frail

National Radio Astronomy Observatory, Box 0, Socorro NM 87801

Abstract. In the last three decades sensitive radio searches have been made for prompt emission accompanying a burst, flaring or fading emission on time scales of days to weeks following a burst, and quiescent radio emission from the host object well after the burst has faded from view. These efforts have recently been rewarded with the detection of the radio afterglow from GRB970508. I will summarize all that has been learned from a continuing study of this source at the Very Large Array (VLA). The absence of radio emission from a large number of well-monitored events similar to GRB970508, suggests that detectable radio emission is not a generic consequence of gamma-ray bursts.

PAST AND PRESENT COUNTERPART SEARCHES

Following Schaefer [34] we divide the search for radio counterparts into three categories: prompt, delayed and quiescent. Prompt searches aim to observe the GRB while it is still ongoing. Recent experiments include FLIRT, operating at 0.07 GHz [3], CLFRT at 0.151 GHz [5], STARE at 0.61 MHz [19] and COBE at 90 GHz [1]. Apart from COBE, these are arrays of antennas operating at relatively low frequencies, providing large field of views ($\sim 10°$) and (in some cases) with rapid steering. The state-of-the-art sensitivity for these systems is 10-100 Jy, far better than earlier all-sky monitor experiments from the mid-1970's (e.g. [2,18]). This sensitivity is not very interesting for detecting incoherent emission from cosmological GRBs. However, should GRBs produce coherent emission, with the high brightness temperatures seen toward pulsars, then it might be possible to use the observed delays in arrival times at different frequencies to constrain GRB distances [27].

The search for delayed radio emission following a GRB is motivated in large part by extragalactic fireball models, which predict a flaring or fading source in the days to weeks after a burst [26,20,40,23]. There are a large number of efforts underway to detect such emission. Past efforts include the CLFRT [15], DRAO [8], WSRT [13] and VLA [28]. Most of these instruments continue to be used for GRB counterpart searches but additional instruments have been added, including the

Molonglo Synthesis telescope (MOST), the Australian Telescope Compact Array (ATCA), the Ryle telescope [30], the Berkeley-Illinois Millimeter Array (BIMA) [36], the Owens Valley Radio Interferometer (OVRO) [35], and the James Clark Maxwell telescope (JCMT) [37]. Sensitivities are typically in the range of 0.1 to 100 mJy, spanning three orders magnitude in frequency from 0.1 to 300 GHz. Most of the instruments in this category are imaging interferometers, capable of synthesizing a field of view whose diameter depends in the size of the individual antennas and a resolution which depends on the maximum separation between antennas. The VLA, for example, has a field of view of $45'/\nu(GHz)$, and depending on the array configuration, the synthesized beam at 20 cm varies from 1.3" to 43" and scales linearly with frequency.

The hosts of GRBs remain an important unsolved problem, which in principal could be revealed by deep radio imaging of their fields. The earliest work in this area was by Hjellming & Ewald [17] for GRB781119, followed up with more fields by Schaefer et al. [33]. In §4 there will be more detailed discussion of the radio source content of GRB error boxes. Summarizing here, it is accurate to say that GRB error boxes are largely devoid of radio sources brighter than 0.25 mJy and certainly except for GRB970508 none of those which have been detected appear to be related to the burst.

BACKGROUND SOURCES AND VARIABILITY

Below a few milliJanskys the classical radio source population of giant ellipticals and quasars gives way to an emerging population of starburst and interacting spiral galaxies [42]. These sources have flatter spectral indices and are extended on average with $\bar{\theta} \simeq 2.6''$ [6]. The classical radio sources do not disappear altogether, but become a diminishing fraction of the source population (10-30%) at these flux levels [7]. A useful rule of thumb is that there are 0.02 background radio sources per square arcminute above a 5-sigma detection in a 10 minute integration with the VLA at 20, 6 or 3.6 cm. More accurate source counts for these wavelengths are given in White et al. [41], Fomalont et al. [6], and Windhorst et al. [42].

Almost all radio sources exhibit modest (~10%) variations in flux density on time scales of months to years. This can be attributed either to extrinsic propagation effects, due to large scale focusing and defocusing of the radio waves by ionized gas in our Galaxy [24], or intrinsic, reflecting the on-going activity of the central engine powering the radio emission. A small class of quasars and BL Lacs undergo rapid flux variations on time scales of a day or less (see Wagner & Witzel [39] and references therein). The most extreme example of these so-called intraday variables (IDVs) is PKS 0405−385, which has a modulation index (defined as a ratio of the rms variation to the mean) approaching 15% at 5 GHz [22]. Most IDVs show variations which are weaker than this [31].

The statistics on radio variability quoted above are based on Jansky radio samples, not the milliJansky population that is of interest in GRB afterglow searches.

The only study of this kind comes from a comparison of multiply observed sources as part of the VLA FIRST survey [16] at 20 cm. For 135,000 sources brighter than 1 mJy which were observed multiple times over several days, only 200 sources varied by more than 25% and only 12 varied by more than a factor of two (or 1 in 10^4). Despite these optimistic projections for the low incidence of background variability, unrelated radio variables continue to be discovered near GRB error boxes with alarming regularity (e.g. [13,10]). Whether this is just due to small number statistics or it is a real source of contamination is unknown at this time. We caution that counterpart searches are exploring flux densities where there is only a cursory knowledge of source variability levels. Until such time as this issue is resolved radio variability should be viewed as an imperfect tool for identifying radio afterglows.

GRB970508 - THE GOLDEN AFTERGLOW

A full account of the detection and the behavior of the radio afterglow from GRB970508 during the first 85 days is given in Frail et al. [11]. Here we will discuss the subsequent behavior of the source out to day 120, and we will summarize the methods used to estimate the size and expansion of the fireball.

The radio light curve at 3.6 cm is shown in Figure 1. The source showed erratic fluctuations during the first few weeks, which subsequently dampened out. The mean flux density reached a broad plateau near 600 μJy but it has since decayed below this level. The behavior of the source at 6 cm is very similar (see Frail et al. [11]) but at 20 cm there are significant differences. GRB970508 was initially undetectable at 20 cm (S<0.1 mJy) but "turned on" at later times increasing in flux density by a factor of four from day 6 to day 80. This behavior has been interpreted as due to a decreasing optical depth from a source that was initially synchrotron self-absorbed. The time evolution of the spectral index (defined as α where $S_\nu \propto \nu^\alpha$) between 6 and 3.6 cm is shown in Figure 2. The large positive and negative swings of α mimic the erratic flux variations seen at early times. However, between day 40 and 80 α converges to a more or less constant value of +0.25, before abruptly changing to more negative values at later times. We suggest that this change marks the start of the decay of the light curve at these frequencies. In the context of the fireball model the decay phase begins when the characteristic energy of the synchrotron electrons shift into the radio band [40,23].

Frail et al. [11] and Shepherd et al. [35] argued that the erratic behavior of the light curve and the spectral index must be ascribed to *external* influences and could not come from intrinsic fluctuations. It is well known that the turbulent, ionized gas in our Galaxy scatters radio waves and produces a phenomena analogous to optical seeing [32]. A plane wave incident on a smooth medium experiences only dispersive effects, but in a clumpy medium the same wave sees a point-to-point variation in the refractive index. This "crinkles" the wavefront and can produce a host of observable phenomena, including rapid flux variations from an otherwise constant source [25,14]. The magnitude and the time scale of the observed flux

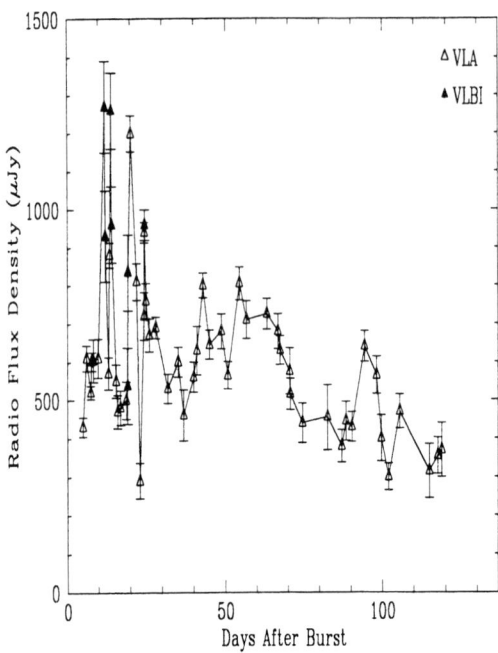

FIGURE 1. The radio light curve of GRB970508 at 3.6 cm.

variations depends on only two fundamental parameters, the Fresnel scale θ_F and the diffractive scale θ_{diff}. The first of these is a purely geometric term. It is the size of the first Fresnel zone at the distance d where most of the scattering is occurring, $\theta_F = \sqrt{2\lambda/\pi d}$. Within θ_F the radiation from the source adds constructively. The diffractive scale θ_{diff}, also known as the spatial coherence scale, is the size of the region over which the rms phase variations amount to no more than one radian. Values of θ_{diff} are known approximately in most directions of our Galaxy [4], so it is possible to estimate the *strength* of scattering toward GRB970508.

In the language of scattering theory, the erratic variations in the flux density and spectral index from GRB970508 are the signature of strong, diffractive scintillation, hitherto only seen toward pulsars [32]. To see the variations in Figure 1 and 2 requires that the source size $\theta_s \leq \theta_{\text{diff}} < \theta_F$. With a factor of two uncertainty we estimate that during the first week GRB970508 has $\theta_s \simeq \theta_{\text{diff}} \simeq 3$ μas. The damping of the amplitude of the observed modulations at later times is interpreted as the expansion of the source beyond the diffractive scale (i.e. $\theta_s > \theta_{\text{diff}}$). In principal, with better sampling, it should be possible to use the scintillation history of radio afterglows to recover the evolution of the fireball radius with time [14]. An independent estimate for the size and expansion of the fireball comes from the flux

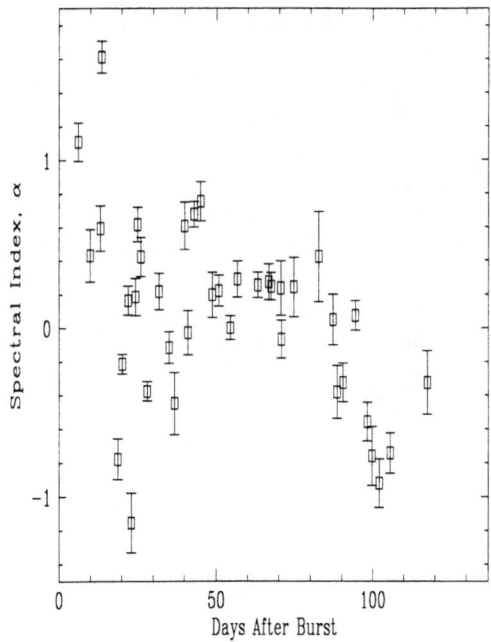

FIGURE 2. A radio spectral index for the afterglow emission from GRB970508 between 3.6 cm and 6 cm. The spectral index, α, is defined as $S_\nu \propto \nu^\alpha$.

density of GRB970508 at 20 cm when it was optically thick. Synchrotron self-absorption has long been used to estimate the angular size of radio sources, since the size at any frequency ν depends only on the source flux S_ν and (weakly) on magnetic field (i.e. $\theta_s^2 \propto B^{1/2} S_\nu / \nu^{5/2}$). Katz & Piran [21] derive a similar expression for a relativistically expanding fireball which is independent of the bulk Lorentz factor and yields a value for θ_s similar to that derived from the scintillation method.

THE NON-DETECTIONS

As exciting and illuminating as the detection of a radio afterglow from GRB970508 has been, the fact remains that no other GRB has been detected at radio wavelengths. There are 17 well-monitored GRBs at centimeter wavelengths [33,8,28,12]. To fall into this category requires that (1) the GRB error region is less than $\sim$10 arcmin^2, (2) that it has been observed between 1 and 10 GHz, (3) with rms sensitivities between 0.01 and 0.1 mJy, and (4) that multiple observations have been made with a range of postburst time scales from days to decades. The great majority of GRB error boxes have no detectable radio sources down to rather sensitive limits. Notable exceptions include GRB781119 [17] and GRB960720 [29]. There is no evidence to suggest that these radio sources occur in GRB fields any more than expected by chance, nor do any of the sources exhibit unusual behavior that would distinguish them from the background population of radio sources [12].

We summarize these non-detections in Figure 3. Each point represents a 3-sigma limit on the absence of any time-variable radio source in a GRB error box. The radio limits presented here represent the most complete temporal sampling of GRB afterglows currently known. On the right hand horizontal axis we convert these flux density values to equivalent R-band magnitude (assuming 0 mag = 3000 Jy). We note that the current formulations of the fireball models predict that the peak flux density of the fireball is approximately the same in all wavelength bands (e.g [40]). Therefore, the results in Figure 3 imply that the existence of afterglows with peak fluxes in excess of 18th magnitude at *any wavelength* must be rare.

If we confine ourselves to the radio band it is certainly true that radio counter-

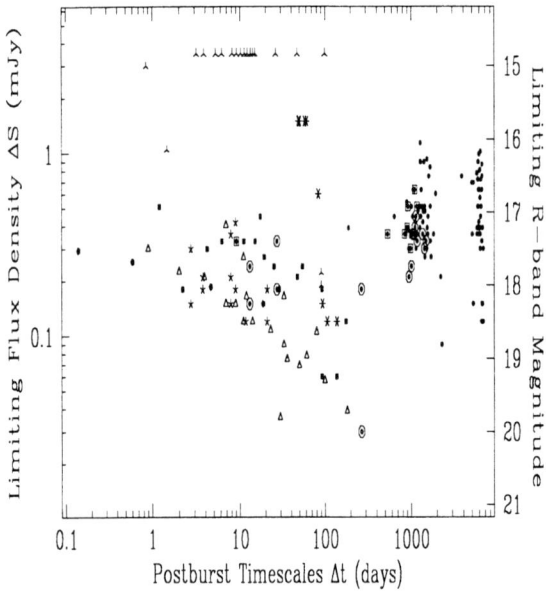

FIGURE 3. Upper limits on the absence of a (time-variable) radio source in the error boxes of 17 well-monitored gamma-ray bursts.

parts to GRBs above 0.1 mJy are rare. An event with a light curve resembling that of GRB970508 would have been detected in 9 of the 17 events followed so far. This approximately 10% detection rate is reminiscent of the fraction of type Ib, Ic and II supernovae detected in the radio band (Montes, *priv. comm.*). In this case, and also possibly for GRBs, the detectability of radio emission depends strongly on the amount of circumstellar material around the progenitor.

Summarizing the current state of our knowledge about the radio counterparts to gamma-ray bursts it should first be said that the detection of the radio afterglow yields unique GRB diagnostics. Unlike optical or X-ray wavelengths one is presented with the possibility of following the full evolution of the fireball emission through all its different stages; first while it is optically thick, then as it slowly rises to a peak flux density and thereafter decays, making a transition from an ultra-relativistic to sub-relativistic shock. Both the scintillation of the radio source and its flux density, when it is synchrotron self-absorbed, allow a determination of the size and expansion of the fireball. At present these values cannot be determined by any other means. In the future, should a sufficiently bright, nearby GRB ($z \simeq 0.1$-0.2) be detected then the technique of VLBI could be used to image the expanding fireball [38]. Finally, we emphasize that bright GRBs ($> 100\ \mu$Jy) in the radio (and perhaps optical and X-rays) are rare, requiring that the largest and most sensitive instruments be used for GRB studies.

ACKNOWLEDGEMENTS

The radio counterpart searches that I have done over the last four years would not have been possible without the help and support of colleagues like Shri Kulkarni, Greg Taylor, and Mark Wieringa. I further appreciate the prompt dissemination of GRB localizations by the BeppoSAX group (lead by Enrico Costa), the IPN group (lead by Kevin Hurley) and the BATSE team (lead by Jerry Fishman).

REFERENCES

1. Ali, S., Schaefer, R. K., Limon, M., & Piccirillo, L., *ApJ* **487**, 114-121 (1997).
2. Baird, G. A. et al., *Astrophys. Space. Sci.* **42**, 69-72 (1976).
3. Balsano, R. et al., these proceedings.
4. Cordes, J. M. et al., *Nature* **354**, 121-124 (1991).
5. Dessenne, C. A. -C. et al., *MNRAS* **281**, 977-984 (1996).
6. Fomalont, E. B. et al., *AJ* **102**, 1258-1277 (1991).
7. Fomalont, E. B. et al., *ApJ* **475**, L5-L7 (1997).
8. Frail, D. A. et al., *ApJ* **437**, L43-L46 (1994).
9. Frail, D. A., & Kulkarni, S. R., *Astrophys. Space. Sci.* **231**, 277-280 (1995).
10. Frail, D. A. et al., *ApJ*, **483**, L91-L94 (1997a).
11. Frail, D. A., Kulkarni, S. R., Nicastro, L., Feroci, M., & Taylor, G.B., *Nature* **389**, 261-263 (1997b).

12. Frail, D. A. et al., *in preparation* (1998).
13. Galama, T. J. et al., *A&A* **321**, 229-235 (1997).
14. Goodman, J., *New Ast* **2**, 449-460 (1997).
15. Green, D. A., et al., *Astrophys. Space. Sci.* **231**, 281-284 (1995).
16. Helfand, D. J., Das, S. R., Becker, R. H., White, R. L., McMahon, R. G. in *Blazar Continuum Variability*, (eds. H. R. Miller, J. R. Webb & J. C. Noble) 214 (Astron. Soc. of the Pacific: San Francisco, 1996).
17. Hjellming, R., & Ewald, S., *ApJ* **246**, L137-L140 (1981).
18. Inzani, P. et al., *Astrophys. Space. Sci.* **56**, 239-243 (1978).
19. Katz, C. A., these proceedings.
20. Katz, J. I., *ApJ* **432**, L107-L109 (1994).
21. Katz, J. I., & Piran, T., *ApJ* **490**, 772-778 (1997).
22. Kedziora-Chudczer, L. et al., *ApJ* **490**, L9-L12 (1997).
23. Mészáros, P., & Rees, M. J., *ApJ* **476**, 232-237 (1997).
24. Mitchell, K. J. et al., *ApJS* **93**, 441-453 (1994).
25. Narayan, R. in *Pulsars as Physics Laboratories* (eds. R. D. Blandford, A. Hewish, A. G. Lyne & L. Mestel) 151-165 (Oxford University Press, Oxford, 1993).
26. Paczyński, B., & Rhoads, J. E., *ApJ* **418**, L5-L8 (1993).
27. Palmer, D. M., *ApJ* **417**, L25-L28 (1993).
28. Palmer, D. M. et al., *Astrophys. Space. Sci.* **231**, 315-318 (1995).
29. Piro, L. et al., *astro-ph/9707215* (1997).
30. Pooley, G., & Green, D. A., *IAU Circ.* **6670** (1997).
31. Quirrenbach, A. et al., *A&A* **258**, 279 (1992).
32. Rickett, B. J., *ARA&A* **28**, 561-605 (1990).
33. Schaefer, B. E. et al., *ApJ* **340**, 455-457 (1989).
34. Schaefer, B. E. in *Gamma-Ray Bursts: The Second Huntsville Symposium*, (eds. G. J. Fishman, J. J. Brainerd, & K. Hurley) 382-391 (AIP: New York, 1994).
35. Shepherd, D. S., Frail, D. A., Kulkarni, S. R., & Metzger, M., *ApJ*, in press (1997).
36. Smith, I. A., Gruendl, R. A., Liang, E. P., & Lo, K. Y., *ApJ* **487**, L5-L7 (1997).
37. Smith, I. A., these proceedings.
38. Taylor, G. B., Frail, D. A., Beasley, A. J., & Kulkarni, S. R., *Nature* **389**, 263-265 (1997).
39. Wagner, S. J., & Witzel, A., *ARA&A* **33**, 163-197 (1995).
40. Waxman, E., *ApJ.* **491**, L19-L22 (1997).
41. White, R. L., Becker, R. H., Helfand, D. J. & Gregg, M. D., *ApJ* **475**, 479-493 (1997).
42. Windhorst, R. A., Fomalont, E. B., Partridge, R. B., & Lowenthal, J. D., *ApJ* **405**, 498-517 (1993).

VLBA Observations of the Radio Counterparts to γ-Ray Bursters

G. B. Taylor*, A. J. Beasley*, D. A. Frail*, & S. R. Kulkarni[†]

National Radio Astronomy Observatory, Box 0, Socorro NM 87801
[†]*Division of Physics, Mathematics and Astronomy 105-24, Caltech, Pasadena CA 91125*

Abstract. The detection of a radio counterpart to GRB970508 by Frail et al. [6] opens the way to applying the powerful technique of Very Long Baseline Interferometry to this intriguing class of objects. We demonstrate the applicability of this technique to GRB970508. Using the Very Long Baseline Array (VLBA), including the phased VLA and the 100-m radiotelescope at Bonn, we were able to image the radio counterpart of GRB970508 at a resolution of 1 mas. This 8.4 GHz image revealed GRB970508 to be very compact (less than 400 micro-arcsecond in angular size). We present the results from an ongoing VLBA monitoring campaign of GRB970508 that commenced 8 days after the initial gamma-ray burst and is still underway 120 days after the burst. We derive a position for the radio counterpart with a standard deviation of 73 micro-arcsecond, and will discuss the limits on proper motions and source expansion obtained so far. Assuming that the radio emission arises from the afterglow of a cosmological fireball, VLBI observations of GRB970508 and future GRBs offer a reasonable hope of directly measuring the expansion.

INTRODUCTION

The Very Long Baseline Array (VLBA) is the first array of antennas devoted full-time to Very Long Baseline Interferometry (VLBI) observations. As such it is well suited to Target of Opportunity projects. On May 16, 1997 UT we initiated a program of observations of the radio counterpart to GRB970508, VLA J065349.4+791619 with the 10-antenna VLBA. The first observation was conducted 7.8 d following the gamma-ray burst and also included the 100-m reflector at Effelsburg, Germany and the Very Large Array (VLA). All the observations were conducted at a central frequency of 8.41 GHz using a bandwidth of 32 MHz. Signals in both senses of circular polarization were recorded on magnetic tapes and processed at the VLBA correlator in Socorro, New Mexico. Results from the first 6 epochs have been published in Taylor et al. [1]. Here we discuss the combined results from the 9 epochs obtained to date.

The atmosphere affects the phase of the signal from the source differently at each of these widely spaced antennas. In order to overcome this and other instrumental

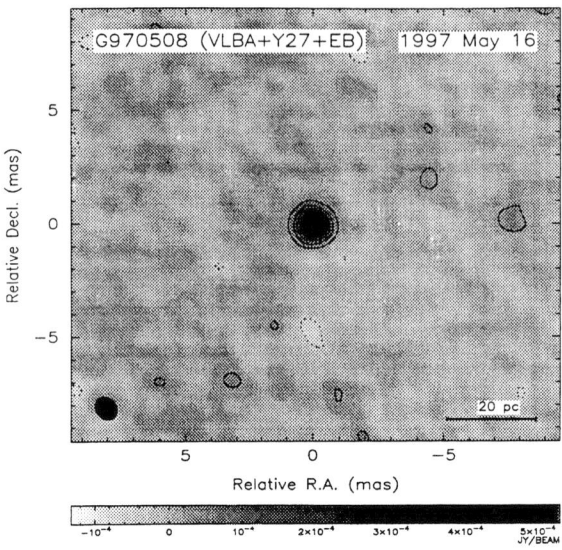

FIGURE 1. The first VLBI observation of VLA J065349.4+791619 on May 16.48 – 16.92 using the VLBA, phased VLA, and 100-m at Effelsburg at 8.4 GHz. Contours are overlaid on a greyscale total intensity image at −60, 60, 120, 180, 240, 300, 360, and 420 μJy/beam.

contributions we alternated observations of VLA J065349.4+791619 with an observation of the strong, compact "calibrator" source B0718+793 located 1.5° away. This technique, known as phase-referencing, is now routinely used for observations of faint sources with the VLBA. The switching time for the first and sixth epochs (involving the phased VLA) was 2 minutes and 1.5 minutes for all other epochs. VLA J065349.4+791619 was clearly detected in all the nine epochs. In all the nine images, VLA J065349.4+791619 is unresolved, although the final resolution of the instrument depends on the antennas involved and the degree to which atmospheric phase fluctuations could be removed (the latter effect is somewhat analogous to the phenomenon of optical "seeing").

Under the best observing conditions (May 28.4 UT) we place a limit on the source size (defined as the full width at half maximum), $\Delta\theta < 0.26$ milliarcsecond (mas). The corresponding lower limit on the brightness temperature $T_B = S\lambda^2/2k\Delta\Omega$ is 1.4×10^8 K; here $\lambda = 3.6$ cm is the wavelength of observation, k, the Boltzmann constant and $\Delta\Omega = \pi\Delta\theta^2/4$ is the solid angle of the source. Within the measurement errors, there is no indication that the source has expanded to a size approaching the resolution of this experiment. The deepest image (May 16.7 UT - Fig. 1.) allows us to limit the extended emission on scales 10 mas – 20 arcsec to less than ∼50 μJy.

From analysis of the positions of VLA J065349.4+791619 over the nine epochs

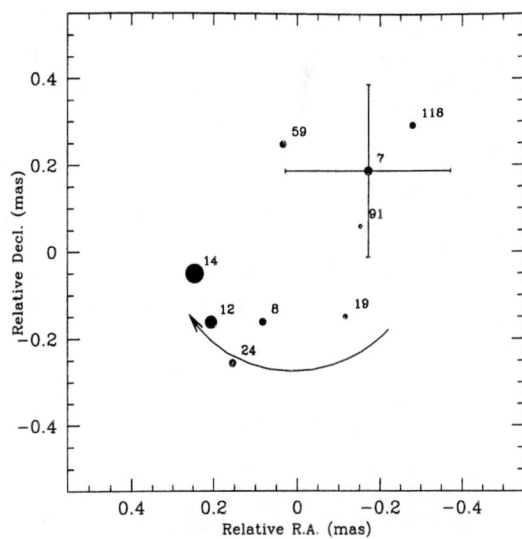

FIGURE 2. Positions of VLA J065349.4+791619 from the 9 VLBI epochs. The zero point was selected to be the fitted position reported in the text. The symbol size is proportional to the upper limit on the source size derived from each observation and the time elapsed in days since GRB970508 was detected by BeppoSAX is shown next to each measurement. The curve shows the portion of the trigonometric parallax ellipse tracked by an object at 3.3 kpc over the 4 months spanned by our observations.

(Fig. 2) we find: (1) The mean position for VLA J065349.4+791619 is $\alpha = 06^h53^m49\overset{s}{.}45133$ and $\delta = +79°16'19\overset{''}{.}51360$. The 1-$\sigma$ error in $\alpha \cos \delta$ is 73 μas and 73 μas in δ. This position is referenced to the absolute coordinates of the calibrator, 0718+793 [2]. (2) We fit a linear model with astrometric parameters α, δ, μ_α, and μ_δ to the data and place a 3σ upper limit on the proper motion of $\mu < 5$ mas yr^{-1}. (3) Assuming no proper motion, our limit on the annual trigonometric parallax is $\Pi < 0.3$ mas. This limit requires the distance of the radio source to be greater than 3 kiloparsec. If both parallax and motion are allowed then the limits are somewhat larger.

The radio emission may originate with the explosive event that produced the γ-ray, X-ray and optical transients, or it may be associated with the host galaxy of GRB970508. Here we compare the radio properties of VLA J065349.4+791619 to those of both "normal" and "active" galaxies, before going on to explore the predictions of cosmological fireball models.

At the sub-mJy level, extragalactic sources are identified with host galaxies having a median optical magnitude V = 23 [3]. The median angular size in the radio of these sources is 2.6 arcsec, and in 90% of the population the radio emission is thought to be a combination of steep-spectrum synchrotron emission from a galactic disk and thermal-bremsstrahlung from large-scale star formation. About 10%

of the sub-mJy population may be dominated by emission from a nuclear starburst or a weak Active Galactic Nucleus (AGN). Our VLBI brightness temperature limit of $>10^8$ K rules out galactic disks and star formation as the source of the radio emission from VLA J065349.4+791619.

Next we consider the possibility that the radio emission from VLA J065349.4+791619 originates from the central engine of an AGN. The radio power of the source at a rest frequency of 17 GHz (for an assumed redshift of 1 and $H_0 = 75$) is 10^{24} W Hz^{-1}. This is comparable to the radio powers of nearby low-luminosity AGN [4]. These FR I type radio sources, however, are generally dominated by emission from extended radio lobes. In the few cases where the cores are sufficiently strong, VLBI observations have revealed considerable structure [5]. In most cases this structure has a surface brightness a factor of 20-80 below that of the peak, so it is difficult to make a direct comparison to VLA J065349.4+791619 as at best we have a peak to rms noise level of 20:1 (Figure 1).

The positive radio spectrum of VLA J065349.4+791619 [6], on average, resembles that of the so-called "GHz-peaked" radio sources (GPS). However, these sources show little variability, rather complicated structure on the parsec scale and have bright optical hosts [7,8]. Gravitationally lensed sources at similar redshifts may have a comparable intrinsic radio power, but these also show much less radio variability than GRB970508, from 0 to 10%. The morphologies of the gravitationally lensed sources are predominantly core+jet [9,10]. The extreme variability of VLA J065349.4+791619 [6] is characteristic of a class of radio sources known as intraday variables (IDVs), but these originate in flat-spectrum radio sources such as BL Lacs and quasars [11].

The radio data on VLA J065349.4+791619 – (*i*) the small angular size and correspondingly high brightness temperature, (*ii*) the steep, positive spectral index and high variability, (*iii*) the absence of any diffuse or jet-like emission around the central source, and (*iv*) the lack of proper motion and parallax – are not commensurate with any known population of radio sources but they are consistent with the expectations of all cosmological fireball models.

Optical observations of the galaxy in which this radio source is located are also relevant. The optical source continues to fade and as yet there are no indications of any host galaxy to a limit of about R=24.5 mag [12]. This rules out the radio source being a GPS, IDV or FR I source since the host galaxy of these objects would be bright. Thus the observations support the idea that the radio source reported here is not a typical extragalactic radio source embedded in the central portions of a host galaxy. The preferred interpretation, consistent with the optical spectroscopy, is that GRB970508 is a fireball located in a low luminosity blue galaxy [12].

Proceeding with the hypothesis that VLA J065349.4+791619 is the radio afterglow from GRB970508 it is interesting to note that our observations rule out significant gravitational lensing on scales of 10 mas – 300 arcsec. This covers the range of lenses discussed in the literature and rules out any substantial flux magnification of the γ-ray burst by gravitational lensing.

The VLBA observations reported here had typically longer durations than the

VLA observations of Frail et al. [6], and thus can be used to probe variability on timescales of hours. During the three observations between May 21 and May 28, significant flux density variability was detected within each 5.6 hour run at the level of 10-20%/hour. These data are plotted in Figure 1 of Frail et al. [6] where it is explained as the result of diffractive interstellar scintillation.

In the future, an afterglow with the same radio luminosity as GRB970508 but at z=0.2 could be 30 times brighter and have an angular diameter twice as large. This would improve the obtainable limits on source size by over a factor of 50 and open the way for using VLBI to study the detailed evolution of a resolved GRB in much the same way as has recently been done for extragalactic supernovae [13]. If θ remains small, then as discussed in [6], there is great potential for using less direct methods such as interstellar scintillation to determine the evolution of the fireball.

ACKNOWLEDGEMENTS

We wish to thank Jean-François Lestrade for useful discussions and assistance in placing limits on the proper motion and parallax. We also thank the VLBA staff for their extraordinary efforts on behalf of this project.

REFERENCES

1. Taylor, G. B., Frail, D. A., Beasley, A. J., & Kulkarni, S. R., *Nature* **389**, 263-265 (1997).
2. Johnston, K. J. et al., *AJ* **110**, 880-915 (1995).
3. Windhorst, R. A., Fomalont, E. B., Partridge, R. B., & Lowenthal, J. D., *ApJ* **405**, 498-517 (1993).
4. Giovannini, G., Feretti, L., & Comoretto, C., *ApJ* **358**, 159-163 (1990).
5. Lara, L. et al., *ApJ.* **474** 179-187 (1997).
6. Frail, D. A., Kulkarni, S. R., Nicastro, L., Feroci, M. & Taylor, G.B., *Nature* **389**, 261-263 (1997).
7. Taylor, G. B., Readhead, A. C. S., & Pearson, T.J., *ApJ* **463**, 95-104 (1996).
8. O'Dea, C. P., Stanghellini, C., Baum, S. A., & Charlot, S., *ApJ* **470**, 806-813 (1996).
9. Campbell, R. M., Lehar, J., Corey, B. E., Shapiro, I. I. & Falco, E. E., *AJ* **110**, 2566-2569 (1995).
10. Porcas, R. W. in *IAU Colloquium 164: Radio Emission from Galactic and Extragalactic Compact Sources* eds. J. A. Zensus, J. M. Wrobel and G. B. Taylor (PASP: San Francisco) submitted (1997).
11. Wagner, S. J., & Witzel, A., *ARA&A* **33**, 163-197 (1995).
12. Pian, E. et al., *ApJ.*, submitted (1997).
13. Marcaide, M. et al., *Science* **270**, 1475-1478 (1995).

The BIMA Search for Millimeter–Wavelength Counterparts to Gamma-Ray Bursts

Robert A. Gruendl[*], Ian A. Smith[†], R. Forester[‡], K. Y. Lo[*||], Edison Liang[†], and M. Leventhal[¶]

[*] Laboratory for Astronomical Imaging, University of Illinois at Urbana-Champaign, Urbana, IL 61801[1]
[†] Rice University, Houston, Texas 77005
[‡] University of California at Berkeley, Berkeley, CA 94720
[||] ASIAA, Taipei, Taiwan 115
[¶] University of Maryland, College Park, MD 20742

Abstract. We present the latest results from our on-going campaign to detect gamma-ray burst counterparts at millimeter-wavelengths using the Berkeley–Illinois–Maryland Association array (BIMA). BIMA is well suited for performing counterpart searches, since it can map the several arc-minute error boxes from initial burst locations, as well as perform deep stares on any better localized counterparts. Here we summarize our observations of GRBs 970111, 970228, 970508, 970815, and 970828 with emphasis on the last three to show the constraints that BIMA can place on the brightness of millimeter–wavelength emission from these bursts.

INTRODUCTION

BIMA is a radio interferometric array operating at mm-wavelengths in Hat Creek, CA [18]. We have observed 5 localized burst errorboxes with BIMA at either 85 GHz (3.5 mm) or 28.5 GHz (9.5 mm) depending on the receivers available at the time of the burst. The resulting 3σ upper limits are summarized in Figure 1. So far we have decreased our response time from one month (for GRB970111) to one week (for GRB970228) [16] and more recently to one day (for GRB970508, GRB970815, and GRB970828) [6]. Currently the fundamental limitation to our response time is that we must interrupt the previously scheduled observations and that we generally try to minimize the impact on array operations. Where telescope time has been available we have also managed a limited monitoring campaign.

---

[1] The Laboratory for Astronomical Imaging is sponsored by the National Science Foundation.

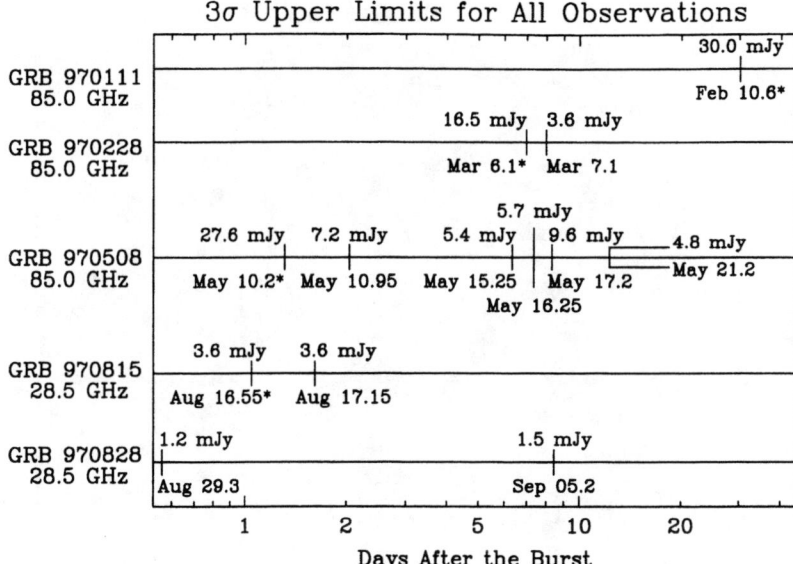

FIGURE 1. Summary of all observations of GRB error-boxes made with BIMA to date. An "*" has been appended to the date for observations which used mosaiced pointings to cover a preliminary error-box.

MOSAICED OBSERVATIONS OF LARGE ERROR BOXES

Our observing strategy has been to exploit the large field of view (2') afforded by BIMA to observe the largish error-box (3–5') generally available for bursts detected by RXTE and BeppoSAX. Thus, we can obtain observations before any localized (<1') counterparts have been detected. To perform a search over these largish areas with BIMA it is necessary to make a mosaiced observation of between 20 and 40 overlapping fields. An example of the final mosaic for GRB970508 observed on May 10.2 UT at 85 GHz (3.5 mm) is shown in Figure 2. This interferometer observation is comprised of 43 overlapping fields which have been combined in a linear mosaic to cover most of the early error box reported by the BeppoSAX team [2]. These observations show no source at the positions later reported for the optical/radio transient [1,3,5,8,17] with a 3σ upper limit of 27.6 mJy [6].

DEEP SEARCH FOR A MM-COUNTERPART TO GRB970508

Once a localized X-ray, optical or radio counterpart has been reported we make much deeper observations that usually require only one pointing. Such observations

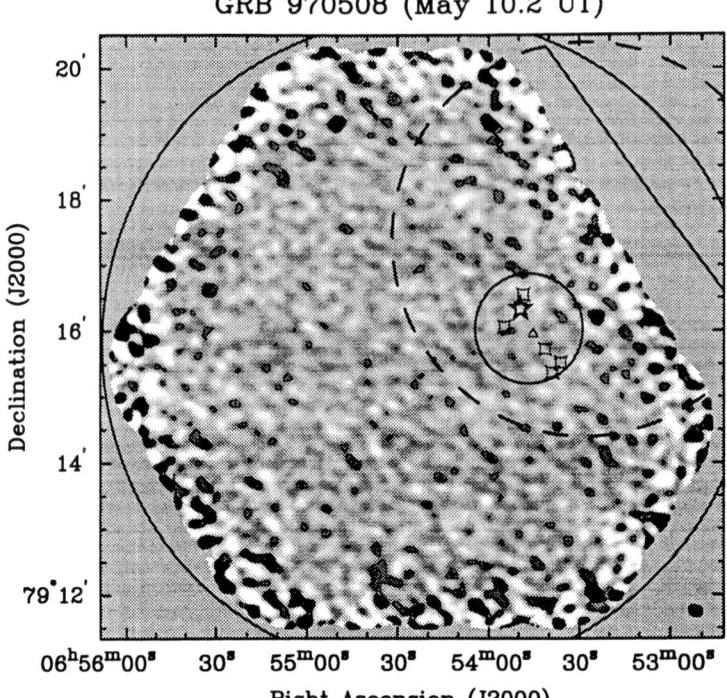

FIGURE 2. Mosaic of the preliminary error box for GRB970508 made at 85 GHz on May 10.2 UT. The grayscale shows the absolute intensity which stretches from -27 mJy (white) to 110 mJy (black) overlaid with contours which show the $2\sigma, 3\sigma, 4\sigma\ldots$ confidence level where 1σ RMS was 9.2 mJy. The large solid circle and intersecting chord are the preliminary BeppoSAX burst error circle [2] and the edge of the IPN arc. The dashed circle is the final error estimate [7] which was announced after this BIMA observation. The small solid circle is the 100″ diameter position for the BeppoSAX X-ray transient [10]. The squares mark the position of optical point sources within this small error-box while the triangle marks the position of a galaxy [13]. The star marks the position of the optical and radio counterpart [1,3,5,8,17].

are much more sensitive (by factor of 4-8 times) compared to the large mosaics because they concentrate all the observing time on a single position. The resulting 3σ upper limits from these pointed or "stare" observations at 85 GHz (3.5 mm) for GRB970508 are summarized in Figure 1 where a modest monitoring campaign was possible. No source was detected in any of these observations at the location of the optical and radio counterparts.

The BIMA upper limits constrain the spectral energy distribution of the counterpart and must be accounted for in any model. The emission at millimeter-wavelengths is lower than predicted by a simple extrapolation from the radio observations [5,11,17] or from the optical spectrophotometry of GRB970508 [3,8].

This will be discussed in more detail in a work which is currently in progress [6].

9.5 MM OBSERVATIONS OF GRB970815 AND GRB970828

The BIMA array has recently added the capability to observe between 27 GHz and 39 GHz through the loan of receivers built by John Carlstrom. This capability is not always available but we have observed GRB970815 (see Figure 3) and GRB970828 (these observations will be detailed in [6]). The main advantage to operating at 28.5 GHz (9.5 mm) with BIMA is that the field of view is much larger ($\sim 6'$) and therefore requires at most two pointings to observe the typical burst error–boxes initially determined by BeppoSAX and RXTE.

The first opportunity to observe a burst counterpart at 28.5 GHz with BIMA was GRB970815. This burst, triggered August 15.5 UT, was seen by the RXTE ASM [14]. We observed this location with BIMA using two overlapping fields (Figure 3a). We made a follow-up observation approximately 36 hours later to confirm the possible 3σ "detection" near the edge of the primary beam in one field but outside the error estimate from RXTE. This second observation (Figure 3b) does not confirm this "source." No obvious counterpart at any wavelength has been reported for this burst. The 3σ upper limit for a counterpart at 28.5 GHz is 3.6 mJy for both of the BIMA observations of GRB970815.

We also observed GRB970828 at 28.5 GHz (not shown here) with BIMA in the D–array approximately 12 hours after the burst was detected by BATSE and RXTE [12]. This configuration has very coarse spatial resolution ($\sim 30''$). The configuration was changed to the B-array (resolution $\sim 10''$) between Sept. 3 and 4. We re-observed this location on September 5th, because the VLA had since noted a source not present on August 29 [4]. We conclude that no mm (or optical or radio) counterparts fall within the final error boxes reported from the RXTE and ASCA observations [9,15]. The 3σ upper limits for a counterpart at 28.5 GHz are 1.2 mJy for the August 29 observation and 1.5 mJy for the September 5 observation.

REFERENCES

1. Bond, H. E., *IAUC* 6654 (1997).
2. Costa, et al., *IAUC* 6649 (1997).
3. Djorgovski, S. G., et al., *Nature* **387**, 876 (1997).
4. Frail, D. A., and Kulkarni, S. R., *IAUC* 6730 (1997).
5. Frail, D. A., et al., *Nature* **389**, 261 (1997).
6. Gruendl, R. A., et al., in preparation.
7. Heise, J., et al., *IAUC* 6654 (1997).
8. Metzger, M. R., et al., *Nature* **387**, 878 (1997).
9. Murakami, T., et al., *IAUC* 6732 (1997).
10. Piro, L., et al., *IAUC* 6656 (1997).

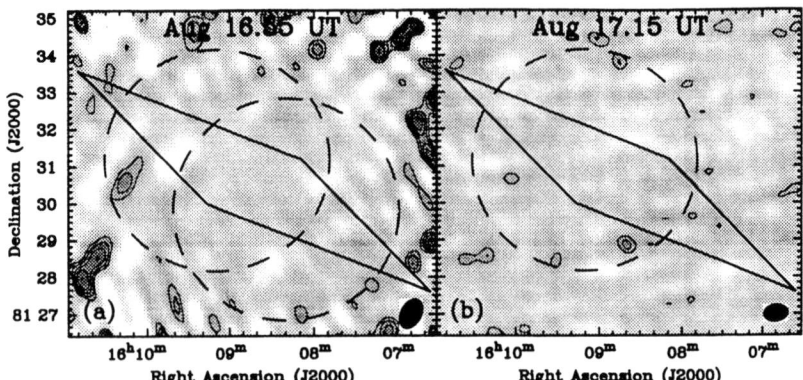

FIGURE 3. BIMA observations of GRB970815 at 28.5 GHz (9.5 mm). In each panel the heavy dashed circle is the BIMA primary beam half-power point and the solid black ellipse in the lower right corner shows the synthesized beam. Contours show the 2σ, 3σ, 4σ ... confidence levels. The observations of GRB970815 in (a) are a linear mosaic of the August 16.55 observation of two overlapping fields and thus were primary beam corrected (resulting in the amplified noise beyond the half-power point). Note that the follow-up observations in (b) do not confirm the 3σ "source" in (a) which falls outside the RXTE error box [14].

11. Pooley, G., and Green, D., *IAUC* 6670 (1997).
12. Remillard, R., Wood, A., Smith, D., and Levine, A., *IAUC* 6726 (1997).
13. Schaefer, B., et al., *IAUC* 6658 (1997).
14. Smith, D., et al., *IAUC* 6718 (1997).
15. Smith, D., et al., *IAUC* 6728 (1997).
16. Smith, I. A., Gruendl, R. A., Liang, E. P., and Lo, K. Y., *ApJ* **487**, L5 (1997).
17. Taylor, G. B., et al., *Nature* **389**, 263 (1997).
18. Welch, W. J., et al., *PASP* **108**, 93 (1996).

The STARE Project: A Progress Report

C.A. Katz*, J.N. Hewitt*, C.B. Moore*, and B.E. Corey[†]

*MIT Department of Physics and Research Laboratory of Electronics
Cambridge, MA 02139
[†]Haystack Observatory, Off Route 40, Westford, MA 01866

Abstract. The Survey for Transient Astronomical Radio Emission (STARE) is a wide-field monitor for transient radio emission at 611 MHz on timescales of fractions of a second to minutes. Consisting of multiple geographically separated total-power radiometers which measure the sky power every 0.125 sec, STARE has been in operation since March 1996. In its first seventeen months of operation, STARE collected data before, during, and after 173 gamma-ray bursts. Seven candidate astronomical radio bursts were detected within ±1 hr of a GRB, consistent with the rate of chance coincidences expected from the local radio interference rates. The STARE data are therefore consistent with an absence of radio counterparts appearing within ±1 hr of GRBs, with 5σ detection limits ranging from tens to hundreds of kJy. The strengths of STARE relative to other radio counterpart detection efforts are its large solid-angle and temporal coverage. These result in a large number of GRBs occuring in the STARE field of view, allowing studies that are statistical in nature. Such a broad approach may also be valuable if the GRBs are due to a heterogenous set of sources.

INTRODUCTION

The Survey for Transient Astronomical Radio Emission (STARE) is a project designed to detect transient radio signals at 611 MHz on timescales of fractions of a second to minutes. Local interference rejection is accomplished by using geographically separated multiple detectors and a coincidence requirement. STARE monitors a large solid angle 24 hours/day, with an operating efficiency of $\sim 95\%$. We present here a brief description of the project and some of the results from the first 17 months of operation.

INSTRUMENTATION

STARE consists of three detectors located at geographically separated sites: Hancock, NH, Green Bank, WV, and Hat Creek, CA. At each site is a total power radiometer, a GPS receiver which provides timing information, and a PC which controls the equipment. Operation is fully automated, and is coordinated over the internet by a computer at MIT, which also receives data from the sites. The organization of the system is shown in Figure 1.

The radiometers are simple single-conversion, dual-polarization receivers. System specifications:

- Center Frequency: 611 MHz,
- Bandwidth : 4 MHz,
- Beam Solid Angle: 1.5 sr,
- Integration Time: $20\mu s$ to $0.125\,s$
- RMS Sensitivity : $\sim 3\,\text{kJy}$ (zenith, $0.125\,s$ averaging).

TRANSIENT DETECTION BY COINCIDENCE REQUIREMENT

Radiofrequency interference (RFI) presents a major obstacle to transient detection at UHF. At these frequencies it is imperative that an experiment have an interference rejection scheme. STARE filters out RFI using a coincidence requirement. To be identified as astronomical, a signal must appear in at least two of the detectors simultaneously. The power of this criterion is apparent from the STARE results: in one year of operation, the Hancock station recorded 78,714 individual events (instances in which the radiometer output exceeded the baseline by at least 5σ) while the Green Bank station recorded 260,407. Out of these, only 138 coincidences were identified, yielding a rejection rate of well over 99%.

With such a scheme, there is always the possibility of coincidences due to chance. Using the mean event rates measured for Hancock ($7\,\text{hr}^{-1}$) and Green Bank ($13\,\text{hr}^{-1}$), the mean time between chance coincidences is found to be about 4 days, though the chance coincidence rate is quite variable since it depends on the (variable) underlying rates of RFI bursts at each site. Adding a third site with a rate identical to that of Green Bank increases the mean time between chance coincidences to about 27 years. For this reason STARE was designed to include three sites. The Hat Creek site, however, has been found to have an RFI environment unsuitable for general transient detection. Work is in progress to remedy this situation in order to give STARE the full interference rejection capability for which it was designed.

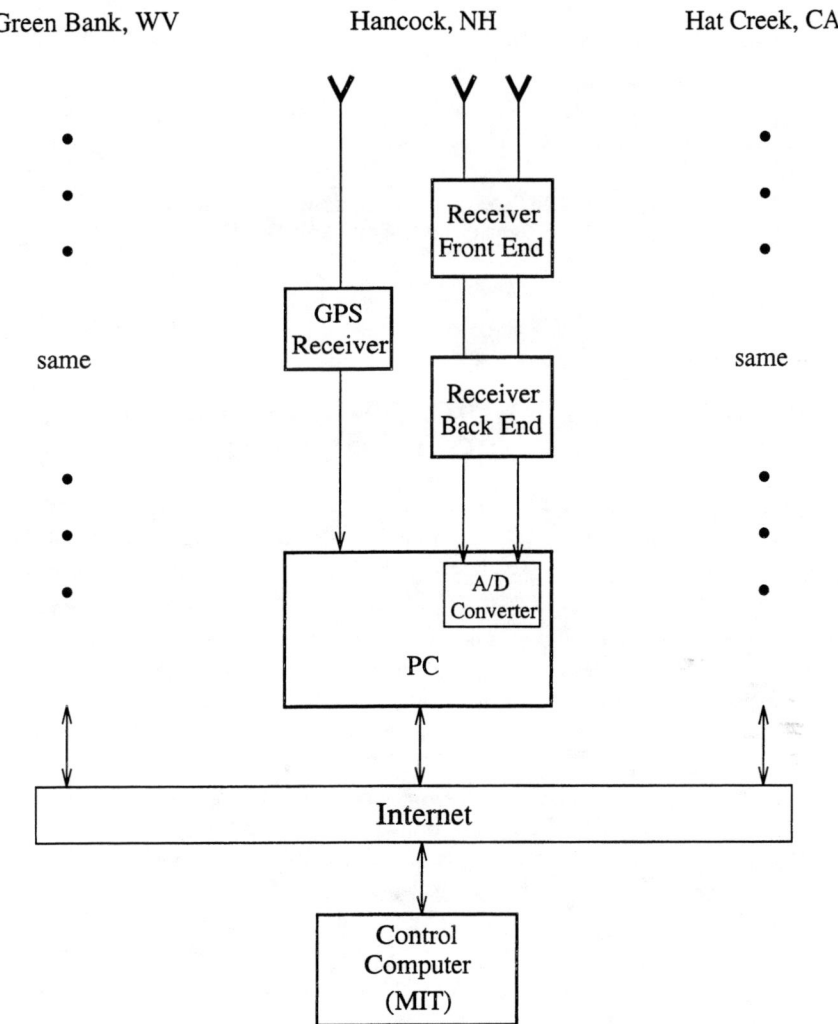

FIGURE 1. STARE System Organization

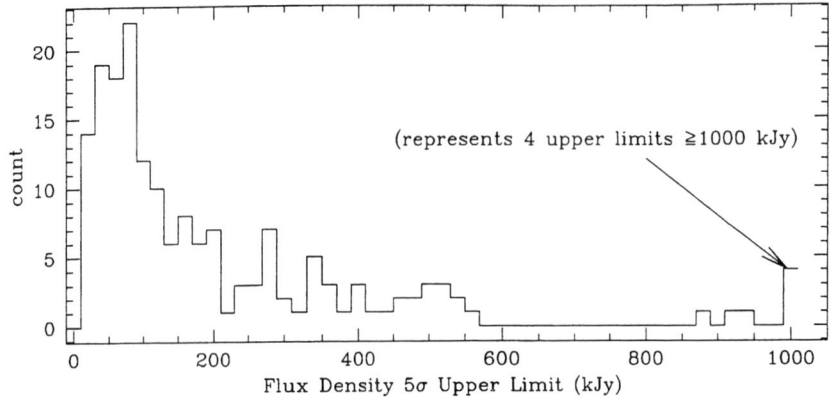

FIGURE 2. Histogram of STARE flux density 5σ upper limits for 173 GRBs

SURVEY FOR RADIO COUNTERPARTS TO GAMMA-RAY BURSTS

A strength of the STARE project in searching for GRB counterparts is that it monitors a large fraction of the sky nearly twenty-four hours per day. It will record from GRBs in its field of view any radio emission occurring after, during, or even before the GRB itself (with intensity above the STARE sensitivity level, of course). The price paid for a large field of view is lower sensitivity than that of a narrow-field detector. However, the large field of view also increases the number of GRBs expected to occur within the antenna beam.

STARE began multi-site operation on 26 March 1996 when the Green Bank station came on-line (Hancock came on-line in August 1995). Between that time and 31 August 1997, 173 GRBs (detected by BATSE) occurred in the field of view (above 20° elevation) of at least one of the STARE sites. For each GRB, the STARE coincidence record was examined for detections within ±1 hr of the gamma-ray event. Seven positive results were found, which is consistent with the number expected from chance. None of the seven occurred simultaneously with a GRB. In addition, for each GRB, the raw STARE data record from each site was examined manually for unusual activity within ±30 min of the GRB. Nothing was found that was not obviously the usual RFI. We conclude that **the STARE data are consistent with an absence of radio counterparts appearing within ±1 hr of GRBs, with 5σ detection limits ranging from tens to hundreds of kJy**. A histogram of the ensemble of upper limits is shown in Figure 2.

FLIRT Update: Rapid Radio Observations of GRBs

R. J. Balsano[1], S. E. Thorsett[1], W. A. Coles[2], B. J. Rickett[2],
P. S. Ray[3], S. Barthelmy[4], P. Butterworth[4], T. Cline[4], N. Gehrels[4]

[1] *Department of Physics, Princeton University, Princeton, NJ 08544-0708*
[2] *Electrical and Computer Engineering Department, University of California, La Jolla, CA 92093-0407*
[3] *Naval Research Laboratory, Washington, DC 20375*
[4] *NASA/Goddard Space Flight Center, Greenbelt, MD 20771*

Abstract. The Fallbrook Low-Frequency Immediate Response Telescope (FLIRT) is a phased radio array located in Fallbrook, CA which is remotely operated from Princeton, NJ. The experiment is designed to detect prompt radio counterparts of gamma-ray bursts (GRBs) at 74 MHz. Observations are triggered by the BATSE locations distributed through the GRB Coordinate Network and begin $\sim$ 10 s after BATSE detects a GRB. FLIRT is capable of detecting a 100 Jy source in a 10 s integration.

A prompt radio burst at 74 MHz would follow a GRB by up to 30 minutes for bursts at cosmological distances. Thus, for cosmological GRBs, FLIRT can probe the pre-GRB radio emission. With even a crude second estimate of the GRB source distance, the dispersion measure can be used to estimate the column density of ionized hydrogen in the intergalactic medium.

INTRODUCTION

GRBs are among the most energetic and enigmatic phenomena known. As events of this past year have shown, rapid follow-up observations of GRBs at lower energies can provide crucial hints about the origins of these enigmatic events. For instance, the most definitive measure of a GRB distance was made for GRB970508 with the detection of redshifted absorption lines in its optical spectrum [10]. Other wavelengths offer opportunities to measure or confirm distances to GRBs as well. In particular, looking for a radio burst coincident with a GRB would measure the delay caused by the interstellar and intergalactic medium [11]. As radio waves pass through an ionized plasma, lower frequencies are delayed relative to higher frequencies, and the delay is proportional to the integrated column density of ionized hydrogen between the source and observer. The dispersion can be used to constrain the distance to the GRB or, given an independent measure of the distance, can measure the density of ionized hydrogen in the intergalactic medium. For a cosmological GRB, the delay between detection of gamma rays and radio waves which were emitted *simultaneously* at the source could be up to 30 *min*. With FLIRT's rapid response, the pre-GRB radio emission can be probed.

After the detection of a radio counterpart to GRB970508 at the ~ 0.5 mJy level [6], one may question whether FLIRT has the sensitivity needed to observe radio emission associated with a GRB. The radio emission observed in this case was presumably the incoherent synchrotron radiation from a cosmological fireball [12]. Since incoherent emission is expected to be quite faint and delayed by days to weeks from the GRB event, a telescope capable of deep imaging is required to search for this type of counterpart. Coherent emission, on the other hand, could be quite bright, as it is in radio pulsars. If GRBs have similar radio flux density to gamma ray flux ratios to those of the 5 radio pulsars seen in gamma rays, they could be detected in the range of $10 - 10^4$ Jy for a 10 s burst. Although other estimates for coherent radio emission which may be relevant to GRBs have been made (e.g. [13], [9]), making quantitative estimates for the strength of coherent emission is extremely difficult and model-dependent. Therefore, we believe whether GRBs also emit measurable radio emission or not is an issue best addressed observationally.

Except for a few GRBs [5], current limits [1,7,2,8] for prompt radio emission associated with GRBs are many kJy. The FLIRT experiment will improve on these limits by at least an order of magnitude.

INSTRUMENT

The Fallbrook telescope is a phased array divided into 8 subarrays, 4 of which are currently functional. Each subarray has two $4 \lambda \times \lambda$ elements in the EW direction and eight elements in the NS direction. The main beam of this configuration is roughly circular with a half-power diameter of 7° giving a total beam area of 45 sq°. Each subarray is independently steered so that a large fraction of the GRB Coordinate Network (GCN) error circle may be surveyed at one time. A description of the telescope's construction can be found elsewhere [4], as can more detailed information on the modifications made for the FLIRT experiment [3].

Radio data from FLIRT are downloaded to UCSD from the remote site in Fallbrook and then moved to Princeton for analysis. In May of 1996, a leased telephone line was installed at the remote site to replace the use of a cellular modem. The change from a 2400 baud cellular modem to a 14.4 kbaud modem used over a dedicated line has reduced the typical delay between the onset of a GRB and the beginning of a FLIRT observation to ~ 10 s (see figure 1). In addition, the higher bandwidth has greatly simplified remote testing and observing.

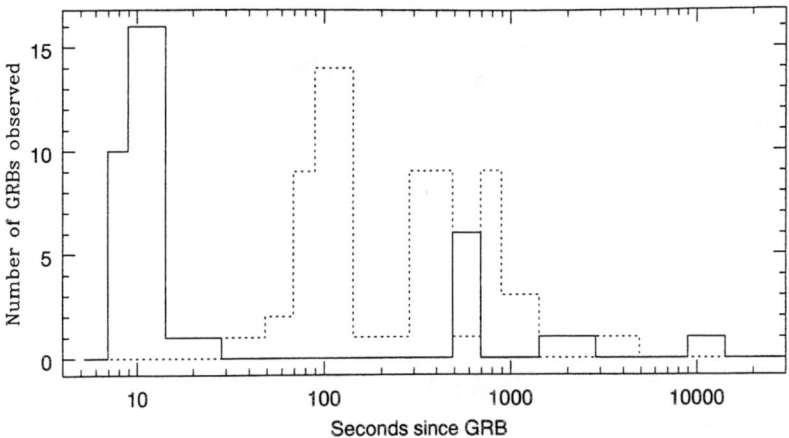

FIGURE 1. Histogram of FLIRT response times for 85 GRBs, defined as the delay between detection of a GRB event by BATSE and the first data records from FLIRT. The dashed and solid lines are histograms for before and after, respectively, the leased line was installed. The peaks at several hundred seconds are due to delayed GRB event notification by GCN during CGRO telemetry gaps.

CALIBRATION

In order to electronically steer the beams, each antenna element must have independent gain and phase control. To calibrate the instrument, differing effective signal paths and gains for each antenna element must be taken into account. Since the telescope is operated remotely, the most efficient way to make these calibrations is to use strong astronomical sources.

The calibration is carried out by pointing 15 of 16 elements in a subarray at a radio source and repeatedly cycling through all phase values on the remaining element. The signal is detected and the resulting interference pattern gives the hardware phase delay and relative gain for the element in question. Each antenna is phased in turn and this process must be iterated until the phase delays measured for all antenna elements converge to 0. Starting from a randomly phased array, convergence to an accuracy of $\pm 3°$ can be achieved in about 4 scans. In practice, however, many hardware and software problems have delayed this process. The calibration is ongoing and will improve pointing accuracy and beam efficiency. Figure 2 shows calibration data taken on the bright supernova remnant Cas A before and after phasing was done.

OBSERVATIONS

Figure 3 shows data for GRB971021. The largest spike is atypical and quite likely terrestrial in origin. The square-wave pattern is present at some level in all data

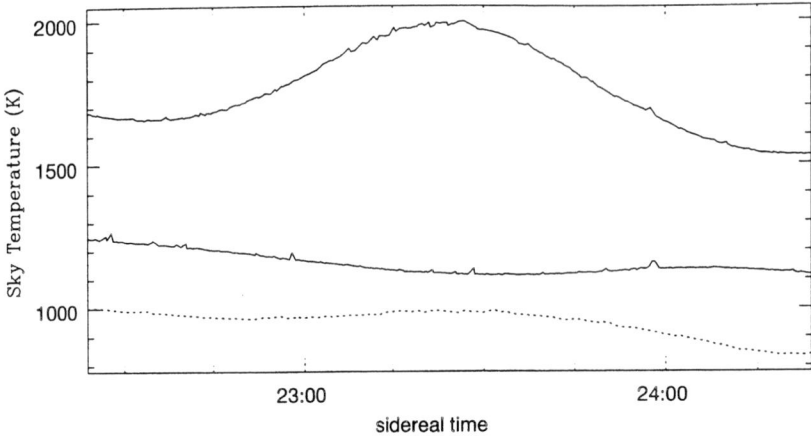

FIGURE 2. The upper two curves are data taken once the array was properly phased. The upper solid line is a drift scan of the 15 kJy source Cas A and the lower solid line is a scan with the array pointed directly overhead. The dotted line shows a scan of Cas A before the array was properly phased. Pointing directly overhead with the array in the unphased state produced essentially the same signal.

taken thus far and is caused by a strong, narrow-band signal at 72.5 MHz. Both hardware and software methods for reducing the amplitude of the interference are being pursued, but neither has been completely effective yet. The greatest challenge to reaching FLIRT's optimal sensitivity will be generating an effective algorithm for finding dispersed signals amidst larger non-dispersed interference.

FUTURE

FLIRT first responded to a GRB notice on July 20, 1995. Many improvements to the radio array and to the remote connection have been made since then. However, numerous important tasks remain to be completed. These include:

- improvement of phase and gain calibration on all 4 subarrays,
- hardware and software filtering schemes to remove terrestrial interference,
- characterization of the spectral response of the system, and
- full-time monitoring for dispersed radio transients.

Once these tasks are completed, FLIRT will provide sensitive and rapid radio followup to a large sample of GRBs.

REFERENCES

1. Baird, G. A. *et al.*, *Astrophys. J.* **196**, L11 (1975).

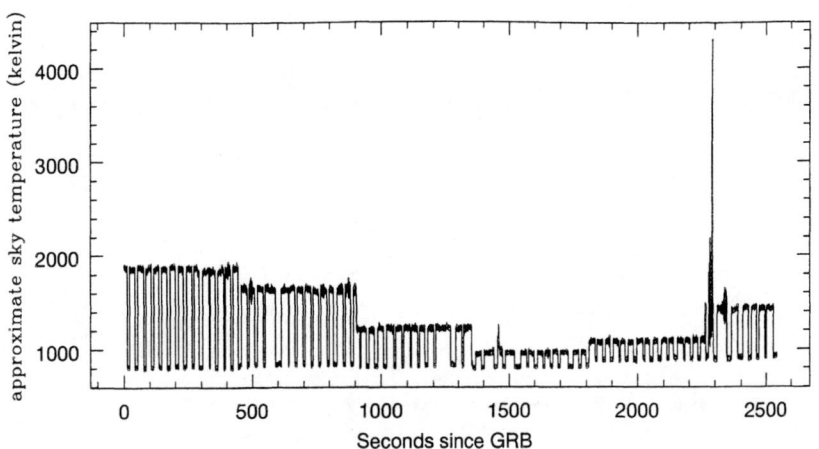

FIGURE 3. FLIRT data for GRB971021 showing typical data. The large spikes at 1460, 2300 and 2350 seconds are non-dispersed signals. The square-wave pattern is caused by a roughly periodic, narrow-band interfering signal. The apparent amplitude of the interference changes when the array is re-phased every 8 minutes.

2. Benz, A. O. & Paesold, G., *Astr. Astrophys.*, in press (1997).
3. Balsano, R. J. et al., in *The 3rd Huntsville Symposium on Gamma Ray Bursts*, p. 575 (1996).
4. Coles, W. A., Frehlich, R. G., and Kojima, M., *Proc. IEEE* **82**, (1994).
5. Dessenne, C. A. C. et al., *MNRAS* **281**, 977 (1996).
6. Frail, D. A., Kulkarni, S. R., & BeppoSAX Team, IAU Circ 6662 (1997).
7. Inzani, P., Sironi, G., Mandolesi, N., & Morigi, G., in *Gamma Ray Transients and Related Astrophysical Phenomena*, ed. R. E. Lingenfelter, H. S. Hudson, & D. M. Worrall, AIP, p. 79 (1982).
8. Katz, C. A., Hewitt, J. N., Moore, C. B., Ellithorpe, J. D., & Rabii, B., in *The 3rd Huntsville Symposium on Gamma Ray Bursts*, p. 651 (1996).
9. Katz, J. I., *Astrophys. J.* **422**, 248 (1994).
10. Metzger, M., et al. IAUC 6655 (1997).
11. Palmer, D., *ApJ* **417**, L25-28 (1993).
12. Paczyński, B. & Rhoads, J., *ApJ* **418**, L5 (1993).
13. Rees, M. J., *Nature* **266**, 333 (1977).

SCUBA Sub-millimeter Observations of Gamma-Ray Burst Counterparts

I. A. Smith*, J. van Paradijs[†,‡], T. J. Galama[†], P. J. Groot[†], L. B. F. M. Waters[†], and C. Kouveliotou[||,¶]

*Department of Space Physics and Astronomy, Rice University, MS-108,
6100 South Main, Houston, TX 77005-1892
[†]Astronomical Institute 'Anton Pannekoek', University of Amsterdam and Center for High-Energy Astrophysics, Kruislaan 403, 1098 SJ Amsterdam, The Netherlands
[‡]Department of Physics, University of Alabama in Huntsville, Huntsville, AL 35899
[||] Universities Space Research Association
[¶]NASA Marshall Space Flight Center, ES-62, Huntsville, AL 35812

Abstract. SCUBA is a new sub-millimeter instrument on the James Clerk Maxwell Telescope. It can use two arrays of bolometers to simultaneously map the same patch of sky ($\sim 2.3'$ in diameter) at 850 and 450 microns. The field of view and sensitivity make it a good instrument to use in the sub-millimeter on arcminute sized gamma-ray burst error boxes. For well-localized radio and optical counterparts, sensitive pointed observations can be made using the central pixel of these arrays, or using the dedicated photometry pixels at 1100, 1350, and 2000 microns. We have an ongoing program of Target of Opportunity GRB observations. As a test run, we observed GRB970508 on May 26 using the 1350 μm pixel. No sources were detected, consistent with other millimeter observations.

INTRODUCTION

An important reason for the long-standing mysteries surrounding the gamma-ray burst (GRB) sources has been the lack of prompt accurate locations to look for quiescent or fading counterparts. This situation changed dramatically this year with the rapid and accurate location of X-ray emission detected during GRBs using the wide-field camera on the Satellite per Astronomia X (BeppoSAX), the All Sky Monitor on the Rossi X-ray Timing Explorer (RXTE), and from the detection of X-ray afterglow in scans of BATSE GRB error boxes with the PCA on RXTE.

These localized X-ray counterparts have led to intense multiwavelength campaigns. So far, two bursts have been associated with optical transients: GRB970228 [17,14,4] and GRB970508 [1,2,12]. GRB970508 was found to have a redshift $0.835 \le z \le 2.3$, while GRB970228 may be related to a faint persistent extended object that is presumably a distant galaxy.

One burst, GRB970508, has been associated with a variable radio source [3,16]. This ∼ mJy source has a *rising* spectral index, and its variability has been attributed to two possible causes: (1) it is intrinsic to the source, and (2) it is caused by interstellar scintillation. These scenarios predict significantly different flux densities in the millimeter region. For example, at 85 GHz case (1) predicts ∼ 10 mJy, while case (2) predicts ∼ 1 mJy. The source was not detected by BIMA at 85 GHz [15,5,6], and the damping of the radio variations is consistent with an expanding source, making scintillation the more likely explanation. Self absorption is also occurring in this case.

The combined observations of GRB970508 show that the peak of the spectrum lies somewhere in the millimeter region. Even before any counterparts were found, two completely separate classes of models had suggested this would be the case: (1) Compton scattering models (e.g. [9,10]), and (2) cosmological fireball models (e.g. [13,8,11]).

To obtain a complete picture of the nature of the burst counterparts, it is clear that one needs to cover the entire spectrum, and sub-millimeter observations with a ∼ mJy sensitivity are needed. This is particularly important since the optical emission can be suppressed by local absorption, and the radio emission can be self absorbed as well as scrambled by interstellar scintillation. Here we discuss our ongoing program of Target of Opportunity observations using SCUBA on the James Clerk Maxwell Telescope.

SCUBA DETAILS

SCUBA is the new sub-millimeter continuum instrument for the James Clerk Maxwell Telescope on Mauna Kea, Hawaii.

It uses two arrays of bolometers to simultaneously observe the same region of sky, ∼ 2.3′ in diameter. The arrays are optimized for operations at 850 and 450 μm. Fully sampled maps of the 2.3′ region can be made by "jiggling" the array. Thus this mode is appropriate for mapping the better localized GRB error boxes as well as those of the X-ray transients.

Deeper photometry using just the central pixel of these arrays can be performed. There are also dedicated photometry pixels for 1100, 1350, and 2000 μm observations (that cannot be used at the same time as the arrays). This photometry mode is appropriate for well localized radio or optical transients.

Scan mapping of larger GRB error boxes using the 450:850 filters is not currently possible (though this mode will eventually be available). This mode will be ideal for mapping the long thin GRB error boxes that are obtained using triangulation between satellites, e.g. 5′ × 0.5′ [7].

The most sensitive measurements will normally be made at 1350 μm, though the sensitivities depend significantly on the weather, particularly at the shorter wavelengths. In photometry mode, for an integration time of 2 hours, we would expect to achieve an rms ≲ 1 mJy at 1350 μm, ∼ 1 mJy at 850 μm, and ∼ 5 − 20 mJy at 450 μm. The jiggle maps give an rms that is a factor $\sim 2-3$ times higher. Having located a SCUBA source, we plan to monitor its decay over the days or weeks that it fades.

OBSERVATIONS OF GRB970508

As a trial run, a 30 minute SCUBA observation of GRB970508 was made on 1997 May 26 using the 1350 μm photometry pixel. The weather conditions were very poor and a preliminary analysis showed that no source was detected with an rms $\sim$ 10 mJy. This is consistent with other mm observations [15,5,6].

ACKNOWLEDGMENTS

The James Clerk Maxwell Telescope is operated by The Joint Astronomy Centre on behalf of the Particle Physics and Astronomy Research Council of the United Kingdom, the Netherlands Organisation for Scientific Research, and the National Research Council of Canada.

We thank the JCMT Director Ian Robson for authorizing the test run on GRB 970508, and Remo Tilanus and Wayne Holland for the preliminary reduction of the data.

This work was supported by NASA grant NAG 5-3824 at Rice University.

REFERENCES

1. Bond, H. E., IAUC 6654 (1997).
2. Djorgovski, S. G., et al., *Nature* **387**, 876 (1997).
3. Frail, D. A., et al., *Nature* **389**, 261 (1997).
4. Galama, T., et al., *Nature* **387**, 479 (1997).
5. Gruendl, R. A., et al., these proceedings.
6. Gruendl, R. A., et al., in preparation.
7. Hurley, K., et al., *ApJ* **485**, L1 (1997).
8. Katz, J. I., *ApJ* **432**, L107 (1994).
9. Liang, E. P., Kusunose, M., Smith, I. A., & Crider, A., *ApJ* **479**, L35 (1997).
10. Liang, E. P., *ApJ*, submitted (1998).
11. Mészáros, P., & Rees, M. J., *ApJ* **476**, 232 (1997).
12. Metzger, M. R., et al., *Nature* **387**, 878 (1997).
13. Paczyński, B., & Rhoads, J. E., *ApJ* **418**, L5 (1993).
14. Sahu, K. C., et al., *Nature* **387**, 476 (1997).
15. Smith, I. A., et al., IAUC 6663 (1997).
16. Taylor, G. B., et al., *Nature* **389**, 263 (1997).
17. van Paradijs, J., et al., *Nature* **386**, 686 (1997).

HOST GALAXIES

Severe Constraints on Possible Cosmological Models Caused by Limits on Host Galaxies

Bradley E. Schaefer

Yale University, PO Box 208101, New Haven, CT 06520-8101

Abstract. Cosmological models of Gamma Ray Bursts (GRBs) generally place them in normal host galaxies at distances corresponding to a luminosity of 6×10^{50} erg s^{-1}. A strong test of this model is the prediction that bursts with small boxes should have galaxies at around seventeenth magnitude. I have obtained data for 26 of the smallest GRB boxes in the U, B, V, R, I, J, H, and K bands which place the limits on the brightest galaxy in the boxes to be considerably fainter than expected. Detailed analysis shows these data to refute the standard model of cosmological bursts. This lack of host galaxies forces all cosmological models to either (1) place GRBs at distances much greater than is accepted by virtually all current models or (2) explain why GRBs always occur outside of normal host galaxies.

INTRODUCTION

The isotropy and inhomogeneity of bursts [6] and the high red shift absorption lines for an associated optical transient [7] have provided strong evidence that GRBs are indeed cosmological. Virtually all published models and expectations place GRBs inside normal host galaxies, such that the host should appear inside burst positional error regions.

The small GRB boxes do not have galaxies sufficiently bright to be normal hosts. This is not to say that the boxes are empty of any galaxies, after all, the Hubble Deep Field demonstrates that there are always galaxies if you look deep enough. The No-Host-Galaxy (NHG) dilemma is that virtually all published cosmological models require host galaxies to appear inside the GRB error regions with magnitudes substantially brighter than the observed limits on the brightest galaxy inside the box.

The NHG dilemma was first posed at the Taos GRB Conference [8], where the hypothesis that bursters reside in galaxies like M31 was strongly rejected. Fenimore et al. [2] and Woods & Loeb [9] examined the same data with more general assumptions, yet they also found that the NHG dilemma could not be explained at the distances deduced from the $LogN - LogP$ curve.

Larson, McLean, & Becklin (LMB) [5] examined six error boxes in the near infrared and noticed galaxies brighter than $K = 15.5$ mag. They claimed that these galaxies solve the NHG dilemma. This claim has severe difficulties: (1) LMB performed a statistical test on the presence of galaxies, however, they counted as successes galaxies that appeared far outside the error box and they counted multiple galaxies as multiple successes for a single box. A galaxy outside the error region is not the host and there cannot be more than one host. Thus, the LMB statistics do not address the NHG question. (2) LMB discuss that the existence of multiple galaxies outside the error region can correlate with the presence of a GRB inside the error region if the burster is inside a galaxy cluster. This is a valid, if weak, method for searching for a host cluster. So LMB are really addressing the No-Host-Cluster problem. To support their claimed solution to the No-Host-Cluster problem, LMB point to an apparent excess of galaxies with $K < 15.5$ mag in the burst fields compared to control fields. However, even their statistical significance is 3.4% (which corresponds to a Gaussian 2.5-sigma probability), such that at face value their claim can only be considered suggestive at best. (3) LMB have found 12 galaxies near GRB fields out of a total area examined equal to 78.8 square arc-minutes, for a density of 548 ± 158 galaxies with $K < 15.5$ per square degree. Yet with their own reference of Gardner, Cowie, & Wainscot [3] and Koo & Kron [4], the background densities range from 252-2600 galaxies with $K < 15.5$ per square degree. Thus, the variation in the background is so large that the real significance of the LMB claim is not even half-a-sigma. (4) In this conference, D. L. Band demonstrated that the error boxes of LMB are sufficiently large and for sufficiently faint bursts that there can be little information content in their data.

These old data and analyses can be substantially improved. I have made extensive searches for host galaxies in the far ultraviolet, optical, and near infrared totaling 255 hours of telescope time. Supplemented with observations reported in the literature, I have made a list of the brightest possible host galaxies in the U, B, V, R, I, J, H, and K bands for each of 26 bursts with the smallest error boxes.

The galactic latitude, error box size, and peak flux are given for all 26 bursts in the table. The sixth column of this table presents the limit on the apparent magnitude for the brightest possible host galaxy inside the three-sigma error box. Each of these magnitudes has been corrected for absorption in our own galaxy. The majority of the values correspond to an actual galaxy of the given magnitude, while the remainder are only limits. The seventh column identifies the band for the limit, which was chosen to be the most restrictive of the available limits.

NO-HOST-GALAXY ANALYSIS

Let us examine a default or minimal cosmological burst model which assumes that (1) GRBs occur inside normal host galaxies and (2) GRBs have a distance scale typified by a peak luminosity of 6×10^{50} erg s^{-1}. This is the 'standard model' as adopted by virtually all published cosmological GRB models.

GRB	Gal. Lat. (°)	Size (sq')	P $erg \cdot cm^{-2} \cdot s^{-1}$	D (Gpc)	m_{lim}	Band	M_{lim}	Fraction
781104	-26	14.0	8.7e-5	0.24	15.94	B	-20.8	0.780
781119	-84	8.0	1.3e-4	0.20	20.16	B	-16.1	0.018
781124	77	48.0	2.3e-5	0.47	20.61	B	-17.3	0.057
790113	-19	78.0	3.6e-5	0.37	18.44	B	-19.0	0.249
790307	14	10.0	4.8e-5	0.32	17.93	I	-19.6	0.139
790313	-25	24.0	2.0e-5	0.50	17.85	B	-20.2	0.581
790325	21	2.0	1.5e-5	0.58	22.08	B	-16.3	0.022
790329	57	41.0	1.3e-5	0.62	20.37	U	-18.1	0.180
790331	-7	20.0	1.5e-5	0.58	13.95	V	-24.8	1.000
790406	-61	0.3	2.4e-5	0.46	24.29	B	-13.6	0.000
790418	-16	2.9	5.6e-5	0.30	22.84	B	-14.3	0.003
790613	38	0.8	3.2e-5	0.40	21.06	B	-16.5	0.027
791105	-50	35.0	1.7e-5	0.54	20.06	B	-18.2	0.128
791116	-76	3.7	6.9e-5	0.27	22.55	B	-14.4	0.003
910122	-31	19.3	2.0e-6	1.58	20.58	B	-19.2	0.292
910219	55	7.3	2.0e-6	1.59	20.98	V	-19.5	0.230
911118	36	12.2	4.1e-6	1.10	20.17	B	-19.2	0.292
920325	-44	2.1	1.7e-6	1.73	19.98	J	-21.3	0.128
920406	-28	0.4	4.6e-6	1.05	23.42	B	-15.8	0.014
920501	2	0.9	3.2e-6	1.26	17.86	J	-22.8	0.393
920711	27	1.4	2.4e-6	1.43	19.95	J	-20.9	0.090
920720	81	1.3	1.5e-6	1.83	18.54	B	-21.4	0.928
920723	8	4.5	6.2e-6	0.90	16.34	K	-24.0	0.649
970228	-18	0.0	4.0e-6	1.12	24.73	V	-15.3	0.004
970402	-9	2.2	1.0e-6	2.24	18.81	U	-20.8	0.909
970508	26	0.0	1.7e-7	5.42	24.82	R	-18.1	0.062

For each of the 26 bursts, the peak flux is known, so the luminosity distance to the burster can be deduced for the standard model (see column five). The corresponding distance modulus is also applicable to any host galaxy, so that the limit on the apparent magnitude can be translated into a limit on the absolute magnitude (see the eighth column of the table). These values have appropriate K and E corrections from Bruzual [1]. We see many examples where the host must have very faint absolute magnitudes. To quantify this statement, we can calculate the position of each limit in the luminosity weighted Schechter luminosity function (column nine). For example, a value of 0.100 implies that the host galaxy must be in the bottom 10% of the luminosity function to satisfy the observed limit.

Let us take as a specific example the burst GRB790406. This region has a galaxy at an apparent magnitude of B=24.33, which corresponds to $B_o = 24.29$ after correction for absorption in our Milky Way. For an error box of this size, the brightest background galaxy will be $B_o = 24.15$ on average, so we see that the observed galaxy is fully consistent with being in the box by chance. For a standard model luminosity and the observed peak flux, the luminosity distance must be 0.46 Gpc and the distance modulus must be 38.31. With a K and E correction of 0.43 mag (appropriate for the distance), the absolute magnitude of any host galaxy must be fainter than -13.6 mag. This is an incredibly faint galaxy. If it indeed exists, then it must reside in the bottom $\sim 0.1\%$ of the luminosity function. While the luminosity function is not well known at these depths, the conclusion remains that the host is improbably faint.

The unlikely chance that the host is incredibly faint is repeated many times. There are four bursts for which the host must reside in the bottom 1%. Of the six bursts with error box sizes less than one square arcminute, the host must be in the bottom 8% of the luminosity function on average. Of the 26 bursts, 14 have a fraction of less than one-sixth, and this is as improbable for normal hosts as rolling 26 dice to get 14 sixes. Alternatively, 20 out of 26 must have the host in the bottom half of the luminosity function, a result which is as probable as tossing 26 coins to get 20 heads. The point is that the limits force any host galaxy (within the standard model) to be unacceptably faint.

In the absence of background galaxies, the fraction should average to 50% if the hosts are drawn from the Schechter luminosity function. With background contamination, the average fraction can only get larger. For large error boxes or for faint bursts, the expectation is that random contamination will produce galaxies inside the box that are not related to the GRB. The limits from such boxes carry little information. To account for this, I have weighted the fractions by the probability that the brightest galaxy in the region will be the host (for some assumed burst luminosity). The weighted average fraction cannot be significantly smaller than 50% for the model to be true.

The observed weighted average fraction is 0.24 ± 0.06 for an assumed luminosity of 6×10^{50} erg s^{-1}. The value expected if there were is 0.23 for background alone, and this shows the galaxy content of GRB boxes to be entirely as expected from background alone. The value expected within the standard model is 0.52 including the background. This result strongly rejects the standard model.

SOLUTIONS

But GRBs are at cosmological distances, so there must be some solution to this NHG dilemma. Possible answers come from denying each of the two tenets of the standard model; either bursters do not reside in normal host galaxies or their distances are much farther than implied by the adopted luminosity.

One possible solution is to place bursters in normal galaxies at substantially greater distances than in the standard model. The NHG analysis can be repeated for increasing luminosities until the observed weighted average fraction is consistent with 0.5. This forces the luminosity to be greater than 10^{55} erg s^{-1}, although the exact value will depend on the details of the cosmology. If a burst model allows for luminosities this large (and survives other severe problems), then it is a solution to the NHG dilemma.

Alternatively, bursters could be placed in active galactic nuclei at great distances. To test this possibility, I have collected x-ray, radio, and infrared limits for the 26 small GRB boxes. For all types of quasars, these limits imply that any such host must have a luminosity distance of typically greater than 100 Gpc. Even for Seyfert galaxies, the limits on the luminosity distance are frequently greater than a Hubble Radius. Thus, GRBs are not associated with AGN, or else their host would already have been spotted.

So what about postulating that GRBs are outside normal host galaxies? Here there are various possibilities, for which the NHG data can provide severe constraints: (A) If bursters are ejected from their birth galaxy, then the observations require that the ejection mechanism be > 50% efficient to satisfy the NHG data. An additional constraint can be obtained by looking in the sky around six high-latitude bright bursts, for which the average transverse distance from the box to the nearest acceptable (former) host is 500 kpc. For an ejection velocity of 500 km s^{-1}, any ejected burst must have an average time delay from ejection to burst of 2×10^9 years. (B) Alternatively, perhaps the probability of a burst is equal from galaxy-to-galaxy. This possibility is an unlikely extreme case since the masses of central black holes and collision rates will scale roughly with galaxy mass and luminosity. Nevertheless, if the fractions are calculated from an unweighted Schechter luminosity function, the standard model is still rejected at the three-sigma level. (C) If bursters occur inside subluminous galaxies, then no more than 50% can be allowed in normal galaxies and the subluminous population must be > 4 magnitudes dimmer than normal galaxies. (D) It is always possible to have GRBs totally unrelated to galaxies.

In conclusion, the NHG dilemma refutes the standard model, yet there are many possible solutions to the NHG dilemma, each of which has significant problems.

REFERENCES

1. Bruzual, G., *Rev. Mex.* **8**, 63 (1983).
2. Fenimore, E. E. et al., *Nature* **366**, 40 (1993).
3. Gardner, J. P., Cowie, L. L., & Wainscot, R. J., *ApJ* **415**, L9 (1993).
4. Koo, D. C. & Kron, R. G., *ARA&A* **30**, 613 (1992).
5. Larson, S. B., McLean, I. S., & Becklin, E. E., *ApJ* **460**, L95 (1996).
6. Meegan, C. et al., *Nature* **355**, 143 (1992).
7. Metzger, M. R. et al., *Nature* **387**, 878 (1997).
8. Schaefer, B. E., in *Gamma-Ray Bursts Observations, Analyses, and Theory*, eds. C. Ho, R. I. Epstein, and E. E. Fenimore (Cambridge: Cambridge Univ. Press), p. 107 (1992).
9. Woods, E. & Loeb, A., *ApJ* **453**, 583 (1995).

Results of an Extragalactic Survey of Gamma-Ray Burst Localizations[1]

Samuel B. Larson and Ian S. McLean

Department of Physics & Astronomy
University of California, Los Angeles
8371 MSA, Los Angeles, CA 90095-1562

Abstract. We present final results of a deep, near-infrared imaging survey of a predetermined, consistent sample of gamma-ray burst localizations. This systematic study, conducted from 1994 - 1996, is the first to be designed specifically to examine the extragalactic content of these regions of sky. JHK images of nine of the smallest error boxes from the third Interplanetary Network together with 70 arcmin^2 of control fields were obtained, reaching levels up to 200 times fainter than previous infrared studies of burst locations. An overabundance of brighter galaxies exists in the gamma-ray burst images at the 98% significance level, indicating that the bursts are not positioned randomly with respect to galaxies. As a group, the brightest galaxy within each error box is consistent with a random population of galaxies, both in brightness and position within the boxes. This suggests that either the true host galaxies are fainter, or the error boxes do not point accurately to the hosts.

INTRODUCTION

Since the discovery by the Burst and Transient Source Experiment that cosmic gamma-ray bursts (GRBs) are distributed isotropically and inhomogeneously about the Earth [1], attempts have been made to examine the extragalactic content of precise GRB localizations for suitable host galaxies [2]. Given our limited understanding of gamma-ray bursts, such analyses are fraught with uncertainties [3]. A different way to test the cosmological interpretation of this phenomenon is to simply compare the galaxy population at GRB localizations with the population seen in the random sky.

Previous quiescent counterpart investigations have searched mainly for peculiarities such as strange colors, variability or high proper motion. Those that did not explicitly ignore extragalactic objects reported only the total number of galaxies in GRB fields [4–6]. These integrated numbers are dominated by faint background

---

[1] This paper has been adapted from Larson and McLean, *ApJ* **491**, in press (1997).

galaxies and will not reveal the modest overabundance that is expected. We developed a new study to analyze the extragalactic content of these regions of sky in much greater detail [7].

PROGRAM DESCRIPTION

More galaxies are likely to be found in the area of a gamma-ray burst if it came from a galaxy than if GRB locations are randomly distributed with respect to galaxies, and this clustering is not limited to the error box. It is unwise to study too large an area of sky, but since infrared arrays are well matched to the typical size of a GRB error box, the whole imaged field is acceptable. Near-infrared is the natural wavelength regime in which to search for extragalactic sources. Galaxies are intrinsically brightest in the near-infrared, they become redder with increasing distance, and near-infrared light is less obscured by interstellar dust than visible light, allowing us to probe very low galactic latitude positions.

Our predetermined targets consist of the smallest localizations from the third Interplanetary Network (IPN^3) [8]. IPN^3 localizations are attractive because nearly all were derived using the same three spacecraft, the allowed timing errors are four times larger than what was adopted for IPN^1 boxes, and the localizations are smaller as a group than those of the IPN^1. Separate control fields were acquired about 20' away from error box centers. These fields provided a good way of checking for systematic errors in the survey, but galaxy number counts from very large area studies [9] were used to determine the expected galaxy brightness distribution since they have much lower statistical errors than could be achieved in this survey.

Deep near-infrared images of nine GRB localizations and 71.67 arcmin^2 of control fields were acquired between March 1994 and August 1996. Limiting magnitudes for 5 σ detection of a point source typically reached $J \sim 20.0$, $H \sim 19.1$, and $K \sim 18.7$, which is 5 to 200 times fainter than previous near-infrared GRB counterpart searches [10,11]. In terms of potential host galaxies, this program is complete to $z \sim 1$ for an $L*$ galaxy. Observational techniques and data reduction steps are detailed elsewhere [12].

GALAXY DISTRIBUTIONS

Figure 1 shows the K' galaxy brightness distribution for both the GRB fields and the control fields over the random galaxy distribution. The control field galaxy distribution agrees remarkably well with the random distribution. This shows that (1) no measurable systematic errors exist between this galaxy survey and the surveys used to construct the random distribution, and (2) the variance in the galaxy counts are within Poisson errors, despite the fact that galaxies have a clustered distribution. The combined data from all large area K-band surveys [9] are similarly well-behaved. Thus, it appears safe to assume a Poisson distribution to estimate the statistical significance of the galaxy content in the GRB fields.

Our null hypothesis assumes that (1) GRBs are positioned randomly with respect to galaxies, (2) the GRB sample is unbiased, and (3) galaxies are distributed randomly. This GRB sample consists of the smallest IPN[3] localizations, with no *a priori* knowledge of their content. IPN localizations are biased toward the brightest bursts, so results can only be applied to bright GRBs. Although this sample is also biased toward the smallest boxes, there is nothing physical about GRBs or the triangulation process that would prefer this bias. Assumption 3 arises only because Poisson statistics are used to calculate the significance of the results, which was shown above to be a good approximation.

An overabundance of galaxies brighter than $K' \sim 16$ exists in the gamma-ray burst fields. A magnitude threshold of $K = 15.5$ was defined as a tool to measure the abundance of brighter galaxies after a galaxy of that brightness was found in the first field. Fifteen galaxies brighter than $K = 15.5$ were found in the GRB fields,

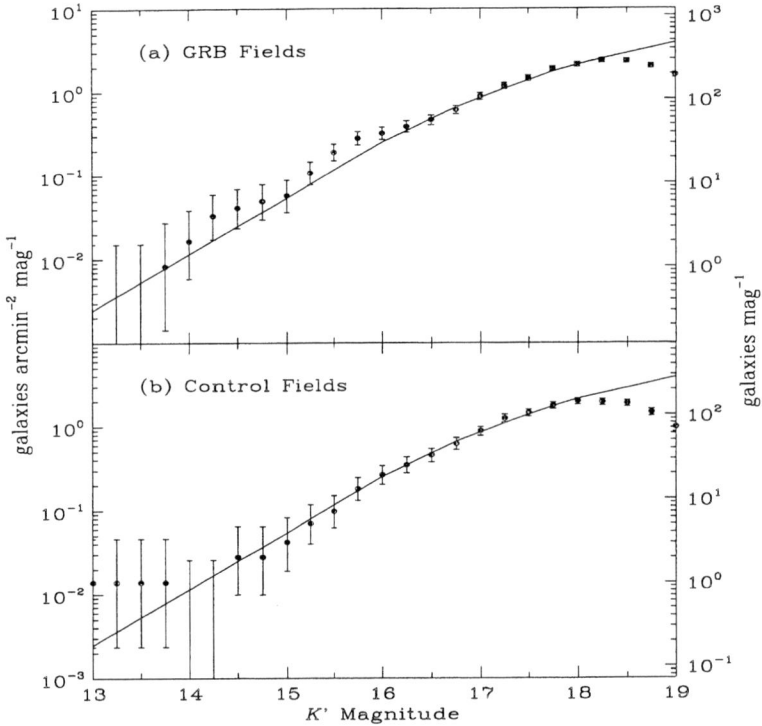

FIGURE 1. K' galaxy distribution for (a) GRB fields and (b) control fields over the expected random galaxy distribution (solid line). The data have been binned four ways to provide measurements every 0.25 mag. Error bars are 1 σ Poisson confidence intervals.

which is 2.6 σ away from the mean of 7.8 for this field of view. The probability of finding at least fifteen by chance is 1.4%. Excluding the first field, which was examined before defining the statistical test, the odds of finding at least 14 galaxies in the eight remaining GRB fields is 2.1%, a 2.4 σ deviation from the mean of 7.4. Individual field analysis shows that eight of the nine fields contain a higher number of galaxies than expected. Thus, this brighter galaxy excess is shared throughout the sample of error boxes and is not dominated by a small number of fields.

BRIGHTEST ERROR BOX GALAXIES

For very small error boxes, the probability that they contain random galaxies brighter than the true host galaxy is comparably small, unless there are no host galaxies, in which case the brightest galaxy is random. Thus, comparisons can be made between the brightest galaxy in each error box and that of a random distribution, without introducing any tenuous assumptions about GRB or host galaxy properties.

We performed two such tests. First, we compared the brightness of these galaxies with what would be expected from the random galaxy brightness distribution. This test was performed using a Monte Carlo analysis by populating each error box in the sample with random galaxies in 0.1 magnitude bins. We ran over 10^5 trials, running a host galaxy mass fraction test [13] on the resulting brightest galaxies after each trial. The measured mass fraction from the real galaxies in these error boxes is only 0.7 σ higher than the Monte Carlo mean.

The second test involved location within the error box. Host galaxies should prefer the middle of the error boxes rather than being randomly distributed inside them. The position of the brightest galaxy in each error box was measured with respect to each of the three annuli that defines the localization. No evidence for clustering or systematic offsets is seen in any individual spacecraft pairing or in the combined distribution.

CONCLUSIONS AND DISCUSSION

With the profusion of galaxies brighter than $K \approx 16$, our null hypothesis has been rejected at the ~98% confidence level using *a priori* statistical tests. As the latter two assumptions of the hypothesis appear to be valid, our conclusion is that bright gamma-ray bursts are not positioned randomly with respect to galaxies.

Although brighter galaxies are overabundant in the GRB fields, the brightest error box galaxies are consistent with a random distribution, both in brightness and position. This combination of results leaves us with three possibilities.

(1) The excess is a statistical fluke. This has a probability of about 2%.

(2) Bright neighbors of fainter host galaxies are causing the excess. This would explain why the unusually bright galaxies are not always located inside the error boxes, but even the ones that are within the localizations are generally not well

centered in them. It is not clear why host galaxies should always have a brighter companion.

(3) The error boxes do not point accurately to the hosts. Bursters may be ejected from their host galaxy [15], or the error boxes themselves may be inaccurate. GRB boxes have changed in the past when arrival times were revised, and comparisons between triangulation and single spacecraft localization techniques have claimed marginal discrepancies [16,17].

ACKNOWLEDGMENTS

We are grateful to John Mulchaey for imaging one of our target fields. We also thank the IPN[3] experimenters for sharing their unpublished data, and David Cline whose suggestions led to this project.

REFERENCES

1. Meegan, C., et al., *Nature* **355**, 143-145 (1992).
2. Schaefer, B. E., in *Gamma-Ray Bursts: Observations, Analyses, and Theories*, 1992, pp. 107-112.
3. Larson, S. B., *ApJ* **491**, in press (1997).
4. Motch, C., Pedersen, H., Ilovaisky, S. A., Chevalier, C., Hurley, K., and Pizzichini, G., *A&A* **145**, 201-205 (1984).
5. Harrison, T. E., McNamara, B. J., and Klemola, A. R., *AJ* **108**, 600-604 (1994).
6. Vrba, F. J., Hartmann, D. H., and Jennings, M. C., *ApJ* **446**, 115-149 (1995).
7. Larson, S. B., Ph.D. Thesis, University of California, Los Angeles (1997).
8. Hurley, K., private communication (1996).
9. Gardner, J. P., Cowie, L. L., and Wainscoat, R. J., *ApJ* **415**, L9-L12 (1993).
10. Apparao, K. M. V., and Allen, D., *A&A* **107**, L5-L6 (1982).
11. Schaefer, B. E., et al. 1987, *ApJ* **313**, 226-230 (1987).
12. Larson, S. B., and McLean, I. S., *ApJ* **491**, in press (1997).
13. Fenimore, E. E., et al., *Nature* **366**, 40-42 (1993).
14. Fenimore, E. E., private communication (1995).
15. Narayan, R., Paczyński, B., and Piran, T., *ApJ* **395**, L83-L86 (1992).
16. McBreen, B., and Metcalfe, L., *Nature* **332**, 234-236 (1988).
17. Graziani, C., and Lamb, D. Q., in *AIP Conf. Proc.* **384**, 382-386 (1996).

A Statistical Treatment of the Gamma-Ray Burst "No Host Object" Issue

David L. Band[1] and Dieter H. Hartmann[2]

[1] *CASS, UC San Diego, La Jolla, CA 92093*
[2] *Dept. of Physics & Astronomy, Clemson University, Clemson, SC 29634*

Abstract. Various burst origin scenarios require a host object in the burst error box or near a well-located position. For example, a host galaxy should be present in the standard cosmological models. We present a methodology which evaluates whether the observed detections and nondetections of potential host objects in burst error boxes are consistent with the presence of the host, or whether all the detections can be attributed to background objects (e.g., unrelated background galaxies). The host object's flux distribution must be modeled. Preliminary results are presented for the "minimal" cosmological model.

INTRODUCTION

In many gamma-ray burst scenarios a host object should be detected when an error box is observed to sufficiently faint fluxes. However, once an error box has been observed, how do we know whether the host object has been detected? Most cosmological models predict that bursts occur in or near galaxies. Since the study of X-ray and optical transients indicate that some and probably all bursts are cosmological, here we will focus on galaxies as the host objects, although the concepts and methodology can be applied to other host object types.

The study of burst error boxes consists of three interrelated aspects. First is the observations of the error boxes, which we assume result in a list of galaxies which are brighter than a limiting flux. These observations can be in any wavelength band in which imaging is possible, although usually optical or infrared images are used. Second is the model for the host object, which guides both the observations and the analysis. For example, the assumption that bursts are cosmological leads the observer to ignore the stars in the error box, although the observer (hopefully) notices any unusual objects in the field. Third is the analysis of the observations in terms of the model. Beyond deciding whether the observations support the model, the analysis methodology also guides the observer as to which error boxes should be searched and to what detection limit. Here we present a new methodology for analyzing burst error box observations, and present preliminary results. A more complete presentation is in press [1].

We emphasize that the analysis of burst error boxes must be made in the context of a model of the expected host. It is nonsensical to ask merely whether galaxies are present in an error box because if one searches deep enough one will find a multitude of faint galaxies. Here we test the "minimal" cosmological model used in most studies of burst ensembles, particularly those analyzing burst error boxes. Bursts are assumed to be standard candles in this model: a basic burst property such as total emitted energy or peak photon luminosity is constant for all bursts, and does not evolve with redshift. Therefore, there is a one-to-one mapping between a burst's redshift and the observed intensity corresponding to the standard candle (e.g., energy fluence for a constant total emitted energy); the redshift-intensity relationship is derived from the intensity distribution under the assumption that the comoving density of burst sources does not evolve. Of course, in this model bursts occur in galaxies. Since a neutron star-neutron star merger is a possible origin of a burst's energy, and the number of compact binary systems is presumably proportional to a galaxy's mass, the burst rate per galaxy is assumed to be proportional to the galaxy's luminosity (for a constant mass-to-light ratio) [2]. Undoubtedly bursts are characterized by a luminosity function, and cosmological density and luminosity evolution is likely, but this "minimal" model has been a reasonable working assumption in the absence of additional data.

B. Schaefer [3] first reported that the galaxies in 8 burst error boxes were fainter than expected. Specifically, Schaefer calculated a large burst energy (up to 2×10^{53} ergs) if the brightest galaxy in an error box was as bright as M31. Fenimore et al. [2] introduced the statistic

$$S = \int_0^{f_{det}} df\, \psi(f) \tag{1}$$

where the brightest galaxy in an error box has a flux of f_{det} and $\psi(f)$ is the flux distribution of the expected host galaxies; S is the fraction of the distribution which is fainter than f_{det}. If f_{det} is indeed the flux of the host galaxy, and $\psi(f)$ is the correct distribution, then S should be distributed uniformly between 0 and 1, with an average of $1/2 \pm (12N)^{-1/2}$ for N error boxes (this test is similar to the V/V_{max} test). Based on the minimal cosmological model, Fenimore et al. found $\langle S \rangle = 0.44 \pm 0.10$ for Schaefer's data. While this $\langle S \rangle$ is consistent with the minimal model, the value of S for a given error box is only an upper limit since the brightest galaxy may be a background galaxy instead of the host galaxy. Similarly, although S. Larson [4–6] reported an overabundance of bright galaxies in his K-band observations of nine IPN[3] boxes, he recognized that many of these galaxies are unrelated background galaxies.

We therefore derived an analysis methodology which includes the unrelated background galaxies in evaluating whether a host galaxy is present. An additional guiding principle was the use of all available information. Thus the method uses all the detected galaxies in the observations. We describe the error box by a probability density $\rho(\Omega)$, where Ω represents the spatial coordinates. Typically it is assumed that $\rho(\Omega) = 1/\Omega_0$ within the 99% contour (a region of size Ω_0), and $\rho(\Omega) = 0$ outside, but more sophisticated treatments are possible. The detection threshold may vary across the error box, e.g., as a result of mosaicing the box with multiple observations of differing quality. Currently we do not include the clustering of background galaxies, which should be a small effect.

METHODOLOGY

Our method is a Bayesian comparison of two hypotheses: H_0—a host galaxy is present in addition to unrelated background galaxies; and H_1—only background galaxies are present. Assume the observations of an error box reveal n_d galaxies with fluxes f_i above a detection limit $f_{\lim}(\Omega)$; these results we represent by the statement D. We set up an odds ratio

$$O(H_0, H_1) = \frac{p(H_0 \mid D)}{p(H_1 \mid D)} = \frac{p(H_0)\, p(D \mid H_0)}{p(H_1)\, p(D \mid H_1)}, \qquad (2)$$

where: $p(H_x \mid D)$ is the probability that hypothesis H_x is true given the observations D; $p(H_x)$ is the "prior," our assessment of the validity of H_x before obtaining the new data D; and $p(D \mid H_x)$ is the likelihood of H_x, the probability of obtaining D if the hypothesis H_x is correct. For simplicity we set the two priors equal to each other, $p(H_0) = p(H_1)$. Therefore, the odds ratio is the ratio of the likelihoods, $O(H_0, H_1) = p(D \mid H_0)/p(D \mid H_1)$.

The likelihoods are calculated by breaking into little bins the three-dimensional space formed by the two spatial dimensions and the flux, and calculating the probability that a host or background galaxy is present or absent in each bin. If the galaxy redshifts are also available, then the redshift can be added as a fourth dimension. Poisson statistics characterize the probability that a background galaxy is found in a given bin. The likelihood for H_0, $p(D \mid H_0)$, is the sum of every possibility for the presence of a host galaxy: either the host is fainter than the limit $f_{\lim}$ or it is one of the detected galaxies. Consequently [1]

$$\frac{p(D \mid H_0)}{p(D \mid H_1)} = \int d\Omega \int_0^{f_{\lim}(\Omega)} df\, \Psi(f)\rho(\Omega) + \sum_{i=1}^{n_d} \frac{\Psi(f_i, z_i)\rho(\Omega_i)}{\phi(f_i, z_i)}, \qquad (3)$$

where $\Psi(f, z)$ is model-dependent host galaxy distribution, $\rho(\Omega)$ is the burst location's probability density across the error box, $\phi(f, z)$ is distribution of background galaxies, and n_d is the number of detected galaxies. If the redshifts of the detected galaxies are unknown, then the z-dependence of Ψ and ϕ should be dropped. In this equation, the first term on the right is the probability that the host galaxy can be hidden below the detection limit, while the sum compares for each detected galaxy the probability that it is the host galaxy to the probability that it is an unrelated background galaxy. Clustering of the background galaxies can be included by multiplying $\phi(f_i, z_i)$ by a function of the distance to the other detected galaxies. This additional factor will usually be of order unity, and will not affect our results qualitatively.

For a database with a number of error boxes the likelihood ratio for the ensemble is the product of the ratios for each box. If the resulting odds ratio is much larger than one, then the presence of host galaxies has been demonstrated. If the ratio is much less than one, then the host galaxy model is incorrect. Finally, if the ratio is of order unity, then the data are insufficient to distinguish between the hypotheses.

By evaluating the likelihood ratio for the expected host and background galaxies, we can determine the method's sensitivity for a given error box. For reasonable assumptions about the distributions, we find that an observation can distinguish between hypotheses if the error box is small enough so that the expected host galaxy is much brighter than average background galaxy. Only then is it clear that a galaxy is the host and not a background galaxy. As currently formulated, this methodology tests a given hypothesis. However, it can easily be modified to fit model parameters.

This methodology was developed for finite size error boxes, such as has been available from the various IPNs. However, the methodology can be readily adapted for other circumstances. Afterglows localize the burst with very small uncertainties (e.g., a fraction of an arcsecond). However, unless the model being tested places the burst in a galactic nucleus, a galaxy within a certain region around the burst would be acceptable as the host; this region can be treated as the error box. Similarly, the burst source might have been ejected from the host galaxy. The distance the source might have traveled before bursting can be used to define the error box around an afterglow; finite size error boxes should be expanded by this distance.

APPLICATIONS

As examples, we apply this methodology to published datasets to test the "minimal" cosmological model described above. In the future we plan to test variants of the cosmological model using a more extensive dataset.

Larson and McLean [6] presented K-band observations of 9 IPN[3] error boxes with an average size of 8 arcmin^2. They listed only the flux of the brightest galaxy in each box, and therefore we use this galaxy as the single galaxy detection and its flux as the detection limit. For all 9 error boxes we find

$$\prod_{j=1}^{9} O_j = 0.25 \qquad (4)$$

which indicates that based on these data we cannot determine whether or not a host galaxy is present. The reason the analysis of these data is inconclusive is that the fluxes of the average expected host galaxy, the detection limit, and the average brightest background galaxy are all comparable; therefore even if the host galaxy is present, the odds ratio will be of order unity.

Schaefer et al. [7] observed 4 small ($\frac{1}{4}$-2 arcmin²) burst error boxes with the *Hubble Space Telescope*. The detection of objects exhibiting bizarre behavior (e.g., proper motion) was the primary purpose of these observations, but our methodology can be applied to the data, nonetheless. Galaxies were detected in 2 of these error boxes. The odds ratio for the four boxes together is

$$\prod_{j=1}^{4} O_j = 2 \times 10^{-6} \quad . \tag{5}$$

This is a clear statement that host galaxies expected by the minimal model are not present.

ACKNOWLEDGEMENTS

We thank B. Schaefer, C. Luginbuhl and F. Vrba for stimulating discussions. This research is supported by the *CGRO* guest investigator program (DLB and DHH) and NASA contract NAS8-36081 (DLB).

REFERENCES

1. Band, D., and Hartmann, H., *Ap. J.* **493**, in press (1998).
2. Fenimore, E. E., *et al.*, *Nature* **366**, 40 (1993).
3. Schaefer, B. E., in *Gamma-Ray Bursts: Observations, Analyses and Theories*, eds. C. Ho, R. I. Epstein, and E. E. Fenimore (Cambridge: Cambridge Univ. Press), 107 (1992).
4. Larson, S. B., McLean, I. S., and Becklin, E. E., *Ap. J.* **460**, L95 (1996).
5. Larson, S. B., *Ap. J.* **491**, in press (1997).
6. Larson, S. B. and McLean, I. S., *Ap. J.* **491**, in press (1997).
7. Schaefer, B. E., Cline, T. L., Hurley, K. C., and Laros, J. G., *Ap. J.*, in press (1997).

Study of the Possible Connection between Gamma-Ray Bursts and Active Galactic Nuclei

J. Gorosabel and A. J. Castro-Tirado

LAEFF-INTA, P.O. Box 50727, 28080 Madrid, Spain.

Abstract. We study the possible correlation of a selected sample of 344 gamma-ray bursts with different subdivisions of the Véron & Véron-Cetty compilation of quasars-AGNs. The intention is to confirm whether the radio quiet quasars are related to GRBs as it was claimed by some authors. On the other hand we analyze the correlation of the radio quiet quasars with the 4B GRBs. We find a 90% correlation between "polygonal-shape" GRBs and radio quiet quasars, however the correlation vanishes for the so called "optimized sample". This fact is supported by the lack of correlation found using WATCH bursts. We detect that the 4B catalog is correlated to radio quiet quasar, however their redshifts are not related to the 4B GRBs fluences, concluding that the found excess could be a statistical fluctuation. On the other hand a 97.6% correlation is found between BL Lac objects and WATCH GRBs, although based on very few coincidences.

INTRODUCTION

The recent optical spectrum of GRB970508 may finally settle the issue of the distance scale in favor of the cosmological models [1]. This view is strengthened by the announced correlation between 3B bursts and radio quiet quasars [2], (hereafter SC97). SC96 reported that GRBs seem to be related to radio quiet quasars (hereafter RQQs) on the basis of a 96.4% confidence level correlation between the 3B bursts with radii $< 1.8°$ and the RQQs of the Véron & Véron-Cetty compilation.

METHOD

The γ-ray burst and quasar-AGN samples

Our selected sample of 344 GRBs is divided into two groups of 311 polygonal and 33 circular error boxes each. The polygonal error boxes have been obtained from different compilations [3-6], whereas the circular boxes are based on the positions provided by WATCH instruments on the *Granat* and *Eureca* satellites [7,8]. We used the 7th version of the "Catalogue of quasars and active nuclei" [9]. In addition to the

TABLE 1. N^l represents the number of sources, C_0^l and P_0^l the number of coincidences of circular and polygonal error boxes, $<C_j^l>$ and $<P_j^l>$ the expected value of coincidences, and finally $PC^l(\%)$ and $PP^l(\%)$ the coincidences excesses.

Main Class	Subsample	N^l	C_0^l	$<C_j^l>$	$PC^l(\%)$	P_0^l	$<P_j^l>$	$PP^l(\%)$
quasar	all	8609	3	3.93	19.5	29	24.17	82.4
quasar	radio quiet	7146	2	2.44	29.2	23	17.57	89.7
quasar	radio quiet-optimized	967	0	0.81	0.0	4	4.88	28.3
quasar	radio-loud	1377	2	1.66	48.8	7	8.24	25.2
quasar	highly polarized	72	0	0.12	0.0	0	0.43	0.0
BL Lac	all	220	2	0.39	94.7	1	1.41	25.8
BL Lac	confirmed	93	1	0.15	86.0	0	0.53	0.0
BL Lac	highly polarized	76	1	0.14	86.1	0	0.52	0.0
BL Lac	radio selected	119	1	0.20	81.4	0	0.83	0.0
BL Lac	X-ray selected	82	0	0.12	0.0	1	0.47	62.3
AGN	AGN	1553	2	1.32	46.5	9	9.39	39.3
AGN	Seyfert 1	888	1	1.06	32.9	3	5.65	8.8
AGN	Seyfert 2	496	0	0.72	0.0	4	3.13	61.4
AGN	suspected Seyfert	97	2	0.11	99.2	1	0.63	52.1
AGN	LINER	71	0	0.09	0.0	1	0.49	55.6
AGN	radio quiet	1346	2	1.52	58.5	8	8.12	42.8
AGN	radio-loud	166	0	0.28	0.0	0	1.21	0.0
	AG	1165	1	0.57	54.3	10	4.46	99.1
	Nucl. H II	116	0	0.12	0.0	0	0.53	0.0

three sections of the catalog (quasars, BL Lac objects, AGNs) several subsamples have been considered depending on their spectral characteristics. We would like to remark upon the so called "optimized RQQs sample", composed by RQQs with redshift $z \leq 1.0$ and absolute magnitude $M_{abs} \leq -24.2$. In the study carried out by SC97 the correlation of this subset reached to a correlation $> 99.7\%$.

The positional coincidences and the correlation estimate

By definition, a positional coincidence occurs when there is a GRB error box containing at least one source. Thus, the number of coincidences of the real circular and polygonal boxes with the 19 subsamples (C_0^l and P_0^l, $l = \{1, 2, .., 19\}$) were calculated and compared to the number of coincidences between 1500 simulated GRBs sets and the 19 quasars-AGNs subsamples (C_j^l and P_j^l, $j = \{1, 2...1500\}$). For the j-th simulated GRB set and for the l-th quasar-AGN subsample, two quantities (hereafter quoted as correlations) T_j^l and M_j^l were introduced: $T_j^l = \sum_{k=1}^{311} \sum_{i=1}^{N^l} t_{ijk}^l$ with $t_{ijk}^l = \frac{-\ln(1-s)}{a_{jk}} \exp(((\frac{x_{ijk}^l}{X_{jk}})^2 + (\frac{y_{ijk}^l}{Y_{jk}})^2)\ln(1-s))$

M_j^l was obtained substituting t_{ijk}^l by a simple Gaussian-like function. For more details see [10]. On the other hand, the same method was applied to the true GRB set providing 19 values of T_0^l and M_0^l. Table 1 and Table 2 display the results.

We also applied the current method to the 4B bursts with radii $< 1.8°$. If the

TABLE 2. N^l represents the number of sources, M_0^l and T_0^l the correlation of circular and polygonal error boxes, $<M_j^l>$ and $<T_j^l>$ the expected value of the correlation, and finally $PM^l(\%)$ and $PT^l(\%)$ the correlation excesses.

Main Class	Subsample	N^l	M_0^l	$<M_j^l>$	$PM^l(\%)$	T_0^l	$<T_j^l>$	$PT^l(\%)$
quasar	all	8609	1.49	4.73	35.4	79.31	52.29	86.7
quasar	radio quiet	7146	0.89	3.49	39.9	75.03	41.88	90.0
quasar	radio quiet-optimized	967	0.00	0.43	0.0	2.44	6.44	48.1
quasar	radio-loud	1377	0.60	0.88	64.3	3.63	7.89	42.9
quasar	highly polarized	72	0.00	0.05	0.0	0.0	0.28	0.0
BL Lac	all	220	1.36	0.14	97.6	0.05	1.71	48.5
BL Lac	confirmed	93	1.26	0.07	98.2	0.0	0.31	0.0
BL Lac	highly polarized	76	0.09	0.06	93.1	0.0	0.28	0.0
BL Lac	radio selected	119	1.26	0.06	98.6	0.0	0.65	0.0
BL Lac	X-ray selected	82	0.00	0.04	0.0	0.05	1.08	57.7
AGN	AGN	1553	0.46	0.99	54.5	4.57	8.98	43.6
AGN	Seyfert 1	888	0.10	0.56	53.9	3.81	5.07	62.1
AGN	Seyfert 2	496	0.00	0.36	0.0	0.74	2.50	51.5
AGN	suspected Seyfert	97	0.35	0.09	96.1	0.01	0.81	75.0
AGN	LINER	71	0.00	0.06	0.0	0.00	0.43	66.3
AGN	radio quiet	1346	0.45	1.83	62.7	4.48	8.58	49.9
AGN	radio-loud	166	0.00	0.13	0.0	0.00	2.73	0.0
	AG	1165	0.10	0.52	65.5	22.45	6.98	95.8
	Nucl. H II	116	0.00	0.03	0.0	0.0	0.87	0.0

gamma fluences of the GRBs (S_γ) were statistically related to the RQQs luminosity distance (d_l) through a law $S_\gamma \propto d_l^{-2}$ it would be an important support for the suggested correlation. Only GRBs containing one RQQ were counted (hereafter single coincidences), discarding the error boxes containing more than one. Once the singular coincidences were calculated, we subtracted the expected value of GRB-RQQ pairs in order to estimate the excess of GRB-RQQ pairs, X. Fig. 1 shows S_γ vs the redshift of the X RQQs that provides highest value of t_{ijk}^l.

RESULTS AND DISCUSSION

Excesses are detected in three sets:

- BL Lacs objects show a significant excess in the coincidences (94.7%) and in the correlation (97.6%) with circular error boxes. However the excess is not noticeable with polygonal boxes. These results do not agree with those obtained by SC97, who found only a 13.5% correlation. A possible explanation could be due to the different sensitivity of the experiments, as WATCH is sampling the strongest bursts and BATSE is detecting a fainter population. It must be stated that the sample of GRBs and the number of coincidences are too small (only two counts) to affirm that they represent two different populations. Thus, the detected correlation is probably a single statistical fluctuation.

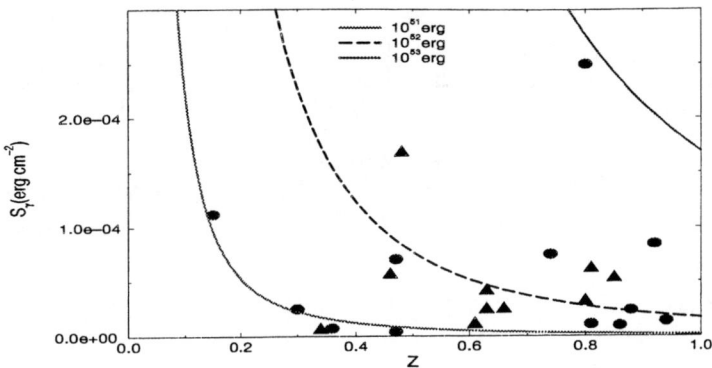

FIGURE 1. The figure represents the fluence of the GRBs versus the redshift of the RQQs. The Lines are different values for the luminosity of the sources whereas the triangles represent the 10 GRB-RQQ pairs with the highest probability of connection.

TABLE 3. Coincidences and Single coincidences between the 4B catalog and RQQs.

Coincidences	N^l	C_0^l	$<C_j^l>$	$PM^l(\%)$	M_0^l	$<M_j^l>$	$PM^l(\%)$
radio quiet	7146	71	63.25	91.40	51.42	52.81	54.02
optimized	967	40	26.85	99.95	9.50	6.78	90.70
Single coincidences	N^l	C_0^l	$<C_j^l>$	$PM^l(\%)$	M_0^l	$<M_j^l>$	$PM^l(\%)$
radio quiet	7146	26	27.33	36.36	2.028	2.039	49.94
optimized	967	26	16.60	97.47	2.164	1.251	99.27

- We detect an excess of RQQs in polygonal error boxes, although this excess is not detected in circular error boxes. The excess seems unreliable because the most important coincidence is based on the quasar found in the GRB781119 error box found during deep surveys, which was included afterwards in the Véron-Cetty quasar catalog [11]. Thus, if the mentioned quasar was extracted from the sample, the correlation would be reduced to 73% and the coincidences excess to 77%, typical of a statistical fluctuation. We would like to point out the few coincidences and low correlation obtained with the "optimized sample", showing no correlation either with WATCH error boxes or with polygonal error boxes. However, as is shown in Table 3 the improved sample is highly correlated with the 4B catalog. Twenty-six single coincidences were detected when only 16.6 ± 4 were expected, quoting an excess of $\sim 10 \pm 4$ single coincidences. For 4 of the 26 GRB-RQQ pairs the 4B catalog does not contain information about the fluences, thus the sample had to be reduced to 22 pairs. As is shown in Fig. 1, there is no clear indication of correlation between the S_γ and Z either for the 22 pairs or for the 10 pairs with higher t^l_{ijk}. The triangles represent the former 10 pairs, and the lines indicate different values for the luminosity of the sources. Thus, if we assume that GRBs are standard candles, we have to conclude that the reported 99.5% correlation between the 4B catalog and RQQ optimized sample is a statistical fluctuation. However, we can not exclude that studies based on a broad luminosity function could reproduce the results.

- AGs and suspected Seyfert galaxies also show an important correlation. The first one is based on AGNs without any morphological type assigned whereas the second one comprises AGNs that are suspected Seyfert galaxies. Any relationship of GRBs with these unclear families of objects seems unreliable.

For the other classes of AGNs and quasars no excess coincidences above random expectation were found.

ACKNOWLEDGMENTS

The authors wish to thank Norbert Schartel for kindly providing the sets of quasars/AGNs. Also Kevin Hurley for giving us the possibility of using the Master GRB list in advance of publication.

REFERENCES

1. Metzger, M. et al., *Nature* **387**, 878 (1997).
2. Schartel, N., Andernach, H. and Greiner, J., *Astron & Astrophys.* **323**, 659 (1997).
3. Golenetskii, S. V., et al., Ioffe Technical Institute, St, Petersbourg, preprint 1026 (1986).
4. Hurley, K., AIP Conf. Proc. **307**, p. 29 (1993).
5. Lund, N., *Ap&SS* **231**, 217 (1995).
6. Hurley, K., private communication (1997).
7. Brandt, S., et al., AIP Conf. Proc. **307**, p. 13 (1994).
8. Sazonov, S.Y., et al. *Astron. & Astrophys.*, in press.
9. Véron-Cetty, M.P., Véron, P., 1996, A catalogue of quasars and active galactic nuclei (7th edition), ESO Scientific Report, in press.
10. Gorosabel, J. et al., *ApJ* **483**, L83 (1997).
11. Pedersen, H., et al., *ApJ* **270**, L43 (1983).

Properties of GRB Host Galaxies

Dieter H. Hartmann[1] and David L. Band[2]

[1] *Department of Physics and Astronomy, Clemson University, Clemson, SC 29634*
[2] *CASS, UC San Diego, La Jolla, CA 92093*

Abstract. The transients following GRB970228 and GRB970508 showed that these (and probably all) GRBs are cosmological. However, the host galaxies expected to be associated with these and other bursts are largely absent, indicating that either bursts are further than expected or the host galaxies are underluminous. This apparent discrepancy does not invalidate the cosmological hypothesis, but instead host galaxy observations can test more sophisticated models.

THE ABSENCE OF THE EXPECTED HOST GALAXIES

Observations of the optical transients (OTs) from GRB970228 [19] and GRB970508 [2] have finally provided the smoking gun that bursts are cosmological. In most cosmological models bursts occur in host galaxies: are these galaxies present, and conversely, what can we learn from them? Underlying any confrontation of theory and data must be a well defined model. Here we show that the host galaxy observations are not consistent with the expectations of the simplest cosmological model, and that these observations can be used to test more sophisticated models.

In the simplest ("minimal") cosmological model the distance scale is derived from the intensity distribution logN–logP assuming bursts are standard candles that do not evolve in rate or intensity. Bursts occur in normal galaxies at a rate proportional to a galaxy's luminosity. This model predicts the host galaxy distribution for a given burst. Are the expected host galaxies present?

For GRB970228 an underlying extended object was found [19], but its redshift and nature have not been established. If the observed "fuzz" is indeed a galaxy at $z \sim 1/4$, it is ~ 5 magnitudes fainter than expected for a galaxy at this redshift. For GRB970508 no obvious underlying galaxy was observed [14] and the nearest extended objects have separations of several arcseconds, but spectroscopy with the Keck telescopes [10] led to the discovery of absorption and emission lines giving a GRB redshift of $z \geq 0.835$. The HST magnitude limit $R_{\lim} \sim 25.5$ [14] for a galaxy coincident with the transient again suggests a host galaxy fainter than expected. Similar conclusions follow from the inspection of IPN error boxes [1,16,17], but see also [6,20]. This absence of sufficiently bright host galaxies is often called the "no-host" problem, which is a misnomer. The point simply is that if galaxies such as the Milky Way provide the hosts to most bursters, and if their redshifts are less than unity, as predicted by the minimal model, we expect to find bright galaxies inside a large fraction of the smallest IPN error boxes.

To demonstrate this quantitatively, consider the apparent magnitude of a typical host galaxy, which we assume has $M_*(B) = -20$ (approximately the absolute magnitude of an L_* galaxy—see discussion below). Using Peebles' notation [13], the apparent magnitude is

$$m = 42.38 + M + 5 \log [y(z)(1+z)] + K(z) + E(z) + A(\Omega, z) + \chi(z) , \quad (1)$$

where $K(z)$ is the usual K-correction, $E(z)$ corrects for the possible evolution of the host galaxy's spectrum, A is the sum of Galactic foreground (position dependent) and intergalactic extinction, and $\chi(z)$ represents any corrections that apply in hierarchical galaxy formation scenarios, where galaxies are assembled through the merger of star forming subunits. The commonly found term $5\log(h)$ is already absorbed in eq. (1). Neglecting potentially large corrections from the K, E, A, and $\chi(z)$ terms, a host like the Milky Way with $M \sim -20$ would have an apparent magnitude $m \sim 22$ for redshifts of order unity. Several small IPN error boxes have no galaxy of this magnitude or brighter. Our simplified treatment agrees with Schaefer's conclusion [16,17] that typical galaxies at the calculated burst distance are absent from burst error boxes.

Thus bursts are farther than predicted from the logN–logP distribution without evolution, or they occur in underluminous galaxies; an extreme limit of the latter alternative is that bursts do not occur in galaxies.

HOST GALAXIES AS A PROBE OF COSMOLOGICAL MODELS

The search for host galaxies is a powerful test of cosmological burst models. From the two above mentioned OTs we conclude that GRBs are cosmological, but the observations have not fixed the distance scale quantitatively, nor have they determined the energy source. While the x-ray, optical, and radio lightcurves (for GRB970508 only) are consistent with the predictions of the basic "fireball afterglow" picture, the fireball's central engine could be the merger of a neutron star binary, the collapse of a massive, rotating star, or the jet produced by accretion onto a massive black hole residing at the center of an otherwise normal galaxy.

The host galaxies found within burst error boxes are a powerful discriminant between different models for the burst energy source. Note that the region within which a host would be acceptable surrounding the sub-arcsecond localizations of an OT is effectively the error box for the host galaxy. Almost all models assume that bursts are associated with galaxies; the issue is the relationship between the burst and the host. In models such as the momentary activation of a dormant massive black hole the burst rate per galaxy is constant. On the other hand bursts are an endpoint of stellar evolution in most models, and therefore to first order we expect the burst rate per galaxy in these models to be proportional to the galaxy's mass and thus luminosity. These two model classes have different host galaxy luminosity functions $\psi(M)$ with different average values of M (the absolute magnitude). In the first case, $\psi(M)$ is proportional to the normal galaxy luminosity function, while in the second case $\psi(M)$ is proportional to the normal galaxy luminosity function weighted by the luminosity $L \propto 10^{-0.4M}$. We approximate the normal galaxy luminosity function with the Schechter function:

$$\Phi(M) = \kappa \; 10^{0.4(M_*-M)(\alpha+1)} \left[\exp\left(-10^{0.4(M_*-M)}\right)\right] \quad , \qquad (2)$$

where κ is the normalization, α is the slope of the faint end, and M_* is the absolute magnitude of an L_* galaxy. Here we use $\alpha = -1$. In the B band $M_*(B) = -19.53$, which corresponds to $L_*(B) = 1.8 \times 10^{10} L_\odot \; h_{75}^{-2} \sim 3 \times 10^{11} \; L_\odot(B) \; h_{75}^{-2}$. In Figure 1 we show the cumulative distributions for the host galaxy magnitudes for the two model classes. As can be seen, the average host galaxy magnitude (i.e., at 0.5) differs by ~ 1.75 magnitudes.

However, we can make better predictions about the host galaxies in cosmological models where bursts are a stellar endpoint. In such models, the burst rate should be a function of the star formation rate (SFR). If there is a substantial delay (e.g., of order a billion years or more) between the GRB event and the star forming activity that created the progenitor, then the burst rate integrates over a galaxy's SFR, and we would not expect the host galaxy to display the signatures of recent star formation. Furthermore, if the progenitor is given a large velocity, then it may travel a large distance from the host galaxy before bursting, and it may become impossible to associate a galaxy with the burst.

In many models the burst occurs shortly after its progenitor star forms (e.g.,

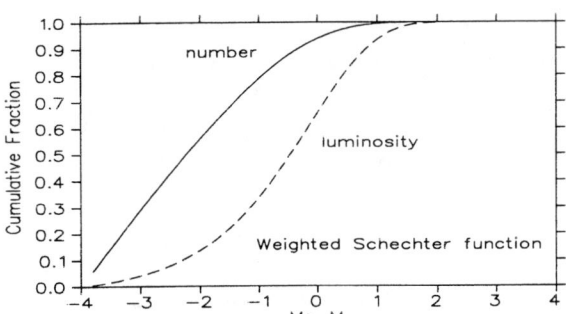

FIGURE 1. The cumulative distribution of host galaxy magnitudes if their luminosity function is weighted by number or luminosity. If GRBs trace light, the 50% point is near M_*. If it is proportional to galaxy number a typical host would be ~ 2 magnitudes fainter.

within a hundred million years or less). We would then expect that on average bursts would occur in galaxies showing evidence of recent star formation. The burst rate should be proportional to the SFR, both for individual galaxies and for a given cosmological epoch.

In particular, the burst rate and the SFR should have the same history, as was recently considered by several groups [12,15,18,21]. Extensive redshift surveys and data from the Hubble Deep Field have reliably determined the cosmic star formation history to $z \sim 5$ [3,4,7–9]. The data clearly suggest a rapid increase in the comoving SFR density with increasing redshift, SFR $\propto (1+z)^4$, reaching a peak rate (at $z \sim 1.5$) about 10–20 times higher than the present-day rate, and decreasing slowly to the present value by $z \sim 5$. This evolution function, $\eta(z)$, enters the differential rate vs. (bolometric) peak flux

$$\partial_P R \propto P^{-5/2} \, \mathrm{E}(z)^{-1} \, \eta(z) \, (1+z)^{-3} \, [(1+z)\partial_z y(z) + y(z)]^{-1} \quad , \qquad (3)$$

where E(z) and $y(z)$ are defined in [13]. For small redshifts the logarithmic slope of this function is Euclidean, i.e. $-5/2$. The solid curve of Figure 2 shows the effects of geometry (bending of logN–logP) and the dashed curve demonstrates how $\eta(z)$ compensates for the geometry out to the redshift at which the cosmic SFR peaks. At larger redshifts the effects of geometry and decreasing SFR then combine and the slope flattens quickly. Comparison with BATSE data suggests that this SFR model deviates from the pseudo-Euclidean slope too abruptly. While several studies [15,18,21] report that the observed SFR generates a brightness distribution consistent with BATSE data, our findings support the different result of Petrosian & Lloyd [12], who suggest that other evolutionary effects must be present in addition to the density evolution described by $\eta(z)$. While a good fit to the data requires a more sophisticated model of source evolution, the basic message is likely to be

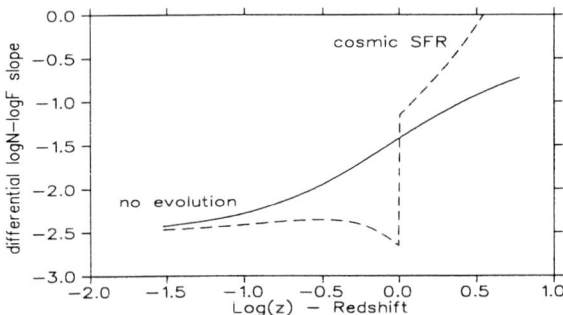

FIGURE 2. The slope of the differential GRB brightness distribution. Non-evolving sources (solid curve) quickly show significant deviation from the Euclidean value $-5/2$ with increasing redshift. If the GRB rate is proportional to the SFR (see text) the apparent Euclidean slope extends to $z \sim 1$ (dashed curve). For greater z the geometry of the universe together with a now decreasing burst rate cause the slope to deviate rapidly from $-5/2$.

the same: the logN–logP distribution does not exclude GRB redshifts much greater than unity.

Therefore, the absence of the host galaxies predicted by the "minimal" cosmological model does not call the cosmological origin of bursts into question. Instead, host galaxy observations will teach us where bursts occur.

REFERENCES

1. Band, D., and Hartmann, D. H., *Ap. J.* **493**, in press (1998).
2. Bond, H. E., IAU Circ. 6654 (1997).
3. Connolly, A. J., *et al.*, astro-ph/9706255 (1997).
4. Ellis, R. S., *ARAA* **35**, 389 (1997).
5. Fuller, G. M., and Shi, X., astro-ph/9711020 (1997).
6. Larson, S. B., *Ap. J.* **491**, in press (1997).
7. Lilly, S., in *Critical Dialogues in Cosmology*, ed. N. Turok, Singapore: World Scientific, 1997, p. 465.
8. Madau, P., *et al.*, *M.N.R.A.S* **283**, 1388 (1996)
9. Madau, P., *et al.*, astro-ph/9709147 (1997).
10. Metzger, M. R., *et al.*, IAU Circ. 6631, 6655, and 6676 (1997).
11. Paczynski, B., astro-ph/9710086 (1997).
12. Petrosian, V., and Lloyd, N. M., these proceedings.
13. Peebles, P. J. E., *Principles of Physical Cosmology*, Princeton: Princeton University Press, 1993.
14. Pian, E., *et al.*, *Ap. J.*, submitted (1997).
15. Sahu, K. C., *et al.*, *Ap. J. Lett.* **489**, L127 (1997).
16. Schaefer, B. E., in *Gamma-Ray Bursts: Observations, Analyses and Theories*, eds. C. Ho, R. I. Epstein, and E. E. Fenimore, Cambridge: Cambridge University Press, 1992, p. 107.
17. Schaefer, B. E., *et al.*, *Ap. J.* **489**, 693 (1997).
18. Totani, T., *Ap. J. Lett.* **486**, L71 (1997).
19. Van Paradijs, J., *et al.*, *Nature* **386**, 686 (1997).
20. Vrba, F. J., *et al.*, these proceedings.
21. Wijers, R.A.M.J., *et al.*, *M.N.R.A.S.*, in press (1997).

The Complex No-Host Problem: Real or Imaginary?[1]

Samuel B. Larson

Department of Physics & Astronomy
University of California, Los Angeles
8371 MSA, Los Angeles, CA 90095-1562

Abstract. The reported lack of galaxies inside of gamma-ray burst (GRB) error boxes has often been used to argue against a cosmological distance scale. However, our limited understanding of GRBs makes it difficult to construct a sensitive test for the presence of suitable hosts without making tenuous assumptions. This difficulty is illustrated by examining the underlying assumptions in the two original host galaxy tests. It is found that these tests are heavily dependent upon poorly known quantities, and many assumptions are already violated by circumstances unrelated to GRBs or host galaxies. As a result, normal galaxies have not been eliminated as the hosts of gamma-ray bursts.

INTRODUCTION

One of the remaining difficulties with the cosmological interpretation of cosmic gamma-ray bursts (GRBs) is the reported lack of potential host galaxies inside their localizations. Obviously, if one looks deeply enough, galaxies will be found inside even the smallest error boxes. The issue is not whether error boxes contain galaxies, but whether they contain *suitable* galaxies. Unfortunately, the lack of knowledge about gamma-ray bursts makes it difficult to define what is suitable. The distance scale, energy production, and underlying source of gamma-ray bursts are unknown, causing host galaxy tests to be underconstrained. Assumptions must be made either about host galaxy luminosities to make inferences about burst energies, or about burst energies to make inferences about host galaxy luminosities. Results are then compared to "expected" values that have been calculated using further assumptions. By the time a conclusion is reached, it is easy to lose sight of what has actually been tested. This paper describes the two host galaxy tests which defined the no-host problem and examines their underlying assumptions.

---

[1] This paper has been adapted from Larson, *ApJ* **491**, in press (1997).

SCHAEFER 1992

The first claim of a host galaxy problem was made by Schaefer [1], who analyzed a collection of moderately deep images, ranging from U- to K-band, of 13 localizations acquired by himself and other researchers. He assumed a standard candle host galaxy, and used the apparent magnitude of the brightest galaxy inside each error box to derive lower limits to the distance and energy of the corresponding bursts.

For each passband, the magnitude of the brightest possible host galaxy in the error box was determined. This magnitude either corresponded to the brightest object in the box or, more often, the detection threshold of the image when no objects were detected. Because sensitivities varied from image to image, the passband imposing the faintest counterpart limit for each error box was chosen. Next, the distances to these brightest galaxies were estimated by assuming they had the intrinsic luminosity of M31 in the chosen passband. These distances were then used to calculate the energy of the corresponding bursts from GRB fluence measurements, assuming isotropic energy release and propagation.

The resulting energies ranged from 2×10^{51} to 5×10^{54} erg, and these were compared to the energy budget of proposed cosmological models. The most plausible models convert some fraction of the binding energy of a neutron star ($\sim 10^{53}$ erg) into gamma rays. Five of the 13 estimated burst energies were greater than 10^{53} erg and two were above 10^{54} erg. In his conclusion and subsequent discussions [2], Schaefer suggested several ways out of this dilemma: GRBs come from subluminous galaxies or do not come from galaxies, GRBs emit more energy than 10^{53} erg, or GRBs are beamed or gravitationally lensed. However, there are other possibilities having nothing to do with bursts or their host galaxies. The null hypothesis in this test consists of the following assumptions.

The sample of localizations is unbiased. Samples consisting of IPN boxes are biased toward the brightest events, so results are only valid for bright GRBs. But this sample is not even a random subset of bright GRBs. Of the 13 targets, four are optical transient fields and one is a soft repeater. These five non-GRBs represent the four highest implied energies. When they are excluded from the sample, only one of eight energies exceeds 10^{53} erg.

Every gamma-ray burst originates inside a galaxy of M31 luminosity and spectral shape. This is a combination of many assumptions: GRBs come from galaxies (which is all we are trying to test), these galaxies are as luminous as M31, and since the deepest passband was always chosen, they have the same spectral shape as M31. Because the outliers of the burst energy distribution are used to invalidate the whole hypothesis, no host galaxies are allowed to be fainter than M31. A real host galaxy population will have a wide luminosity distribution, and furthermore, M31 is not even representative of a "normal" galaxy. M31 is an $L*$ galaxy. If GRBs trace all luminous matter in the universe, then the median host galaxy luminosity will be closer to $0.5L*$. This not only causes an overestimation of the burst energies, it induces a large variance in them, so it is incorrect to use the outliers of this

GRB energy distribution to make inferences about the whole sample. In addition, consistently choosing the most restrictive passband for each position will cause a systematic underestimation of a galaxy's true bolometric luminosity.

Every host galaxy is located inside its error box. If the boxes are inaccurate, as their occasional movement might suggest, or if bursters are ejected from their galaxies [3], this assumption is violated. Interestingly, the smallest boxes—those given highest priority—might be least likely to contain the host galaxy.

There are no galaxies brighter than the adopted limit in each error box. Many magnitude limits were set by the detection threshold of an image. These are not $5\,\sigma$ detection limits, i.e., galaxy detection is not complete to these levels. It is conceivable that galaxies one magnitude brighter than these limits are present in these fields but escaped detection. For example, Schaefer's magnitude limit for field GRB 781124 is $R = 20.0$, yet Harrison, McNamara, & Klemola [4] report finding a galaxy of $V = 18.7$ and $I = 17.8$ inside the box.

No gamma-ray bursts can release more energy than 10^{53} erg, and GRB energy is released and propagated isotropically. These are model dependent assumptions that are used to compare the estimated burst energies with what can be supplied by cosmological models. Models that can supply more energy than 10^{53} erg or can beam or lens the radiation, even for a few events, are exempted from these limits.

Excluding the non-GRBs from the sample and accounting for the galaxy magnitude underestimations, the inferred GRB energies range from 5×10^{50} to 5×10^{52} erg, and the null hypothesis is not violated. These energies are entirely consistent with cosmological models that liberate neutron star binding energies.

FENIMORE ET AL. 1993

Instead of assuming a standard-candle host galaxy to estimate the distance to—and energy of—the bursts, this study [5] assumed a standard-candle burst to estimate the distance to—and luminosity of—the host galaxies. Burst distances were estimated by fitting a nonevolving standard-candle burst of peak flux 6×10^{50} erg s^{-1} to the observed GRB brightness distribution, which predicts $z \sim 0.8$ for the faintest BATSE bursts. Using these burst distances, and the galaxy limiting magnitudes of Schaefer [1], they calculated the inferred intrinsic luminosity of the brightest galaxy in each of eight GRB error boxes.

The resulting host galaxy luminosity distribution was compared to a mass-weighted galaxy luminosity function that assumes a constant mass-to-light ratio throughout the whole distribution. For each host galaxy luminosity, the fraction of stellar mass contained by galaxies fainter than itself was integrated from this function. If GRBs trace the luminous matter in the universe, these two galaxy distributions should be the same, and the mass fractions will be evenly distributed between 0 and 1 with an average of 0.5. An average less than 0.5 indicates a lack of host galaxies: as a group, the nearest galaxy in each error box is farther away than the burst.

The average mass fraction for this sample of eight GRB sources was 0.42. For a sample of eight measurements, this is 0.8 σ away from 0.5 and is therefore consistent. Fenimore et al. derived an upper limit of $M_V = -18$ for the mean absolute magnitude of host galaxies, conclusively ruling out active galaxies and allowing for "normal" galaxies only if the true galaxies in these error boxes were no more than one magnitude fainter than Schaefer's limits. The null hypothesis in this test, which they were unable to reject, is built on the following assumptions.

The GRB sample is unbiased. The brightness bias is still present, but the non-GRBs in Schaefer's sample were removed. This appears to be a random sample of bright GRBs, although the selection criteria have never been explained.

GRB peak fluxes are 6×10^{50} erg s^{-1} nonevolving standard candles. This inference is used as a statistical GRB distance indicator. In this test, outliers in the host galaxy distribution are not used to make any inferences about the whole population; only the average of the mass fraction distribution is used. Therefore, random deviations in these distances for particular bursts will not cause a rejection of the null hypothesis, although they may lead to an overestimate of the statistical significance of any results. The main problem here is the possibility of a systematic error in the distance estimates. Underestimating the distances would cause smaller mass fractions and violate the null hypothesis. Mao & Paczyński [6] find that the faintest BATSE bursts are at $z \sim 1.5$, almost twice as distant as the estimates used here.

Every host galaxy has the same spectral profile as M31, they are all located inside their error boxes, and there are no galaxies brighter than the quoted limit in each box. These three assumptions result from using Schaefer's magnitudes and are subject to the same problems.

GRBs come from galaxies whose luminosities are distributed according to a certain mass-weighted luminosity function. Even if GRBs do come from normal luminous matter, it is very difficult to come up with the proper galaxy distribution to use for comparison. Galaxy luminosity functions are not well known and depend on evolution and environment. Practically any mass fraction can be obtained by adjusting the luminosity function in the desired way. This freedom, combined with the wide distribution of acceptable average mass fractions that comes from using such a small number of measurements, makes this assumption extremely difficult to violate.

DISCUSSION

The results of both of the original host galaxy tests are consistent with the presence of suitable host galaxies. This does not mean that GRBs do come from normal galaxies. The brightest galaxies inside these error boxes may have no association with the bursts; in fact, independent tests indicate that they are consistent with a random galaxy distribution [7,8].

Both samples of GRBs contain somewhat large error boxes, which are more likely to contaminate a host galaxy test with random bright galaxies. However, eliminating these boxes from the sample comes at the expense of the statistical significance of the results. Host galaxy tests are already challenged by the small number of localizations. More importantly, selecting a subsample of a larger sample of targets after studying their content destroys the *a priori* validity of the test. For a given sample of events, it is always possible to define a hypothesis and a subsample of those events that violates the hypothesis.

ACKNOWLEDGMENTS

I thank Ian McLean, Eric Becklin, Ned Wright, and Fred Vrba for many useful discussions which led to this study.

REFERENCES

1. Schaefer, B. E., in *Gamma-Ray Bursts: Observations, Analyses, and Theories*, 1992, pp. 107-112.
2. Schaefer, B. E., in *AIP Conf. Proc.* **307**, 1994, pp. 382-391.
3. Narayan, R., Paczyński, B., and Piran, T., *ApJ* **395**, L83-L86 (1992).
4. Harrison, T. E., McNamara, B. J., and Klemola, A. R., *AJ* **108**, 600-604 (1994).
5. Fenimore, E. E., et al., *Nature* **366**, 40-42 (1993).
6. Mao, S., and Paczyński, B., *ApJ* **388**, L45-L48 (1992).
7. Larson, S. B., and McLean, I. S., *ApJ* **491**, in press (1997).
8. Band, D., Hartmann, D., Vrba, F., these proceedings.

The USNO Deep Optical Survey of Small IPN[3] Localizations: Where are the Hosts?

F.J. Vrba[1], C.B. Luginbuhl[1], M.C. Jennings[2,3], D.H. Hartmann[4],
K.C. Hurley[5], C. Kouveliotou[6], C.A. Meegan[6], G. Fishman[6],
T.L. Cline[7], M. Boër[8]

[1] *U.S. Naval Observatory Flagstaff Station, P.O. Box 1149, Flagstaff AZ 86002*
[2] *P.O. Box 66, Corona del Mar, CA 92625*
[3] *Visiting Scholar, IGPP, University of California Riverside, Riverside, CA 92521* [4] *Department of Physics and Astronomy, Clemson University, Clemson SC 92634-1911*
[5] *Space Sciences Laboratory, University of California, Berkeley, CA 94720*
[6] *NASA/Marshall Space Flight Center, Huntsville, AL 35812*
[7] *NASA/Goddard Space Flight Center, Greenbelt, MD 20771*
[8] *Centre d'Etude Spatiale des Rayonnments, Toulouse, FRANCE*

Abstract. We present a summary of our deep optical photometric surveys of five small (< 8 arcmin^2) IPN^3 GRB localizations at high Galactic latitude ($b^{II} > 25°$). All of the localizations contain galaxies ranging in brightness from V$\sim$18 to our survey limits at about V$\sim$23. Without model-dependent assumptions of acceptable GRB distances, these galaxies, which are several magnitudes brighter than the galaxy associated with the GRB970228 optical transient, are viable GRB host candidates. We find a QSO candidate in only one of the localizations, and are therefore unable to support the hypothesis, based on larger IPN^1 localizations, that QSO/AGNs or blue galaxies may be associated with GRBs [1,2]. We examine possible reasons for this negative result.

INTRODUCTION

With the advent in the late 1970's of IPN[1] gamma-ray burst (GRB) localizations of a few to tens of arcmin^2 [3], many researchers anticipated quick progress on the GRB problem with the discovery of obvious low-energy counterparts or hosts. Not only did this prove not to be the case, but even IPN[1]-size localizations are typically too large to allow definitive correlations to be made with more mundane and common potential hosts such as stars, galaxies, QSOs, etc. The overwhelming evidence of GRB isotropy [4] points strongly to an extragalactic origin for GRBs. Yet, ubiquitous fainter galaxies have been rejected by some as potential hosts for the sources of GRBs on the basis of apparently reasonable, but probably naive assumptions about intrinsic host brightness and GRB maximum energy, isotropic propagation, *etc.* [5]. The so-called "no-host problem" thus created may be more a problem of underlying assumptions than a lack of potential hosts [6]. For example, simply relaxing the demand for isotropic propagation seems to allow faint galaxies as potential hosts, even though the smallest current GRB localizations remain much too large to clinch such an identification.

In this paper we summarize the results of a deep photometric UBVI survey of five high Galactic latitude GRB localizations produced with the Third Interplanetary Network (IPN[3]) [7]. This survey is a follow-on of our earlier work [1] on larger error boxes produced by IPN[1]. Portions of this work have been presented in preliminary form elsewhere [8,9].

SUMMARY OF FIELDS AND DATA

Observations for each field consist of UBVI–filtered CCD images taken at the U. S. Naval Observatory, Flagstaff Station 1.0-m and CTIO 1.55-m telescopes during 1993–1995. Integrations were typically 100/80/60/60 minutes through the U/B/V/I filters, though in several cases the exposures were shorter. These data are therefore less homogeneous than those our earlier survey [1].

The fields summarized here are at $|b^{II}| > 25°$, where Galactic extinction is expected to be lower than $A_V \sim 0.6$ mag. Table 1 shows the details and summary results for each field. The first seven columns list the GRB event's name, Galactic latitude, error box area in arcmin^2, and UBVI limiting magnitudes (defined at S/N = 5), respectively.

TABLE 1. GRB Field Summaries

GRB	b^{II}	area	U_{lim}	B_{lim}	V_{lim}	I_{lim}	V_{QSO}	V_{galaxy}
910219	+55.3	7.29	22.5	23.7	23.6	22.2	20.7	19.8
920406	−25.9	1.64	21.5	24.7	24.3	22.3	(20.2)	19.8
920525	−30.7	3.54	19.6	23.0	24.0	22.2		22.3
920711	+26.2	2.50	22.8	23.5	22.8	21.5	(21.8)	19.0
920720	+77.1	3.78	22.3	22.5	23.0	21.9		18.0

The eighth column lists the V magnitude of any QSO candidates within (simple number) or near (enclosed in parentheses) the IPN^3 error box. These candidates are identified based on color criteria [1]; about 2/3 are expected to be QSOs. The last column lists the V magnitude of the brightest resolved galaxy found within each error box.

DISCUSSION

Our earlier result [1] suggesting an association of QSO candidates with IPN[1] localizations is supported by the discovery of a QSO candidate within the GRB790325 localization [2]. Using these combined data and recent QSO statistics [11], the overall association probability is approximately 0.94 [10]. The presence of a QSO candidate within only one of our surveyed IPN[3] localizations (see Figure 1) is inconsistent with the IPN[1] results and must be due either to the lack of an actual

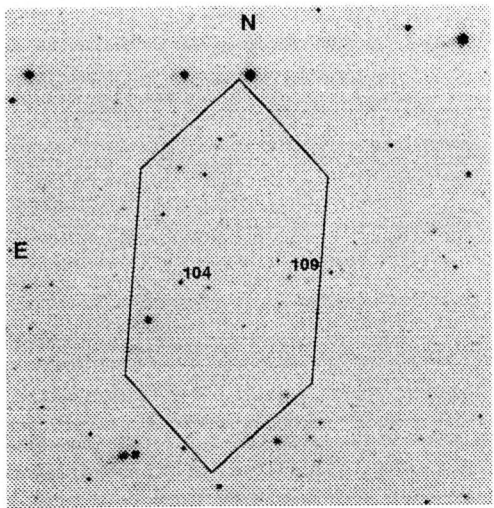

FIGURE 1. GRB910219 (5.7x5.7 arcmin field). Object 109 is a QSO candidate; 104 is the brightest resolved galaxy in the error box.

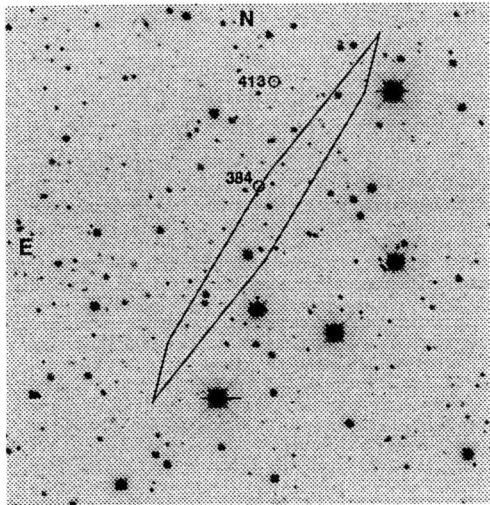

FIGURE 2. GRB920406 (6.1x6.1 arcmin field). Object 413 is a QSO candidate; 384 is the brightest resolved galaxy in the error box.

QSO/GRB association or to some limitation of the new IPN^3 data. If the former, our suggestive IPN^1 results [10] are simply a statistical fluke. It must be kept in mind that the QSO candidates identified in our IPN^1 studies are identified only on the basis of broad-band colors; none are yet verified spectroscopically. Although our statistics take into account an empirical 'hit rate' of verified QSOs based on similar color-based searches, if fewer than the expected number are in fact QSOs, the statistical significance of the IPN^1 association will decrease.

If the latter, the following possibilities suggest themselves. These data do not go as deep, in several cases, as the data for the IPN^1 survey [1]; this could leave QSO candidates unidentified even though they may exist in each box. Another possibility is that some IPN^3 localizations may have unrecognized small ($\sim$1 arcmin) positional errors. For the generally larger IPN^1 localizations, such errors would be relatively immaterial. For the smaller IPN^3 localizations, such errors could be sufficient to move the localizations off the GRB sources. This possibility may be supported by the presence of QSO candidates near but just outside the localizations in two fields (GRB920406 and GRB920711; see Figures 2 and 3). Independent verifications of the IPN localizations for SGR1806-20 [12], the bursting pulsar J1744-28 [13] and several BeppoSAX localizations [14–16] do not generally verify the IPN timing annuli to better than about 2-5 arcmin, and utilize different suites of spacecraft for the timing data than were used for the localizations searched in our IPN^3 survey.

All of the error boxes surveyed contain galaxies, often relatively bright ones. Without imposing model-dependent assumptions about GRB distances [5], such galaxies are potential hosts for the GRB sources. The brightest galaxies, as listed

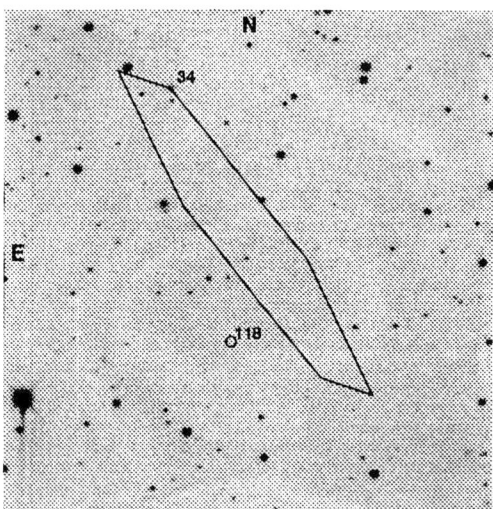

FIGURE 3. GRB920711 (5.7x5.7 arcmin field). Object 118 is a QSO candidate; 34 is the brightest resolved galaxy in the error box.

in Table 1, are much brighter and presumably nearer than the faint probable galaxy associated with the GRB970228 optical transient, which in the "no-host" scenario is too faint to be the host for the GRB associated with it, and must therefore be a chance superposition. The data reported here cannot lend any independent support to the identification of galaxies as GRB hosts, since the error boxes are still too large for a statistically significant correlation. However, it does seem safe to say that these data do not support the alternative hypothesis, that there are no potential host galaxies present within the localizations – the classic "no-host problem."

REFERENCES

1. Vrba, F. J., Hartmann, D. H., and Jennings, M. C., *Astrophys. J.* **446**, 115 (1995).
2. Vrba, F. J., Luginbuhl, C. B., Hartmann, D. H., and Jennings, M.C., these proceedings.
3. Cline, T. L., et al., *Astrophys. J.* **237**, L1 (1980).
4. Meegan, C., et al., these proceedings.
5. Schaefer, B. E., these proceedings.
6. Larson, S. B., and McLean, I. S., *Astrophys. J.* **491**, in press (1997).
7. Cline, T. L., et al., in Gamma-Ray Bursts, eds. Paciesas, W.S. and Fishman, G., AIP Conf. Proc. **265**, 72, (AIP, New York, 1992).
8. Luginbuhl, C. B. et al., *Astrophys. & Sp. Sci.* **231**, 289 (1995).
9. Luginbuhl, C. B., Vrba, F. J., Hudec, R., Hartmann, D. H., and Hurley, K., *Gamma-Ray Bursts*, New York: AIP Press, eds. Kouveliotou, C., et al., **384**, pg. 676, (1996).
10. Vrba, F. J., Luginbuhl, C. B., Jennings, M. C., and Hartmann, D. H., in preparation (1998).
11. Boyle, B. J., Jones, L. R., and Shanks, T., *M.N.R.A.S.* **251** 482 (1991).
12. Hurley, K., et al., *IAU Circular* 6512 (1997).
13. Hurley, K., et al., *IAU Circular* 6286 (1996).
14. Hurley, K., et al., *IAU Circular* 6571 (1997).
15. Hurley, K., et al., *IAU Circular* 6594 (1997).
16. Cline, T.L., et al., *IAU Circular* 6593 (1997).

Deep Imaging of the IPN[1] Localization of GRB790325 (GBS 1810+31)

Frederick J. Vrba[1], Christian B. Luginbuhl[1],
Dieter H. Hartmann[2], and Mark C. Jennings[3,4]

[1] *U.S. Naval Observatory, Flagstaff Station, P.O. Box 1149, Flagstaff, AZ 86002*
[2] *Department of Physics and Astronomy, Clemson University, Clemson, SC 92634-1911*
[3] *P.O. Box 66, Corona Del Mar, CA 92625*
[4] *Visiting Scholar, IGPP, University of California, Riverside, Riverside, CA 92521*

Abstract. The IPN[1] localization of GRB790325 is one of the smallest GRB localizations, but it has been poorly surveyed due to the nearby bright star 104 Her. We have used a new camera employing a Tek 2048^2 CCD with a 9 magnitude ND apodizing region to block the light of 104 Her and obtain deep UBVI imaging of the localization. The observations, obtained at the USNO 1.55-m telescope, are the deepest yet obtained of this localization, revealing 20 objects clearly in or at the edge of the error box. Among these is a QSO candidate which is unlikely to be located randomly within the localization based on known QSO statistics.

INTRODUCTION

While optical counterpart radiation has now been detected from GRBs 970228 [1] and 970508 [2], the nature of GRB sources remains far from resolved. Deep optical and infrared surveys of the smallest GRB localizations are still valuable tools in searching for the sources of bursts [3-5]. One of the smallest localizations from the IPN[1] network is that of GRB790325 (GBS 1810+31) at 1.52 arcmin^2 [6]. Unfortunately, under normal observing conditions this small localization is largely obscured by the glare from the bright (V = 4.97) M3III star 104 Her, which lies approximately 1.2 arcmin from the nearest edge and 1.8 arcmin from the center of the localization. This circumstance has allowed only shallow or incomplete surveys of the localization's contents [7,8]. More recently [9], the HST has been used to obtain B, U, and UV imaging of the entire localization. However, the ground based U, B, V, I CCD imaging observations reported here appear to be the deepest yet obtained of this localization.

TABLE 1. Limiting magnitudes for apodized imaging of GRB790325 (GBS 1810+31) localization.

Filter	Exposure Time (min)	Limiting Mag. (20% Phot)	Limiting Mag. (Detection)
U	120	20.8	21.5
B	60	23.0	23.8
V	60	22.2	22.9
I	25	21.2	21.6

OBSERVATIONAL RESULTS

Since December 1995 a special camera (designated ND9) employing a Tek 2048^2 CCD detector with a 9 magnitude apodizing region has been used routinely in the USNO parallax program to obtain astrometry of bright individual stars with respect to a reference frame of much fainter stars [10]. The 3-mm diameter Inconel attenuation spot is mounted 1-mm in front of, and centered on the CCD, producing, at the USNO 1.55-m telescope, a 40 arcsec diameter apodized region in an 11.1 x 11.1 arcmin^2 FOV. The out-of-focus image of 104 Her produced by light reflected from the Inconel spot was moved away from the localization by tilting the photometric filters located a few inches above the CCD camera. Preceeding the ND9 observations, we obtained normal CCD observations at the USNO 1.0-m telescope of an area just East of the IPN[1] localizaion in order to set up local photometric standards within the FOV of ND9 at the 1.55-m.

Deep UBVI imaging of the localization, employing ND9 at the 1.55-m telescope, was obtained on UT 1997 June 4 and 5. Table 1 gives the total exposure times of the stacked images taken on these two nights along with the limiting magnitudes for 20% photometry and detection in each filter. Figure 1 shows approximately 25% of the full ND9 I-band frame over which the IPN[1] localization is drawn. Since 104 Her is very red, contaminating residual scattered light within the IPN box is by far the worst in the I-band image shown, with essentially no contamination in B and U. Within this 4.8 x 5.8 arcmin^2 region, excluding the small area covered by 104 Her and the thin slice of sky to its east, photometry was obtained of every object within the remaining 27 arcmin^2 to the limiting magnitudes listed in Table 1. In the surveyed region 181 objects were detected. Of these, 7 appear to be barely resolved galaxies, at our resolution, between $20.6 \leq V \leq 21.5$.

Figure 2 displays the (U−V) vs. (V−I) color-color diagram for all objects having precise UVI photometry in the surveyed region. Most of the objects fall along the unreddened stellar main sequence (MS) which is also plotted. The two objects slightly, but significantly, to the upper left of the MS are resolved galaxies, while the object marked '91' has colors consistent with a QSO.

Within the localization we detect 20 objects, 5 of which are close to the localization boundaries. These objects are identified in Figure 3 which employs our full–survey numbering system. Most of these objects appear to be normal stars.

However, object '70' is a V = 20.6 resolved galaxy also identified as such by [9], and object '91', near the center of the box, is the V = 21.3 QSO candidate mentioned above. The QSO was also detected by [9] in the B-band, but no particular mention of it is made since it was not detected in their U or UV-band HST observations. The fact that we detected it easily at U = 20.7 while the HST observations [9] did not detect it with a quoted detection limit of U = 22.4 may indicate that it is significantly variable in U, consistent with it being a QSO. However, we also point out that while we detected 20 objects within the error box, the HST UV-B imaging [9] only detected 8 objects. Several other of our 20 objects should have been detected at U by the HST with their stated 5σ detection limit of U = 23.0.

DISCUSSION

A weak correlation of QSO candidates (flat-spectrum objects identified by their positions in the UBVI color-color diagrams) with 5 of the smallest, high b^{II} IPN[1] localizations has been reported previously [3]. The discovery reported here of a V = 21.3 QSO candidate in the smallest IPN[1] localization (1.52 arcmin^2) is of some statistical significance by itself. Applying a conservative correction factor of 0.65 actual QSOs per color-selected candidate (based on experience in identifying QSOs from UBVI photometry), and using the commonly accepted QSO areal density [11] corrected for the fainter limiting magnitude of these observations, the occurence

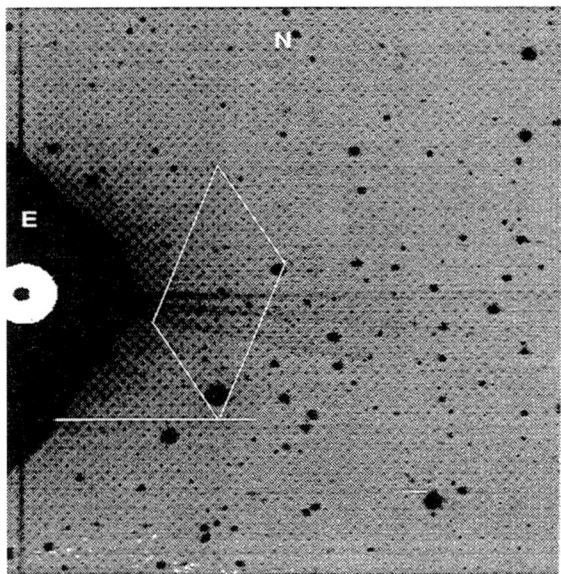

FIGURE 1. The ND9 I-band image of the GRB790325 region surveyed photometrically in this study. The IPN[1] localization is shown. The apodized image of 104 Her is at the left.

of a QSO candidate in the GRB790325 localization is significant at approximately 96.4% confidence.

In a forthcoming paper [12] we will present the full photometric database for our survey of this localization and discuss the overall QSO/GRB statistics for our combined IPN[1] localization surveys. These suggestive results call for the testing of the QSO/GRB correlation hypothesis by systematic surveys of additional, even smaller, GRB localizations. We report preliminary results of our observations for small, high b^{II} IPN[3] localizations elsewhere in these proceedings [13].

REFERENCES

1. Groot, P. J., et al., *IAU Circular* No. 6584 (1997).
2. Bond, H. E., *IAU Circular* No. 6654 (1997).
3. Vrba, F. J., Hartmann, D. H., and Jennings, M. C., *Astrophys. J.* **446**, 115 (1995).
4. Larson, S. B., and McLean, I. S., *Astrophys. J.* **491**, in press (1997).
5. Schaefer, B. E., Cline, T. L., Hurley, K. C., and Laros, J. G., preprint (1997).
6. Laros, J. G., et al., *Astrophys. J.* **290**, 728 (1985).
7. Ricker, G. R., Vanderspek, R. K., and Aihar, E. A., *Adv. Sp. Sci.* **6**, 75 (1986).
8. Schaefer, B. E., et al., *Astrophys. J.* **313**, 226 (1987).
9. Schaefer, B. E., Cline, T. L., Hurley, K. C., and Laros, J. G., *Astrophys. J.*, in press (1997).

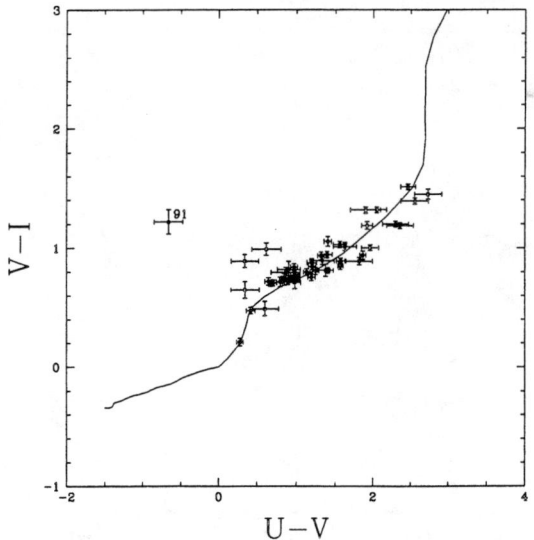

FIGURE 2. The (U−V) vs. (V−I) color-color diagram for all stars in the surveyed region with precise UVI photometry. Object 91, which lies within the IPN[1] localization, is a QSO candidate.

10. Dahn, C. C., *Fundamental Stellar Properties: The Interaction between Observation and Theory*, IAU Symp. No. 189, eds. T. R. Bedding et al., pg. 19 (1997).
11. Boyle, B. J., Jones, L. R., and Shanks, T., *M.N.R.A.S.* **251** 482 (1991).
12. Vrba, F. J., Luginbuhl, C. B., Jennings, M. C., and Hartmann, D. H., in preparation (1998).
13. Vrba, F. J., Luginbuhl, C. B., Jennings, M. C., Hartmann, D. H., Hurley, K. C., Kouveliotou, C., Meegan, C. A., Fishman, G., Cline, T. L., and Boër, M., these proceedings.

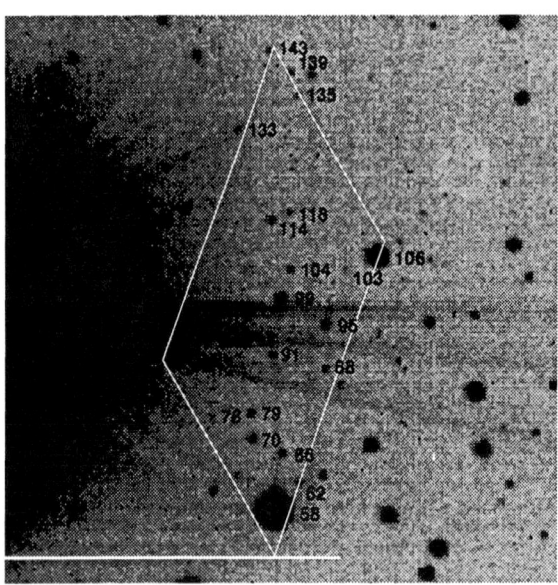

FIGURE 3. An enlargement of the ND9 I-band image indicating the 20 objects within the IPN[1] localization identified in this survey.

NIR Imaging of Gamma-Ray Burst Error Boxes

S. Klose[1,2], B. Stecklum[2], J. Eislöffel[2,3,4], J. L. Hora[3] and R. Tuffs[1,5]

[2] *Thüringer Landessternwarte Tautenburg, D-07778 Tautenburg, Germany*
[3] *Institute for Astronomy, University of Hawaii, Honolulu, HI 96822*
[4] *Max-Planck-Institut für Astronomie, Königstuhl 17, D-69117 Heidelberg, Germany*
[5] *Max-Planck-Institut für Kernphysik, Saupfercheckweg 1, D-69117 Heidelberg, Germany*

Abstract. We report on a test of the cosmological model of gamma-ray bursts (GRBs) based on K-band statistics of galaxies in six GRB error boxes. Although our database is small and inhomogeneous, we can use it to draw some general conclusions about the nature of the GRB hosts.

INTRODUCTION

Over the last years various attempts have been made by several groups to tackle the problem whether all GRBs are cosmological in origin or not. A typical approach is to search for an overabundance of a special class of astronomical objects in GRB error boxes. The conclusions drawn from such studies seem to be, however, controversial. In particular, some B-band surveys point to a lack of appropriate host galaxies in GRB error boxes [1,2], whereas K-band surveys of other boxes [3,4] do not show this "no-host problem".

We have extended these near-infrared surveys and observed five GRB error boxes. These boxes were selected either because of their small size (IPN[1]) or because of their very strong burst (IPN[3]). As a sixth box, we also have included here GRB970228, in order to check whether in this case K-band statistics of galaxies would give a result in agreement with the former cases, or whether it would clearly fail. Basic data from K-band number-count statistics of galaxies (in units of number deg^{-2} mag^{-1}) were taken from [5].

The reason for performing a K-band survey instead of a B-band survey was twofold. First, the colour B–K of galaxies increases with increasing redshift z. At $z = 0$ this colour is around 4 on average (cf. [6]). Secondly, if GRBs prefer galaxies

---

[1]) Visiting astronomer, German-Spanish Astronomical Centre, Calar Alto, operated by the Max-Planck-Institut für Astronomy, Heidelberg, jointly with the Spanish National Commission for Astronomy.

CP428, *Gamma-Ray Bursts:* 4th Huntsville Symposium
edited by C. A. Meegan et al.
© 1998 The American Institute of Physics 1-56396-766-9/98/$15.00

TABLE 1. Surveyed GRB error boxes.

GRB	fluence[a]	reference	box size[b]	reference	remarks
790418	6.5	[13–15]	2.9	[19]	3 σ box; 100%[c]
791105b	1.3	[13,15]	41	[19]	3 σ box; 100%[c]
791116	16.0	[13,1]	4.5	[19]	3 σ box; 100%[c]
940217b	66 ± 27	[16]	360[d]	[16]	1 IPN annulus[d]; 60%[c]
960924	26	[17]	117	[20]	1 IPN annulus[e]; 35%[c]
970228	0.2	[18]	3	–	assumed, see text; 100%[c]

[a] in units of 10^{-5} erg cm^{-2}
[b] error box size in units of arcmin^2
[c] fraction of error box surveyed
[d] The observed field is centred at the midpoint of the area covered by the intersection of the $\sim 6'$ wide 3 σ IPN annulus, the 95% *EGRET* contour, the 2 σ *COMPTEL* contour, and the 1 σ *BATSE* error circle: $\sim 360'^2$ (see [16,21]).
[e] Area of the 3 σ IPN annulus within the 1 σ *BATSE* error circle [20]. The survey was centred on the most probable burster position [20,22].

with violent star formation, then K-band surveys should better reveal these galaxies than optical surveys.

OBSERVATIONS AND RESULTS

Most observations reported here were performed between 24 and 27 September 1996, using the Calar Alto 2.2-m telescope equipped with the MAGIC near-infrared camera [7]. MAGIC was used in its wide-field mode with $1''\!.62$ pixels. This survey was supplemented by observations with the Calar Alto 3.5-m telescope (MAGIC; GRB790418 [8], GRB970228 [9]) and by observations on 26 April 1997 with the University of Hawaii 2.2-m telescope on Mauna Kea (Quick Infrared Camera [10]; GRB970228). All observations reported here were performed in the K'-band. The $K' - K$ conversion followed the procedure applied in [8]. Photometry was done using infrared standard stars from Casali & Hawarden [11] and Elias *et al.* [12]. The observed error boxes are listed in Table 1, the results obtained are summarized in Table 2.

Explanations to Table 2: Here, two probabilites, p_1 and p_2, are given. Probability p_1 applies if the galaxy of magnitude K under consideration lies within the 3 σ error box (IPN^1 boxes) or within only one IPN annulus (IPN^3 boxes). It is the probability of finding a galaxy of magnitude $K \pm 0.5$ mag in an area equal to the area of the 3 σ error box (IPN^1 boxes) and, respectively, in an area defined by the width of the IPN annulus and a length $2r$, where r is the distance of the galaxy from the most probable burster position. Probability p_2 is used if the galaxy lies very close to, but outside, the 3 σ error box (IPN^1 boxes) and, respectively, very close to, but outside, the IPN annulus (IPN^3 boxes). It is the probability of finding a galaxy of magnitude $K \pm 0.5$ mag within an area equal to a circular area of radius r, where r is the distance of the galaxy from the centre of the 3 σ error box (IPN^1

TABLE 2. K-band statistics of brightest galaxies.

GRB	K-magn.[a]	B, R-magn.[a]	redshift[b]	RA[c]	DEC[c]	p_1	p_2
790418	16.3 ± 0.2	21.6, 19.5[d]	1.1	5:54:22.4	-6:58:56	–	0.9
	17.6 ± 0.3	22.2, 20.2[d]	1.3	5:54:21.3	-6:59:42	1.0	–
791105b	14.6 ± 0.2	20.6, 17.7	0.7	22:54:02.9	-2:11:14	1.0	–
	15.1 ± 0.2	21.2, 18.7	0.8	22:53:56.6	-2:09:34	–	1.0
791116	14.9 ± 0.2	20.3, 18.2[e]	0.6	0:12:43.7	-15:43:11	–	0.7
	> 16	–	–	–	–	1.0	–
940217b	13.4 ± 0.1	17.4, 15.2	0.05	1:58:20.6	3:54:22	0.6	–
	14.1 ± 0.2	17.8, 16.5	0.04	1:58:17.7	3:53:02	1.0	–
960924	12.9 ± 0.1	17.3, 14.2	0.06	2:23:58.0	1:13:39	0.07	–
	13.6 ± 0.1	18.3, 16.1	0.12	2:24:26.7	1:07:57	–	0.4
970228	16.1 ± 0.1	–	–	5:01:46.8	11:46:16	1.0	–
	17.0 ± 0.2	–	–	5:01:48.7	11:47:13	1.0	–

[a] if not stated otherwise, magnitudes are according to [28]
[b] estimated based on Fig. 9 in [29] after a self-consistent correction of colour according to redshift, assuming late type galaxies and a $q_0 = 1/2$ universe
[c] J 2000, $+/- 1''$; [d] V, I magnitude [19]; [e] B, I magnitude [19]

boxes) and, respectively, the distance of the galaxy from the most probable burster position (IPN^3 boxes).

Remarks to the observed fields (Table 2):

- GRB790418: The K-brightest galaxy lies only $5''$ outside of the 3 σ error box. Therefore, we assumed it lies within the box, calculated p_1 and set $p_2 = p_1$.

- GRB791105b: No source is visible at the position of the optical transient reported in [23]. Any such source is fainter than $K = 16$.

- GRB791116: A bright galaxy lies $\sim 50''$ outside the 3 σ error box and $r \sim 2'$ away from its centre. Galaxies within the error box must have $K > 16$ and, therefore, could not be identified by their morphology.

- GRB940217b: The K-brightest galaxy lies within the 3 σ IPN annulus (width $6'$ [16]) and $r \sim 8.5'$ away from the most probable burster position given in [24]. For the second brightest galaxy $r = 7.5'$.

- GRB960924: The K-brightest galaxy lies within the IPN annulus (width $1'$) only $r \sim 13'$ away from the most probable burster position [20,22], see Figure 1. A further K-bright galaxy lies $\sim 30''$ outside the IPN annulus and $r \sim 4'$ away from the most probable burster position.

- GRB970228: We applied the procedure of K-band statistics of galaxies in a circular area of $1'$ radius centred on the reported celestial position of the optical transient [25].

CONCLUSIONS

Based on only a handful of surveyed GRB error boxes of very different sizes, including three boxes of relatively large size, we cannot draw statistically well-founded conclusions. This concerns a possible overabundance of K-bright (and extended) galaxies in GRB error boxes, as well as the question whether there is a monotonic relation between the fluence of a burst and the magnitude of the K-brightest galaxy found in the corresponding error box. The former seems to be indicated by GRB960924 (Figure 1), but more such cases are needed. The latter is not in strong conflict with GRBs 960924, 940217b, and 791105b, and also not in conflict with GRB970228: this was the weakest burst in our sample and with respect to the other surveyed boxes K-band statistics in an arcmin-sized field centered on the burster should not show a bright galaxy. If the latter applies, however, then the lack of K-bright galaxies within the GRB790418 and GRB791116 error boxes (Table 2) indicates that GRB sources can be located several arcsec away from their hosts. Conversely, if GRBs do not occur several kpc away from their host galaxies (for instance, within the hypernova scenario [26]), then GRB790418 and GRB791116 imply that either 1) GRBs are not standard candles (at least with respect to their gamma-ray fluence), or 2) the distance scale of GRBs is larger than previously thought ([27], see also these proceedings), or 3) GRBs mostly occur in galaxies (around $z = 1$) which are not only under-luminous in the B-band [1,2], but also in the near-infrared (or a combination of all three points).

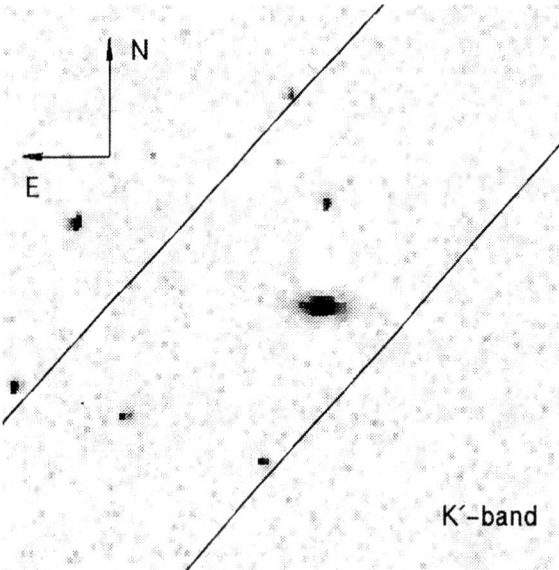

FIGURE 1. A bright anonymous galaxy lies within the 1′ wide IPN annulus of GRB960924.

REFERENCES

1. Schaefer, B. E., in *Gamma-Ray Bursts - Observations, Analyses and Theories*, Eds. C. Ho, R. I. Epstein, and E. E. Fenimore, Cambridge University Press, 1992, p. 107.
2. Schaefer, B. E. et al., *Astrophys. J.* **489**, 693 (1997).
3. Larson, S. B., McLean, I. S., and Becklin, E. E., in *Gamma-Ray Bursts*, Eds. C. Kouveliotou, M. F. Briggs, and G. J. Fishman, AIP Conf. Proc. **384**, Part Two, New York: AIP, 1996, p. 666.
4. Larson, S. B., McLean, I. S., and Becklin, E. E., *Astrophys. J.* **460**, L 95 (1996).
5. Djorgovski, S. et al., *Astrophys. J.* **438**, L13 (1995).
6. Songaila, A. et al., *Astrophys. J. Suppl. Ser.* **94**, 461 (1994).
7. Herbst, T. M., and Rayner, J. T., in *Infrared Astronomy with Arrays*, Ed. I. S. McLean, Dordrecht: Kluwer, 1993, p. 515.
8. Klose, S., Eislöffel, J., and Richter, S., *Astrophys. J.* **470**, L93 (1996).
9. Klose, S., Stecklum, B., and Tuffs, R., *IAU Circ.* 6611 (1997).
10. Hodapp, K.-W. et al., *New Astronomy* **1**, 177 (1996).
11. Casali, M., and Hawarden, T., JCMT-UKIRT Newsletter **4**, 33 (1992).
12. Elias, J. H. et al., *Astron. J.* **87**, 1029 (1982).
13. Schaefer, B. E. et al., *Astrophys. J.* **340**, 455 (1989).
14. Boer, M. et al., *Astron. & Astrophys.* **249**, 118 (1991).
15. Baity, W. A., Hueter, G. J., and Lingenfelter, R. E., Ed. S. Woosley, AIP Conf. Proc. **115**, New York: AIP, 1984, p. 434.
16. Hurley, K. et al., *Nature* **372**, 652 (1994).
17. Meegan, C. A. et al., Current *BATSE* Gamma-Ray Burst Catalog, 1997, see http://www.batse.msfc.nasa.gov/data/grb/catalog/
18. Palmer, D. et al., *IAU Circ.* 6577 (1997).
19. Vrba, F. J., Hartmann, D. H., and Jennings, M. C., *Astrophys. J.* **446**, 115 (1995).
20. Hurley, K., private communication (1997).
21. Tokanai, F. et al., *PASJ* **49**, 207 (1997).
22. Klose, S. et al., *Astron. J.*, submitted (1997).
23. Schaefer, B. et al., *Astrophys. J.* **286**, L1 (1984).
24. Winkler, C. et al., *Astron. & Astrophys.* **302**, 765 (1995).
25. van Paradijs, J. et al., *Nature* **386**, 686 (1997).
26. Paczyński, B., *Astrophys. J.*, submitted (1997).
27. Wijers, R. A. M. J. et al., *MNRAS*, submitted (1997).
28. Monet, D. et al., USNO-SA1.0, U.S. Naval Observatory, Washington DC, 1996.
29. Boselli, A. et al., *Astron. & Astroph. Suppl.* **121**, 507 (1997).

On The Possible GRB/QSO and GRB/Abell Cluster Correlations

K. Hurley[1], D. Hartmann[2], C. Kouveliotou[3], C. Meegan[4], G. Fishman[4], T. Cline[5]

[1] *UC Berkeley Space Sciences Laboratory, Berkeley, CA 94720-7450*
[2] *Clemson University, Clemson, SC 29634-1911*
[3] *NASA Marshall Space Flight Center, Universities Space Research Association, Huntsville, AL 35812*
[4] *NASA Marshall Space Flight Center, Huntsville, AL 35812*
[5] *NASA Goddard Space Flight Center, Greenbelt, MD 20771*

Abstract. A study by Kolatt and Piran [7] has presented evidence that gamma-ray burst sources are correlated with Abell clusters at the 95% confidence level, based on analyses of bursts in the BATSE 3B catalog. Similarly, evidence has recently been presented by Schartel et al. [9] that the positions of BATSE 3B bursts are correlated with radio-quiet quasars. This association is significant at the 99.8% confidence level, and is also supported by an analysis of several smaller error boxes from the 1st Interplanetary Network. Using more precise localization information from the 3rd Interplanetary Network, we have reanalyzed these possible correlations. We find that most of the Abell clusters which are in the relatively large 3B error circles are not in the much smaller IPN/BATSE error regions. We have also found that none of the QSOs which are in the 3B error circles are in the smaller IPN/BATSE error regions. We repeated our analysis using the bursts in the BATSE 4B catalog, with similar results.

INTRODUCTION

Kolatt and Piran [7] (hereafter KP) analyzed the BATSE 3B catalog [8] in conjunction with data on Abell clusters [1], and concluded that these gamma-ray bursts were correlated with them at the 95% confidence level. In their study, they selected the 3B bursts with error circle radii <2.77°, and bursts and clusters with $|b| > 30°$. They then calculated the number of burst-cluster pairs, N(θ), whose separation was smaller than a given angle θ, for θ=1 – 6°. Comparing this number to the numbers found for randomly generated catalogs, they found a number for θ=4° that was significant at the 95% confidence level.

Schartel et al. [9] analyzed the 3B catalog in conjunction with data on quasars and active nuclei [10], and concluded that these gamma-ray bursts were correlated with them at the 99.8% confidence level. In their study, they selected the 80 3B bursts with error circle radii <1.7°, and considered radio-loud and radio-quiet quasars, BL Lac objects, and AGN's, among others. We focus here only on the radio-quiet quasar (RQQ) association. They constructed an optimized sample of bright RQQs, and found that 20 3B error circles contained at least one RQQ. Comparing this number to the numbers found for randomly generated burst catalogs, they concluded that this association was significant at the 99.8% confidence level.

If these findings could be confirmed, they would constitute statistical evidence for counterparts to GRB sources, and would indicate that many GRBs are at cosmological distances. We have attempted to confirm these results by subjecting them to more stringent tests. The positions of those Abell clusters or RQQ's which appear to be related to GRBs, based on their locations within BATSE error circles, must be consistent with all known localization information for the bursts in question. In particular, they must also lie within the annuli or the error boxes of the 3rd Interplanetary Network (IPN3) for those bursts. The area covered by these precise positions is over an order of magnitude smaller than the BATSE error circles, leading to a stronger test.

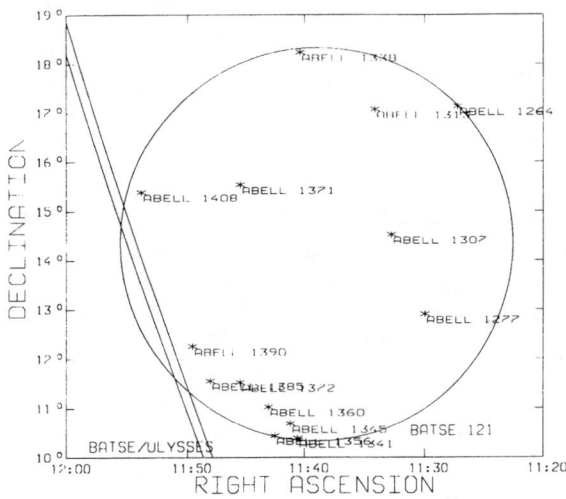

FIGURE 1. The field around BATSE burst 121. The BATSE 4° radius error circle and the Ulysses/BATSE triangulation annulus are shown. None of the Abell clusters in this field lies within the triangulation annulus.

IPN DATA

For the period covered by the 3B catalog, IPN3 consisted of the Ulysses spacecraft, at distances up to 6 AU [3], BATSE, and, until mid-1992, Pioneer Venus Orbiter (PVO). When only Ulysses and BATSE observed a burst, the resulting

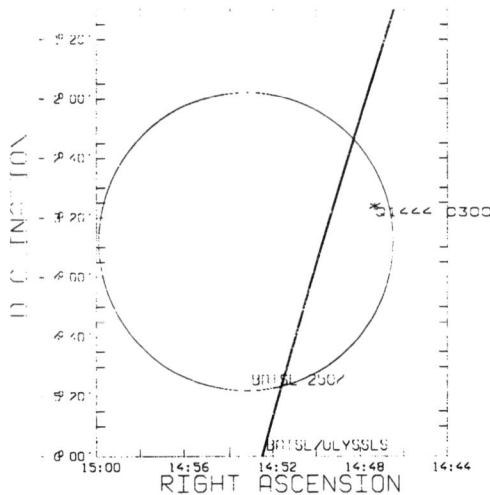

FIGURE 2. The BATSE error circle for burst 2507 (radius 1.7°), and the corresponding IPN annulus (width 40″). The radio-quiet quasar Q1444-0300 lies in the BATSE error circle, but not in the IPN annulus.

localization was an annulus generally crossing the BATSE error circle. When PVO observed the burst also, an error box resulted with dimensions in the several arcminute range in the best cases. Examples may be found in Hurley et al. [4] and Cline et al. [2]. The annulus widths varied from about 10″ to 1000″, with an average of 5.7′. Descriptions of the data set have appeared in Hurley et al. [5,6].

Of the 136 "accurately" localized BATSE bursts selected by KP on the basis of their error circle size and galactic latitude, 91 have IPN3 annuli associated with them, and 12 have IPN3 error boxes. Of the 80 "accurately" localized BATSE bursts selected by Schartel et al. on the basis of their error circle size, 78 have at least one IPN3 annulus associated with them, and 20 have IPN3 error boxes. The total areas covered by the IPN/BATSE locations are less than the areas covered by the corresponding BATSE error circles alone, by factors of ~10 to 30 in these two studies, respectively. This substantially reduces the probability of chance correlations.

ANALYSIS

The all-sky catalog of Abell clusters [1] contains 4073 rich clusters, each having at least 30 bright members, and covers redshifts $z < 0.2$. Following KP, a latitude cut of $|b| > 30°$ was applied to both cluster and burst catalogs. We began by verifying the numbers of bursts and Abell clusters in the KP study, 136 (after latitude and error circle size cuts are applied) and 3616 (after the latitude cut), respectively. For each cluster, we then checked each of the BATSE error circles to see whether their positions were consistent. If it was, we finally checked for consistency with any IPN3 location information. Since, in the KP study, Abell clusters within 4° of a BATSE position were considered to be correlated with the burst, we assigned an error radius of 4° to the BATSE bursts, and also used this to define the BATSE/IPN3 error box. An example is shown in Figure 1.

We found that 16 Abell clusters lie within the BATSE/IPN3 error boxes, while 15.8 are expected by chance. This indicates that the overwhelming majority of the Abell clusters which are in BATSE error circles are there by chance. We reanalyzed the data using the standard 1σ BATSE error circle radii, defined as

$$\sqrt{\sigma_{sys}^2 + \sigma_{stat}^2}$$

where σ_{sys} is the systematic error, 1.6°, and σ_{stat} is the statistical error, given in the BATSE 3B catalog for each burst. In this case, 5.5 Abell clusters should fall in the BATSE/IPN error circles by chance. Six were found to actually lie in them.

The catalog of Véron-Cetty and Véron [10] contains a total of 8609 QSO's. Here we confirmed the numbers of bursts and QSO's in the Schartel study, 80 (after error circle size cuts are applied) and 7146 (after the cut for radio-quiet quasars only), respectively. For each RQQ, we then checked each of the BATSE error circles to see whether their positions were consistent. If it was, we finally checked for consistency with any IPN3 location information. We found that 0.3 RQQs would be expected to lie within the BATSE/IPN error boxes by chance. None were actually found to lie within them, indicating that the RQQs which are in BATSE error circles are there by chance. Figure 2 shows a typical result.

Since the time of the KP and Schartel et al. studies, the BATSE 4B catalog has been released, and we repeated our analyses using it.

If, as in the original KP study, clusters within 4° of a BATSE burst are counted as associated with the burst, we would expect 27.8 clusters to fall in BATSE/IPN annuli by chance; 24 were actually found to lie within the annuli. If we use the normal definition of the error radii for the BATSE bursts, we would expect 8.6 to fall in BATSE/IPN annuli by chance. Eight were actually found to lie within the annuli.

For the Schartel et al. study, applying the same cut as before (error circle radii <1.7°), we have 124 BATSE bursts. Of them, 120 have one or more IPN annuli associated with them, and 20 have two annuli. We would have expected 0.5 RQQs to lie in the BATSE/IPN error boxes by chance. Again we found that none of the RQQs actually lay in the BATSE/IPN error boxes.

CONCLUSIONS

It seems likely that the two correlations, between Abell clusters and GRBs, and between radio-quiet quasars and GRBs, are simply statistical fluctuations. We stress that our analysis is a preliminary one, because the IPN localizations are being refined and updated. However, in most (although not all) cases, this results in narrower annuli, which should not change the results.

ACKNOWLEDGMENTS

KH is grateful for Ulysses support under JPL Contract 958056, and IPN3 support under NASA NAG 5-1560. DH acknowledges support from NASA through the COMPTON Observatory guest investigator program. We are indebted to T. Kolatt, T. Piran, and N. Schartel for many useful discussions about this study.

REFERENCES

1. Abell, G. O., Corwin, H. G., and Olowin, R. P., *Ap. J. Supp.* **70**, 1 (1989).
2. Cline, T. et al. in *Gamma-Ray Bursts*, eds. W. Paciesas and G. Fishman (New York: AIP), 1992, p. 72.
3. Hurley, K. et al. *Astron. Astrophys. Suppl. Ser.* **92(2)**, 401 (1992).
4. Hurley, K. et al. *Astron. Astrophys. Suppl. Ser.* **97(1)**, 39 (1993).
5. Hurley, K. et al. in *Gamma-Ray Bursts - Second Workshop*, eds. G. Fishman, J. Brainerd, and K. Hurley (New York: AIP), 1994 p. 27.
6. Hurley, K. et al. in *3rd Huntsville Symposium on Gamma-Ray Bursts*, eds. C. Kouveliotou, M. Briggs, and G. Fishman, AIP Conf. Proc. 384 (New York: AIP), 1996 p. 422.
7. Kolatt, T., and Piran, T., *Ap. J.* **467**, L41 (1996).
8. Meegan, C. et al., *Ap. J. Supp.* **106**, 45 (1996).
9. Schartel, N., Andernach, H., and Greiner, J., *Astron. Astrophys.*, in press (1997).
10. Véron-Cetty, M-P., and Véron, P., ESO Scientific Report, in press, (CDS catalog 188, available by anonymous ftp to adc.gsfc.nasa.gov) (1996).

SHOCKS

Theoretical Models of Gamma-Ray Bursts

P. Mészáros[1]

Department of Astronomy & Astrophysics, 525 Davey Lab,
Pennsylvania State University, University Park, PA 16802

Abstract. Models of gamma ray bursts are reviewed in the light of recent observations of afterglows which point towards a cosmological origin. The physics of fireball shock models is discussed, with attention to the type of light histories and spectra during the gamma-ray phase. The evolution of the remnants and their afterglows is considered, as well as their implications for our current understanding of the mechanisms giving rise to the bursts.

INTRODUCTION

The discovery of X-ray, optical and radio afterglows of gamma-ray bursts (GRB) amounts to a major qualitative leap in the type of independent observational handholds on these objects. Together with existing γ-ray signatures, these provide significantly more severe constraints on possible models, and may indeed represent the light at the end of the tunnel for understanding this long-standing puzzle of astrophysics.

The report of long wavelength observations of GRB970228 over time scales of days to weeks at X-ray (X), and months at optical (O) wavelengths [4] was the most dramatic recent development in the field. In this and subsequent IAU circulars, it was pointed out that the overall behavior of the long term radiation agreed with theoretical expectations from the simplest relativistic fireball afterglow models published in advance of the observations ([27]; see also [58]). A number of theoretical papers were stimulated by this and subsequent observations (e.g. [53,60,43,64] among others), and interest has continued to grow as new observations provided apparently controversial evidence for the distance scale, possible variability and the candidate host [48]. New evidence was added when the optical counterpart to the second discovered afterglow (GRB970508) yielded a redshift lower limit placing it at a clearly cosmological distance [30], and this was strengthened by the detection of a radio counterpart [11,54] as well as evidence for the constancy of the associated diffuse source and continued power law decay of the point source [12].

---

[1]) also Center for Gravitational Physics and Geometry, Pennsylvania State University

This new evidence reinforces the conclusions from previous work on the isotropy of the burst distribution which suggested a cosmological origin (e.g. [13]). Observational material on this is provided chiefly by a superb data base (currently of over 1700 bursts in the 4B catalog) which continues being accumulated by the BATSE instrument, complemented by data from the OSSE and Comptel instruments on CGRO, as well as Ulysses, KONUS and other experiments. At gamma-ray energies, much new information has been collected and analyzed, relevant to the spatial distribution, the time histories, possible repeatability, spectra, and various types of classifications and correlations have been investigated. At the same time, investigations of the physics of fireball models of GRB have continued to probe the γ-ray behavior of these objects, as well as the afterglows. Much of the recent theoretical work has concentrated on modeling the time structure expected from internal and external shock models, multi-wavelength spectra, the time evolution and the spectral-temporal correlations.

THE DISSIPATIVE FIREBALL SCENARIO

The dissipative (or shock) fireball model is a fairly robust astrophysical scenario, independent of the particular type of progenitor, based only on the fact that it must inject the inferred large amount of energy inside the very small volume required by causality and the timescales characteristic of GRB [44,22]. The observational and physical motivation for this generic scenario of GRB has been described in detail elsewhere (e.g. [26]). The very large energy deposition inside a small volume leads to characteristic photon energy densities which lead to an optically thick $\gamma e^{\pm}$ fireball that is highly super-Eddington. The resulting expansion must be highly relativistic ($\Gamma \sim 10^2 - 10^3$), in order to avoid having the observed 0.1-10 GeV photons degraded by photon-photon interactions, and to yield the right timescales and energies. The fireball initially is thermal, and converts most of its radiation energy into kinetic energy (bulk motion). This kinetic energy of motion must be tapped via some dissipation mechanism, which is most likely to be shocks, and these probably occur after the fireball becomes optically thin, as suggested by the nonthermal spectra.

The plasma, MHD and radiation physics involved in the fireball shock scenario are familiar, being used in a number of other astrophysical situations. The basic ingredients, such as a high Γ outflow, collisionless shocks, magnetic field generation at some fraction of equipartition, acceleration of electrons to a power law, efficient energy exchange between protons and electrons, etc. are common features (or common problems, to varying degrees) in AGN, pulsars winds and supernova remnants. In AGN and possibly pulsar winds, conditions qualitatively similar to those in GRB seem to obtain, and these sources are known to have in many cases efficiencies of at least tens of percent in converting bulk kinetic energy into nonthermal particles and radiation. As in those sources, for GRB fireballs it is assumed that the fluid approximation is valid whenever the usual plasma kinetic theory criteria are satisfied, e.g. that the dimensions of the region are much larger than the proton gyroradius or the proton Debye length. The shocks serve to reconvert the kinetic energy of the outflow into random energy, and to accelerate relativistic particles which can radiate a power-law spectrum. The cosmological fireball shock model appears to have received strong confirmation from the afterglow observations, and from the fact that many of the basic gamma-ray signatures can be understood within the framework of the model without undue parameter twisting.

The generic nature of this scenario stems from the fact that it is largely *independent* of the detailed nature of the primary energy release mechanism, whether it be a binary compact object merger (NS/NS or NS/BH, e.g. [33]), a "failed Supernova Ib" [65], a young ultrastrongly magnetized pulsar [56], a "hypernova" [36], etc. This is because the primary mechanism is initially enshrouded in an optically thick pair fireball, which washes out most of the details, the observed radiation being produced outside the pair photosphere. Some information, however, may be carried through, e.g. in the details of the light curve (especially if this is due to internal shocks, see below).

A major theoretical question is how the very large bulk Lorentz factors inferred from observations are produced. Neutrino-antineutrino annihilation leading to pairs [8] have been proposed. Since the merger would lead also to enormous radiation pressure, a baryon rich outflow would however pollute the $e^{\pm}, \gamma$ fireball, but a clean fireball might be achieved if tidal heating and annihilation occur before merger, or if enough annihilations occur around the centrifugally evacuated binary rotation axis [21]. Numerical simulations using Newtonian potentials [47] indicate that this may not be straightforward, although effects like turbulent convection and magnetic fields could improve the pair luminosities. Matthews, et al. [20] use a general relativistic hydro code and conclude that both neutron stars collapse to black holes before merger, producing enough heating to power a pair luminosity comparable to required estimates. The disagreement between numerical simulation results is debated, and further refinements in models involving neutrino annihilation should be forthcoming.

On the other hand, magnetic fields may be responsible for a large or even dominant fraction of the relativistic stress tensor in the fireball. Super-strong magnetic fields are probably generated during the collapse of the rapidly rotating configuration [56,21,32,55,58], and this may contribute significantly to the energy density of the fireball. Such fields could in fact be dynamically dominant around the rotation axis, especially if the central object collapses to a black hole, leading to a Poynting dominated outflow which could be almost baryon-free [28]. Magnetic fields would, of course, also ensure a high radiation efficiency. MHD numerical simulations are difficult, as in pulsar winds and AGN jets, and have not so far been done. In any case, it is worth stressing that the motivation for high bulk Lorentz factor ($\Gamma \gtrsim 10^2$) outflows in GRB is largely observational, in particular the observation of 0.1-10 GeV photons, which are hard to explain otherwise (e.g. [16]).

GAMMA RAY TEMPORAL PROPERTIES AND SPECTRA

Two types of fireball models have been discussed, both of which produce the nonthermal spectrum via shocks occurring after the fireball has become optically thin, as inferred from the nonthermal nature of the spectrum. These involve different explanations for the typical duration of the burst, and predict different time variabilities. In the first type (a) (e.g. [22]) the shocks are those caused by interaction of the gaseous fireball ejecta with an external medium. In this case the typical duration is given by the Doppler delayed arrival of the light from the beginning and end of the ejecta shell, or from the delay between surface elements within the light cone. This assumes that any "intrinsic" burst duration is shorter than the above duration (impulsive approximation). Any "intrinsic" short time variability is washed out by the fact that radiation is received from a light cone and a finite width over which Γ varies by at least a factor 2. (The afterglows discussed below are well fitted, in their overall average features, by the late stages of an external shock).

In the second type of model (b) [45,35] the shocks leading to γ-rays are those which may be expected within the outflow itself, e.g. internal shocks caused by the catching up of faster portions with slower portions of the flow. These, if they occur, tend to do so at smaller radii than the previous external shocks, and the duration is likely to be given by the intrinsic duration of the energy release (since the Doppler delayed light arrival or light cone duration is likely to be shorter than the latter). One of the two stated purposes for introducing this model is that it *does* specifically allow arbitrarily complicated light curves [45], which are expected to reflect any "intrinsic" variability injected at the base of the outflow. These models are also referred to as wind models, or central engine models.

Detailed kinematical calculations [9] show explicitly some of the constraints imposed by observations on external shock light curves. Sari & Piran [49] showed analytically that external shocks in a blobby external medium would not be able to reproduce very complicated light curves with many subpulses, unless the efficiency is very low, $\lesssim 1\%$. Nonetheless, as shown by Panaitescu & Mészáros [37], if magnetic inhomogeneities are present or develop in the ejecta, and there is some pre-beaming in the comoving frame, one can get up to 5-10 peaks with good efficiency in an external shock light curve, and the spectral-temporal correlations are close to the observed values. For bursts with a very large number of subpulses, simulations of bolometric internal shock light curves [7,19] are in good qualitative agreement with the observations.

The nonthermal radiation spectrum of GRB is likely to be due to synchrotron or inverse Compton (IC) radiation of electrons or positrons accelerated in the optically thin shocks described above. Particles accelerated, e.g. by a Fermi type mechanism, in the presence of modest magnetic fields lead to nonthermal photon spectra similar to those observed (e.g. [24] for impulsive shock spectra, and [40] for wind spectra). Basically, two types of spectra are possible: those where the observed "break" in the 50 KeV - 2 MeV range is due to the synchrotron characteristic energy, or those where it is due to the IC upscattering of a lower energy break (typically at optical energies) which itself is due to synchrotron. The latter requires smaller magnetic fields and smaller electron minimum energies γ_m. Above γ_m shock acceleration is assumed to provide the electron power law responsible for the flattish νF_ν spectrum characteristic of bursts: an electron index $p \sim -3$ reproduces this well. The burst spectra can satisfy the X-ray paucity condition (i.e. the observation that generally $F_x \lesssim 3 \times 10^{-2} F_\gamma$ during the γ-ray burst), since below the break one expects a spectrum $\nu F_\nu \propto \nu^{4/3}$. On the other hand, spectra flatter than this can be easily obtained in an inhomogeneous magnetic field, or for a spatially varying bulk Lorentz factor, so that "soft excess" bursts can also be produced. In addition, one expects significant simultaneous emission at GeV energies, and modest but detectable simultaneous X-ray and optical emission [23,24,17,40]. In addition, if GRB occur inside galaxies where the external medium has an appreciable density, one would expect internal shock bursts to be followed by external shock bursts, which can be relevant for, e.g. delayed GeV emission [25]. In general one expects different spectral and temporal properties for such compound bursts. A study of the properties of internal shocks followed by external shocks [41] provides constraints on the internal parameters of the outflow (duration, variability timescale and luminosity) in different external environments.

One of the signs of the development of the subject is that γ-ray observations of GRB have become sufficiently detailed and extensive that they are beginning to probe questions of the internal physics of the models, such as the shock acceleration, the magnetic field equipartition fraction, and the radiative efficiency involved. As far as the specific radiation mechanism, Tavani [52] has presented detailed synchrotron spectra calculated numerically with a distribution of shocked electrons produced by a specific diffusive acceleration mechanism, and these were fitted to a variety of observed γ-ray spectra. Another investigation [3] aimed at testing the synchrotron hypothesis uses the fact that an electron distribution with a low-energy cutoff would produce a low-frequency asymptotic intensity spectrum with a slope of 1/3, while the time integrated high-frequency slope would be expected to be –1/2, compatible with a sample of BATSE spectra studied by these authors. This issue remains under discussion, e.g. [5]. The problem of a relatively high radiative efficiency during the γ-ray event is, clearly, one of the requirements of a fireball shock, or indeed of any other model. In particular, one needs to ensure that much of the energy carried in protons (if these are present and energetically dominant) is shared with the radiating electrons or pairs. Specific mechanisms have been proposed for this [1]. A high radiative efficiency is natural in models where magnetic fields are prominent (e.g. [32,57,55]). The electron-proton exchange would also be obviated in the reverse shock for scenarios involving Poynting dominated outflows where the inertia is mainly due to pairs [28], although in the late stages of deceleration the blast wave pushed ahead of the ejecta will unavoidably include baryons.

Another area of contact between models and observations is in the area of spectral-temporal correlations in the gamma-ray range. Fenimore & Sumner [10] find that the observed spectral break energies decay in time faster than predicted from single shell analytic models. Crider, et al. [6] argue that the evolution of the spectral break with integrated photon flux may be a restriction on simple models. Numerical hydrodynamic simulations [37] of external shock models indicate that many of the commonly observed correlations are well-reproduced; among these are a brightness and hardness correlation, hardness and duration anti-correlation, a hard to soft evolution outside of intensity pulses, a break energy increasing with intensity during a pulse, the break energy decreasing with increasing fluence, earlier pulses being harder, pulses being narrower at higher energies, etc. This is a continuing area of activity.

THE IMPLICATIONS OF AFTERGLOWS

The breakthrough Beppo-SAX observation of GRB970228 provided both a study of the long-term X-ray decay (extending over days) and an accurate localization permitting subsequent optical follow-ups (extending to months). The X-ray and optical sources are both point-like, as expected for a fireball at cosmological distances (dimension $\sim 0.1 - 1$ pc after $\sim$ months). The spectra are nonthermal, as expected from the simplest model based on synchrotron radiation of shock-accelerated electrons from a decelerating shell interacting with an external medium [27], and as predicted, it decays as power law in time with an index close to the expected value (see also [58,60,43,50]. Furthermore, a fuzzy extended source was identified around the point source. While the initial optical magnitude of the point source decayed from ~ 20 to ~ 24, there was uncertainty as to whether the diffuse source remained constant and whether it showed any proper motion [2]. However, the September 1997 HST images [12] have largely dispelled such doubts, indicating that the diffuse source remained at $m_R \sim 25.5$ with negligible proper motion, being compatible with a distant ($z \sim 1$), faint (possibly irregular) galaxy, while the optical point source is still present at $m_R \sim 28$, right along the extrapolation of the earlier power law.

A major highlight was the detection of the afterglow of GRB970508, which largely followed the pattern of GRB970228, but which added considerable excitement because of new, even if not entirely unexpected, features. The most significant of these is that a redshift limit was obtained [30] of $0.835 \leq z \lesssim 2.3$, based on several systems of absorption lines. Another previously unobserved phenomenon was that the optical flux of the point source initially rose as a power law, followed by a decay similar to that of GRB970228. This, in fact, is what one expects from a cloud where the spectrum has a peak initially above optical frequencies that shifts downwards during its expansion [17], and is in agreement with the Mészáros & Rees [27] simplest model. Another previously unobserved feature was the detection of a radio afterglow in GRB970508, about a week after the outburst, peaking after weeks and then decaying slowly. With a self-absorption frequency around 5-10 GHz (overestimated in early calculations), this is also compatible with the 'simplest' model (e.g. [61,18]). Furthermore, scintillations in the radio spectrum, predicted by Goodman [14], were also observed [11], providing a nice double-check on the physical dimension of the source of ~ 0.1 pc. An interesting, and less expected result is that the scintillation is of large amplitude and broad band, suggesting it is diffractive [63]. This requires a small size, which comes from the fact that the intensity is concentrated in a ring of radial extent substantially smaller than the radius of the visible disk [62,38,51]. This is because for equal observer times one sees an egg-shaped region of the outflow and the portion around the edges corresponds to a younger, hence hotter and higher field, stage of the remnant. Another unexpected feature is that the optical light curve appears to have been steady or decaying for a brief (few days) initial period before it started to rise and then decay [42]. One explanation for an initial decay could be that it is due to emission from a central jet, which later becomes overwhelmed by emission from a more energetic isotropic outflow at large angles [29].

One issue is whether the fireball, as it slows down by sweeping increasing amounts of external matter, evolves with $\Gamma \propto r^{-3/2}$ as expected in the adiabatic limit, or as $\Gamma \propto r^{-3}$ as expected if the remnant is in the radiative regime [44]. This would have consequences for the evolution of the afterglow [59,18]. In the latter case, the remnant would evolve faster, and could reach the nonrelativistic regime sooner, even if after some time it becomes adiabatic, as it should. Physically, however, for the remnant dynamics to be 'radiative' implies that most of the kinetic energy in protons and fields has to be radiated in less than a dynamic time [29]. This would require field reconnection, as well as efficient mechanisms for protons to re-energize electrons whose cooling timescale is shorter than the dynamical time in the cooling region (behind the shock front and throughout the remnant), and it is far from being understood how this would happen. In fact, the optical power law of GRB970228 is unbroken so far, after 8 months, arguing for an early onset of the adiabatic regime, and indicating that the nonrelativistic regime has not been reached yet (which is strong indication for a cosmological distance [64]). Another observational constraint comes from the radio scintillations in GRB970508: this requires a relatively small size $\lesssim$ few 10^{17} cm, after a time of several weeks. The longitudinal size is $r \sim 4(2n+1)\Gamma_o^2 ct$ where $n = (3/2, 3)$ for an (adiabatic, radiative) remnant, but the ring structure of the remnant emitting region reduces somewhat the coefficient in front [38]. The adiabatic behavior seems more in accord with observations [62]. However numerical calculations of the light curve and spectral evolution (e.g. [39]) are needed in order to address this issue more thoroughly.

A question is why some bursts (e.g. GRB970111) are detected in γ-rays but not in X/O, despite being in in the field of view of Beppo-SAX. One reason may be that the γ-ray emission could be due to internal shocks (leaving essentially no afterglows [27]) and the environment has a very low density, so the external shock occurs at larger radii and over longer times than in "canonical" afterglows, resulting in a sub-threshold X-ray intensity. This may be the case if GRB arise from compact binaries which are ejected to considerable distances from the host galaxy, where the external density may be much lower than the typical ISM values. Low density environments may also occur if the GRB goes off inside a pulsar cavity inflated by one of the precursors in the binary. This gives rise to a deceleration shock months after the GRB with a much lower brightness. Conversely, an interesting consequence of anisotropic models [64,46,29] is that there could be a large fraction of detectable afterglows for which no γ-rays are detected. The outflow at large angles is certain to be more baryon-loaded, and therefore of lower Γ, so that the shocks would occur later and would be at longer wavelengths.

It is also possible that some bursts arise in unusually high density environments (such as a star-forming region, where failed supernova or hypernova progenitors may reside [36]. This could lead to a more rapid onset of the deceleration leading to the X-ray phase, and it would also imply an increased neutral gas column density and optical depth in front of the source. A special case is that of GRB970828, where X rays have been observed, but no optical radiation down to faint levels [15]. The presence of a significant column density of absorbing material has been inferred from the low energy turnover of the X-ray spectrum [31], and the corresponding dust absorption may in fact be sufficient to cause the absence of optical emission (Wijers & Paczyński, private comm.). The difference between the low density and high density environments cases could be tested if future observations of afterglows reveal a correlation with the degree of galaxy clustering or with individual galaxies.

ACKNOWLEDGEMENTS

I am grateful to M.J. Rees for stimulating collaborations in this subject, as well as to H. Papathanassiou, A. Panaitescu and R. Wijers. This research has been supported in part by NASA NAG5-2857.

REFERENCES

1. Bykov, A. & Mészáros, P., *ApJ(Lett)* **461**, L37 (1996).
2. Caraveo, P., et al., these proceedings.
3. Cohen, E., et al., preprint (astro-ph/9703120) (1997).
4. Costa, E., IAU Circ. 6572; *Nature* **387**, 783 (1997).
5. Crider, A., Liang, E. & Preece, R., "Confronting Synchrotron Shock and Inverse Comptonization Models with GRB Spectral Evolution", these proceedings.
6. Crider, A., Liang, E. & Preece, R., "Testing the Invariance of Cooling Rate in Gamma-Ray Burst Pulses", these proceedings.
7. Daigne, F. & Mochkovich, R., *MNRAS*, in press (1997); these proceedings.
8. Eichler, D. et al., *Nature* **340**, 126 (1989).
9. Fenimore, E., Madras, C. & Nayakshin, S., *Ap.J.* **473**, 998 (1996).
10. Fenimore, E. & Sumner, C., in *Proc. All-Sky Observations in the Next Decade*, (RIKEN, Japan), in press (1997).
11. Frail, D., et al., *Nature*, in press (1997).

12. Fruchter, A. et al., IAU Circ. 6747 (1997).
13. Fishman, G. & Meegan, C., *A.R.A.A.* **33**, 415 (1995).
14. Goodman, J., *New Astronomy*, in press (1997).
15. Groot, P. J., et al., *ApJ*, submitted (1997).
16. Harding, A.K. and Baring, M.G., in *Gamma-ray Bursts*, ed. G. Fishman, et al., p. 520 (AIP 307, NY) (1994).
17. Katz, J., *ApJ(Lett.)* **432**, L109 (1994).
18. Katz, J. & Piran, T., *ApJ*, in press (1997).
19. Kobayashi, T., Sari, R. & Piran, T., *ApJ*, in press (1997).
20. Matthews, G., et al., preprint (astro-ph/9710229); J. Wilson, these proceedings.
21. Mészáros, P. & Rees, M.J., *ApJ* **397**, 570 (1992).
22. Mészáros, P. & Rees, M.J., *ApJ* **405**, 278 (1993).
23. Mészáros, P. & Rees, M.J., *ApJ(Lett.)* **418**, L59 (1993).
24. Mészáros, P., Rees, M.J. & Papathanassiou, *ApJ* **432**, 181 (1994).
25. Mészáros, P. & Rees, M.J., *MNRAS* **269**, L41 (1994).
26. Mészáros, P., in *Proc. 17th Texas Symp. Relat. Astrophys*, (N.Y. Acad Sci., NY) 759, 440 (astro-ph/9502090) (1995).
27. Mészáros, P. & Rees, M.J., *ApJ* **476**, 232 (1997).
28. Mészáros, P. & Rees, M.J., *ApJ* **482**, L29 (1997).
29. Mészáros, P., Rees, M.J. & Wijers, R., *ApJ*, submitted (1997); (astro-ph/9709273).
30. Metzger, M. et al., *Nature* **387**, 878 (1997).
31. Murakami, T., et al., these proceedings.
32. Narayan, R., Paczyński, B. & Piran, T., *Ap.J.(Lett)* **395**, L83 (1992).
33. Paczyński, B., *ApJ(Lett)* **308**, L43 (1986).
34. Paczyński, B. & Rhoads, J., *ApJ(Lett)* **418**, L5 (1993).
35. Paczyński, B. & Xu, G., *ApJ* **427**, 708 (1994).
36. Paczyński, B., these proceedings.
37. Panaitescu, A & Mészáros, P., *ApJ*, in press (1997); (astro-ph/9703187).
38. Panaitescu, A & Mészáros, P., *ApJ(Lett)*, in press (1997); (astro-ph/9709284).
39. Panaitescu, A. & Mészáros, P., these proceedings.
40. Papathanassiou, H. & Mészáros, P., *ApJ(Lett)* **471**, L91 (1996).
41. Papathanassiou, H., Ph.D. thesis, Pennsylvania State University (1997).
42. Pedersen, H., preprint (astro-ph/9710322) (1997).
43. Reichart, D., *ApJ*, in press (1997).
44. Rees, M.J. & Mészáros, P., *MNRAS* **258**, P41 (1992).
45. Rees, M.J. & Mészáros, P., *ApJ(Lett)* **430**, L93 (1994).
46. Rhoads, J., preprint (1997); these proceedings.
47. Ruffert, M., these proceedings.
48. Sahu, K., et al., *Nature* **387**, 476 (1997).
49. Sari, R. & Piran, T., preprint (astro-ph/9701002) (1997).
50. Sari, R., *ApJ(Lett)* **489**, L37 (1997).
51. Sari, R., preprint (astro-ph/9709300) (1997).
52. Tavani, M., *ApJ* **466**, 768 (1996).
53. Tavani, M., *ApJ(Lett)* **483**, L87 (1997).
54. Taylor, G.B., et al, *Nature*, in press (1997).

55. Thompson, C., *MNRAS* **270**, 480 (1994).
56. Usov, V.V., *Nature* **357**, 472 (1992).
57. Usov, V.V., *MNRAS* **267**, 1034 (1994).
58. Vietri, M., *ApJ(Lett)* **478**, L9 (1997).
59. Vietri, M., *ApJ*, submitted (1997).
60. Waxman, E., *ApJ(Lett)* **485**, L5 (1997).
61. Waxman, E., *ApJ(Lett)* **489**, L33 (1997).
62. Waxman, E., *ApJ(Lett)* in press (1997); (astro-ph/9709190).
63. Waxman, E., Kulkarni, S. & Frail, D., *ApJ(Lett)*, submitted (1997); (astro-ph/9709199).
64. Wijers, R., Rees, M.J. & Mészáros, P., *MNRAS* **288**, L51 (1997).
65. Woosley, S., *ApJ* **405**, 273 (1992).

Kinematic Arguments Against Single Relativistic Shell Models for GRBs

E. E. Fenimore, E. Ramirez, and M. C. Sumner

NIS-2, MS D436, Los Alamos National Laboratory, Los Alamos NM 87545

Abstract. Two main types of models have been suggested to explain the long durations and multiple peaks of gamma-ray bursts (GRBs). In one, there is a very quick release of energy at a central site resulting in a single relativistic shell that produces peaks in the time history through its interactions with the ambient material. In the other, the central site sporadically releases energy over hundreds of seconds forming a peak with each burst of energy. We show that the average envelope of emission and the presence of gaps in GRBs are inconsistent with a single relativistic shell. We estimate that the maximum fraction of a single shell that can produce gamma-rays in a GRB with multiple peaks is 10^{-3}, implying that single relativistic shells require 10^3 times more energy than previously thought. We conclude that either the central site of a GRB must produce $\sim 10^{51}$ erg s^{-1} for hundreds of seconds, or the relativistic shell must have structure on a scales the order of $\sqrt{\epsilon}\Gamma^{-1}$, where Γ is the bulk Lorentz factor ($\sim 10^2$ to 10^3) and ϵ is the efficiency.

INTRODUCTION

Two classes of models have arisen that explain different (but not all) aspects of the duration of GRBs. In the "external" shock model [1], the release of energy is very quick and a relativistic shell forms that expands outward for a long period of time (10^5 to 10^7 sec). At some point, interactions with the external medium (hence the name) cause the energy of the bulk motion to be converted to gamma rays. The alternative theory is that a central site releases energy in the form of a wind or multiple shells over a period of time commensurate with the observed duration of the GRB [2]. The gamma rays are produced by the internal interactions within the wind, hence these scenarios are often referred to as internal shock models.

In Fenimore, Madras, & Nayakshin [3], we used kinematics to demonstrate that a single relativistic shell has extreme difficulties explaining the observed GRB time structure. We have made direct comparisons to the observations for three of the most potent arguments: the average envelope [4,5], gaps in the time history [3], and the maximum active fraction of the shell [6]. In this paper, we summarize those arguments.

ARGUMENT 1: AVERAGE ENVELOPE

If a single relativistic shell with high bulk Lorentz factor (Γ) expands outward from a central site towards an observer, the observed time structure is dominated by two effects. First, although the shell might produce gamma rays for a long period of time (say t_0 to $t_{\max}$), the shell keeps up with the photons such that they arrive at a detector over a short period of time. If the shell has velocity $v = \beta c$ such that the Lorentz factor, Γ is $(1-\beta^2)^{-1/2}$, then photons emitted over a period t arrive at a detector over a much shorter period, $T = (1-\beta)t \approx t/(2\Gamma^2)$. Second, the curvature causes regions of the shell off-axis to arrive later at the detector. The additional distance that photons must travel is $\sim R(1-\cos\theta)$ where R is the radius of the shell ($\sim ct$). At a typical observable angle of $\theta = \Gamma^{-1}$, the delay due to the curvature is the same order as the time scale of arrive for on-axis photons: $t/(2\Gamma^2)$. In [3,4], we showed that a single symmetric shell produces a "FRED"-like shape (fast rise, rapid decay):

$$V(T) = V_0 \frac{T^\omega - T_0^\omega}{T^{\alpha+1}} \qquad \text{if } T_0 < T < T_{\max}$$
$$= V_0 \frac{T_{\max}^\omega - T_0^\omega}{T^{\alpha+1}} \qquad \text{if } T > T_{\max} \quad , \qquad (1)$$

where V_0 is a constant, $\omega = \alpha + 3 - \nu$, α is the spectral number index (e.g., 1.5), and ν is a power law index for the intrinsic variation of the shell's emissivity as a function of time. The expansion effects occur in the rise of the envelope and the curvature dominates in the fall. We have also shown [4] that during the decay phase, the spectra should evolve as T^{-1}.

To test this, we have added together 32 bright BATSE bursts with durations between 16 and 40 s. We align each burst by scaling it to a standard duration defined to be $T_{100} = (T_{90} + T_{50})/0.7$ where T_{90} and T_{50} are the durations that contain 90% and 50% of the counts. Figure 1 is from reference [5] which should be consulted for compete details. The average envelope *and* the average spectral evolution are linear whereas a single relativistic shell predicts that they should be power laws with indices $-\alpha-1$ and -1, respectively. We conclude that the average envelope of GRBs is not consistent with a single relativistic shell.

ARGUMENT 2: GAPS IN TIME HISTORY

Gaps or precursors in GRBs produce the strongest evidence against a single relativistic shell. The sharp rise in the average profile indicates that the shell emits for a short period of time (i.e., t_0 to $t_{\max}$ is short relative to the duration of the event), so that the shape of the overall envelope is dominated by photons delayed by the curvature. During the decay phase, the region that can contribute photons to a given section of the time history is an annulus oriented about the line of sight to the observer (see Fig. 2). Gaps in the time history indicate that some annuli

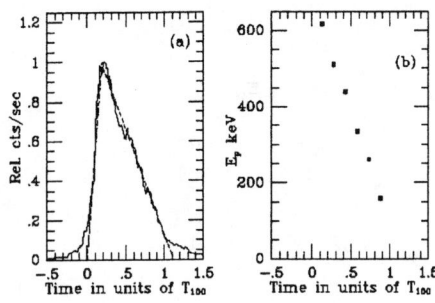

FIGURE 1. The average temporal and spectral evolution of bright events with intermediate durations (T_{90} between 16 and 40 s) based on the BATSE MER data. (a) The average time history. The decay phase starting 20% after the beginning of the T_{100} period is linear rather than the expected power law. (b) The average evolution of the the peak of the νF_ν distribution. The peak energy is also a linear function rather than the expected T^{-1}. These patterns are inconsistent with that expected from a single relativistic shell.

emit while others do not. These annuli are causally disconnected, making it difficult to achieve this large scale coherence. (See Figure 7 in [3] for attempts to fit the emission of shells to bursts with gaps and precursors.)

ARGUMENT 3: ACTIVE FRACTION OF SHELL

Each dot in Fig. 2 is a causally connected region. Note region 3 has more dots so it produces a smoother time history (the intensity is less because the emission is off axis so fewer photons are beamed towards the observer). We have shown [7] that the volume of the annulus that contributes at any time is a constant so all sections of a time history should have about the same smoothness. We assume that the "peaks" in a time history represent Poisson fluctuations in the number of entities contributing at any time. We determine the total number of entities ($N_N = \mu_N(T/\Delta T)$) up to time T by determining the rate of entities: $\mu_N = N^2/\delta N^2$ where N and δN are the mean and root-mean-square of the profile. The fraction of the shell that became active is $\epsilon = N_N A_N / A_S$. Here, A_N is the size of each entity ($= \pi c^2 \Gamma^2 \Delta T^2 / k$, where k is 13 for entities arising from entities that grow at the speed of sound and is 1 for interactions with interstellar matter (ISM) clouds, see [6]). The total area of the shell is $A_S = 4\pi c^2 \Gamma^2 T^2 f$ where f is the fraction of the shell out to $\theta = \Gamma^{-1}$ that contributes up to time T. For FRED-like bursts, Eq. 1 usually fits the profile such that f is unity at $T = 0.8 T_{50}$. For non-FRED-like bursts, we simply assume that $f = 1$ at $T = 0.8 T_{50}$. Figure 3 gives the efficiency for 6 FRED-like bursts and 46 bright, long complex BATSE bursts based on

$$\epsilon = N_N \left[\frac{\Delta T}{2T}\right]^2 \frac{1}{kf} = \frac{N^2}{(\delta N)^2} \frac{\Delta T}{3.2 k T_{50}} \quad , \tag{2}$$

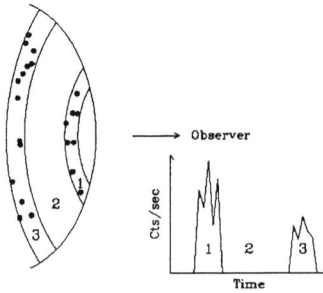

FIGURE 2. Schematic of the relationship between the emission on a shell and the observed time history. The curvature delays the photons from off-axis regions such that at any one time, the observer sees photons from an annulus oriented around the line of sight. The perpendicular size of the shell is $\sim \Gamma T$ whereas a causally connected entity (represented by the dots) is only $\Gamma \Delta T$. Here, T and ΔT are the time in the time history and a typical time scale of variation. Gaps imply that entire causally disconnected regions do not emit (e.g., region 2 produces gap 2 in the time history). The number of entities in each annulus determines the variability of the time history.

where we have used the case of shocks growing at the speed of sound. For complete details see reference [6]. Thus, the spikiness of GRB time histories implies that only $\sim 10^{-3}$ of the surface of a shell becomes active. This is lower than previously estimated [3,8], and implies that models require $\epsilon^{-1} \sim 10^3$ times more energy than previously thought. Of course, reducing the fraction of the sky into which each shell expands can compensate for low efficiency for the small price of requiring a higher density (by ϵ^{-1}) of GRBs in the universe.

A common misconception is that one can just use ISM clouds that cover most of the shell's surface. Each cloud could cause a relatively large peak while efficiently utilizing the area of the shell. This does not work because the curvature of the expanding shell prevents the shell from engaging the cloud instantaneously. Rather, the portion of the shell at $\theta \sim \Gamma^{-1}$ requires $R(1 - \cos\theta)/v$ longer before it reaches the cloud. Even if the cloud happens to have a concave shape such that the shell reaches the cloud simultaneously over a wide range of angles, the resulting photons at $\theta \sim \Gamma^{-1}$ must travel farther to the detector resulting in emission that is delayed by $R(1 - \cos\theta)/c$. Since the speed is weakly dependent on Γ or the ambient material, there is no reason to believe variations in the ambient material could cause the shell to develop into a plane wave oriented towards the observer such that the photons produced by an interaction with an ISM cloud or a shock would arrive as a short flare. Only the instantaneous interaction of two plane parallel surfaces oriented perpendicular to the line of sight can produce a short peak from large surfaces.

These three arguments make a strong case against single, symmetric relativistic shells that undergo variations either due to shocks or interactions with the ISM. There are two alternative explanations. First, one can accept the internal shock

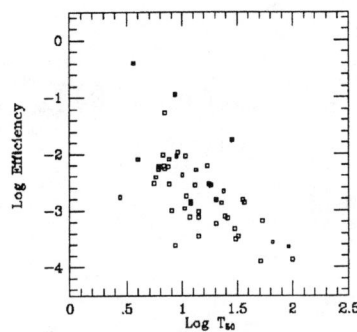

FIGURE 3. Typical values of the fraction of a relativistic shell that becomes active during a GRB as a function of the duration of the emission (T_{50}). The six solid squares are FRED-like BATSE bursts for which direct estimates of the size of the shell can be made. The 46 open squares are long complex BATSE bursts where we estimate the size in a manner similar to the FRED-like estimate. Under most conditions, the efficiency is $\sim 0.1 \Delta T/T$. These low values imply that either only a small fraction of the shell converts its energy into gamma-rays or that GRBs consist of very fine jets with angular sizes much smaller than Γ^{-1}.

models [2]. These models have two weaknesses: there is a concern that internal shocks are rather inefficient, and the long, complex time history of a GRB requires the central site to produce 10^{51} erg s^{-1} for hundreds of seconds. Second, one can retain the quick energy release associated with the single shell but break the spherical symmetry of the shell by having the emitting material confined to fine jets with angular width the order of $\sim \sqrt{\epsilon}\Gamma^{-1}$.

REFERENCES

1. Mészáros, P., and Rees, M. J., *ApJ* **405**, 278 (1993).
2. Rees, M. J., and Mészáros, P., *ApJ* **430**, L93 (1994).
3. Fenimore, E. E., Madras, C. D., and Nayakshin, S., *ApJ* **473**, 998 (1996) astro-ph/9607163.
4. Fenimore, E. E., and Sumner, M. C., *All-Sky X-Ray Observations in the Next Decade*, Eds. M. Matsuoka, N. Kawai, in press, (1997), astro-ph/9705052.
5. Fenimore, E. E., *ApJ*, submitted.
6. Fenimore, E. E., Cooper, C., Ramirez, E., and Sumner, M. C., *ApJ*, to be submitted.
7. Sumner, M. C., and Fenimore, E. E., these proceedings, astro-ph/9712302.
8. Sari, R., and Piran, T., *ApJ*, in press.

The Internal-External GRB-Afterglow Model

Tsvi Piran[1] and Re'em Sari[2]

Racah Institute of Physics, The Hebrew University, Jerusalem 91904, Israel

Abstract. Most current GRB models are based on the conversion, via shocks, of the energy of an ultra-relativistic flow to the observed radiation. We show that *external shocks* cannot produce the highly variable temporal structure observed in most GRBs. *Internal shocks*, on the other hand can produce this structure, provided that the *inner engine* that produce the relativistic flow generates an unsteady and irregular "wind". However, internal shocks can convert only a fraction of the total energy to radiation. Most of the kinetic energy of the flow must be eventually dissipated via external shocks. We suggest that this is the origin of the observed afterglow. According to this *Internal-external* model the GRB and afterglow are not produced by the same process and hence they do not scale directly to each other. Furthermore, under certain conditions the initial phase of the afterglow might radiate in gamma rays and in these cases this signal will overlap as a smooth component with the variable signal produced by the internal shocks, leading to a combined scenario of the GRB itself.

INTRODUCTION

According to the current generic picture [1] GRBs arise from a three stage process: (i) A compact inner engine produces a relativistic energy flow (relativistic particles or electromagnetic Poynting flux – but not the observed photons) which (ii) transport the energy outwards to an optically thin region where (iii) it is converted to the observed radiation. Step (iii) could occur due to external shocks [2] resulting from the deceleration of the energy flow by some external medium (e.g. the ISM). Alternatively, internal shocks [3,4] that would arise in an irregular flow with non-uniform velocities could convert the kinetic energy to radiation. We show here that internal shocks rather than external shocks convert the energy in GRBs. Since internal shocks can convert only a fraction of the energy to gamma rays [5–8], we suggest that the remaining energy produces the afterglow via external shocks. As the afterglow is produced by a different process than the main GRB its properties do not scale directly to the properties of the GRB.

---

[1] tsvi@nikki.fiz.huji.ac.il
[2] sari@nikki.fiz.huji.ac.il

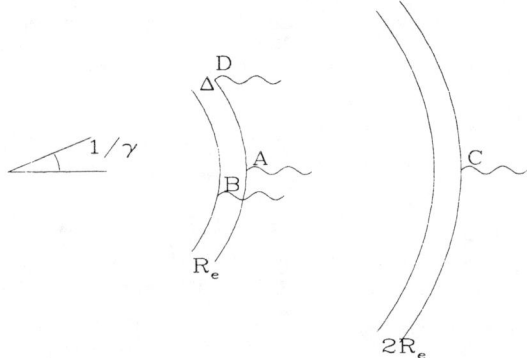

FIGURE 1. *Time Scales in Shocks:* (1) $T_R = T_C - T_A$ (where $T_{A,B,C,D}$ are the arrival times of photons from the events A, B, C, D respectively); (2) $T_{\text{angular}} = T_D - T_A$; (3) $T_\Delta = T_B - T_A$.

TEMPORAL STRUCTURE AND THE ANGULAR SPREADING PROBLEM

Most bursts have a highly variable temporal profile with a rapid variability, on a time scale $\delta T \ll T$, T being the burst's duration. It is useful to define a parameter $\mathcal{N} \equiv T/\delta T \sim 100$ which is a measure of this variability.

Consider a relativistic shell that converts its energy to radiation. Let Δ be the width of the shell and let the conversion take place between R_E and $2R_E$. The emitting material moves with a Lorentz factor γ_e. There are three generic time scales (see Fig. 1).

- The radial time scale, T_R: The difference in arrival time between two photons emitted at R_E and $2R_E$. We have, up to numerical factors of order unity:

$$T_R \approx R_E/\gamma_e^2 c. \tag{1}$$

- The angular time scale, T_{angular}: The difference in arrival time between two photons emitted along the line of sight and at an angle θ from the line of sight. Because of relativistic beaming an observer detects radiation only from an angular scale γ_e^{-1} around the line of sight. Thus, the angular size of the observed region always satisfies $\theta \leq \gamma_e^{-1}$:

$$T_{\text{angular}} \approx R_E \theta^2 / c \leq R_E/\gamma_e^2 c. \tag{2}$$

- The shell crossing time, T_Δ: The light crossing time of the shell corresponds to the time difference between the photons emitted from the shell's front and from its back. This equals to:

$$T_\Delta = \Delta/c. \tag{3}$$

Quite generally a fourth time scale, the cooling time scale, is shorter than all those scales [10,13].

Comparison of Eqs. 1 and 2 reveals that if the system is "sufficiently spherical" (that is if $\theta > \gamma_e^{-1}$) then, due to relativistic beaming, effectively $\theta \approx \gamma_e^{-1}$ and $T_R \approx T_{\text{angular}}$ [1,11,12]. This leads to the angular spreading problem. Blending of emission from regions from an angle γ_e^{-1} from the line of sight leads to smoothing of the signal on a time-scale: $T_{\text{angular}} \approx T_R$. Therefore, unless $T_\Delta > T_R \approx T_{\text{angular}}$ there will be a smooth single peak burst with $\delta T \approx T$. The condition for the angular spreading problem is, therefore,:

$$\Delta < R_E/\gamma_e^2. \tag{4}$$

This condition is always satisfied if the emission is due to external shocks[3].

To produce a variable burst with $\delta T \ll T$, within the external shock scenario, one must break the spherical symmetry on scales smaller than γ_e^{-1}. Detailed analysis [12] shows that the emitting regions must have an angular size smaller than $(\gamma_e \mathcal{N})^{-1} \leq 10^{-4}$ to produce such a variability. A sufficiently narrow jet can satisfy this restriction. But it is not clear how produce such a jet. Hydrodynamic acceleration, for example, cannot produce a jet whose angular width is smaller than γ_e^{-1}. A second possibility is that the emission emerges from numerous small size emitting bubbles. To produce a variable burst the number of bubbles (emitting regions) should be smaller than $\mathcal{N}$, otherwise the contribution from different bubbles will average out to a smooth signal. The maximal solid angle of each bubble is $(\gamma_e \mathcal{N})^{-2}$. Therefor the total solid angle of all bubbles is smaller than $(\gamma_e^2 \mathcal{N})^{-1}$, which is only $\mathcal{N}^{-1}$ of the observed solid angle. This leads to an intrinsic inefficiency in conversion of energy to radiation of magnitude $\mathcal{N}^{-1}$ which is less than 1%.

Let's turn now to a thick shell which we call a "wind". If the wind is irregular, internal shocks could form and those would produce the observed GRB. Internal shocks take place at $R_E \approx \delta \gamma^2$ (where δ is the length scale of variability of the wind – $\delta \leq \Delta$ and γ is the initial Lorentz factor). In this case the condition $T_{\text{angular}} \approx T_R < T_\Delta = \Delta/c$ is always satisfied. This will produce a burst whose overall duration is Δ/c and the observed variability scale is[4] $\delta T = \delta/c \approx T_{\text{angular}} \approx T_R$. The variability scale could be much shorter than the duration. Both the duration and the variability are determined now by the activity of the inner engine and not by the emitting regions. The observed temporal structure reflects the activity of

---

[3] Strictly speaking this was shown only for hydrodynamic shocks [12,9].
[4] This is provided, of course, that the cooling time is shorter than T_{angular} [13].

the inner engine, which must be producing a relatively long and highly irregular wind. Kobayashi, Piran & Sari [6,8] and Mochkovich & Daigne [7] have shown that internal shocks can actually reproduce the temporal structure observed in GRBs.

PREDICTIONS AND OBSERVATIONS

Internal shocks can convert only a fraction of the total energy to radiation [5–8]. A few months before the discovery of the afterglow by BeppoSAX, Sari & Piran [12] pointed out that since at least half of the initial kinetic energy remains in the flow after it has produced a GRB via internal shocks the flow will interact later via an external shock with the surrounding medium. This shock will produce a signal that will follow the GRB – an afterglow. The idea of an afterglow in other wavelengths was suggested earlier [14–16] but it was suggested as a follow up of the, then standard, external shocks scenario. In this case the afterglow would have been a direct continuation of the GRB activity and its properties would have scaled directly to the properties of the GRB. According to internal-external model (internal shocks for the GRB and external shocks for the afterglow) of Sari & Piran [12] the afterglow is not produced by the same mechanisms as the GRB and therefore its features should not be scaled directly to the properties of the GRB. This was in fact the case in the recent afterglow observations. In all models of external shocks the observed time $t \propto R/\gamma_e^2$ and the typical frequency $\nu \propto \gamma_e^4$. Since most of the emission takes place at practically the same radius and all that we see is the variation of the Lorentz factor we expect quite generally [17]:

$$\nu \propto t^{2\pm\epsilon}. \tag{5}$$

The small parameter ϵ reflects the variation of the radius and it depends on the specific assumptions made in the model. We would expect that $t_X/t_\gamma \sim 50$ and $t_{opt}/t_\gamma \sim 300$. The observations of GRB970508 show that $(t_{opt}/t_\gamma)_{observed} \approx 10^4$. This is in a clear disagreement with the single external shock model for both the GRB and the afterglow while it is in agreement with the internal-external model.

CONCLUSIONS AND FURTHER PREDICTIONS

We have seen that external shocks cannot produce variable GRBs (or at least they cannot do this efficiently). The only viable alternative is production via internal shocks, which can produce the observed variable structure. This has several implications. First, since internal shocks can convert only a fraction of the kinetic energy to radiation, a significant amount of energy remains and this could produce later afterglow via external shocks. Since the afterglow is produced by external shocks it must be smooth. A clear prediction of this model is that there should be no rapid temporal variability in the afterglow.

The initial external shock signal might also be in the gamma-ray range, depending on various parameters of the model. In fact the observations of BeppoSAX suggested that in GRB970228 the initial external shock signal was in the hard X-ray – soft gamma-ray range. In this case it is possible that this initial phase will overlap the gamma-ray activity of the internal shocks [18], adding a smooth component that overlaps the variable component produced by the internal shocks. This smooth component will vary from hard to soft, as is generally the case in external shocks (see Eq. 5), on a time scale comparable to the time scale of its emission. This may explain the softening generally observed in GRBs.

Finally we point out that the internal shock scenario restricts the inner engines that could power the relativistic flow. The observed temporal structure in this scenario reflects directly the activity of the inner engine [6,8]. This engine must operate for a long duration, up to hundreds of seconds in some cases, and it must produce a highly variable wind to form internal shocks. This directly rules out all explosive inner engine models and severely constrains most others.

ACKNOWLEDGMENTS

We thank Ramesh Narayan and Jonathan Katz for helpful discussions. The research was supported by a US-ISRAEL BSF grant 95-328 and by NASA grant NAG5-3516.

REFERENCES

1. Piran, T., in *Some Unsolved Problems in Astrophysics*, Eds. Bahcall, J. N. and Ostriker, J. P., Princeton University Press, 1997.
2. Mészáros, P., & Rees, M. J., *MNRAS* **258**, 41p (1992).
3. Narayan, R., Paczyński, B., & Piran, T., *Ap. J.* **395**, L83 (1992).
4. Rees, M. J., & Mészáros, P., *Ap. J.* **430**, L93 (1994).
5. Mochkovitch, R., Maitia, V., & Marques, R., in *Towards the Source of Gamma-Ray Bursts, Proceeding of 29th ESLAB Symposium*, eds. Bennett, K. & Winkler, C., 531 (1995).
6. Kobayashi, S., Piran, T., & Sari, R., *Ap. J.* **490**, 92 (1997).
7. Mochkovitch, R, & Daigne F., these proceedings.
8. Kobayashi, S., Piran, T., & Sari, R., these proceedings.
9. Sari, R., & Piran, T., *Ap. J.* **455**, L143 (1995).
10. Sari, R., Narayan, R., & Piran, T., *Ap. J.* **473**, 204 (1996).
11. Fenimore, E. E., Madras, C., & Nayakshin, S., *Ap. J.* **473**, 998 (1996).
12. Sari, R., & Piran, T., *Ap. J.* **485**, 270 (1997).
13. Sari, R., & Piran, T., *MNRAS* **287**, 110 (1997).
14. Paczyński, B. and Rhodas, J., *Ap. J. Lett.* **418**, L5 (1993).
15. Katz, J. I., *Ap. J.* **422**, 248 (1994).
16. Mészáros, P., & Rees, M. J., *Ap. J.* **476**, 232 (1997).
17. Katz, J. I., & Piran, T., *Ap. J.* **490**, 772 (1997).
18. Sari. R., *Ap. J. Lett.* **489**, L37 (1997).

Internal Shock Models for Gamma-Ray Bursts: Temporal and Spectral Properties

Robert Mochkovitch and Frédéric Daigne

Institut d'Astrophysique de Paris
98 bis, Bd Arago 75014 Paris, France

Abstract. We follow the evolution of a relativistic wind when the initial distribution of the Lorentz factor is highly non-uniform. We use an approach where the wind is modelled by a large number of layers which can collide and form internal shocks. We suppose that the magnetic field and the electron Lorentz factor reach equipartition values in these shocks. Synchrotron photons emitted by the relativistic electrons have typical energies in the gamma-ray range in the observer frame. Synthetic bursts are constructed as the sum of the contributions from many elementary internal shocks and their temporal and spectral properties are compared to the observations.

INTRODUCTION

Some progress have been made recently in the theoretical understanding of gamma-ray bursts in parallel to the observation of X-ray, optical and radio afterglows which have confirmed that bursts are indeed located at cosmological distances. One of the most popular model involves internal shocks in a relativistic wind [1] produced after the coalescence of two neutron stars [2] or the collapse of a massive star [3,4]. Only internal shocks can explain the complex temporal profiles seen in most observed bursts [5] and external shocks are more likely responsible for the afterglow. We present in this work the results of a simple model for the evolution of a relativistic wind with a non-uniform distribution of the Lorentz factor. This model allows to compute both the temporal and spectral properties of bursts produced by internal shocks.

SIMPLE MODEL OF THE RELATIVISTIC WIND

We study the evolution of the relativistic wind using a very simple method where a succession of layers are emitted every 2 ms with a varying Lorentz factor, during a total time t_w. We follow the layers as the wind expands and when a rapid one (of mass m_1 and Lorentz factor Γ_1) catches up with a slower one (m_2, $\Gamma_2 < \Gamma_1$) they collide and merge to form a single shell of resulting Lorentz factor Γ_r. If the dissipated energy

$$e = m_1 c^2 \Gamma_1 + m_2 c^2 \Gamma_2 - (m_1 + m_2) c^2 \Gamma_r \,, \tag{1}$$

is radiated in the gamma-ray range on a time scale shorter than the shell expansion time, the burst profile will be made by a succession of elementary contributions of duration (in the observer frame)

$$\Delta t = \frac{r}{2c\Gamma_r^2} \,, \tag{2}$$

r being the shell radius. We estimate Γ_r by considering that most of the energy available in a collision is already released when the less massive of the two layers has swept up a mass comparable to its own mass in the other layer. Then

$$\Gamma_r \simeq \sqrt{\Gamma_1 \Gamma_2} \,, \quad e = mc^2(\Gamma_1 + \Gamma_2 - 2\Gamma_r), \quad \text{with} \quad m = \min(m_1, m_2) \,. \tag{3}$$

The arrival time of each of the elementary contributions from internal shocks is calculated relative to a signal which would have travelled at the speed of light from the source to the observer. It is given by

$$t_a = t_e - \frac{r}{c} \,, \tag{4}$$

where t_e is the emission time. The evolution of the system is followed until all the layers are ordered with Γ decreasing from the front to the back of the wind. The efficiency of the dissipation process is then obtained as

$$f_d = \frac{\sum_s e_s}{\sum_i m_i c^2 \Gamma_i} \,, \tag{5}$$

where the e_s are the energies released in the internal shocks and the m_i, Γ_i are the initial masses and Lorentz factors of the layers.

This very simple approach is naturally very crude because it neglects all pressure waves propagating throughout the wind. Nevertheless, we expect that it can reproduce the basic behavior of the real phenomenon.

EMISSION PROCESSES

The process by which the dissipated energy is finally radiated depends on the energy distribution of protons and electrons in the shocked material and on the values of the comoving density and magnetic field. The average energy which is dissipated per proton in a shock between two layers of equal mass is

$$\epsilon = (\Gamma_{\text{int}} - 1) m_p c^2 \,, \tag{6}$$

where

$$\Gamma_{\text{int}} = \frac{1}{2}\left[\left(\frac{\Gamma_1}{\Gamma_2}\right)^{1/2} + \left(\frac{\Gamma_2}{\Gamma_1}\right)^{1/2}\right], \qquad (7)$$

is the Lorentz factor for internal motions in the shocked material. We then compute the characteristic Lorentz factor Γ_e for the electrons with the expression proposed by Bykov & Mészáros [6] who have considered acceleration by turbulent magnetic field fluctuations. They get

$$\Gamma_e \sim \left[\left(\frac{\alpha_M}{\zeta}\right)\left(\frac{\epsilon}{m_e c^2}\right)\right]^{1/(3-\mu)}, \qquad (8)$$

where α_M is the fraction of the dissipated energy which goes into magnetic fluctuations, ζ the fraction of the electrons which are accelerated and μ the index of the fluctuation spectrum. With $1.5 \leq \mu \leq 2$, $\alpha_M = 0.1 - 1$, $\zeta \sim 10^{-3}$ and $\epsilon/m_e c^2 \sim 500$ (corresponding to $\Gamma_1/\Gamma_2 = 4$) values of Γ_e in the range $10^3 - 10^4$ can be obtained.

The equipartition magnetic field is given by

$$B_{\text{eq}} \sim (8\pi\alpha_M n\epsilon)^{1/2}, \qquad (9)$$

where n is the comoving proton number density

$$n \sim \frac{\dot{E}}{4\pi r^2 \bar{\Gamma}^2 m_p c^3}. \qquad (10)$$

In Eq.(10) $\dot{E}$ is the power injected in the wind and $\bar{\Gamma}$ its average Lorentz factor. Assuming $\dot{E} = 10^{52}$ erg s^{-1}, $\bar{\Gamma} = 300$, and $\alpha_M = 1/3$, the equipartition magnetic field at a radius $r \sim ct_{\text{var}}\bar{\Gamma}^2$ where most of the collisions take place is $B_{\text{eq}} \sim (10^2 - 10^3)$ G for $t_{\text{var}} = 1$ s (t_{var} being the characteristic time scale for the variations of the Lorentz factor). From the values of Γ_e, B_{eq} and $\bar{\Gamma}$ the typical synchrotron energy can be calculated

$$E_{\text{syn}} = 500 \left(\frac{\Gamma_r}{300}\right)\left(\frac{B}{1000\text{G}}\right)\left(\frac{\Gamma_e}{10^4}\right)^2 \text{ keV}. \qquad (11)$$

and the synchrotron spectrum radiated by the electrons is therefore

$$\frac{dn(E)}{dE} \propto \frac{e}{E_{\text{syn}}}\left(\frac{E}{E_{\text{syn}}}\right)^{-x}, \qquad (12)$$

with $x = 2/3$ for $E < E_{\text{syn}}$ and $2 < x < 3$ for $E > E_{\text{syn}}$.

RESULTS AND DISCUSSION

For a given initial distribution of the Lorentz factor in the wind it is now possible, using the simple model decribed above, to obtain both the temporal profiles and spectra of synthetic bursts. Figure 1 shows a typical example where the initial distribution of the Lorentz factor is made of a rapid component (with an average Γ equal to 400 to which rapid fluctuations of maximum amplitude ± 20 %) and of some slower layers (with Γ between 100 and 300). The total mass in the slow layers has to be comparable to the mass in the rapid component in order to keep the efficiency at a reasonable level. It can be seen that the synthetic profile is made of a series of pulses which show the characteristic FRED shape found in observed bursts [7].

The overall spectrum is the sum of all the elementary contributions (12) from internal shocks. The spectrum corresponding to the profile of Fig. 1 is represented in Fig. 2. In the four BATSE bands, between 20 keV and a few MeV, it can be well fitted with Band's formula [8] with the parameters $\alpha = -1.33$, $\beta = -2.31$ and $E_0 = 544$ keV.

The general spectral properties of our synthetic bursts are also in good agreement with the observations (see [9] for details). Short bursts are found to be harder than long bursts, the transition in the duration-hardness relation occurring for $t_{90} \sim$ a few seconds. This is a natural consequence of the model since in short bursts dissipation takes place closer to the source where the magnetic field and the synchrotron energy are larger. Hardness and count rate are correlated with the hardness usually preceding the count rate. Finally, pulses become narrower at high energy, their (half maximum) width following a relation $W(E) \propto E^{-p}$. The value of p is close to 0.4 for pulses of 2 – 10 s but decreases to 0.2 for shorter pulses in the range 0.1 – 1 s.

The comparison of the temporal and spectral properties of our synthetic bursts to the observations therefore appears quite encouraging. A potential problem is how-

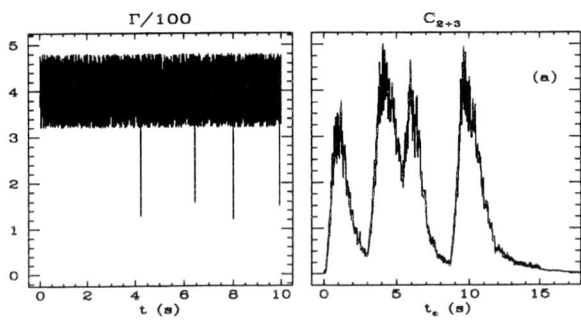

FIGURE 1. Example of burst profile with four intensity pulses two of which partially overlap.

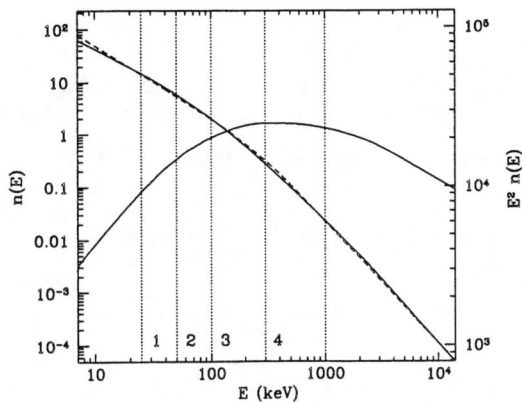

FIGURE 2. Spectrum of the burst corresponding to the profile of Fig.1. The number of photons per energy interval $n(E)$ and the product $E^2 n(E)$ are shown in arbitrary units. The dashed line is a fit of the spectrum with Band's formula in the interval 10 keV – 10 MeV. The product $E^2 n(E)$ is maximum at the peak energy $E_\mathrm{p} = 365$ keV.

ever the rather low efficiency for the conversion of wind kinetic energy to gamma-rays. The value of f_d (Eq. 5) is typically of the order of 10% which has to be multiplied by the fraction of the dissipated energy which is finally radiated by the electrons. This leads to a global efficiency which does not exceed a few percent. Since the total energy available by accretion after the coalescence of two neutron stars is probably less than 10^{53} erg [2], beaming (or more energetic sources such as hypernovae [4]) will be necessary to account for cosmological gamma-ray bursts.

REFERENCES

1. Rees, M.J., and Mészáros, P., *ApJ* **430**, L93 (1994).
2. Ruffert, M., Janka, H.T., and Schaefer, G., *A&A* **319**, 122 (1997).
3. Woosley, S.E., *ApJ* **405**, 273 (1993).
4. Paczyński, B., preprint astro-ph/9706232 (1997).
5. Sari, R., and Piran, T., *MNRAS* **287**, 110 (1997).
6. Bykov, A., and Mészáros, P., *ApJ* **461**, L37 (1996).
7. Norris, J.P., et al., *ApJ* **459**, 393 (1996).
8. Band, D., et al., *ApJ* **413**, 281 (1993).
9. Daigne, F., and Mochkovitch, R., to appear in *MNRAS*.

Can Internal Shocks Produce the Variability in GRBs?

Shiho Kobayashi[1], Tsvi Piran[2] and Re'em Sari[3]

Racah Institute of Physics, Hebrew University, Jerusalem 91904, Israel

Abstract. We discuss the possibility that gamma-ray bursts result from internal shocks in ultra-relativistic matter. Using a simple model we calculate the temporal structure and we estimate the efficiency of this process. In this model the ultra-relativistic matter flow is represented by a succession of shells with random values of the Lorentz factor. We calculate the shocks that take place between those shells and we estimate the resulting emission. Internal shocks can produce the highly variable temporal structure observed in most of the bursts provided that the source emitting the relativistic flow is highly variable. The observed peaks are in almost one-to-one correlation to the activity of the emitting source. A large fraction of the kinetic energy is converted to radiation. The most efficient case is when an inner engine produces shells with comparable energy but very different Lorentz factors. It also gives the most preferable temporal structure.

INTRODUCTION

The only known mechanism for producing cosmological GRB is decelerations of ultra-relativistic matter having a Lorentz factor $\gamma \geq 100$ [3]. The kinetic energy of the ultra-relativistic matter is converted into internal energy by relativistic shocks. These shocks can be due to the interstellar medium (ISM), "external shocks", or inside the shell itself due to non-uniform velocity, "internal shocks". Electrons are heated by the shocks and the internal energy is then radiated via synchrotron emission and inverse Compton scattering.

Most of bursts have a highly variable temporal profile with a time scale of variability significantly shorter than the overall duration. Clearly any GRB model must be able to explain this complex temporal structure. Two of us have shown that the efficiency of the external shocks process is less than 1% in order to produce the time variability of GRB [5]. In view of the difficulties within the external shock scenario it is worthwhile to consider the question whether internal shocks could

---

[1] shiho@astro.huji.ac.il
[2] tsvi@shemesh.fiz.huji.ac.il
[3] sari@shemesh.fiz.huji.ac.il

actually produce the temporal structure observed in GRBs and what would be the efficiency of this process.

TWO SHELL INTERACTION

Internal shocks arise in a relativistic wind with a non-uniform Lorentz factor and convert a portion of the kinetic energy to a radiation. We represent the irregular wind by a succession of relativistic shells [2]. A collision of two shells is the elementary process in our model. A rapid shell (denoted by the subscript r) catches up a slower one (s) and the two merge to form a single one (m). Using conservation of energy and momentum we calculate the Lorentz factor of the merged shell to be

$$\gamma_m \simeq \sqrt{\frac{m_r \gamma_r + m_s \gamma_s}{m_r/\gamma_r + m_s/\gamma_s}}, \qquad (1)$$

where $\gamma_i (\gg 1)$ and m_i are the Lorentz factors and the masses of these shells. The internal energy of the merged shell is the difference of kinetic energy before and after the collision:

$$E_{int} = m_r c^2 (\gamma_r - \gamma_m) + m_s c^2 (\gamma_s - \gamma_m). \qquad (2)$$

The emitted radiation will be observed as a pulse with a width δT. Three time scales, the cooling time, the hydrodynamic time, and the angular spreading time determine δT. In most of cases, the cooling time scale of electrons is much shorter than the hydrodynamic time scale, so we can neglect the cooling time. The hydrodynamic time scale is the time that the shock crosses the shell. For simplicity, we estimate the emission time scale by the time that the reverse shock crosses the rapid shell. If both shells have Lorentz factor of order γ, this observed time scale is of order l_r/c where l_r is the width of the rapid shell. If the collision of the shells takes place at a large radius R, angular spreading effects the width of the pulse. The separation of the shells is L, the effect of angular spreading on the pulse width is $\sim L/c$. If the separation L is larger than the width of the shells l, the pulse width δT is determined by angular spreading. The shape of the pulse become asymmetric with a fast rise and a slower decline which GRBs typically show.

MULTIPLE SHELL MODEL

For multiple shells, numerous collisions take place. Each collision produces a pulse. To construct the temporal structure in a given model we calculate the time sequence of the two shell collisions until there are no more collisions, i.e., until all the shells have merged to form a single shell or until the shells are ordered with increasing values of the Lorentz factor. We superimpose the resulting pulses from each collision.

In order to reduce the number of parameters describing the wind we adopt the following simplifications. We assume a constant initial width l and a constant initial separation L. In other words, we assume that the "inner engine" operates for a fixed period l/c and then it is quiet for a fixed period L/c.

We assume that the Lorentz factor of each shell is uniformly distributed between $\gamma_{\min}$ and $\gamma_{\max}$. For the distribution of the density we have two kinds of models. In the first kind we consider models in which the density, the mass or the energy of each shell is a random variable which is uniformly distributed between 1 and some maximal value $X_{\max}$. In the second kind the density is correlated with the Lorentz factor as $\rho_i \propto \gamma_i^{\eta-1}$ so that the mass and kinetic energy of each shell are proportional to γ_i^{η} and $\gamma_i^{\eta+1}$ respectively. For $\eta = 1$ the shells have equal density, for $\eta = 0$ the shells have equal mass and for $\eta = -1$ the shells have equal energy. The random density/mass/energy models give almost the same efficiency as the corresponding constant model, and the dependence on $X_{\max}$ is low. We therefore present the result for the correlation models which are more simple.

TEMPORAL STRUCTURE

The calculated temporal structure is a superposition of the pulses from the elementary two shell collisions. Two typical temporal profiles are presented in Figs. 1a, b. These depict the luminosity for a model with $N = 100$ and $L/l = 5$ as a function of the observed time for shells with a constant energy (Fig. 1a) or a constant density (Fig. 1b). During the evolution of the N shells, almost N collisions happen and almost N pulses are produced. For the constant energy shells, the internal energy of the merged shell E_{int} takes almost the same value for most of the collisions so most of the peaks have comparable amplitude (see Fig. 1a). Thus, we practically observe the "shell structure" produced by the inner source as almost all two shell collisions produce an observable peak. If on the other hand for the constant density shells (see Fig. 1b) a few collisions produce significantly higher peaks than others and only those are observed. Thus the number of the observed peaks is much less than N, the observed temporal structure corresponds only weakly to the activity of the inner source. Visual inspection suggests that

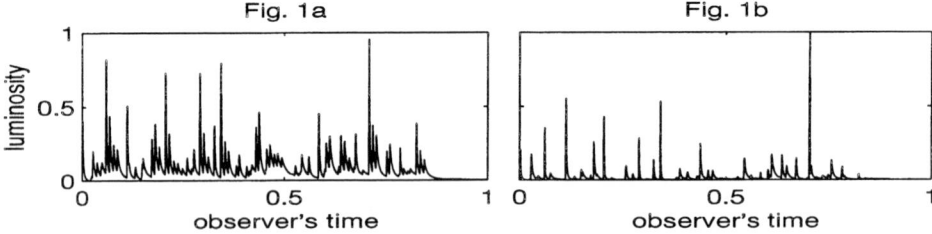

FIGURE 1. The luminosity versus the observer's time for different models

TABLE 1. Efficiency with standard deviation for different models

N	η	γ_{min}	γ_{max}	efficiency [%]
100	1	20	1000	10.9 ± 1.4
100	1	50	1000	10.0 ± 1.3
100	1	100	1000	8.5 ± 1.1
100	1	200	1000	6.1 ± 0.7
100	1	500	1000	1.7 ± 0.2
100	0	20	1000	19.2 ± 2.9
100	0	50	1000	16.0 ± 2.3
100	0	100	1000	12.4 ± 1.6
100	0	200	1000	7.7 ± 0.9
100	0	500	1000	1.8 ± 0.2
100	-1	20	1000	25.0 ± 3.5
100	-1	50	1000	19.5 ± 2.4
100	-1	100	1000	14.1 ± 1.6
100	-1	200	1000	8.2 ± 0.9
100	-1	500	1000	1.8 ± 0.2
20	1	100	1000	7.6 ± 2.5
20	-1	100	1000	11.0 ± 3.3
1000	1	100	1000	8.7 ± 0.3
1000	-1	100	1000	15.1 ± 0.5
1000	-1	10	10000	39.3 ± 2.6

the temporal structure produced by the constant energy shells resembles better the observed temporal structure.

EFFICIENCY

We have shown that internal shocks could produce the highly variable temporal structure. We ask now what is the efficiency of this process? The relativistic shells collide with each other and merge into more massive shells. The overall efficiency of conversion of kinetic energy in internal shocks can be calculated from the initial and final kinetic energies as:

$$\epsilon = 1 - \frac{\Sigma m_i^{(f)} \gamma_i^{(f)}}{\Sigma m_i^{(i)} \gamma_i^{(i)}} \qquad (3)$$

where the superscript (f) and (i) represent the initial and final values, respectively. For each choice of the parameters of the model, we have evaluated the efficiency for 100 realizations. The mean efficiency and its standard deviation are listed in Table 1.

The efficiency is only a few percent if the spread in the Lorentz factor is relatively low (a factor of 2-3). This agrees with results of Mochkovitch, Maitia and Marques

[2] who concluded that the efficiency of this process is less than 10%. However, higher efficiency could be reached if the spread in γ is larger. The spread required, for example, for 10% is $\gamma_{max}/\gamma_{min} \sim 6$ for $N = 100$ and $\eta = -1$ (see ref. [1] for a full detail of our model).

CONCLUSIONS

Using a simple model, we have shown that internal shocks can produce the highly variable profile observed in most GRBs. The temporal structure does not reflect the exact activity of the source, but it gives the time scale of the inner engine.

We have shown that the number of peaks is almost the same as the number of shells that the inner engine emitted. The separation between the peaks corresponds to the duration for which the inner engine was quiet. The variability of peaks height can tell as whether all shells have comparable energy (low variability in observed peak heights) or not.

The efficiency of this process is low (less than 2%) if the initial spread in γ is only a factor of two. However the efficiency could be much higher. The most efficient case is when the inner engine produces shells with comparable energy but with very different Lorentz factors. In this case ($\eta = -1$, and spread of Lorentz factor $\gamma_{max}/\gamma_{min} > 10^3$) the efficiency is as high as 40%. For a moderate spread of Lorentz factor $\gamma_{max}/\gamma_{min} = 10$, with $\eta = -1$, the efficiency is 20%.

ACKNOWLEDGEMENTS

We thank J.I.Katz for many useful discussions. S.K. gratefully acknowledges the support by the Golda Meir Postdoc fellowship. S.K. also thanks a grant from the Israeli Ministry of Science. This work was supported in part by a US-Israel BSF grant and by a NASA grant.

REFERENCES

1. Kobayashi, S., Piran, T., and Sari, R., *Ap. J.*, in press (1997).
2. Mochkovitch, R., Maitia, V., and Marques, R., *in Towards the Source of Gamma-Ray Bursts, Proceeding of 29th ESLAB Symposium*, eds. Bennett, K. and Winkler, C., TIDC Reproduction, 1995, 531.
3. Piran T., *in Unsolved Problems in Astrophysics*, eds. Bahcall. J. and Ostriker J.P., Princeton University Press, 1996, ch.18, 343.
4. Sari, R. and Piran T., *Ap. J.* **455**, L143 (1995).
5. Sari, R. and Piran T., *Ap. J.* **485**, 270 (1997).

Hydrodynamical Study of Internal Shocks in a Relativistic Wind

Frédéric Daigne and Robert Mochkovitch

Institut d'Astrophysique de Paris
98 bis, Boulevard Arago 75014 Paris France

Abstract. In the context of internal shock models for GRBs a 1D lagrangian hydrocode has been developed to follow the evolution of a relativistic wind with a very inhomogeneous distribution of the Lorentz factor. We suppose that the energy dissipated in the internal shocks which form in the wind is efficiently radiated so that the shocks can be considered as quasi-isothermal. We present the first results obtained with this code. They are compared to a simple model where all pressure waves are suppressed in the wind so that layers with different velocities only interact by direct collisions.

INTRODUCTION

Many models for cosmological GRBs involve a relativistic wind which dissipates a fraction of its kinetic energy in the interaction with the interstellar medium (external shock models) or in shocks taking place within the wind (internal shock models). The complex temporal profiles of most observed bursts can be explained in internal shock models where they reflect the activity of the central engine which is responsible for the initial very non-uniform distribution of the Lorentz factor in the wind, whereas it is very difficult to have both a complex temporal profile and a high efficiency of the emission process in external shock models [1].

To study internal shocks we have first used a very simple method where the wind is made of a large number of individual layers with different Lorentz factors which interact by direct collisions only (all pressure waves are neglected). The typical energy dissipated per proton in such shocks reaches about 100 MeV. Assuming equipartition between the magnetic field and the populations of protons and accelerated electrons leads to typical values for $B \sim 10^3 - 10^4$ G and $\Gamma_e \sim 10^4$ [2]. The dissipated energy is then radiated on a synchrotron time scale which is much shorter than the shell expansion time.

The temporal and spectral properties of synthetic bursts obtained in this way are in reasonable agreement with the observations [3,4] but our very crude description of the dynamics of the wind can naturally be questioned. We have therefore begun to develop a relativistic 1D lagrangian hydrocode with the aim of comparing the results of detailed hydrodynamical calculations to our simple model. We do not discuss the details of the emission processes behind the shocks and we simply assume that the dissipated energy is radiated sufficiently fast to consider that the shocks are quasi-isothermal.

We present below the equations of the relativistic flow and the method we have used to solve the Riemann problem in the isothermal case. The first results of our code are then compared to the simple, pressureless model.

RELATIVISTIC LAGRANGIAN HYDRODYNAMICS – RIEMANN PROBLEM

Lagrangian equations of relativistic hydrodynamics: we write the equations of mass, momentum and energy conservation in a fixed frame ("the observer frame") in spherical symmetry

$$\begin{cases} \frac{\partial(\frac{1}{D})}{\partial t} - \frac{\partial(x^2 v)}{\partial \xi} = 0 , \\ \frac{\partial S}{\partial t} + x^2 \frac{\partial P}{\partial \xi} = \begin{cases} 0 & \text{if adiabatic} \\ -vL & \text{with radiative losses} \end{cases} , \\ \frac{\partial E}{\partial t} + \frac{\partial(x^2 P v)}{\partial \xi} = \begin{cases} 0 & \text{if adiabatic} \\ -L & \text{with radiative losses} \end{cases} . \end{cases} \quad (1)$$

The following quantities appear in the system (1): P is the pressure in the fluid rest frame, v is the fluid velocity in the observer frame and D, S and E are the mass, the specific momentum and the specific energy (including mass energy) densities in the observer frame. D, S and E can be related to quantities in the fluid rest frame: $D = \rho\Gamma$, $S = h\Gamma v$ and $E = h\Gamma - \frac{P}{\Gamma\rho}$, where $\Gamma = \frac{1}{\sqrt{1-v^2}}$ is the Lorentz factor, ρ the rest-mass density and $h = 1 + \epsilon + \frac{P}{\rho}$ is the specific enthalpy density (ϵ being the specific internal energy density) ; $L = -\frac{d\epsilon}{d\tau}$ is the power lost by radiation (per unit mass) in the fluid rest frame (τ is the proper time) ; $\xi = \int_{x_0}^{x} Dx^2 dx$ is the lagrangian mass coordinate. We complete the system (1) by the equation of state $P = (\gamma - 1)\rho\epsilon$, γ being a constant.

The Riemann problem: both in the newtonian and the relativistic cases, an initial configuration consisting in a discontinuity separating two constant states L and R gives rise to three elementary nonlinear waves, two of which being either shocks or rarefaction waves propagating towards L and R from two new states L_* and R_* and the third one being a contact discontinuity separating L_* and R_*. The common values of the pressure p_* and the velocity v_* of the two intermediate state L_* and R_* are determined implicitly by

$$v_* = v_{L_*}(p_*) = v_{R_*}(p_*) \text{ with } \begin{cases} v_{S_*}(p_* \leq p_S) = \mathcal{R}_S(p_*) & \text{(rarefaction)} \\ v_{S_*}(p_* > p_S) = \mathcal{S}_S(p_*) & \text{(shock)} \end{cases} , \quad (2)$$

$\mathcal{S}_S(p_*)$ (resp. $\mathcal{R}_S$) giving the velocity of the state S_* connected to S by a shock (resp. a rarefaction wave) as a function of the physical quantities defining S.

For an adiabatic evolution, the expressions of $\mathcal{S}_S$ and $\mathcal{R}_S$ have been given by Martí and Müller [5]: in particular, $\mathcal{S}_S$ is derived from the relativistic Rankine-Hugoniot relations. In our code, we modify $\mathcal{S}_S$ considering that the post-shock state S_{**} (connected to S by a discontinuity following the Rankine-Hugoniot conditions) is radiating very efficiently so that it reaches on a short length scale a new state S_* whose temperature equals the pre-shock value. In the shock frame the stationary flow of the shocked fluid is then governed by

$$\rho \Gamma' v' = C^{ste} = -j, \text{ (constant mass flux accross the shock)}, \qquad (3)$$

$$\begin{cases} \frac{\partial(\rho h \Gamma'^2 v'^2 + P)}{\partial x'} = -\rho \Gamma' v' L \\ \frac{\partial(\rho h \Gamma'^2 v')}{\partial x'} = -\rho \Gamma' L \end{cases} \text{ leading to } \Gamma'^{\frac{\gamma}{\gamma-1}} P \pm j \int \frac{\Gamma'^{\frac{1}{\gamma-1}}}{\sqrt{\Gamma'^2 - 1}} d\Gamma' = C^{ste}, \qquad (4)$$

(for $\gamma = \frac{4}{3}$, (4) gives $\Gamma'^4 P \pm j \left(\sqrt{\Gamma'^2 - 1} + \frac{1}{3}(\sqrt{\Gamma'^2 - 1})^3 \right) = C^{ste}$) and

$$T_* = T_S \iff P_{S*}/\rho_{S*} = P_S/\rho_S \text{ ("isothermal" condition)}. \qquad (5)$$

These three new equations connecting S_* and S_{**} allow us to express v_* as a function of p_* and S only, just like in the adiabatic case, so that the procedure to solve the Riemann problem remains unchanged.

RESULTS AND DISCUSSION

The code we have developed is an extension of the piecewise parabolic method (PPM) of Colella and Woodward [6] for 1D lagrangian problems in relativistic hydrodynamics. Such an extension in the eulerian (and planar) case has already been presented by Martí and Müller [7] and we have used a very similar approach. We have successfully tested our code against several classical problems of numerical relativistic hydrodynamics, both in the adiabatic and radiative cases.

The results of the hydrocode have been compared to the simple model (without pressure waves) in Fig. 1. The adopted initial distribution of the Lorentz factor is a step function where a discontinuity separates a slow part ($\Gamma_1 = 10$) emitted during the first 20% of the total duration ($\tau = 10$ s) and a rapid part ($\Gamma_2 = 40$) emitted during the remaining 80%. The energy injection rate is constant so that the masses in the slow and rapid parts are equal. This initial distribution of Γ leads to the formation of a forward (FS) and a reverse shock (RS) which propagate from the original discontinuity until they reach the limits of the wind. The adopted values for the Lorentz factors are too small for a realistic burst model (an average Γ of a few 10^2 is required to avoid photon-photon annihilation along the line of sight) but the problem can then be handled in a reasonable computing time (improvements to

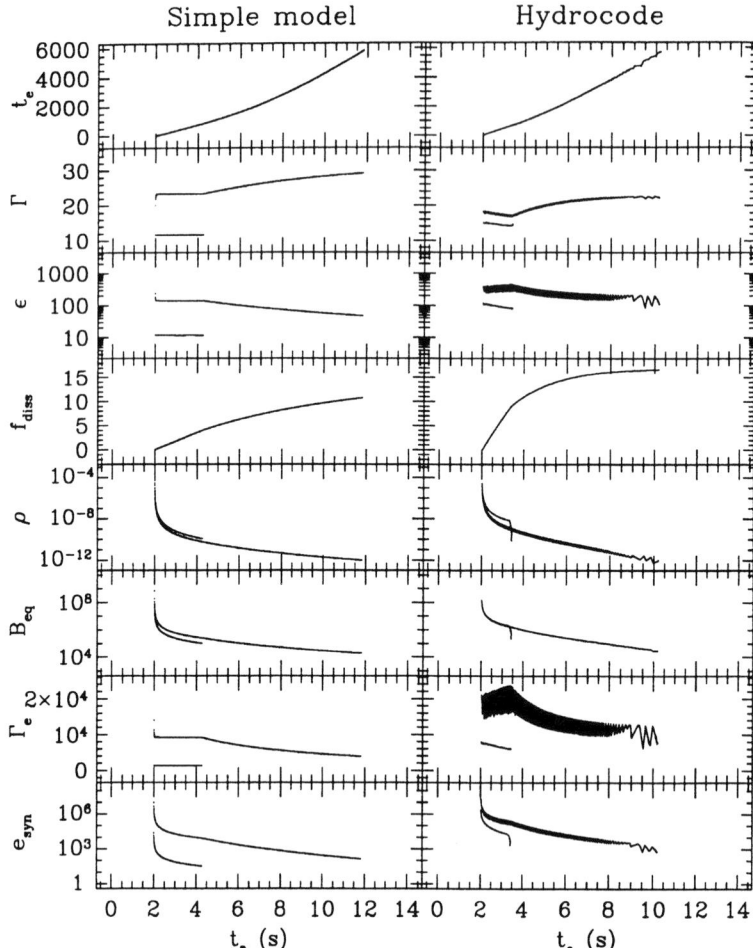

FIGURE 1. Wind parameters obtained both with the simple model (left column) and the hydrocode (right column). As a function of the arrival time t_a of the signal to the observer we have represented: (1) the corresponding emission time t_e in a fixed frame (in s), (2) the Lorentz factor Γ behind the shocks (with two branches for the FS and RS), (3) the dissipated energy ϵ (in MeV per proton), (4) the percentage f_{diss} of the total energy which has been dissipated, (5) the comoving density ρ (in g cm^{-3}) behind the shocks, (6) the equipartition magnetic field $B_{\text{eq}} \sim (8\pi\alpha_B \frac{\rho}{m_p}\epsilon)^{1/2}$ with $\alpha_B = 1/3$ (in G), (7) the typical Lorentz factor of the electrons calculated with the Bykov and Mészáros [2] formula $\Gamma_e \sim \left[\left(\frac{\alpha_M}{\zeta}\right)\left(\frac{\epsilon}{m_e c^2}\right)\right]^{1/(3-\mu)}$ with α_M (fraction of the dissipated energy which goes into magnetic fluctuations) $=1/3$, ζ (fraction of the electrons which are accelerated) $=10^{-3}$ and μ (index of the fluctuation spectrum) $=1.75$ and (8) the synchrotron energy e_{syn} in the observer frame (in keV).

the code should soon allow to consider values of Γ one order of magnitude larger), so that a comparison of the detailed calculations to the results of the simple model can already be made to estimate the errors caused by the suppression of all pressure waves in the wind. We will obtain however excessive values of $e_{\rm syn}$ as a consequence of these too-small Lorentz factors: they lead to an early dissipation in a region where the density and equipartition magnetic field are larger.

The results of Fig. 1 show an overall similarity between the detailed calculations and the simple model in spite of the very crude approximations which have been made. The qualitative dynamical behavior of the wind is comparable and only quantitative differences can be noted. The Lorentz factor behind the shocks is slightly smaller in the detailed calculations and the evolution is 20% shorter. The total efficiency reaches 16% while it is only 11% in the simple model. The dissipated energy per proton is larger, especially in the FS where it is close to 100 MeV instead of 10 MeV. The post-shock density is also larger because the compression behind the shock is not properly taken into account in the simple model. These differences are amplified in $e_{\rm syn}$ (since $e_{\rm syn} \propto \rho^{1/2}\epsilon^{2.1}$) which is typically a factor of 10 larger in the RS and up to 300 times larger in the FS. It should be remembered however that $e_{\rm syn}$ is a by-product of the calculation contrary to other quantities which appear in Fig.1 such as t_e, Γ, ϵ or ρ which are entirely determined by the hydrodynamics of the wind. The synchrotron energy depends on parameters (ζ, α_B) which are still poorly constrained. It is then possible (even if it is not satisfactory) to adjust these parameters in order to have $e_{\rm syn}$ in the correct energy range.

REFERENCES

1. Sari, R., Piran, T., *MNRAS* **287**, 110 (1997).
2. Bykov, A. & Mészáros, P., *ApJ* **461**, L37 (1996).
3. Mochkovitch, R. & Daigne, F., these proceedings.
4. Daigne, F. & Mochkovitch, R., to appear in *MNRAS*.
5. Martí, J. & Müller, E., *J. Fluid Mech.* **258**, 317 (1994).
6. Colella, P. & Woodward, P.R., *J. Comput. Phys.* **54**, 174 (1984).
7. Martí, J. & Müller, E., *J. Comput. Phys.* **123**, 1 (1996).

Relativistic Shocks and Gamma-Ray Burst Afterglow Lightcurves

Tomasz Bulik[1,2] and Marek Sikora[1]

[1] *Nicolaus Copernicus Astronomical Center, Bartycka 18, 00716 Warszawa, Poland*
[2] *Department of Astronomy and Astrophysics, University of Chicago, 5640 South Ellis Avenue, Chicago, IL 60637*

Abstract. We consider the evolution of a relativistic blast wave propagating through a low density medium. We calculate the deceleration of such a wave as it collides with external matter and consider the energy dissipation rate as a function of the observers time and the bolometric flux from such a wave. We discuss the results in the light of the optical gamma-ray burst afterglows GRB970228 and GRB970508.

INTRODUCTION

The discovery of gamma-ray afterglow emission in the X-rays [1] and then in the optical band [2] has lead to a breakthrough in understanding GRBs. Although only a few such events have already been seen it seems that the X-ray emission correlates with the original gamma-ray burst. The ratio of the optical afterglow flux to the gamma-ray flux in the burst varies by more than three orders of magnitude between bursts. An explanation of this may be that the optical afterglow comes from a different physical regime than the GRB event itself. For example the gamma emission may be related to the internal shocks while the optical afterglow to shocking with external matter.

Here we consider propagation through an external medium, and energetics of a relativistic blast wave with various radiative efficiencies and calculate the bolometric flux that can be observed. We discuss the physical properties of the wave given the observed optical afterglow lightcurves.

PROPAGATION OF A RELATIVISTIC SHELL

We consider a spherical shell with an initial Lorentz factor Γ_0 expanding in a medium with a uniform density ρ, and calculate the dynamical evolution of such shell as it collides with the external medium. Some details of the simulations are described in [3]. The evolution depends crucially on the details of the radiative losses in a shell. Two extreme cases can be distinguished: when radiation is very efficient

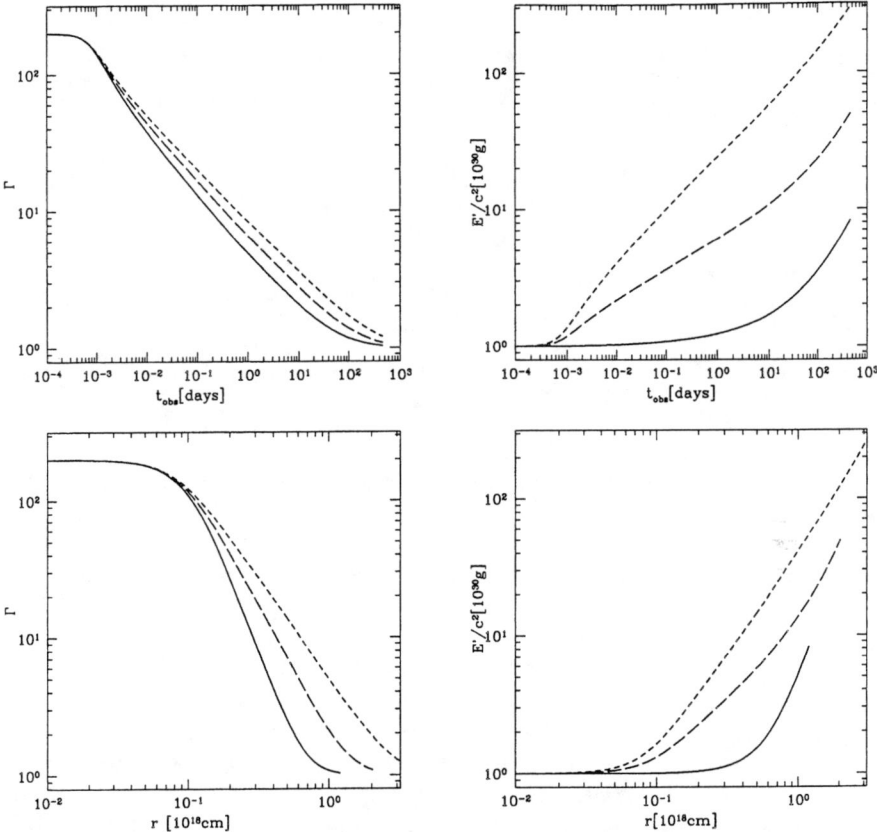

FIGURE 1. Evolution of a relativistic shell. The thick lines correspond to the radiatively efficient case $f = 0$ the short dashed lines correspond to the radiatively inefficient case $f = 1$, while the long dashed lines represent the intermediate case $f = 0.5$. The left panels show Γ, and the right panels the mass (energy) on the observers time (top), and the distance traveled (bottom).

and all the heat generated due to collisions is radiated away instantaneously, and when the radiative mechanisms are inefficient and the energy released remains in the shock (thus increasing its internal mass-energy). In general, one can define a family of solutions parameterized by $0 < f < 1$, where f describes the fraction of the shock energy that stays in the shock and is not radiated away. As an illustrative example we present evolution of a shock with initial $\Gamma_0 = 200$, propagating in a medium with $\rho = 10^{-24} \mathrm{g\, cm^{-3}}$, for three characteristic values of f. After an initial phase of motion with a constant Γ the shock begins to decelerate having swept up a mass $\approx M_0 \Gamma_0^{-1}$. This happens after traversing a distance of a few times 10^{16}cm. Then the mass (energy) E'/c^2 contained in the shock becomes to grow,

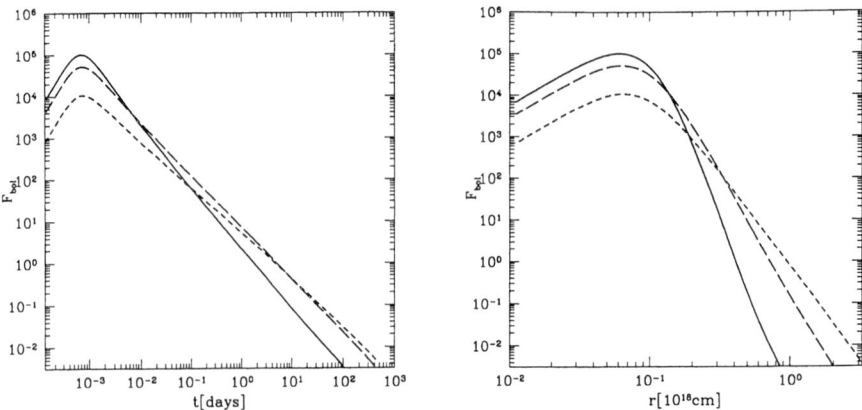

FIGURE 2. The observed bolometric flux from a relativistic blast wave. The solid line corresponds to the radiatively efficient case $f = 0$, the long dashed line to the intermediate case $f = 0.5$, and the short dashed line represents a nearly adiabatic case $f = 0.9$.

and the growth is the fastest in the adiabatic case. The decrease of the Lorentz factor as the shock decelerates is the fastest in the radiatively effective case $f = 0$, and proceeds at the slowest rate for the adiabatic $f = 1$ case. It is worth noting that in the radiatively efficient case the dependence of Γ on the time is not well approximated by a single power law, the power-law index of the decay continuously decreases after the initial quick slowdown. The mass (energy) of the radiatively efficient shock is nearly constant during the deceleration and starts to grow as r^3 when it becomes nonrelativistic. The mass (energy) of the shock in the adiabatic case grows as $r^{3/2}$ during the entire deceleration phase, and in the intermediate case $f = 0.5$ there are two regimes, initially adiabatic-like, and later similar to the radiatively efficient case, see the right panel in Figure 1.

An interesting quantity to calculate is the bolometric flux seen by an observer $F_{\rm bol}$. It can be related to the total energy released in the shock per unit comoving time $d\epsilon'/dt'$, $F_{\rm bol} \propto (d\epsilon'/dt')(\Omega_{\rm D}/\Omega_{\rm out})\Gamma^4 \propto \Gamma^2 d\epsilon'/dt' \propto r^2\Gamma^4$, provided that $\Omega_{\rm D} < \Omega_{\rm out}$, where $\Omega_{\rm D} = \pi/\Gamma^2$ is the Doppler angle, $\Omega_{\rm out}$ is the angular extent of the outflow. The latter result is general and does not depend on the details of the radiative efficiency in the flow.

We calculate the bolometric flux $F_{\rm bol}$ as a function of the observers time and as a function of the distance traveled. The results are presented in Figure 2. It is characteristic that all the curves have maxima at about the same moment of time, namely when the shock begins effective deceleration having swept $\approx M_0 \Gamma_0^{-1}$ from the ambient medium. The maximum of the bolometric flux is independent of the radiative efficiency in the shock. Note that the rate of decay of the energy radiated as a function of time (or r) varies as the parameter f is changed, and is close to t^{-1} (r^{-4}) for the adiabatic case and about $t^{-1.5}$ (r^{-7}) in the radiatively efficient case.

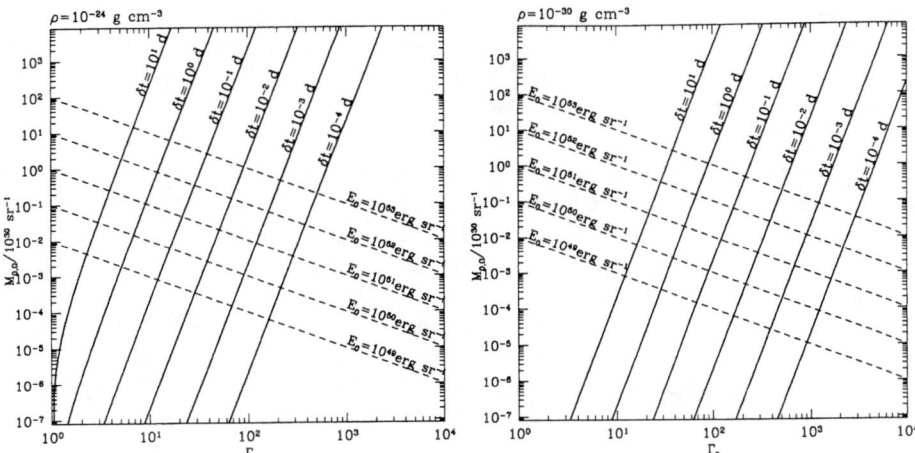

FIGURE 3. Maxima of the bolometric flux $F_{\rm bol}$ as a function of the initial Lorentz factor and the mass of the shell per unit solid angle. the left panel corresponds to the medium with the density of 10^{-24} g cm^{-3} and the right panel to the density 10^{-30} g cm^{-3}.

RESULTS

We calculate the time at which the bolometric flux is maximized for a broad range of initial values of the Lorentz factor and masses of the expanding shell per unit solid angle. We considered two cases: propagation in the medium with density of 10^{-24} g cm^{-3}, typical for the interstellar medium, in the medium with the density of 10^{-30} g cm^{-3}, like that in the intergalactic space. We present the results in Figure 3.

Early limits set by the Bologna University telescope [4] show that the afterglow of GRB970228 began about 18 hours after the gamma-ray burst. At later times the optical light curve of GRB970228 afterglow followed a $\sim t^{-1.1}$ decay law [5]. Here we assume that the turn-on of the optical afterglow is connected with the maximum of the kinetic energy dissipation rate. For a spherical wave $\Omega = 4\pi$ propagating in ISM with the initial energy $E_0 \sim 10^{51}$ ergs, we find that the initial Lorentz factor must have been ~ 15 (see Figure 3). The optical light curve of GRB970508 showed a maximum about ~ 2 days after the burst [6]. For such a time delay the initial Lorentz factor must have been ~ 7. These low Lorentz factors contradict the requirement $\Gamma \geq 100$ from the optical thinness of the source to photon-photon pair creation [7] during the gamma-ray burst.

This difficulty can be overcome in a number of ways. A high Lorentz factor can be reconciled with the observed time delays by imposing a strong collimation ($\Omega \ll 4\pi$), and/or assuming that the blast wave propagates in an extremely low density environment (see Figure 3). The very low density environment can be found e.g. in globular clusters, which are known to have a large population of binary systems [8]. An alternative explanation of the problem shown above lies in assuming a two phase deceleration scenario. In the first phase the blast wave starts with a large Lorentz factor $\Gamma_0 \geq 100$ and is quickly decelerated to a $\Gamma \approx 10$ by colliding with a shell of increased density. Such a shell can exist e.g. due to precursor activity of the binary system. In the case of GRB970508 this phase may be related to the X-ray afterglow, and the initial decay of the optical light a few hours after the burst [9]. The second phase of deceleration is located at much larger distances and is related to interaction with the ISM. The wave enters this phase with Lorentz factor $\Gamma \sim 10$. The kinetic energy dissipation rate peaks about ~ 2 days after the gamma-ray burst. In this model the observed rise and decrease of the optical light about 1-2 days after the GRB event itself is not related to the spectral evolution and transition of the peak luminosity through the optical range, as proposed by [10]. The rise and subsequent decrease is connected with the variations of the bolometric flux, which is followed by modulation in the optical band. Such a model allows to overcome the problem (requirement of a very hard spectrum for energies below the maximum) posed by a very short "turn-on" time of the optical afterglow, see e.g. [9].

ACKNOWLEDGMENTS

The authors wish to acknowledge the support of the following grants KBNP03D00911, KBNP03D01113, NASA NAG 5-4509 and NASA NAG 5-28 (TB) and KBN2P03D01209 (MS).

REFERENCES

1. Costa, E., et al., IAU Circ. 6572 (1997).
2. Groot, P. J., et al., IAU Circ. 6584 (1997).
3. Bulik, T. & Sikora, M., in *Proceedings of the Workshop on Jets in AGNs*, Kraków 1997, eds. Ostrowski, M., Sikora, M., Madejski, G., Begelman, M.
4. Guarnieri, A., et al., IAU Circ. 6582 (1997).
5. Reichart, D., *ApJL* **485**, 57 (1997).
6. Galama, T. J., et al., IAU Circ. 6665 (1997).
7. Piran, T., preprint astro-ph/9507114 (1997).
8. Ziółkowski, private comunication (1997).
9. Pedersen, H., these proceedings.
10. Mészáros, P. & Rees, M., *ApJ* **476**, 232 (1997).

AFTERGLOWS

What Have We Learned From GRB Afterglows?

J. I. Katz* and T. Piran[†]

*Department of Physics and McDonnell Center for the Space Sciences
Washington University, St. Louis, MO 63130
[†]Racah Institute of Physics
Hebrew University, Jerusalem, 91904, Israel

Abstract. The discovery by BeppoSAX and coordinated ground-based observations of persistent X-ray, visible and radio counterparts to GRBs has successfully concluded a search begun in 1973, and the observed redshifted absorption lines have proved that GRB are at cosmological distances. The problem of explaining the mechanisms of GRB and their persistent counterparts remains. There are two classes of models: 1) GRB continue weakly for days at all frequencies; 2) GRB emission shifts to lower frequencies as relativistic debris sweeps up surrounding gas (in an "external shock") and slows. Class 1 predicts that the visible afterglow should be accompanied by continuing gamma-ray emission, as hinted by the high energy emission of GRB940217 and the "Gang of Four" bursts of October 27–29, 1996. It also suggests that the persistent emission will fluctuate. Behavior of this sort may be found in "internal shock" models. Class 2 has been the subject of several theoretical studies which disagree in assumptions and details but which predict that at each frequency the flux should rise and then decline, with the maximum coming later at lower frequencies. Some of this behavior has been observed, but data from GRB970508 show that its afterglow cannot be simply extrapolated from its gamma-ray emission. It is likely that both classes of processes occur in most GRB. Comparisons between GRB show that they are not all scaled versions of the same event. These results suggest that most gamma-ray emission is the result of "internal shocks" while most afterglow is the result of "external shocks", and hint at the presence of collimated outflows. Self-absorption in the radio spectrum of GRB970508 permitted the size of the radiating surface to be estimated, and in future GRBs it may be possible to follow the expansion of the shell in detail and to construct an energy budget.

INTRODUCTION

The visible, infrared and radio afterglows of GRB970228 and GRB970508 have taught us many things. Some of them are obvious: The absorption redshift $z = 0.835$ of GRB970508 established the cosmological distance scale of GRB beyond any reasonable doubt, confirming the very strong case made on statistical grounds [1] from BATSE data. It is also evident that afterglows are very faint. This suggests that the simultaneous visible counterparts to GRBs, as yet unobserved, will also be faint, so that experiments designed to detect them will have to be very sensitive. The loss of the original HETE, carrying an insensitive ultraviolet imager, may therefore have been fortunate, for it will be replaced by HETE-2 which instead will carry a soft X-ray CCD which may yield important spectral information.

Observations of afterglows can answer a number of harder questions too:

1. What is the relative importance of internal *vs.* external shocks?

2. Where does a GRB end and its afterglow begin?

3. Does gamma-ray activity last as long as the afterglow, and could it be an inseparable part of the afterglow?

These questions are central to the understanding and interpretation of afterglows.

INTERNAL *VS.* EXTERNAL SHOCKS

External shocks have been widely considered for GRB since they were suggested by Rees and Mészáros [2]. They predict a hard-to-soft evolution of the spectrum. In many models the duration or elapsed time t is related to a characteristic emission frequency ν_c by a power law $t \propto \nu_c^{-\alpha}$; the exponent α is model-dependent [3–5] but is usually close to 1/2. This is in remarkably good agreement with X-ray and soft gamma-ray observations [6] of the single-peaked GRB960720 in which $\alpha = 0.46$. However, the fact that GRBs with a wide range of durations (from tenths to hundreds of seconds) have comparable ν_c argues against external shock models, which generally predict that these two quantities should vary roughly reciprocally; it is difficult for ν_c to be in the soft gamma-ray range if the duration is minutes.

There must be more to GRBs than external shocks. Fenimore, *et al.* [7] and Sari and Piran [8] showed that an external shock can produce only smoothly varying time-dependent emission, not the spiky multi-peaked structure found in many (but not all) GRBs. Such complex variation must reflect variation in the supply of energy; it cannot be explained solely by interaction with a heterogeneous medium, however complex. A variable outflow may radiate when different fluid elements interact with each other. This process is generally called an internal shock because it does not involve an external medium, although there need be no shock in the hydrodynamic sense of a discontinuity in pressure, density and temperature between two fluids each in thermodynamic equilibrium. Kinematic constraints require that there be inelastic interaction between streams of matter emitted with widely differing Lorentz factors in order that radiation be produced with reasonable efficiency.

Afterglows have, so far, been observed to have a smooth single-peaked time dependence in visible light (their complex time dependence at radio frequencies results from interstellar scintillation [9]), and therefore are naturally explained as a consequence of external shocks. This model predicted [3,4] both the existence of afterglows and the gradual rise to maximum which was observed in both GRB970228 [10] and GRB970508 [11]. However, there is no evidence that internal shocks could not produce the required time dependence, and they are therefore possible explanations for smooth single-peaked GRBs and for afterglows, as well as for the multi-peaked GRBs for which they are required.

Is it meaningful to distinguish an afterglow from the GRB itself? This question first arises for X-rays, whose photon energy range overlaps that traditionally assigned to GRBs. It has no good answer, and should be considered a matter of nomenclature rather than of physics. Rather, we should be concerned with deciding which emission is produced by internal shocks and which by external; the taxonomist may wish to label the former the GRB and the latter the afterglow.

There are two possible limits [5]. In one all the radiation is the product of an external shock. This is excluded, at least for multi-peaked GRBs. It may describe single-peaked GRBs and their afterglows. External shock models successfully predicted [4] the delayed maximum of visible afterglow brightness observed in GRB970228 [10] and GRB970508 [11], and are likely to explain such afterglow emission even if they cannot explain gamma-ray emission. These models also successfully predicted [3,5] self-absorption in the radio afterglow, as was observed [12] from GRB970508.

In the other limit all the emission of a GRB is the product of internal shocks, which may continue for days [13] beyond the nominal gamma-ray duration; the afterglow is a continuation, at lower intensity, of the gamma-ray emitting phase. This has been suggested for GRB970228, where it agrees with the observed instantaneous spectrum [14], and the predicted continuing X-ray emission (beyond the decaying afterglow) has been observed from GRB970508 [15]. This hypothesis is consistent with the hours-long gamma-ray emission of GRB940217 [16], and may explain the "Gang of Four" bursts of October 27–29, 1996 [17] as a single GRB.

HOW TO INTERPRET THE DATA

Early work on GRBs and their afterglows attempted to construct analytic and numerical models of the entire process. Unfortunately, several essential quantities are poorly known (and unlikely to be the same for all GRBs or uniform within a single GRB), such as the distribution of energy among the electrons, ions and magnetic field, the surrounding density and magnetic field, the efficiency of radiation and the degree of collimation. As is so often the case, the real world has turned out to be more complex than theorists could imagine, and a more phenomenological approach is required.

Theories of radiation mechanisms generally predict instantaneous spectra, but observations of afterglows are rarely simultaneous across the spectrum; X-ray, visible and radio observers face different observational constraints. It is possible to fit multiparameter models [18,19] to non-simultaneous data, but the phenomenologist would like a more direct comparison of the emission in different frequency bands. For example, in each band of interest the peak spectral intensity can be measured, or at least estimated, if measurements are obtained close to and straddling the maximum. The resulting function $F_{\max}(\nu)$ is not an instantaneous spectrum, but makes it possible to compare emission mechanisms in different spectral bands.

In the case of GRB970508 $F_{\max}(\nu)$ can be estimated at GHz, visible (and near-IR), hard X-ray and soft gamma-ray energies (the maximum of the soft X-ray intensity was not observed), and is shown in Figure 1. A single mechanism is unlikely to explain emission in all these bands, because $F_{\max}(\nu)$ is not consistent with a single power law, as would be predicted by a single mechanism which does not have a characteristic frequency to define a spectral break. The visible data form a "hinge"; on purely phenomenological grounds it is not possible to say whether this emission is produced by the same mechanism as the GHz emission or the hard X-rays and soft gamma-rays. Neither of the limiting cases discussed in the previous section is satisfactory. A plausible hypothesis suggests that GHz and visible emission result from external shocks and hard X-rays and soft gamma-rays (the classical GRB) from internal shocks. This explains why an early model [4] underestimated the time to maximum and overestimated the brightness of the visible afterglow of GRB970508: it extrapolated from the gamma-ray emission of an assumed bright GRB, but this extrapolation was invalid (even if scaled to the lesser flux of GRB970508) because the gamma rays were produced by a different mechanism.

It is also possible to compare [5] the properties of different GRBs and their afterglows by comparing the burst-to-burst ratios of $F_{\max}(\nu)$ in different spectral bands. The afterglow of GRB970508 was roughly 20 times brighter than that of GRB970228 in soft X-rays and visible light, when scaled to their soft gamma-ray brightnesses. The afterglow of GRB970508 was at least four times brighter than that of GRB970111 at 1.43 GHz, and at least 1000 times brighter than that of GRB970828 in visible light when similarly scaled (in these last two cases the ratios are lower limits because only upper limits exist to the fluxes of GRB970111 and GRB970828 in these bands). These ratios quantify the conclusion that GRB970508 had a remarkably bright afterglow, compared to the the other GRBs for which useful data exist, at all frequencies at which comparisons are possible.

This leads to the important (and unexpected) conclusion that afterglows are "all different". They are not all scaled versions of the same event, and any single simple model must fail. It argues against models in which all GRBs and afterglows are related by a single scaling parameter, such as the ambient density, energy or distance, or even some combination of these. A natural interpretation is that (at least in these GRBs) the hard X-rays and gamma rays are produced by internal shocks, and the lower frequency afterglow by an external shock. If so, there need be no close correlation between the brightnesses of GRBs and their afterglows, partly because the mechanisms are different, but also because internal shock properties depend on the detailed temporal and spatial structure of the outflow and may well be very different in different GRBs.

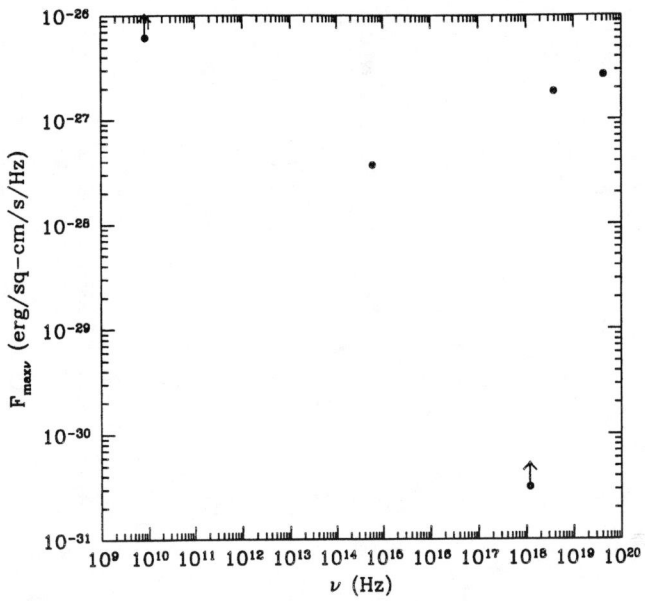

FIGURE 1. $F_{max}(\nu)$ for GRB970508

A GENERIC AFTERGLOW MODEL

A generic afterglow model is based on the assumption of an external shock, which permits specific predictions to be made. The interaction between a relativistic debris shell and the surrounding medium, or among various elements of an outflowing wind, has been the subject of many papers, but the basic physics is not understood. Even the essential collisionless shock is largely a matter of speculation, although recent calculations [20] have begun to attack the problem. Still, a few features common to most afterglow models are independent of assumptions as to the mechanism of entropy production. The asymptotic ($\nu_s \ll \nu \ll \nu_c$) instantaneous spectrum [4] has the form $F_\nu \propto \nu^{1/3}$ between a self-absorption frequency ν_s and a characteristic synchrotron frequency ν_c. Below ν_s the spectrum $F_\nu \propto \nu^2$, while above ν_c the flux falls off with a slope reflecting the high energy "tail" to the particle distribution function; a power law is typically observed in GRBs, but its slope is unpredictable, and is observed to differ from burst to burst and with time in a given burst. The instantaneous spectra $F_\nu(t)$ are bounded from above by the function $F_{\max}(\nu)$. In simple analytic models $F_{\max}(\nu) \propto \nu^{-\beta}$, with the exponent β typically [4,5] close to 0; any function of ν other than a power law would define a characteristic break frequency and require a more complex model.

As time progresses both ν_s and ν_c decrease. At a given frequency ν_0 the flux F_{ν_0}

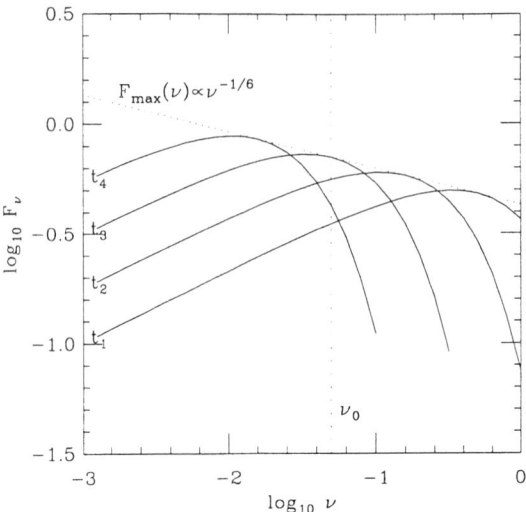

FIGURE 2. Instantaneous GRB spectra for a model in which $F_{\max}(\nu) \propto \nu^{-1/6}$; $t_1 < t_2 < t_3 < t_4$.

rises until ν_c declines to $\sim \nu_0$, after which F_{ν_0} decreases. This predicted behavior has been observed, at least qualitatively, in the afterglows of GRB970228 [10] and GRB970508 [11]. The predicted rate of the initial rise $F_\nu(t) \propto t^\delta$, where the preceding expressions for $F_{max}(\nu)$ and $t(\nu_c)$ and the instantaneous F_ν lead to

$$\delta = \frac{\beta + 1/3}{\alpha}.$$

Typically, δ is estimated to be slightly less than unity; in one model [4] $\delta = 4/5$ while in another [5] $\delta = 6/7$. Figure 2 shows the evolution of the instantaneous spectrum. F_{ν_0} rises until $t = t_3$ and then declines.

TESTING THE GENERIC AFTERGLOW MODEL

The predictions of the generic external shock afterglow model can be tested with data from the two observed afterglows. The predicted rise to maximum has been observed. Some data [21] suggested agreement ($\delta \approx 0.7$) with the predicted rate of rise for GRB970508, but more complete data for GRB970228 [10] and GRB970508 [11] suggest a much steeper rise, with $\delta \approx 2$–3.

These large values of δ appear to disagree with the model. Sari [22] pointed out that in the early stages of external shock models, when the relativistic debris shell has not yet been significantly slowed by the ambient medium, the luminosity should rise $\propto t^2$ at all frequencies ($\delta = 2$), a simple consequence of the increasing ($\propto r^2$) area of the shell. If this is the explanation then the rapid rise should level out after a time

$$t \approx \left(\frac{3E}{4\pi\rho c^2}\right)^{1/3} \frac{1}{2c\gamma^{8/3}},$$

when deceleration becomes important (the earlier stages of the rising flux could be hidden under steady emission by other processes, such as internal shocks). This leads to an estimate of the ambient density ρ:

$$\rho \approx 10^{-33} \left(\frac{E}{10^{52}\,\text{erg}}\right) \left(\frac{\gamma}{10^2}\right)^{-8} \left(\frac{t}{10^5\,\text{s}}\right)^{-3} \text{g/cm}^3.$$

If the model is correct this implies an extraordinarily low ambient density out to a radius $\sim 2c\gamma^2 t \sim 20(\gamma/10^2)^2(t/10^5\,\text{s})$ pc, or a surprisingly low value of $\gamma \sim 10$. Such a low density would be remarkable, even in the intergalactic medium (although a pre-coalescence pulsar wind could create a bubble), and such a low γ is inconsistent with gamma-gamma pair production constraints (although in a long duration internal shock model a low γ wind could follow a brief high γ wind which produces the gamma-ray emission).

If the afterglow is produced by internal shocks it is probably not possible to predict its time dependence. However, the instantaneous spectrum is still predictable. There is some spectral evidence that the first several hours of afterglow in GRB970228 were the product of internal shocks [14]. Internal shocks are a possible explanation of disagreements between the observed time dependence and predictions of the generic external shock model.

This early (3–8 hours after the GRB) period of roughly constant visible intensity in GRB970508 poses another problem. In either internal or external shock models, if the electron synchrotron cooling time is short compared to the duration of observation then the observed spectrum is that integrated over the electrons' cooling history [23]: $F_\nu \propto \nu^{-1/2}$. Comparison of the visible and soft X-ray [11] fluxes during this period shows a deficiency of X-rays compared even to this spectrum, suggesting that ν_c lies within or below the X-ray band. This is consistent with emission by an internal shock with a low value of γ, as suggested by both the near constant visible intensity and the long-delayed onset of the rapid rise.

These conclusions are based on limited data from two afterglows. The outline of the external shock afterglow model is supported by the data, but the detailed interpretation, especially of the rise of the visible intensity to maximum, must await more data from more afterglows.

THE INSTANTANEOUS SPECTRUM

The predicted [4] instantaneous asymptotic spectrum $F_\nu \propto \nu^{1/3}$ for $\nu \ll \nu_c$ should be applicable to both internal and external shocks. It is based on several plausible assumptions: incoherent synchrotron radiation, no cooling (radiative or adiabatic) and a phase space argument for the electron distribution function produced by a relativistic shock which heats the entire electron distribution function, rather than just a "tail" of suprathermal particles. It should apply to the synchro-Compton spectrum too, but with a different coefficient.

Like all theoretical predictions, it is only speculation until empirically tested. So far, the data are inconclusive and not completely consistent. Some X-ray and soft gamma-ray observations of GRBs support the prediction [23,24], but others disagree [25,26]. The inconsistency may result from the difficulty of extracting quantitative spectral information from NaI scintillator data, which have low intrinsic spectral resolution, especially at photon energies < 100 KeV where the asymptotic spectrum is expected. This difficulty may also account for the long-standing controversy over the reality of line features in GRB radiation, and may only be resolved when data from detectors of intrinsically higher resolution become available.

Visible data [27] lead to an exponent 0.25±0.25 in the pre-maximum phase of the afterglow of GRB970508, in agreement with the predicted 1/3 (agreement is not expected at and after maximum, because then the frequency of observation exceeds ν_c). Radio data [12] from the same afterglow lead to an exponent ≈ 0.2, also in agreement with prediction. The results are encouraging, but not yet conclusive.

If a GRB were to be detected in visible light during its initial brief phase of gamma-ray emission the spectral exponent could be determined quite accurately by comparing fluxes at frequencies separated by a factor of more than 10^4. Such simultaneous detection remains the holy grail of GRB visible counterpart research.

SELF-ABSORPTION

Self-absorption of GRB afterglows at GHz frequencies was predicted [3,5] and confirmed [12]. This was no great surprise; the physics is elementary, though a little different (the spectral exponent below ν_s is predicted to be 2 rather than 2.5) than in the usual case of synchrotron self-absorption by a power law distribution of electron energies. Self-absorption is important because it may lead to a measurement of the emission radius r as a function of time with few uncertain assumptions. The flux for $\nu \ll \nu_s$ is

$$F_\nu = 2\pi \nu^2 m_p \zeta (1+z) \frac{r^2}{D^2},$$

where D is the distance, z the cosmological redshift, m_p the proton mass and ζ an equipartition factor defined as $k_B T_e / \gamma m_p c^2$; $\zeta = 1/9$ in the case of complete electron-ion-magnetic equipartition. The shock Lorentz factor (assumed $\gg 1$) drops out; this is fortunate, for it is poorly known and likely to remain so.

If $r(t)$ were inferred from measurements of the self-absorbed flux then it would be possible to reconstruct the expansion history and slowing down of the relativistic debris in an external shock model. This would permit determination of the ambient density and the efficiency of radiation as the initial kinetic energy is radiated or shared with swept-up matter. A preliminary attempt [5] to construct such an energy budget (based on one inferred value of r) for the afterglow of GRB970508 led to the conclusion that the total kinetic energy after seven days was only $\sim 10^{49} n_1$ erg, where n_1 is the ambient particle density. This is much less than the $\sim 3 \times 10^{51}$ erg inferred from the gamma-ray radiation, assuming isotropic emission.

There are three possible implications of this result: $n_1 \gg 1$ cm^{-3}, as might be found in a dense star-forming region; strong beaming of the gamma rays (and therefore of the initial relativistic outflow); or a radiation efficiency > 99% in the first seven days of the event. Any or all of these are possible, and any or all would be important.

A THEORIST'S WISH LIST

We have compared the observed afterglows to theoretical predictions based on simple models, and have generally found qualitative agreement. In order to test the theory, particularly that of relativistic shocks, in more detail, more difficult measurements will have to be performed. Here is a list of ambitious goals a theorist might set for his observational colleagues:

1. Accurate X-ray spectroscopy at moderate resolution (~ 30) to test the predicted instantaneous $F_\nu \propto \nu^{1/3}$ ($F_\nu \propto \nu^{-1/2}$ if synchrotron cooling is important). This may be performed by ZnCdTe or Ge detectors, or by X-ray CCD.

2. Simultaneous spectral measurements across several decades of frequency, with $\nu < \nu_c$ throughout (hence the intensity must not have reached its maximum anywhere in this range). This could be achieved by observation of the visible counterpart to a GRB during its strong gamma-ray activity, or by simultaneous measurements of visible and radio afterglow before the visible maximum (within the first day for GRB970228 and GRB970508).

3. Measurement of the intensity as a function of time in the self-absorbed regime of radio afterglow.

Each of these measurements is likely to be difficult. Locating and measuring a visible counterpart to a GRB within tens of seconds is much harder than doing it with several hours of imaging X-ray observations. The only afterglow observed at radio frequencies was not strong enough to be detected until long after the visible maximum, and self-absorption reduces the strength of the radio emission even more. The first goal is probably the most feasible, and would test relativistic shock theory; the last is probably the one which would lead to the greatest understanding of how GRBs and their afterglows really work.

ACKNOWLEDGEMENTS

We thank R. Sari for discussions. JIK thanks NSF 94-16904 for support, Washington University for the grant of sabbatical leave and the Hebrew University for hospitality and a Forchheimer Fellowship.

REFERENCES

1. Meegan, C. A., *et al.*, *Nature* **355**, 143 (1992).
2. Rees, M. J., and Mészáros, P., *MNRAS* **258**, 41p (1992).
3. Katz, J. I., *Ap. J.* **422**, 248 (1994).
4. Katz, J. I., *Ap. J. Lett.* **432**, L107 (1994).
5. Katz, J. I., and Piran, T., *Ap. J.* **490**, 772 (1997).
6. Piro, L., *et al.*, *Astron. Ap.* in press (astro-ph/9707215) (1997).
7. Fenimore, E. E., Madras, C. D., and Nayakchin, S., *Ap. J.* **473**, 998 (1996).
8. Sari, R., and Piran, T., *Ap. J.* **485**, 270 (1997).
9. Goodman, J., *New Astron.* **2**, 449 (1997).
10. Guarnieri, A., *et al.*, *Astron. Ap.* **328**, L13 (1997).
11. Pedersen, H. *et al.*, *Ap. J.* in press (astro-ph/9710322) (1998).
12. Frail, D. A., *et al.*, *Nature* **389**, 261 (1997).
13. Katz, J. I., *Ap. J.* **490**, 633 (1997).
14. Katz, J. I., Piran, T., and Sari, R., *PRL* submitted (astro-ph/9703133) (1997).
15. Piro, L., *et al.*, *Astron. Ap.* submitted (astro-ph/9710355) (1998).
16. Hurley, K., *et al.*, *Nature* **372**, 652 (1994).
17. Connaughton, V., *et al.*, *Proc. 18th Texas Symp. Rel. Ap.* in press (1997).
18. Mészáros, P., and Rees, M. J., *Ap. J.* **476**, 232 (1997).
19. Wijers, R. A. M. J., Rees, M. J., and Mészáros, P., *MNRAS* **288**, 51p (1997).
20. Usov, V. V., and Smolsky, M. V., *PRE* in press (astro-ph/9704152) (1997).
21. Djorgovski, S. G., *et al.*, *Nature* **387**, 876 (1997).
22. Sari, R., *Ap. J. Lett.* **489**, L37 (1997).
23. Cohen, E., *et al.*, *Ap. J.* **488**, 330 (1997).
24. Schaefer, B. E., *et al.*, *Ap. J.* submitted (1997).
25. Preece, R. D., *et al.*, "The 'Line of Death' for the Synchrotron Shock Model: Confronting the Data," these proceedings.
26. Crider, A., Liang, E., and Preece, R., "Restrictions and Predictions from GRB Spectral Evolution," these proceedings.
27. Groot, P., *et al.*, "Optical and Radio Follow-up of GRB970508," these proceedings.

Afterglows as Diagnostics of Gamma-Ray Burst Beaming

James E. Rhoads

Kitt Peak National Observatory[1]
950 N. Cherry Ave., Tucson, AZ 85719

Abstract. If gamma-ray bursts are highly collimated, radiating into only a small fraction of the sky, the energy requirements of each event may be reduced by several (up to 4-6) orders of magnitude, and the event rate increased correspondingly. The large Lorentz factors ($\Gamma \gtrsim 100$) inferred from GRB spectra imply relativistic beaming of the gamma rays into an angle $\sim 1/\Gamma$. We are at present ignorant of whether there are ejecta outside this narrow cone.

Afterglows allow empirical tests of whether GRBs are well-collimated jets or spherical fireballs. The bulk Lorentz factor decreases and radiation is beamed into an ever-increasing solid angle as the burst remnant expands. It follows that if gamma-ray bursts are highly collimated, many more optical and radio transients should be observed without associated gamma rays than with them. In addition, a burst whose ejecta are beamed into angle ζ_m undergoes a qualitative change in evolution when $\Gamma \zeta_m \lesssim 1$: Before this, $\Gamma \propto r^{-3/2}$, while afterwards, $\Gamma \propto \exp(-r/r_\mathrm{r})$. This change results in a potentially observable break in the afterglow light curve.

Successful application of either test would eliminate the largest remaining uncertainty in the energy requirements and space density of gamma-ray bursters.

The ejecta from gamma-ray bursts must be highly relativistic to explain the spectral properties of the emergent radiation [4,1]. The gamma rays we observe are therefore only those from material moving within angle $1/\Gamma$ of the line of sight, and offer no straightforward way of determining whether the bursts are isotropic emitters or are beamed into a small angle. (Here Γ is the bulk Lorentz factor of expansion.)

Afterglow emission at longer wavelengths is expected to arise later in the evolution of the burst than the original gamma rays. It therefore offers at least two ways of testing the burst beaming hypothesis.

---

[1] Kitt Peak National Observatory is part of the National Optical Astronomy Observatories, operated by the Association of Universities for Research in Astronomy.

Burst and Afterglow Event Rates

First, because Γ is lower at the time of afterglow emission than during the GRB itself, the afterglow cannot be as collimated as the GRB can. This implies that the afterglow event rate should exceed the GRB event rate substantially if bursts are strongly beamed. Allowing for finite detection thresholds,

$$\frac{N_{12}}{N_2} \leq \frac{\Omega_1}{\Omega_2} \leq \frac{N_1}{N_{12}} \;, \tag{1}$$

where N_1, N_2 are the measured event rates above our detection thresholds at our two frequencies; N_{12} is the rate of events above threshold at both frequencies; and Ω_1, Ω_2 are the solid angles into which emission is beamed at the two frequencies.

A full derivation of this result and discussion of its application is given in [6]. Rather than reproduce it, I will refer the reader to that paper and will here discuss the second test more fully than was possible in [6].

Dynamical Calculations: Numerical Integrations

The second test is based on differences between the dynamical evolution of beamed and isotropic bursts. We explore the effects of beaming on burst evolution using the notation of [5]. Let Γ_0 and M_0 be the initial Lorentz factor and ejecta mass, and ζ_m the opening angle into which the ejecta move. The burst energy is $E = \Gamma_0 M_0 c^2 \zeta_m^2 / 4$, where we assume a unipolar jet geometry. Let r be the radial coordinate in the burster frame; t, t_{co}, and $t_\oplus$ the time from the event measured in the burster frame, comoving ejecta frame, and terrestrial observer's frame; and f the ratio of swept up mass to M_0.

The key assumptions in our beamed burst model are that (1) the energy and mass per unit solid angle are constant at angles $\theta < \zeta_m$ from the jet axis and zero for $\theta > \zeta_m$ (see [2] for an alternative model); (2) the energy in the ejecta is approximately conserved; (3) the ambient medium has uniform density; and (4) the cloud of ejecta + swept-up material expands in its comoving frame at the sound speed $c_s = c/\sqrt{3}$ appropriate for relativistic matter. The last of these assumptions implies that the working surface of the expanding remnant has a transverse size $\sim \zeta_m r + c_s t_{co}$. The evolution of the burst changes when the second term dominates over the first.

The full equations describing the burst remnant's evolution are then

$$f = \frac{1}{M_0} \int_0^r r^2 \Omega_m(r) \rho(r) dr \;, \tag{2}$$

$$\Omega_m = \pi(\zeta_m + c_s t_{co}/ct)^2 \approx \pi(\zeta_m + t_{co}/\sqrt{3}t)^2 \;, \tag{3}$$

$$\Gamma = (\Gamma_0 + f)/\sqrt{1 + 2\Gamma_0 f + f^2} \approx \sqrt{\Gamma_0/2f} \;, \tag{4}$$

$$t = r/c \ , \qquad t_{\rm co} = \int_0^t dt'/\Gamma \ , \quad {\rm and} \quad t_\oplus = \int_0^t dt'/2\Gamma^2 \ . \qquad (5)$$

These equations can be solved by numerical integration to yield $f(r)$, $\Gamma(r)$, and $t_\oplus(r)$. Figure 1 shows $\Gamma(r)$ from such integrations for an illustrative pair of models (one beamed, one isotropic).

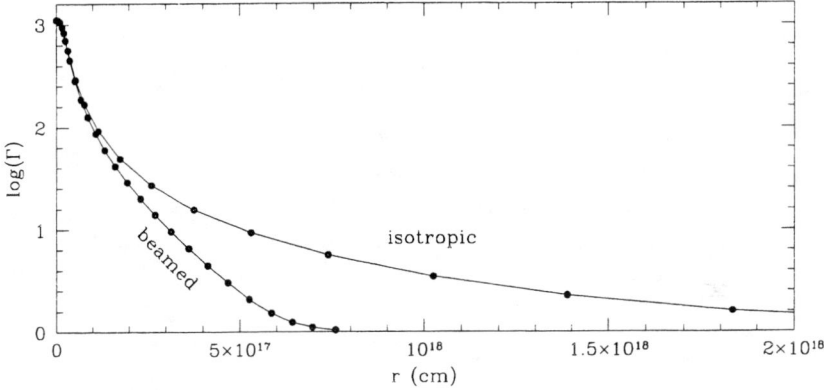

FIGURE 1. Dependence of the bulk Lorentz factor Γ on the burst expansion radius for an isotropic burst and a burst beamed into an opening angle $\zeta_{\rm m} = 0.01$ radian. Both bursts follow a $\Gamma \propto r^{-3/2}$ evolution initially, but the beamed burst changes its behavior at $\Gamma \approx 100 \approx 1/\zeta_{\rm m}$, beyond which its Lorentz factor decays exponentially with radius.

The emergent synchrotron radiation can also be calculated if we assume an electron energy spectrum and assume that electrons and magnetic fields have constant fractions of the equipartition energy density. For illustrative purposes, we again follow the assumptions in [5]. The electron energy spectrum is $N(\mathcal{E}) \propto \mathcal{E}^{-2}$, i.e. a power law with equal energy per decade, so that the synchrotron spectrum peaks where $\tau = 0.35$, rising as $\nu^{5/2}$ at low (optically thick) frequencies and falling as $\nu^{-1/2}$ at high (optically thin) frequencies [3]. The relevant equations are a straightforward modification of equations 11–20 of [5]. Figure 2 shows the peak flux density as a function of observed frequency for the models used in Figure 1. We caution the reader that more recent electron energy spectra grounded in observations (e.g. [7]) may be more reliable. We hope to incorporate such spectra in our calculations in future.

Dynamical Calculations: Analytic Integrations

The most interesting dynamical change introduced by beaming is a transition from a power law $\Gamma \propto r^{-3/2}$ to an exponentially decaying regime $\Gamma \propto \exp(-r/r_{_\Gamma})$. This can be derived by considering the approximate evolution equations for the

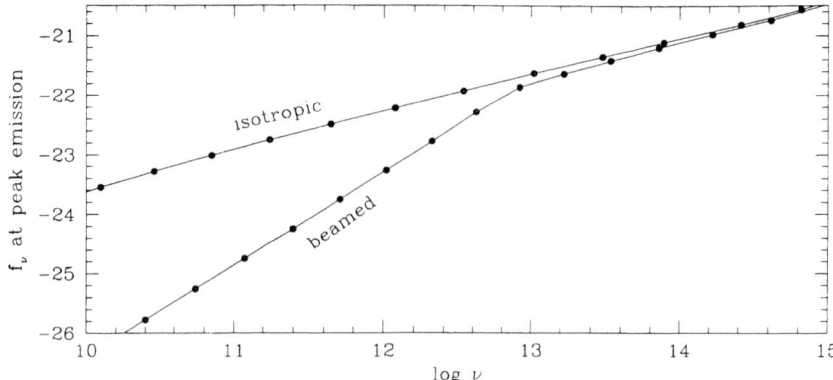

FIGURE 2. The dependence of the peak flux density f_ν on observed frequency ν for the same pair of bursts. The electron spectrum follows the model of Paczyński & Rhoads [5]. The peak in the synchrotron emission for this model occurs at the frequency where optical depth effects become important. The predicted break in the power law caused by beaming should be observable. Similar breaks occur in the dependence of f_ν and $\nu_{\rm peak}$ with time, and are expected to be a generic feature of beamed GRB afterglow models.

regime where (a) $1/\Gamma_0 \lesssim f \lesssim \Gamma_0$, so that $\Gamma \approx \sqrt{\Gamma_0/2f}$; and (b) $c_s t_{\rm co} > \zeta_m r$ (corresponding to $f \gtrsim 9\Gamma_0 \zeta_m^2$):

$$df/dr \approx \frac{\pi}{M_0} c_s^2 t_{\rm co}^2 \rho \quad , \quad dt_{\rm co}/dr \approx \sqrt{\frac{2f}{c^2 \Gamma_0}} \quad , \quad dt_\oplus/dr \approx \frac{f}{c\Gamma_0} \quad . \tag{6}$$

It follows that

$$\sqrt{f} df = \frac{\pi}{\sqrt{2}} \frac{c\, c_s^2 \rho \sqrt{\Gamma_0}}{M_0} \times t_{\rm co}^2 dt_{\rm co} \approx \frac{\pi}{3\sqrt{2}} \frac{c^3 \rho \sqrt{\Gamma_0}}{M_0} \times t_{\rm co}^2 dt_{\rm co} \quad . \tag{7}$$

This is easily integrated to obtain

$$f^{3/2} = \left(\frac{\pi \sqrt{\Gamma_0} c\, c_s^2 \rho}{\sqrt{8} M_0} \right) t_{\rm co}^3 + const \quad . \tag{8}$$

The constant of integration becomes negligible once $c_s t_{\rm co} \gg \zeta_m r$, so that equation 8 becomes $f \propto t_{\rm co}^2$. It is then clear from equations 6 that f, Γ, $t_{\rm co}$, and $t_\oplus$ will all behave exponentially with r in this regime. Retaining the constants of proportionality, we find

$$f \propto \exp(2r/r_\Gamma) \quad \text{where} \quad r_\Gamma = \left[\frac{1}{\pi} \left(\frac{c}{c_s} \right)^2 \frac{\Gamma_0 M_0}{\rho} \right]^{1/3} \quad . \tag{9}$$

Further algebra yields $\Gamma \propto \exp(-r/r_\Gamma)$ and $t_\oplus \propto f \propto \exp(2r/r_\Gamma)$, so that $\Gamma \propto t_\oplus^{-1/2}$. Thus, while the evolution of $\Gamma(r)$ changes from a power law to an exponential at $\Gamma \sim 1/\zeta_m$, the evolution of $t_\oplus(r)$ changes similarly. The net result is that $\Gamma(t_\oplus)$ has a power law form in both regimes, but with a break in the slope from $\Gamma \propto t_\oplus^{-3/8}$ when $\Gamma > 1/\zeta_m$ to $\Gamma \propto t_\oplus^{-1/2}$ when $\Gamma < 1/\zeta_m$.

Of course, Γ is not directly observable, and we ultimately want to predict observables like the frequency of peak emission ν_m, the flux density $F_{\nu,m}$ at ν_m, and the angular size θ of the afterglow. With the electron energy spectrum described above, the relevant power law scalings before beaming becomes dynamically important are $\nu_m \sim t_\oplus^{-2/3}$, $F_{\nu,m} \sim t_\oplus^{-5/12}$, and $\theta \sim t_\oplus^{5/8}$. At late times, $\nu_m \sim t_\oplus^{-1}$, $F_{\nu,m} \sim t_\oplus^{-3/2}$, and $\theta \sim t_\oplus^{1/2}$. Our numerical integrations confirm these relations, though the transition between the two regimes is quite gradual for ν_m.

Combining these scalings with the spectral shape yields predictions for the light curve at fixed observed frequency. The most dramatic feature is in the light curve shape for $\nu > \nu_m$, which changes from $F_{\nu,\oplus} \sim t_\oplus^{-3/4}$ to $F_{\nu,\oplus} \sim t_\oplus^{-2}$. These exponents are generally sensitive to the assumed electron energy distribution in the blast wave.

Conclusions

Establishing whether or not gamma-ray bursts are beamed will be valuable in understanding source populations and burst mechanisms. There are at least two potentially observable consequences of beaming.

(1) The event rate for afterglows should exceed that for bursts substantially if bursts are strongly beamed. A quantitative comparison of rates at two frequencies yields quantitative limits on the ratio of beaming angles.

(2) The dynamical evolution of a beamed burst remnant changes qualitatively when $\Gamma < 1/\zeta_m$. The resulting changes in the light curves could be observed.

REFERENCES

1. Goodman, J., *ApJ* **308**, L47 (1986).
2. Mészáros, P., Rees, M. J., & Wijers, R. A. M. J., astro-ph/9709273 (1997).
3. Pacholczyk, A. G., *Radio Astrophysics*, San Francisco: W. H. Freeman (1970).
4. Paczyński, B., *ApJ* **308**, L43 (1986).
5. Paczyński, B., & Rhoads, J. E., *ApJ* **418**, L5 (1993).
6. Rhoads, J. E., *ApJ* **487**, L1 (1997).
7. Waxman, E., *ApJ* **485**, L5 (1997).

Measuring the GRB Parameters from Afterglow Observations

Re'em Sari and Tsvi Piran

Racah Institute of Physics, Hebrew University, Jerusalem 91904, Israel

Abstract. We examine the relativistic decelerating shell model for γ-ray burst afterglow, and show how one can infer the model's free parameters from afterglow observations.

INTRODUCTION

An essential ingredient of all cosmological GRB models is a bulk relativistic motion, which is dissipated to give the observed γ-rays (see [1]). Sari & Piran [2] have shown that deceleration on an external medium can not give rise to the variability observed in the bursts, while internal shocks can produce efficiently the observed variability [3] if the "inner engine" has considerable fluctuations. These considerations predict that after the main GRB event, which is produced by internal shocks, the ejecta will decelerate due to interaction with the ISM, emitting radiation at longer and longer wavelengths [4–6,2]. This emission has been detected recently for several GRBs following an accurate determination of their position. While the details of the internal shocks model contain too many free parameters (dozens of shells, each with different Lorentz factor and different separations), the afterglow is a relatively clean problem. In addition, the quality of the afterglow data is considerably superior to that of the main burst (broad band spectrum, orders of magnitude evolution in time and flux). These conditions make the γ-ray bursts afterglow an excellent candidate for deducing some of the parameters of the relativistic shock model.

We divide the parameters of the afterglow models into three groups, those that are inherited from the main GRB, those that describe the surrounding matter, and those that describe the micro-physics of particle acceleration by the shock.

- **GRB inherited parameters:** The most important parameter in this group is the explosion energy in the burst E (under the assumption of a spherical explosion). Additional parameters, such as the width of the shell and its initial Lorentz factor γ_0 influence only the very beginning of the interaction with the ISM [7]. The opening angle of the wind θ is observable only at the very late stage when the Lorentz factor of the shell falls below $\gamma < 1/\theta$. If these parameters can be deduced from the afterglow, they would provide valuable direct information on the GRB itself.

- **Surrounding medium parameters**: In the simplest model the ISM is described by a single parameter n, the density of particles. Typical values can be 1 cm^{-3} if the burst is inside a galaxy or 10^{-6} cm^{-3} if the burst is in the inter-galactic medium. More complex models consider a density profile or inhomogeneous ISM and therefore have additional parameters. We ignore this possibility in this paper.

- **Micro-physics parameters**: Throughout the relativistic stage, it is reasonable to assume that a constant fraction of the shock's energy ϵ_e is given to the electrons, and a constant fraction ϵ_B is given to the magnetic field. Shock acceleration usually creates a power law distribution of the particle energy (electrons in our case) $N(\gamma_e) \propto \gamma_e^{-p}$. In contrast to parameters from the two previous groups, these parameters could be calculated from first principles, if a complete theory of particle acceleration in a relativistic shock would have existed. These parameters reflect, therefore, our ignorance in this basic physical process. It is also reasonable to assume that the micro-physics within the burst itself (internal shocks) is the same as that of the afterglow (external shocks).

AFTERGLOW MODELS

The basic model that we consider is the "internal-external" model [8] in which a relativistic shell, after going through internal shocks and producing the GRB itself, encounters the ISM and decelerates via external shocks, giving rise to the afterglow. Although the number of parameters is relatively small and well defined, there are a few possible regimes for the model. The hydrodynamical evolution can be either radiative or adiabatic, the electrons can be cooling fast or slow, and the emission process might be synchrotron or inverse Compton or both. The system can even switch during its evolution from one regime to the other. However, for a given set of parameters it is possible to find out which regime is valid at any given moment. For example, given ϵ_e and ϵ_B it is possible to estimate the electron cooling time and to check if these can cool on the dynamical time of the system or not.

Some models are internally inconsistent. Radiative evolution is possible only if a considerable fraction of the energy of the system is given to the electrons, i.e., $\epsilon_e \to 1$, and the electrons cooling time is shorter than the dynamical time of the system (fast cooling). Otherwise the evolution is adiabatic. Radiative hydrodynamical evolution in which the electrons carry only a small fraction of the energy ($\epsilon_e \ll 1$) is impossible. Alternatively, if $\epsilon_e \to 1$, adiabatic evolution during a fast cooling stage is impossible. A more complete treatment of the possible situations is given in [9,10].

THE OPTICAL DATA

We consider the optical data of GRB970508 (similar data, however more sparse, are also available for GRB970228). As time evolves more and more mass is collected by the shock, it decelerates and radiates in lower and lower frequencies. Consider a single fixed frequency. Initially the typical synchrotron frequencies are above the observed one. As the shock decelerates the flux at the observed frequency increases. Then, when the typical frequency goes below the observed frequency, the flux begins to decrease. This behavior is in qualitative agreement with the observed afterglow. Such a theoretical light curve is presented in Figure 1 (see [9] for detailed derivation).

We focus on three clear features of the observed light curve of GRB970508 in the R-band (where most of the observations where done): the peak flux $F_{\text{peak}} = 40\mu J$, the time in which the peak is obtained $t_{\text{peak}} = 2$ days and the index $\alpha \cong 1.2$ of the long power law decrease that follows the peak: $F_\nu \propto t^{-\alpha}$.

We begin with the third observable, the decay power law index α. The theoretical value of this power law reflects the power law distribution of electrons accelerated by the shock. Fitting this decrease we can measure the power p of the electron distribution. The relation between p and α depends on the model (see the figure for the different possible relations), leading to values of p between 2.3 and 2.6. However since there is no theory for a-priori calculating p these relations can not be used to distinguish between models. This data is therefore just evidence for the existence of a power-law distribution of electrons in the shock.

The rise and decay before and after the peak are very shallow in the high frequency case, where high frequencies are defined as frequencies higher than the synchrotron typical frequency ν_0 at the time t_0 of transition to slow cooling. While at the low frequency light curves, the peak is sharper. We therefore deduce that the low frequency light curve is the relevant one for the R-band, i.e. the transition from fast to slow cooling occurred *before* the typical synchrotron frequency has reached the R-band. In this case the equations (ignoring IC and cosmological red-shift) for the peak flux and the time in which it is observed are

$$F_{\text{peak}} = 1.1 \times 10^5 \epsilon_B^{1/2} E_{52} n_1^{1/2} D_{28}^{-2} \ \mu J = 40 \ \mu J \text{ (for GRB970508)},$$
$$t_{\text{peak}} = 0.69 \epsilon_B^{1/3} \epsilon_e^{4/3} E_{52}^{1/3} \nu_{15}^{-2/3} \text{ days} = 2 \text{ days (for GRB970508)}. \tag{1}$$

We have scaled the results for a total energy of $E = E_{52} \times 10^{52}$erg and ISM density of $n_1 \text{cm}^{-3}$. These are two equation with four unknowns n_1, E_{52}, ϵ_e and ϵ_B ($D_{28} = 1$, as suggested by the observed redshift of this afterglow, and $\nu_{15} = 0.4$ for the R-band).

Using $n_1 = 1$, the solution to these equation yields $\epsilon_B^{1/2} \epsilon_e^4 \cong 10^4$. As both ϵ_e and ϵ_B are by definition less than 1, this is an impossible solution. Furthermore, if by adding some fudge factors we manage to achieve a solution with $\epsilon_e \sim \epsilon_B \sim 1$, then the cooling would be fast at 2 days, leading (with such a large value of ϵ) to a radiative evolution and to the high frequency light curve. We therefore find

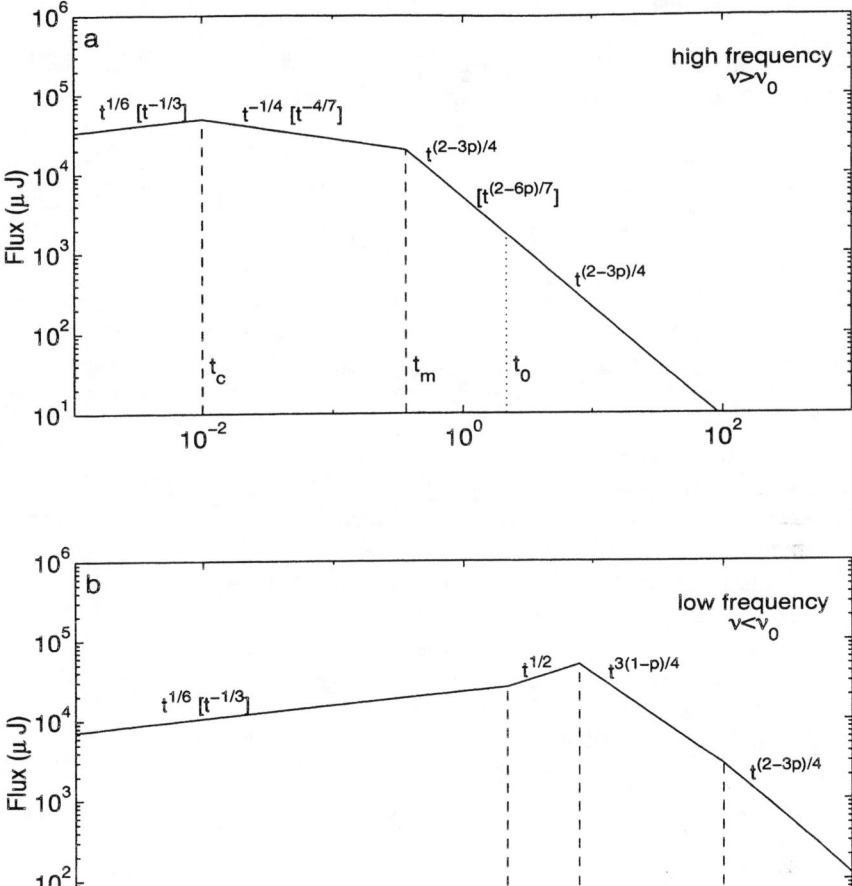

FIGURE 1. Light curve due to synchrotron radiation from a spherical relativistic shock, ignoring the effect of self-absorption. (a) The high frequency case ($\nu > \nu_0$). (b) The low frequency case ($\nu < \nu_0$). The light curves have four segments, separated by the critical times, t_c, t_m, t_0. Time t_m is the time in which the typical synchrotron frequency crosses the observed frequency and t_c is the time from which the electron radiating into the observed frequency is cooling fast. The scalings within square brackets are for radiative evolution (which is restricted to $t < t_0$) and the other scalings are for adiabatic evolution. The break points are shown for adiabatic evolution with $\epsilon_e = 0.5$, $\epsilon_B = 0.2$, $n = 1$ cm^{-3}, $E_{52} = D_{28} = 1$, $p = 2.5$. Panel (a) corresponds to $\nu = 3 \times 10^{14}$ Hz and panel (b) to $\nu = 3 \times 10^{12}$ Hz.

that there is no consistent solution with $n = 1 \text{cm}^{-3}$. The addition of cosmological red-shift effects and IC processes does not change this conclusion.

If instead we consider $n_1 = 10^{-6}$ we find a solution with $\epsilon_e \cong \epsilon_B \cong 0.3$. With $n_1 \cong 10^{-6}$ we are guaranteed to be in the slow cooling scenario, which is a consistency check. The solution $n \sim 10^{-6} \text{cm}^{-3}$ provides an acceptable fit for the afterglow data.[1]

DISCUSSION

We have derived the predicted light-curve of synchrotron models from an expanding shell decelerating on the ISM. We have shown that a qualitative agreement with the data is possible. With $n = 1$ cm^{-3} the theoretical peak is too early, and significantly brighter than that observed. Seeking quantitative agreement with the observations we find the surrounding density is more likely to be around 10^{-6} cm^{-3} than the usual ISM value.

Probably, the main puzzle in GRB afterglow is why there is no optical afterglow in some well-localized GRBs. The previous discussion suggests that in case of a GRB inside a galaxy, with the usual ISM density of $n = 1$ cm^{-3}, the cooling will be fast. This might lead to a rapid and significant loss of energy in the first few hours, making the optical afterglow appear on a short time scale and undetectable after few days.

We thank Ramesh Narayan and Jonathan Katz for helpful discussions. The research was supported by a US-ISRAEL BSF grant 95-328 and by NASA grant NAG5-3516. Re'em Sari thanks The Clore Foundations for support.

REFERENCES

1. Piran, T., in *Unsolved Problems in Astrophysics* (eds J. N. Bahcall and J. P. Ostriker), 343-377 (Princeton, 1996).
2. Sari, R. & Piran, T., *ApJ* **485**, 270 (1997).
3. Kobayashi, S., Piran, T. & Sari, R.,submitted, astro-ph/970513 (1997).
4. Paczyński, B. & Rhoads, J., *ApJ* **418**, L5 (1993).
5. Katz, J., *ApJL* **432**, L107 (1994).
6. Mészáros, P. & Rees, M., *ApJ* **476**, 232 (1997).
7. Sari, R., *ApJL* **489**, L37 (1997).
8. Piran, T. & Sari, R., these proceedings.
9. Sari, R., Piran, T. & Narayan, R., *ApJ*, submitted (1997).
10. Sari, R., Piran, T. & Narayan, R., in preparation (1997).

---

[1]) The possibility of surrounding density of 10^{-6} cm^{-3} might encounter a problem with the inferred size of the fireball from radio scintillation; we will address this problem in detail elsewhere.

Theory of GRB Afterglows

M. Tavani

*Columbia University, 538 West 120th St., New York, NY
and IFCTR-CNR, via Bassini, 15, I-20133 Milan, Italy*

Abstract. Relativistic shock models predict that GRB afterglows are produced only if an energetic population of particles are accelerated and evolve at late times. Strong X-ray afterglows are **not** expected by the brightest GRBs if severe cooling during the burst suppresses the energetic lepton population. Single (prompt) and/or multiple acceleration episodes during and after the bursts can explain the X-ray and optical observations. BeppoSAX and ASCA observations of X-ray afterglows are inconsistent with neutron star cooling models and spindown energy dissipation of magnetized compact stars.

INTRODUCTION

The recent discovery of GRB X-ray afterglows lasting hours/days by BeppoSAX, RXTE and ASCA [3,8,14,27] was a breakthrough both observationally and theoretically. Follow-up observations in the optical and radio bands can be carried out within a reasonable response timescale, and the search for GRB counterparts may resolve the problem of the origin of these enigmatic events. At this time, two GRB error boxes showed the existence of prominent optical transients, GRB970228 [23] and GRB970508 [1]. The nature of the optical source related to GRB970228 is controversial [2,12], and a faint galaxy might be its host [23,18]. The optical transient associated with GRB970508 (of 'anomalous' X-ray/optical/radio properties compared to other GRBs) shows absorption [13] as well as OII emission lines indicating the existence of a foreground galaxy at redshift $z = 0.835$ [13]. These data are encouraging, even though more detections of optical transients within GRB error boxes are needed to confirm the extragalactic origin of GRBs. Fig. 1 shows the GRBs detected by BeppoSAX on the 4B logN-logP distribution function.

However, irrespective of the ultimate origin of GRB sources (i.e., whether they are extragalactic, Galactic or a combination of the two), recent afterglow detections are challenging the theory of GRB emission in many ways. *Contrary to pre-SAX models of GRB emission* characterized by a single bulk Lorentz factor ($\Gamma \simeq 10^2 - 10^3$) of expanding shocks (e.g., [9,10]), a substantial fraction of the GRB energy is emitted at delayed times (at least $\sim 10-20\%$ of the visible energy in the 2-10 keV band for the GRB970228 afterglow [3]). Radiative dissipation at delayed times of order days/weeks requires special conditions of the energetic outflow and surrounding medium as we recently pointed out [20]. The GRB phenomenon is complex, indicating that the particle acceleration and efficient radiation occur during a timescale much longer than the burst itself. A second remarkable observational feature is the variety of temporal decays of the GRB X-ray afterglows. Three out of five error boxes searched within $\lesssim 2$ days after a GRB show fading X-ray sources with a temporal behavior of the flux $f_X \propto t^{-\alpha}$, with $\alpha \simeq 1.3 - 1.5$ (GRB970228 [3], GRB970402 [16], GRB970828 [14]).

This temporal behavior is in contradiction with reverse shock relativistic models (as first shown in [20]), and it might be reconciled with forward shock emission of single-Γ flows only for peculiar and very hard initial particle energy distribution functions [20,26]. The brightest burst detected by BeppoSAX (GRB970111) does not show a detectable afterglow ~ 16 hours after the event, clearly showing a steep temporal decay with $\alpha \gtrsim 2$. The 1997 May 5th burst is different from all the others: it shows an X-ray lightcurve that does not fit any of the power-law decays observed for other GRBs (except maybe during the initial phase of the afterglow, see [17]), and is associated with a *late-time rising* optical transient [1] and variable radio source [4].

PHYSICS OF EMISSION

The energy dissipated in GRB afterglows is a substantial fraction of the total (at least $\sim 10\%$ of the total GRB970228 energy is observed 1–2 days later by BeppoSAX as an X-ray afterglow in the 2–10 keV band [3]). This is surprising for models [10] that predicted a relatively rapid evolution of the burst emission into the X-ray ($\lesssim 10 - 100$ sec) and optical ($\lesssim 10^3 - 10^4$ sec) ranges. This fact is now established in four out of five GRBs with rapid ($\lesssim 2$ days) afterglow searches. It is then clear that the GRB phenomenon persists during a timescale much longer than the burst detected at energies larger than ~ 30 keV. Different models for GRB afterglows can be considered, as we did in the preliminary study of reference [20]. Table 1 of reference [20] summarizes the timing behaviors of GRB afterglows for the most plausible models, and Fig. 2 shows a specific calculation of delayed X-ray and optical emission from forward shock expanding and radiating in a constant density medium [20]. The non-thermal spectra observed for the GRB970228 [5] and GRB970828 [27] X-ray afterglows are consistent with the prediction of a highly relativistic distribution of particles radiating in the 2–10 keV band and evolving from the initial impulsive event. *Cooling neutron star models (see Fig. 2 of reference [20]) and spindown energy dissipation of magnetized compact objects (predicting a flux decay of the type t^{-2}) are excluded by the crucial BeppoSAX and ASCA spectral results.* Accretion phenomena onto the remnant compact object (presumably a black hole in the case of neutron star coalescence) might produce a fading X-ray flux as observed. However, the time decay of the X-ray afterglow, $t^{-1.3}$, observed in three GRBs is not natural in accretion models. Our analysis [20] and others

[26,25,24] showed that relativistic forward shocks may explain the data for special conditions of particle acceleration.

If observable radio-optical-X-ray GRB afterglows are produced by 'radiative remnants' of relativistic outflows, a number of important consequences follows. The possibility of detecting X-rays at delayed times is strongly coupled with the hardness and energy range of the particle distribution function during the decay phase of the impulsive acceleration phase. We can consider single and multiple acceleration scenarios for an expanding relativistic flow. If the burst provides the single acceleration of the simplest models, we can easily calculate the late evolution of the expanding and radiative population of energized particles for different assumptions about their distributions as done in reference [20]. As remarked above, reconciling a time decay behavior of the type $t^{-1.3}$ would require a very hard particle distribution [26] which is *not* consistent with the hard-to-soft pulse spectral evolution detected in the majority of GRBs [21].

The emission of visible afterglows can then be traced back to the existence of a hard component of the particle distribution that does **not** *necessarily contribute to the main GRB pulse emission, and only later becomes radiation efficient.* A dramatic manifestation of this 'decoupling' is the interpretation of the 'delayed' gamma-ray emission observed by EGRET for GRB970217 [7]. This emission can now be interpreted as originating from a very energetic particle population with its typical energy shifted towards energies substantially larger than shown during the main GRB pulse decay [22].

We also note that in cooling models strong X-ray afterglows are not necessarily associated with the most intense GRBs. On the contrary, afterglow emission can be suppressed by strong radiative cooling expected to operate for the most intense GRBs [22]. This might be the case for the very bright GRB970111 with no strong X-ray afterglow 16 hrs after the event [28,6]. This burst is characterized by a relative faintness of the high-energy emission above hundreds of keV (classifiable as 'no-high-energy' (NHE) GRB pulse according to reference [15]). Lack of prompt particle acceleration may fail to produce a prominent non-thermal particle distribution during the main part of the GRB pulse. The 'quasi-Maxwellian' GRB spectrum obtained from the shock synchrotron model may describe successfully this situation [19]. NHE pulses may then be left without the energetic population of particles required by impulsive models of GRB afterglows. More data are required to test the hypothesis of single-acceleration models for GRB afterglows. Strong (weak) X-ray afterglows at late times are expected to be produced by HE (NHE) GRB pulses. Alternately, multiple-acceleration processes during the shock wave expansion may operate leading to the required particle energization. In this case, no strong correlation between prompt and delayed GRB emission is expected, and only the late time evolution (determined by the external environment) matters.

The coupling between late acceleration and radiation of GRB pulses and afterglow emission required in simple models [26] is most likely not consistent with BeppoSAX and ASCA observations. New alternatives need to be considered. Multiple-Γ fireball models with radiative/adiabatic expansion are plausible (but not natural) scenarios. The expanding fireball might be characterized by a multiple-Γ expansion, and different radiative components can become visible as the corresponding bulk kinetic energy is dissipated (e.g., [11]). This model breaks the simplicity of symmetric relativistic expansion, and strongly constrains the properties of a central source required for a complex outflow. Furthermore, the absence of strong X-ray precursors expected by low-Γ fireballs as they become optically thin is problematic for multiple-Γ models.

REFERENCES

1. Bond, H.E., IAU Circ. no. 6654 (1997).
2. Caraveo, P.A., et al., these proceedings.
3. Costa, E., et al., *Nature* **387**, 783 (1997).
4. Frail, D., et al., *Nature* **389**, 261 (1997).
5. Frontera, F., et al., *ApJL*, in press (1997).
6. Galama, T.J., et al., *ApJ* **486**, L5 (1997).
7. Hurley, K., et al., *Nature* **372**, 652 (1994).
8. Marshall, F.E., et al., IAU Circ. no. 6727 (1997).
9. Mészáros, P., Rees, M.J. & Papathanassiou, H., *ApJ* **432**, 181 (1994).
10. Mészáros, P. & Rees, M.J., *ApJ* **476**, 232 (1997).
11. Mészáros, P., these proceedings.
12. Metzger, M.R., et al., IAU Circ. no. 6631 (1997).
13. Metzger, M.R., et al., *Nature* **387**, 878; IAU Circ. no. 6676 (1997).
14. Murakami, T., et al., IAU Circ. nos. 6729, 6732 (1997).
15. Pendleton, G., et al., *ApJ* **489**, 175 (1997).
16. Piro, L., et al., IAU Circ. no. 6617 (1997).
17. Piro, L., et al., IAU Circ. no. 6656 (1997); *A&A* in press.
18. Sahu, K., et al., *Nature* **387**, 476 (1997).
19. Tavani, M., *ApJ* **480**, 351 (1997).
20. Tavani, M., *ApJ* **483**, L87 (1997).
21. Tavani, M., in the Proc. of the NATO ASI *The Many Faces of Neutron Stars*, eds. R. Buccheri, J. van Paradijs (Dordrecht: Kluwer), in press (1998).
22. Tavani, M., in preparation (1998).
23. van Paradijs, J., et al., *Nature* **386**, 686 (1997).
24. Vietri, M., preprint (1997).
25. Waxman, E., preprint (1997).
26. Wijers, R.A.M.J., Rees, M.J., & Meszaros, P., *MNRAS* **288**, L51 (1997).
27. Yoshida, A., et al., these proceedings.
28. in 't Zand, J., et al., IAU Circ. no. 6569 (1997).

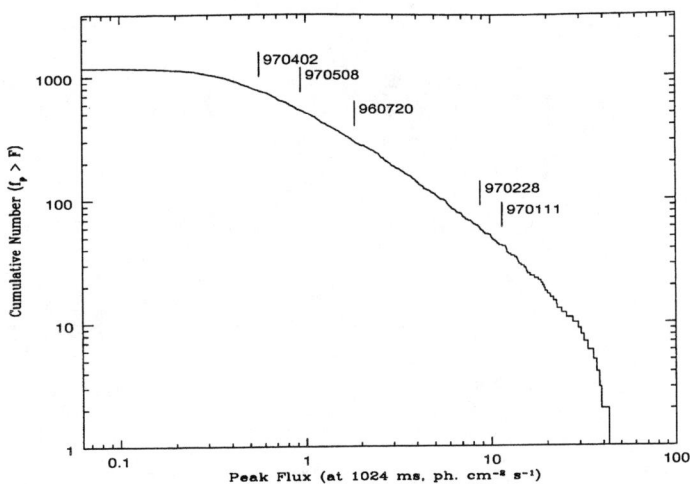

FIGURE 1. GRBs detected by BeppoSAX plotted on the 4B catalogue brightness distribution.

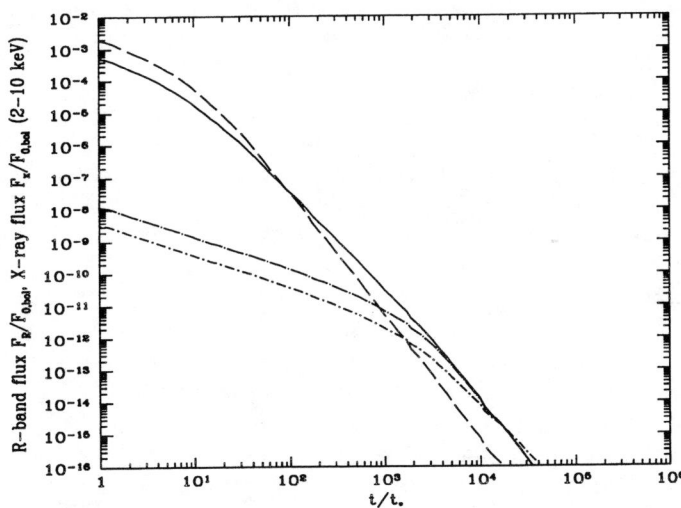

FIGURE 2. Calculated evolution of X-ray and optical (R-band) fluxes for an adiabatically expanding forward shock with no re-acceleration (from [20]). The temporal axis scale is in units of the afterglow radiation timescale t_*, with the synchrotron critical frequency evolving as $\nu_* \sim (t/t_*)^{-3/2}$. *Solid curve:* X-ray (2-10 keV) flux (in units of the initial bolometric flux) for a spectral form $F_\nu \sim \nu^{-1}$ above ν_*; *long-dashed curve:* the same for $F_\nu \sim \nu^{-3/2}$ above ν_*. *Long-dashed-dotted curve:* R-band flux (in units of the initial bolometric flux) for $F_\nu \sim \nu^{-1}$ above ν_*; *short-dashed-dotted curve:* the same for $F_\nu \sim \nu^{-3/2}$ above ν_*.

Limit on the Distance to Gamma-Ray Bursts from Parallax

Tomasz Bulik[1,2] and Bożena Czerny[1]

[1] *Nicolaus Copernicus Astronomical Center, Bartycka 18, 00716 Warszawa, Poland*
[2] *Department of Astronomy and Astrophysics, University of Chicago, 5640 South Ellis Avenue, Chicago, IL 60637*

Abstract. Recent breakthrough discoveries of gamma-ray burst afterglows in X-rays and in the optical band allow the application of parallax distance estimation to this exciting phenomenon. We give a short outline of the method, in which we estimate the amount and direction of parallactic motion for a given distance, direction on the sky, and time of year. We apply this method to the Hubble Space Telescope observations of the probable afterglow from GRB970228, and to the observations of the proposed radio counterpart of GRB970508. We obtain 3σ distance lower limits of 5 pc for GRB970228 and 316 pc for GRB970508, which excludes the Oort Cloud and all local sites for these bursts.

INTRODUCTION

The breakthrough in the field of GRBs came with the detection of GRB970228 on February 28, 1997 by the Italian-Dutch satellite BeppoSAX [7]. A fast and accurate measurement of the source position with the X-ray camera yielded an error box only ~ 3 arcmin in radius [8]. This was followed by numerous optical observations, which led to a detection of an optical variable [11] which has been associated with the burst. This source was observed by HST on March 26 [15] and April 7 [16]. A comparison of the two Hubble observations led to a disputed claim that the point source has a significant proper motion of ≈ 540 mas year^{-1} [4], with a displacement between March 26 and April 7 of 18 ± 4.5 mas in the SE direction. These results have been questioned by Sahu, et al. [17], who found no evidence for proper motion; indeed, their result was a displacement of about 6 mas (at approximately 1.5 to 2σ level) in the direction opposite to that found by Caraveo, et al. [4]. However, the analysis by Butler & Shearer [3] supports the early results of Caraveo, et al. [4]. Butler & Shearer [3] compared the first HST observation with the results of the second in two spectral filters V and I, and found the displacement of 14.3 ± 3.2 mas and 13.3 ± 3.6 mas, respectively, in the same direction as first reported by Caraveo, et al. [4] (we will denote these results as Butler & Shearer a, and Butler & Shearer b, respectively). A similar analysis has

TABLE 1. Lower limits on the distance to GRB970228.

$\delta\phi$ [mas]	Significance	$D >$ [pc]	Reference
12	3σ	6.3	Sahu et al. [17]
30	3σ	2.5	Castander [6]
18	4σ	4.2	Caraveo et al. [4]
12.8	4σ	5.9	Butler and Shearer a [3]
14.4	4σ	5.2	Butler and Shearer b [3]

been performed by other groups; Castander [6] found a conservative upper limit of 10 mas on the displacement.

Two months later another burst with a probable low-energy afterglow was found; GRB970508 discovery by the BeppoSAX satellite [9] was followed by a massive observational campaign. First an optical variable was found [1], and then a variable radio source was discovered at a location consistent with the optical source. The radio source was monitored for at least three weeks [18,19], and VLBA observations allowed it to be located with a very high accuracy of 0."001.

EXPECTED PARALLAX

Every object on the sky has a periodic angular velocity due to the motion of the Earth around the Sun. For a source located at (δ, α) at a given time t_{obs}, this velocity is

$$\omega = 2\pi D^{-1}(1 - u^2)^{1/2} \text{rad yr}^{-1} \qquad (1)$$

where D is the distance to the source in astronomical units, $u = \cos\tau(\cos\delta\sin\alpha\cos\epsilon + \sin\delta\sin\epsilon) - \cos\delta\cos\alpha\sin\tau$, ϵ is the obliquity of the ecliptic, $\tau = 2\pi(t_{\text{obs}} - t_{\text{Vernal}})/1\text{yr}$, and t_{Vernal} is the date of the vernal equinox. The velocity on the sky is in the direction described by the angle θ (measured counterclockwise) to the local direction North:

$$\cos\theta = \frac{\sin\tau\sin\delta\cos\alpha - \cos\epsilon\cos\tau\sin\delta\sin\alpha + \sin\epsilon\cos\tau\cos\delta}{(1-u^2)^{1/2}} \qquad (2)$$

$$\sin\theta = \frac{\sin\alpha\sin\tau + \cos\epsilon\cos\tau\cos\alpha}{(1-u^2)^{1/2}}. \qquad (3)$$

If an object, like a GRB afterglow, is observed for a time $\delta t \ll 1$ yr the expected displacement is

$$\Delta\phi = 2\pi D^{-1}(1-u^2)^{1/2}\frac{\delta t}{1\text{yr}}, \qquad (4)$$

which may allow a distance determination provided that the proper velocity of the object is small enough. In general, a longer observation of a trajectory on the

TABLE 2. Estimates of the distance to GRB970228 with 1σ errors.

$\Delta\alpha$ [mas]	$\Delta\delta$ [mas]	$\delta\phi$ [mas]	θ [deg]	D [pc]	Reference
13.5	-11.7	18 ± 4.5	70	$12.86^{+43}_{-5.6}$	Caraveo et al. [4]
13.5	-5.9	14.3 ± 3.2	54	$9.0^{+4.5}_{-2.3}$	Butler and Shearer a [3]
9.45	-9.45	13.3 ± 3.6	75	$21.7^{+268}_{-10.5}$	Butler and Shearer b [3]

sky leads to the determination of both the parallax and the proper velocity. This, however, is difficult for a fading source, such as a GRB afterglow.

If instead no displacement is found, the distance to the source can be constrained

$$D > \frac{2\pi}{\delta\phi}(1-u^2)^{1/2}\frac{\delta t}{1\text{yr}}, \qquad (5)$$

where $\delta\phi$ is the upper limit on the displacement.

APPLICATION OF THE PARALLAX METHOD

We first illustrate the method using the Hubble Space Telescope observations of the GRB970228 optical afterglow [17]. The results of these observations are shown in Figure 1. The expected direction of the parallactic motion at the time of observation was approximately towards the North-East, and is shown by the two arrows corresponding to the date of the first and the last observation. The measured displacements are shown as circles.

We first consider all the measurements as upper limits on the displacement, which leads to a lower limit for the distance to GRB970228. From equation (5) we obtain in this case: $D > 6.26 \times 10^{-3} \frac{\delta t}{1 \text{ day}} \frac{1"}{\delta\phi}$. The results for the two HST observations ($\delta t = 12$ day) are presented in Table 1. These estimates vary when considering results of different data analysis teams. It can be safely said that the distance to GRB970228 is higher than approximately 5 to 6 pc.

The distance estimates based on the assumption that the displacements are not upper limits but detections are summarized in Table 2, where we list the measured displacement, the angle between the displacement, the expected direction of the parallactic motion, and the estimate of the distance.

As a second example we apply the parallax method to the observations of the possible radio afterglow from GRB970508 [18,19]. The accuracy of the initial discovery by Frail, et al. [10] with the VLA was 0.1"; however, a later VLBA observation yielded an impressive accuracy of 0.001", and the extent of the object has been established to be smaller than 0.0003". Here equation (5) becomes

$$D > 1.58 \times 10^{-2} \frac{\delta t}{1 \text{ day}} \frac{1"}{\delta\phi} \qquad (6)$$

Taking into account just the first observation [18], which lasted 0.23 day, imposes a limit $D > 3.6$ pc, similar to the one obtained with HST for GRB970228 over

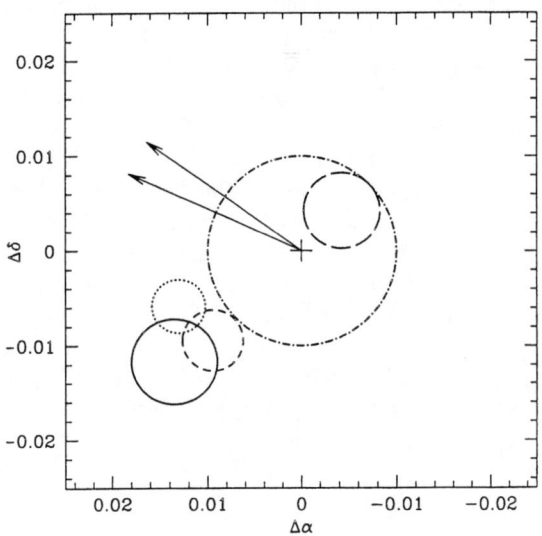

FIGURE 1. Position of the probable afterglow from GRB970228 in the HST observation on April 7 compared to the position from HST observation on March 26. Each circle represents a 1σ contour. The long dashed circle is the measurement of Sahu et al. [17], the solid circle is the result of Caraveo et al. [4,5], and the short-dashed and the dotted circles are the results of Butler and Shearer [3], for the two sets of data they analyzed. The dot-dashed circle is the conservative estimate of the positional uncertainty by Castander [6]. The arrows show the direction of the expected parallactic motion at the time of the two observations: March 26 (upper arrow) and April 7 (lower arrow).

almost two weeks. Given that the source has not moved by more than 0."001 in 20 days [19] this limit goes up to $D > 316$ pc.

DISCUSSION

We have applied a method based on expected parallactic motion to the HST observations of the GRB970228 afterglow. We find a lower limit on the distance of about 5 pc, and can not constrain the distance from above. This is because we deal either with a limit on the displacement between two observations, or with a detection of motion in a direction that is at a large angle to that expected for parallax. It should be noted that we have assumed that the component of the displacement along the expected parallax is entirely due to the parallax. This assumption does not have to hold, and if the object is close, say several pc away, then its proper velocity will very likely lead to confusion. The limit presented here clearly rules out the Oort cloud as the site of GRB970228, and also rules out all local origins for GRB970508. The limit presented above is much stronger than the estimate based on triangulation; Hurley [12] used the IPN together with the highly accurate optical position to obtain a distance limit of 12000 AU, i.e. 0.05 pc. IPN distance estimates based on the measurement of flatness of the wave front are much smaller, typically a few hundred AU.

We have not been able to constrain the distance to GRB970228 from above. If the displacement measured is real, then the proper motion leads to some constraints on the distance from other physical considerations, e.g. comparison of the velocity with the speed of light, or energetics of a relativistic shock [2].

If all GRBs come from a single physical mechanism then the existing evidence favors the cosmological distance scale. However, GRBs are extremely heterogeneous: their durations span at least four orders of magnitude and show bimodality [13,14], the light curves of bursts have many shapes, there is evidence of two classes of bursts based on their spectra [20], and even the properties of afterglows are widely disparate. Hence, there may actually be several distinct classes of GRBs, just as earlier in this century the distinction between novae and supernovae was unclear until improved observations demonstrated that their energy releases are different by ten orders of magnitude. Thus, even if redshifts are determined for some GRBs, distance limits to other GRBs are still valuable.

ACKNOWLEDGMENTS

The authors wish to acknowledge the support of the following grants KBN-P03D00911, KBN-P03D01113, NASA NAG 5-4509 and NASA NAG 5-2868 (TB), and KBN-2P03D00410 (BC). TB wishes to thank the Aspen Center for Physics for hospitality, Mr. and Mrs. Anuszkiewicz of Cambridge, Ontario for kindly allowing me to use a laptop, and Cole Miller for many helpful remarks on the manuscript.

REFERENCES

1. Bond, H., IAU Circ. 6654 (1997).
2. Bulik, T., & Sikora, M., in *Proceedings of the Workshop on Jets in AGNs*, Kraków 1997, eds. M. Ostrowski, M. Sikora, M. Begelman, G. Madejski.
3. Butler, R. F., & Shearer, A., *A&A*, submitted.
4. Caraveo, P.A., Mignani, R., & Bignami, G.F., IAU Circ. 6629 (1997).
5. Caraveo, P.A., et al., *A&A*, submitted (1997).
6. Castander, F., private communication (1997).
7. Costa, E., et al., IAU Circ. 6572 (1997).
8. Costa, E., et al., IAU Circ. 6576 (1997).
9. Costa, E., et al., IAU Circ. 6649 (1997).
10. Frail, D. A., Kulkarni, S., and the SAX Team, IAU Circ. 6662 (1997).
11. Groot, P. J., et al., IAU Circ. 6584 (1997).

12. Hurley, K., *ApJL*, submitted (1997).
13. Kouveliotou, C., et al., *ApJL* **413**, L101 (1993).
14. Lamb, D. Q., Graziani, C., & Smith, I. A., *ApJL* **413**, 11L (1993).
15. Sahu, K., Livio, M., Petro, L., & Macchetto, F. D., IAU Circ. 6606 (1997).
16. Sahu, K., Livio, M., Petro, L., Macchetto, F. D., van Paradijs, J., Kouveliotou, C., Fishman, G., & Meegan, C., IAU Circ. 6619 (1997).
17. Sahu, K., et al., *Nature* **387**, 486 (1997).
18. Taylor, G. B., Beasley, A. J., Frail, D. A., & Kulkarni, S. R., IAU Circ. 6670 (1997).
19. Taylor, G. B., Frail, D. A, Beasley, A. J., & Kulkarni, S. R., *Nature*, in press (1997).
20. Pendleton, G. N., in *Gamma Ray Bursts, Proceedings of the Third Huntsville Symposium*, (New York: AIP), 1993, eds.: C. Kouveliotou, M. F. Briggs, G. J. Fishman.

Afterglow Hydrodynamics

Re'em Sari[1]

Racah Institute of Physics, Hebrew University, Jerusalem 91904, Israel

Abstract. We explore the interaction of a relativistic shell with a uniform interstellar medium (ISM). At a very early stage (few seconds), the observed bolometric luminosity increases as t^2. On longer time scales (more than ~ 10 s), the luminosity drops as t^{-1}. If the main burst is long enough, an intermediate stage of constant luminosity will form. In this case, the afterglow overlaps the main burst, otherwise there is a time separation between the two. The effects of deceleration and accumulation of ISM mass on the relation between the observed time, the shock radius, and its Lorentz factor is discussed. We show that even if only a small fraction of the energy is given to the radiating electrons, most of the energy can be radiated over time.

THE AFTERGLOW RISE AND THE MAIN BURST

All cosmological models of GRBs include relativistic motion, which is then dissipated to give the observed γ-rays (see [1]). Sari & Piran [2] have shown that deceleration on the ISM could not give rise to the variability observed in the bursts, while internal shocks could produce efficiently the observed fluctuations [3] if the "inner engine" has considerable fluctuations. These considerations predict that after the main GRB event, which is produced by internal shocks, the ejecta decelerate due to interaction with the ISM, emitting radiation at longer and longer wavelengths [4–6,2]. This emission has been detected recently for several GRBs due to an accurate determination of their position.

The hydrodynamical evolution is determined by four parameters: the initial shell's Lorentz factor η, the energy of the shell $E = E_{52}10^{52}$ erg, the width of the shell (observer frame) $\Delta = \Delta_{12} \times 10^{12}$ cm and the ISM density $n = n_1$ cm^{-3}. The basic details of the interaction between the shell and the ISM where given in [7,8]. In the following we review these results and estimate the bolometric luminosity at each stage from the rate of internal energy production, i.e.,

$$L(t) \sim 4\pi R^2 \frac{dR}{dt}\gamma^2 n m_p c^2 \sim 8\pi R^2 \gamma^4 n m_p c^3 \sim 32\pi \gamma^8 t^2 n m_p c^5 \qquad (1)$$

Where we used $R = 2\gamma^2 ct$ for the second and third expressions (this is approximate, see the discussion in the next section).

---

[1] sari@nikki.fiz.huji.ac.il

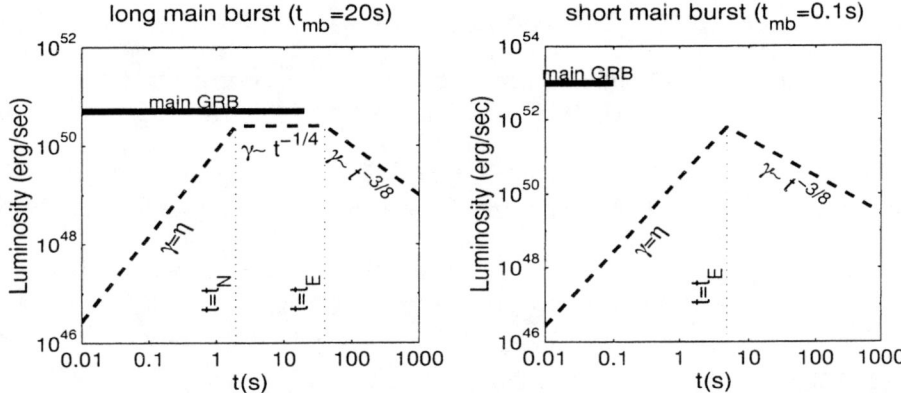

FIGURE 1. Relativistic (left frame) and Newtonian (right frame) cases are drawn in dashed lines. At the early stage, the Lorentz factor is constant and the luminosity increases due to the increase in shell area. When the ISM has energy comparable to the total energy ($t = t_E$), a self-similar solution is established and the luminosity drops as t^{-1}. If the shell is thick (typical for long main bursts) the reverse shock becomes relativistic at t_N, before the self-similar solution is established and some deceleration begins, leading to constant luminosity. The solid line gives the luminosity of the main GRB. Both frames use $E = 10^{52}$ erg and $\eta = 300$.

When the shell encounters the ISM, two shocks are formed: a forward shock accelerating the ISM and a reverse shock decelerating the shell. The forward shock is always highly relativistic since the initial Lorentz factor $\eta \gg 1$. As the density ratio between the shell and the ISM is initially very large, the reverse shock is initially Newtonian and the shell's Lorentz factor equals its initial value $\gamma = \eta$. However due to the increase in the area of the shell, it produces internal energy at an increasing rate of $L = 2.5 \times 10^{50} \gamma_{300}^8 n_1 t_s^2$ erg/s, where t_s is the observed time in seconds.

This initial stage of constant Lorentz factor will continue until either the shell has given the ISM an energy comparable to its initial energy at

$$t_E = 5\ E_{52}^{1/3} \eta_{300}^{-8/3} n_1^{-8/3}\ \text{s}, \tag{2}$$

or until the reverse shock is no longer Newtonian at

$$t_N = 1.5 E_{52}^{1/2} \Delta_{12}^{-1/2} \eta_{300}^4 n_1^{-1/2}\ \text{s}, \tag{3}$$

whichever comes first. Defining

$$\xi \equiv \frac{t_N}{t_E} = 0.3 E_{52}^{1/6} \Delta_{12}^{-1/2} \eta_{300}^{-4/3} n_1^{-1/6}, \tag{4}$$

then for $\xi > 1$, the energy is dissipated to internal energy before the Lorentz factor of the shell is reduced considerably. If $\xi < 1$, then the reverse shock turns relativistic before the kinetic energy of the shell was extracted. In this case the Lorentz factor decreases with time according to $\gamma(t) = 280\ E_{52}^{1/8}\Delta_{12}^{-1/8}n_1^{-1/8}t_s^{-1/4}$, independent of its initial value η [8]. The relation $\gamma \propto t^{-1/4}$ leads through Eq. (1) to $L(t) = cE/(2\Delta)$ independent of time. This stage will continue until the shell has given the shocked ISM energy comparable with its own at $t_E = 2\Delta/c$. This behavior is illustrated in Figure 1.

The relation of the afterglow with the main burst can now be established assuming that the latter is due to internal shocks. In the internal shocks scenario, the width of the shell, Δ, can be inferred directly from the observed main burst duration $\Delta = ct_{mb}$ [9.3]. For long bursts, $t_{mb} \sim 20$ sec and $\Delta \sim 6 \times 10^{11}$ cm while for short bursts $t_{mb} \sim 0.1$ sec and $\Delta \sim 3 \times 10^9$ cm. Using Eq. (4) and $\eta > 100$ we find that the reverse shock is likely to be Newtonian for short bursts and might be relativistic for long bursts. Both cases are therefore of physical interest.

If the reverse shock is relativistic, then the observed peak of the afterglow emission is flat and overlaps the observed GRB emitted by internal shocks. Both end after an observed time of $\sim \Delta/c$. If the reverse shock is Newtonian, then the afterglow peaks on t_E which is longer by a factor of $3\xi^2/2 > 1$ than the main burst duration Δ/c. The duration and luminosity of the main burst and the afterglow rise are shown for the Newtonian and relativistic cases in Figure 1.

In both cases, the properties of the main burst (from IS) and the afterglow are very different. The main burst is usually highly variable (depending on the internal structure of the shells) while the afterglow, which is due to external shocks is expected to be smooth [2]. The afterglow spectrum should peak, in the beginning, around 30KeV-100MeV depending on the initial Lorentz factor η and on the fraction of internal energy in electrons and magnetic field [10]. If the peak energy is too high it might not be observed in the first stage by the BATSE equipment. However, later, as the ejecta decelerates, the emission frequency decreases in time and should cross the soft γ-ray region.

SELF-SIMILAR STAGE

When most of the energy has been given to the ISM, and assuming that radiation losses are small, the energy in the shocked ISM is proportional to $M\gamma^2 \sim R^3\gamma^2$. This equals the constant initial kinetic energy E, so that $\gamma \propto R^{-3/2}$. The same scaling applies to the shock Lorentz factor $\Gamma = \sqrt{2}\gamma$. This implies a quantitative change in the relation between t, R and γ. Photons that were emitted from the shock while it has propagated a small distance δR will be observed on a timescale of $\delta t \sim \delta R/2\Gamma^2 c$. Integrating this over time using $\Gamma \propto R^{-3/2}$ we have $t = R/16\gamma^2 c$, compared with the commonly used expression $R/2\gamma^2 c$. More detailed work concerning the angular spreading in photon arrival time show that the effective relation is $t = R/c_t\gamma^2 c$ with c_t in the range of 3 to 7 [11–13].

Blandford and McKee [14] have described an analytical hydrodynamical profile for the case in which the scaling $\gamma \propto R^{-3/2}$ applies. Using this solution, it is possible to get the exact coefficient relating R, γ, n and E as

$$E = 16\gamma^2 R^3 n m_p c^2 / 17. \tag{5}$$

Together with an approximate relation of $t = R/4\gamma^2 c$, we have

$$R(t) = 2.3 \times 10^{16} E_{52}^{1/4} n_1^{-1/4} t_d^{1/4} \text{ cm} , \quad \gamma(t) = 440 E_{52}^{1/8} n_1^{-1/8} t_s^{-3/8} \tag{6}$$

The time for which the flow becomes sub-relativistic is therefore given by $t = 120 E_{52}^{1/3} n_1^{-1/3}$ days.

The assumption of constant energy can not be strictly correct since some radiation is emitted. We define ϵ_e to be the fraction of the internal energy that is radiated, and lost from the system. As long as the cooling is fast, this is the fraction of energy given by the shock to electrons and is estimated to be larger than 10% [15,16]. This number seems to be negligible, and therefore the energy loss was neglected by previous analyses. Using $dE = -\epsilon_e 4\pi R^2 dR \gamma^2 n m_p c^2$ and Eq. (5) we get $E(t) \propto t^{-17\epsilon_e/16}$. This is a good approximation as long as $\epsilon_e \ll 1$, as it used the relation (5) which is valid with no energy loss. The effect of radiation losses on the scaling (6) can be estimated by replacing $E \to E(t/t_0)^{-17/16\epsilon_e}$.

DISCUSSION

We have explored the early evolution of the interaction of a relativistic shell with the ISM. We have shown that at the initial stage, where the Lorentz factor of the shock is still constant, the luminosity increases as t^2. This stage lasts for about ~ 10 s. For long bursts it is followed by a constant luminosity phase which overlaps (in the observer frame) the main bursts. On the other hand, if the main GRB is short enough, separation is expected between the main burst and the afterglow luminosity peak. This behavior might be detectable in BATSE's data.

The effects of deceleration and collection of mass on the observed timescale were derived. For radiation emitted from the shock vicinity on the line of sight connecting the source and the observer, the observed time scale is given by $t = R/16\gamma^2 t$. Photons arriving from points that are not on the line of sight arrive later leading to a smaller numerical coefficient in this relation.

The role of energy loss due to the radiation was found to be non-negligible even if the part of the internal energy that is radiated at each time is small. Since the self-similar stage begins after ~ 10 s and the cooling is fast up to few days, radiation losses can reduce the energy in the system by a factor of ~ 3 if $\epsilon_e = 0.1$ or a factor of ~ 30 if $\epsilon_e = 0.3$.

ACKNOWLEDGMENTS

The research was supported by a US-ISRAEL BSF grant 95-328 and by NASA grant NAG5-3516. I thank The Clore Foundations for support.

REFERENCES

1. Piran, T., in *Unsolved Problems in Astrophysics* (eds J. N. Bahcall and J. P. Ostriker) 343-377 (Princeton, 1996).
2. Sari, R. & Piran, T., *ApJ* **485**, 270 (1997).
3. Kobayashi, S., Piran, T. & Sari, R., astro-ph/970513 (1997).
4. Paczyński, B. & Rhoads, J., *ApJ*, **418**, L5 (1993).
5. Katz, J., *ApJ* **432**, L107 (1994).
6. Mészáros, P. & Rees, M., *ApJ* **476**, 232 (1997).
7. Sari, R. & Piran, T., *ApJ* **455**, L143 (1995).
8. Sari, R., *ApJ* **489**, L37 (1997).
9. Sari, R. & Piran, T., *MNRAS* **287**, 110 (1997).
10. Sari, R., Narayan, R. & Piran, T., *ApJ* **473**, 204 (1996).
11. Waxman, E., submitted to *ApJ Letters*; astro-ph/9709190 (1997).
12. Panaitescu, A. & Mészáros, P., submitted to *ApJ*; astro-ph/9709284 (1997).
13. Sari, R., submitted to *ApJ Letters* (1997).
14. Blandford, R. D. & McKee, C. F., *Phys. of Fluids* **19**, 1130 (1976).
15. Waxman, E., *ApJ*, in press; astro-ph/9704116 (1997).
16. Waxman, E., *Nature*, submitted; astro-ph/9705229 (1997).

THEORY

Radio Constraints on Shell Afterglow Theories

J. J. Brainerd

Physics Department, University of Alabama in Huntsville, Huntsville, AL 35899

Abstract. The radio afterglow from GRB970508 places strong constraints on the gamma-ray burst energetics. In this article, I discuss the constraint on the total energy in the gamma-ray burst if the afterglow is optically thick synchrotron emission from a relativistic shell that sweeps up interstellar gas. It is found that if the shell cools rapidly as it expands, then the gamma-ray burst must contain of order a solar rest mass energy or more to produce the radio emission. If none of the energy is radiated, then the energy in the gamma-ray burst is much larger than the observed gamma-ray fluence of 2×10^{51} ergs, the source diameter must be larger than the inferred 3 μarcsec, and > 20% must end up in the electrons at the time of the radio emission. Some of these limits can be circumvented by allowing a short period of efficient cooling.

SYNCHROTRON LIMITS

The gamma-ray burst GRB970508 is unique in having a measured redshift [3] and a radio afterglow [2]. These properties permit us to place a lower limit on the thermal energy stored in the electrons producing the observable radio emission [1]. These limits are

$$E_e > 4.14 \times 10^{60} \text{ ergs} \left[\frac{(p+2)^3 G(p, x_0)}{H^3(p, x_0)} \right] \left(\frac{r}{h} \right)^2 \left(\frac{\Gamma + U}{1+z} \right)^{-5} F_{\text{mJy}}^4 \nu_{\text{GHz}}^{-7} \theta_{\mu\text{arcsec}}^{-6}. \quad (1)$$

In this equation, h is the Hubble constant in units of $100 \text{ km s}^{-1} \text{ Mpc}^{-1}$, and r is the coordinate distance $r = \left[q_0 z + (q_0 - 1) \left(\sqrt{1 + 2q_0 z} - 1 \right) \right] / q_0^2 (1+z)^2$, where q_0 is the expansion coefficient and z is the redshift. The dimensionless bulk momentum of the radio source is U, and the bulk Lorentz factor is $\Gamma = \sqrt{1 + U^2}$. The flux of the source in milliJanskys is F_{mJy}, the observed frequency in GHz is ν_{GHz}, and the source diameter in μarcsec is $\theta_{\mu\text{arcsec}}$. The term in brackets represents convolutions of the electron distribution function with the synchrotron emissivity function. The electron distribution is a power law of index p and low energy cutoff γ_0. The parameter x_0 is the dimensionless synchrotron frequency, which is defined as

$$x_0 = \frac{2\epsilon}{3\gamma_0^2 \bar{B} \sin \psi} \frac{1+z}{\Gamma + U}, \quad (2)$$

where ϵ is the photon energy in units of $m_e c^2$ and $\bar{B} = B/B_{cr}$, with $B_{cr} = 4.414 \times 10^{13}$ G. The synchrotron cooling timescale gives a lower limit on the value of x_0. The term in brackets in equation (1) is $\propto x_0^{\frac{p-2}{2}}$ for $x_0 \gg 1$, and it is $\propto x_0^{2/3}$ for $x_0 \ll 1$.

To make effective use of this equation, one needs some way to relate the bulk Lorentz factor to the observed size of the radio source. For an isotropic source expanding at a constant velocity, one has the relationship

$$U = 8.66 \frac{r(1+z)}{h} \frac{\theta_{\mu\text{arcsec}}}{T_{\text{day}}}, \qquad (3)$$

where T_{day} is the time in days that has passed since the gamma-ray burst. In the discussion that follows, $T_{\text{day}} = 25$, $z = 0.835$, $q_0 = 0$, $h = 0.8$, $\nu_G = 1.43$, $F_{\text{mJy}} = 0.1$, and $p = 3$.

One finds that the limits on energy found by combining equations (1) and (3) are very high, with the lowest limits occurring at the value of x_0 for which the cooling timescale equals the time of the observation [1]. For $\theta = 3\,\mu\text{arcsec}$, one finds $E_e > 10^{51}$ ergs, and for $\theta = 6\,\mu\text{arcsec}$, $E_e > 10^{49}$ ergs. One also finds that the source must be $> 24\,\mu\text{arcsec}$ in diameter.

SHELL THEORIES

The predominate theory of afterglow emission is that a relativistic shell of baryonic material propagates into the interstellar medium, and the kinetic energy of the shell is converted into thermal energy by a shock [5]. The velocity of the shell is determined by the amount of material swept up.

The total energy in the baryonic shell can be related directly to the thermal energy of the electrons responsible for the radio emission. The initial energy in the shell is $E_s = M_s c^2 \Gamma_s$. Because of the relativistic motion, the fraction of the shell that is observed in the radio is $2/\Gamma(\Gamma + U)$. The thermal energy in the electrons responsible for the observed radio emission is $E_e = m_e c^2 \gamma_0 N_e (p-1)/(p-2)$, where N_e is the number of electrons producing the observed radio emission. Assuming that the atoms in the shell and the interstellar medium are hydrogen, this can be written as $E_e = 2 m_e c^2 \gamma_0 M / m_p \Gamma(\Gamma + U)$, where $M = M_s + m$, and m is the mass of the interstellar medium that is swept up. The total initial energy in the shell is then

$$\frac{E_s}{E_e} = \frac{m_p \Gamma_s (p-2)}{m_e \gamma_0 (p-1)} \frac{1}{2} \Gamma(\Gamma + U) \frac{M_s}{M}. \qquad (4)$$

The shell theory has two regimes that define the maximum and minimum values of M [4]. In the first, the kinetic energy created as mass is swept up by the shell is immediately radiated away; this is cooling expansion. In the second, radiative and transfer processes are inefficient, so that essentially no kinetic energy is lost as the system expands; this is non-radiative expansion.

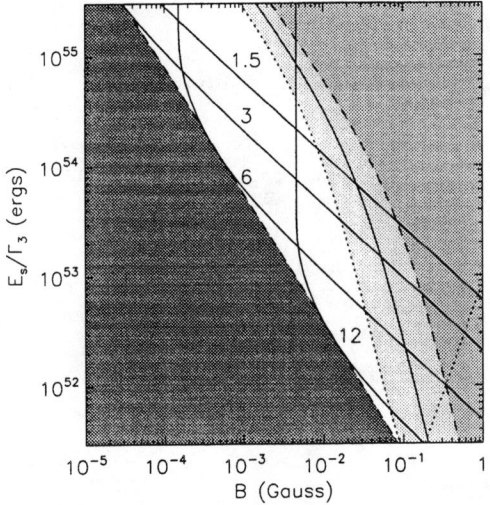

Fig. 1—Minimum E_s/Γ_3 versus magnetic field strength for cooling expansion. Here, $\Gamma_3 = \Gamma/10^3$. The solid lines are lower limits on E_s/Γ_3 for source diameters of 1.5, 3, 6, and 12 μarcsec. The shaded region to the lower left is the region with no solution. The shaded regions to the right are the synchrotron cooling limits for timescales of, from right to left, 1/10 yr, 1 yr, and 10 yr.

Cooling Expansion

For cooling expansion, the dimensionless bulk momentum for $U_s \gg 1$ is given by

$$\frac{U}{U_s} = \frac{2M_s(M_s + m)}{(M_s + U_s m)(2M_s + m)}. \tag{5}$$

In this equation, U_s is the initial momentum, and m is the rest mass swept up by the shell. Inverting this equation for the total rest mass M in the limit of $U \ll U_s$, one finds

$$\frac{M}{M_s} = 1 + \frac{1}{U} + \frac{1}{U(U + \Gamma)}. \tag{6}$$

Combining equations (1), (4), and (6), one finds the lower limits on E_s. These limits are shown in Figure 1 as functions of magnetic field for four different values of $\theta_{\mu arcsec}$.

In this figure, one sees that the energy in the gamma-ray burst must be greater than 10^{54} ergs for the 3 μarcsec source diameter inferred from a ISM scintillation origin of the radio variability. Even a size as large as 6 μarcsec requires a source energy that is greater than 10^{53} ergs.

Non-radiative Expansion

Non-radiative expansion sweeps up much more gas than cooling expansion, because the thermal energy contributes to the rest mass energy of the shell. If the mass and momentum of the shell at the onset of adiabatic expansion are M_a and

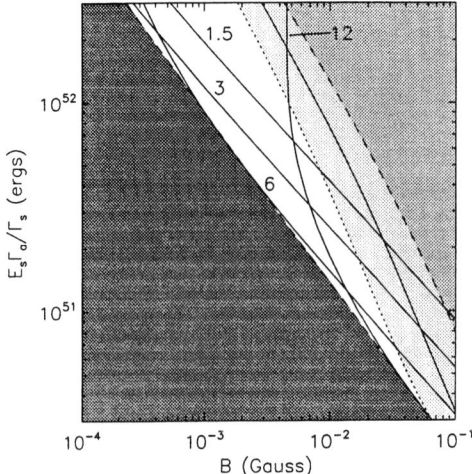

Fig. 2—Minimum shell energy versus magnetic field strength for a non-radiatively expanding shell. The initial bulk Lorentz factor is Γ_s and the bulk Lorentz factor at which cooling expansion ends is Γ_a. The curves are the same as in Figure 1.

U_a, then the dependence of U on the rest mass m that is swept up is

$$\frac{U}{U_a} = \frac{M_a}{\sqrt{M_a^2 + 2\Gamma_a M_a m + m^2}}. \tag{7}$$

Inverting for the total mass, one finds

$$\frac{M}{M_a} = 1 + \frac{U_a^2 - U^2}{U(\Gamma_a U + U_a \Gamma)}. \tag{8}$$

For the gamma-ray energy in GRB970508, a non-radiative expansion with $U_a \approx 10^3$ requires a local interstellar medium of density $> 10^3$ cm^{-3}.

When there is rapid cooling as the shell expands from U_s to U_a, and adiabatic expansion from U_a to U, virtually all of the rest mass at U is from the adiabatic expansion. In this case, one can set $M_a = M_s$, so that equations (1), (4), and (8) give lower limits on $E_s \Gamma_a / \Gamma_s$. These are shown in Figure 2 for four different values of $\theta_{\mu arcsec}$. When interpreting this figure, one must keep in mind that the fluence in the gamma-ray band is $\approx 2 \times 10^{51}$ ergs if the source emits isotropically, so the total energy in the gamma-ray burst must be much greater than this value if $\Gamma_a = \Gamma_s$. If $\Gamma_a \ll \Gamma_s$, so that the gamma-ray fluence can be interpreted as the gamma-ray burst energy, the limits in Figure 2 rapidly rise above 2×10^{51} ergs.

An important aspect of non-cooling expansion is that some solutions have more energy in the radio emitting electrons than in the initial shell. For $\Gamma_a \approx U_a \gg 1$, equation (4) can be written as

$$\Psi = \frac{\Gamma(\Gamma+U)}{2E_s} E_e \Gamma = \frac{m_e \gamma_0 (p-1)}{m_p (p-2)} \frac{\Gamma}{U(\Gamma+U)} \frac{\Gamma_a}{\Gamma_s}. \tag{9}$$

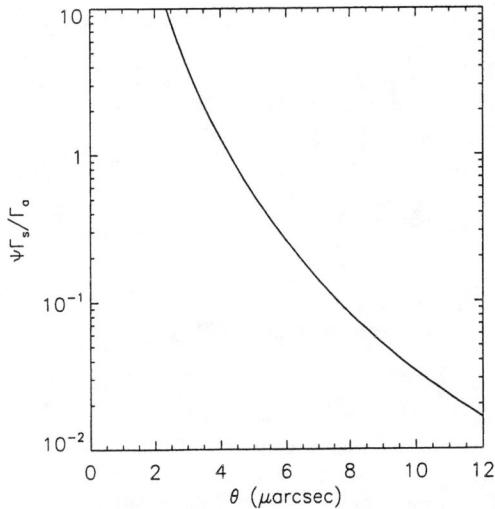

Fig. 3—Ratio of electron energy to gamma-ray burst energy. From equation (9).

Maximum values of γ_0 [1] are $\gamma_0 = 3787$ for $\theta = 3\,\mu\text{arcsec}$ and $\gamma_0 = 533$ for $\theta = 6\,\mu\text{arcsec}$, so the thermal and kinetic energy in electrons, $E_e\Gamma$, is approximately the initial kinetic energy in the fraction of the shell that produces the radio afterglow, $2E_s/\Gamma\,(\Gamma+U)$. The fraction of the energy in the electrons as a function of source diameter is shown in Figure 3. One clearly requires $\theta_{\mu\text{arcsec}} > 3$. Even with $\theta_{\mu\text{arcsec}} = 6$, one requires 0.2 of the initial shell energy to end up as electron thermal and kinetic energy.

A small area of parameter space is available for a 2×10^{51} ergs gamma-ray burst, but this requires a short interval of rapid cooling. For $\Gamma_s/\Gamma_a > 5$, the energy is 2×10^{51} ergs for $\theta = 6\,\mu\text{arcsec}$.

REFERENCES

1. Brainerd, J. J., *ApJ*, submitted (1997).
2. Frail, D. A., Kulkarni, S. R., Nicastro, L., Feroci, M., & Taylor, G. B., *Nature* **389**, 261 (1997).
3. Metzger, M. R., et al., *Nature* **387**, 878 (1997).
4. Rees, M. J., & Mészáros, P., *MNRAS* **258**, 41P (1994).
5. Wijers, R., *MNRAS*, in press (1997).

Spectral Properties of Gamma-Ray Bursts in the 1 GeV - 1 TeV Range

Matthew G. Baring[†] and Alice K. Harding

NASA/Goddard Space Flight Center, Greenbelt MD 20771
[†] *Compton Fellow, Universities Space Research Association*

Abstract. The properties of gamma-ray burst spectra in the 1 GeV - 1 TeV range may elucidate our understanding of these sources. If bursts are indeed at cosmological distances, it is indisputable that they possess relativistic bulk motion with large Lorentz factors. Calculations of pair production transparency in bursts usually assume an infinite power-law source spectrum for simplicity. However, photons above the EGRET range can potentially interact with sub-MeV photons in such calculations. The observed breaks and spectral curvature seen by BATSE around 1 MeV in many bursts is thus important to include in the pair opacity internal to the source. We present our recent determinations of gamma-ray burst pair opacity using such non-power law spectral forms. The comparative depletion of photons below 1 MeV can generate broad absorption troughs (corresponding to significant opacity) in the 1 GeV - 1 TeV range for sources located in the galactic halo. These distinctive signatures can differentiate between a cosmological or galactic origin of a given source. For cosmological bursts, spectral turnovers in the super-GeV range are possible, with their shape depending on source parameters, notably the distribution of photons within the source. Instruments like INTEGRAL, GLAST, Whipple and MILAGRO will play an important role in discriminating between these possibilities.

INTRODUCTION

High-energy gamma rays have been observed for six gamma-ray burst sources by the EGRET experiment on board the Compton Gamma-Ray Observatory (CGRO), including the emission of an 18 GeV photon by the GRB940217 burst [9]. Taking into account EGRET's field of view, these detections indicate that emission in the 1 MeV–10 GeV range is probably common among bursts, if not universal. One implication of GRB observability at energies around or above 1 MeV is that, at these energies, two-photon pair production ($\gamma\gamma \rightarrow e^+e^-$) is not producing any significant spectral attenuation in the source. Attenuation by pair creation in the context of GRBs was first explored by Schmidt [13], who assumed isotropy of radiation in the source, and concluded that the (then) detection of photons around 1 MeV limited bursts to distances less than a few kpc.

The observation by EGRET of emission above 100 MeV has led to the popular suggestion that GRB photon angular distributions are highly beamed and produced by a relativistically moving or expanding plasma (e.g. [7,10]). Various determinations of the bulk Lorentz factor Γ of the medium supporting the GRB radiation field have been made in recent years, mostly concentrating(e.g. [10,4]) on the simplest case where the angular extent of the source was of the order of $1/\Gamma$, with an infinite power-law burst spectrum. This paper presents our recent refinements of pair production opacity calculations, treating the effects of spectral curvature in the BATSE energy range. We find that the influence of such curvature on the spectra and inferred bulk motions for bursts of cosmological origin depends on the distribution of photons within the source. For sources at galactic halo distances, we observe that for realistic parameters of the bulk motion, source opacity may arise only in a portion of the 1 GeV – 1 TeV range, with transparency returning in the super-TeV range, resulting in the appearance of distinctive, broad absorption troughs. Such features may be seen by current and future ground-based initiatives such as Whipple, MILAGRO and HEGRA, and space missions such as GLAST.

PAIR OPACITY AND SPECTRAL CURVATURE

Recent authors have invoked relativistic beaming in GRBs with emission detected above 10 MeV. This hypothesis builds on the property that $\gamma\gamma \to e^-e^+$ has a threshold energy E_1 that is strongly dependent on the angle Θ between the photon directions: $E_1 > 2m_e^2 c^4/[1 - \cos\Theta]E_2$ for target photons of energy E_2. Hence radiation beaming associated with relativistic bulk motion of the underlying medium can dramatically reduce the optical depth, $\tau_{\gamma\gamma}$, *internal* to sources at enormous distances from earth, suppressing γ-ray spectral attenuation turnovers.

Usually opacity calculations (e.g. [10,2,4]) yield $\Gamma \sim 10$ for galactic bursts and $\Gamma \sim 10^2 - 10^3$ for cosmological ones, and assume an infinite power-law spectrum $n(\varepsilon) = n_\gamma \varepsilon^{-\alpha}$, where ε is the photon energy in units of $m_e c^2$, for which the pair creation optical depth assumes the form $\tau_{\gamma\gamma}(\varepsilon) \propto \varepsilon^{\alpha-1}\Gamma^{-(1+2\alpha)}$ for $\Gamma \gg 1$. Despite this expedient approximation, most bursts detected by BATSE show significant spectral curvature in the 30 keV–500 keV range (e.g. [1]). Furthermore, BATSE sees MeV-type (i.e. 500 keV–2 MeV) spectral curvature with significant frequency in bright bursts, including EGRET sources (e.g. see [12]). Such curvature could, in principle, reduce the opacity of potential TeV emission from these sources. The effects of a depletion of low energy photons in the BATSE range relative to the EGRET quasi-power-law spectra can quickly be determined by approximating spectral curvature with a power-law broken at a dimensionless energy ε_B ($= E_\text{B}/0.511\,\text{MeV}$), and possessing a low energy cut-off at ε_c:

$$n(\varepsilon) = n_\gamma \varepsilon_\text{B}^{-\alpha_h} \begin{cases} 0, & \text{if } \varepsilon \leq \varepsilon_c, \\ \varepsilon_\text{B}^{\alpha_l}\varepsilon^{-\alpha_l}, & \text{if } \varepsilon_c \leq \varepsilon \leq \varepsilon_\text{B}, \\ \varepsilon_\text{B}^{\alpha_h}\varepsilon^{-\alpha_h}, & \varepsilon > \varepsilon_\text{B}. \end{cases} \quad (1)$$

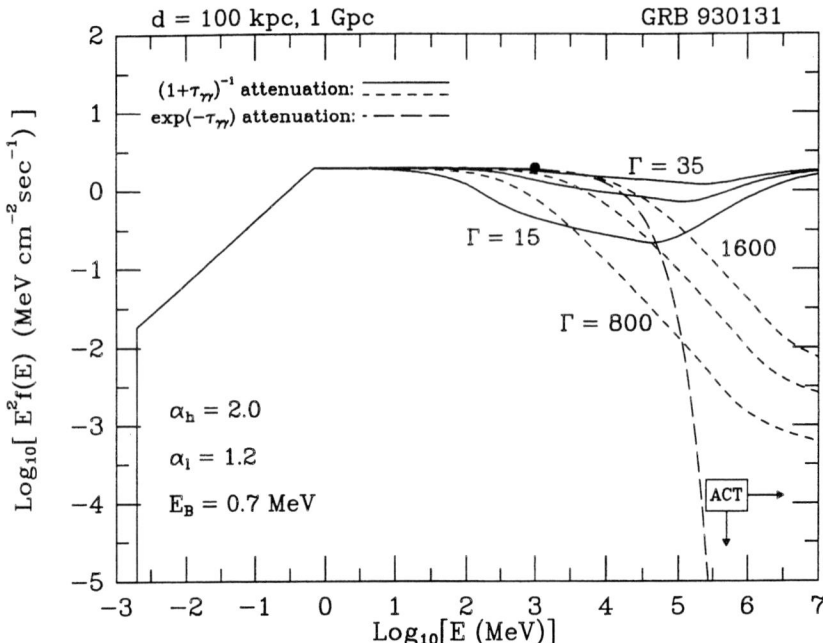

FIGURE 1. The attenuation, internal to the source, of a broken power-law spectrum for GRB930131 at source distances typical of galactic halo (solid curves, $\Gamma = 15, 25, 25$) and cosmological (short dashed curves, $\Gamma = 800, 1200, 1600$) origin, and different bulk Lorentz factors Γ for the emitting region. The spectra, plotted in the $E^2 f(E)$ (i.e. νF_ν) format, are attenuated by the factor $1/(1 + \tau_{\gamma\gamma})$ (except for the $\Gamma = 1600$, long dashed line case) for optical depths whose form is in [5]. The source spectrum was modeled with a power law broken at $E_B = 0.7$ MeV, with spectral indices $\alpha_l = 1.2$ and $\alpha_h = 2.0$. A low-energy cutoff at 2 keV was used to mimic X-ray paucity. The filled circle denotes the highest energy EGRET photon at 1000 MeV (see [14]). The current threshold and sensitivity for ACT observations of bursts is indicated by the "ACT" box.

The optical depth $\tau_{\gamma\gamma}(\varepsilon)$ for such a distribution is presented in [5] and is easily obtained from results determined in [8] and [3] for truncated power-laws. The most notable feature of $\tau_{\gamma\gamma}(\varepsilon)$ is that it is no longer necessarily a monotonically increasing function of energy ε. Two parameters, $\eta_B = \max\{1, \sqrt{\varepsilon_B \varepsilon}/\Gamma\}$ and $\eta_c = \max\{1, \sqrt{\varepsilon_c \varepsilon}/\Gamma\}$, govern the importance, or otherwise, of spectral curvature effects; when these equal unity, curvature effects are negligible and the pure power-law result $\tau_{\gamma\gamma}(\varepsilon) \propto \varepsilon^{\alpha-1} \Gamma^{-(1+2\alpha)}$ emerges. Conversely if η_B or η_c exceeds unity, the shape of the BATSE spectrum becomes crucial to optical depth estimates.

This important consideration was discussed by [5], who modeled GRB spectral curvature using broken power-laws and simple exponential attenuation as a first approximation. They found that the presence of such curvature generally has minimal influence on the spectra (below 1 TeV) and inferred bulk motions for bursts of cosmological origin. This result followed as a consequence of there being a plentiful supply of target photons (at $E_1 \sim m_e^2 c^4 \Gamma^2/E_2$) above the BATSE range in sources at extragalactic distances. In contrast, for galactic halo sources, they observed that source opacity may arise only in a portion of the 1 GeV – 1 TeV range, with transparency returning in the super-TeV range, resulting in the appearance of distinctive, broad absorption troughs. Such features would provide a unique identifier for bursts in halo locales.

In this paper, we have modeled the attenuation function in two extreme cases: by an exponential factor $\exp\{-\tau_{\gamma\gamma}\}$, arising from a delta-function photon spatial distribution, and by a factor of $1/(1+\tau_{\gamma\gamma})$, as is appropriate for source photons being distributed in a roughly spatially-uniform manner (i.e. including "skin effects"). If Γ were chosen large enough to permit emission out to TeV energies, the spectra would also suffer attenuation due to the *external* supply of infrared background photons ([15,11]). Attenuation external to the source will always produce an exponential turnover of the spectrum.

Attenuation of spectra appropriate to the "Superbowl" burst GRB930131 are depicted in Fig. 1 for different Γ. Most of the curves depicted are for attenuation by a factor of $1/(1+\tau_{\gamma\gamma})$. The contrast between spectral shapes for cosmological and galactic halo bursts is striking. Absorption troughs appear in the 100 kpc cases and are quite distinct from the broken power-law structure in cosmological scenarios. They become more pronounced as Γ decreases, resulting from the non-monotonic behaviour of the optical depth with energy: $\tau_{\gamma\gamma}$ drops below unity around the TeV range due to the "depleted supply" of interacting photons in the low energy BATSE portion of the spectrum. These distinctive features arise only for large spectral breaks (i.e. $\delta\alpha = |\alpha_h - \alpha_l| \gtrsim 1.3$). The spectral indices above 1 GeV in the 1 Gpc examples are defined uniquely in terms of those at lower energies, a property that can distinguish this internal absorption from the external photon absorption. For both situations, GRB930131 would have been easily detectable by air Čerenkov telescopes (ACTs) if it had been observed in a slew search. An exponential attenuation case (for source distance $d = 1$ Gpc) is also illustrated in the figure: similar examples are given in [5]. This corresponds to substantial spatial confinement of the target photons, and produces sharp cutoffs in cosmological scenarios that would render bursts undetectable by ACTs; halo bursts would still be detectable, however their absorption troughs would be much more pronounced. A diversity of such spectral shapes (e.g. troughs, shelfs and turnovers) might be anticipated for GRBs. Clearly future observations and/or upper limits by ACTs will play a prominent role in constraining burst scenarios and model parameters such as Γ.

CONCLUSION

The importance of the spectral attenuation results presented here is immediately obvious. The absorption troughs in Fig. 1 cannot be produced for large source distances and are unambiguous markers of a galactic halo burst population; they are consequently a potentially powerful observational diagnostic. Observations by the air Čerenkov detectors Whipple, MILAGRO and HEGRA in the 300 GeV–10 TeV range, combined with a probing of the 100 MeV – 100 GeV range by future space instrumentation such as GLAST could confirm or deny the existence of such absorption features. Note that the observation of apparently sharp cutoffs would not distinguish between cosmological or galactic burst hypotheses, or between intrinsic and extrinsic absorption. However, the broken power-law attenuation signatures in cosmological burst spectra, shown in Fig. 1, would clearly require instrinic absorption and would reflect the distribution of photon density within the source. Since the shape and position of these features are strongly dependent on the source flux below 10 keV, coupled X-ray and hard gamma-ray observations are clearly warranted.

REFERENCES

1. Band, D., et al., *Astrophys. J.* **413**, 281 (1993).
2. Baring, M. G., *Astrophys. J.* **418**, 391 (1993).
3. Baring, M. G., *Astrophys. J. Supp.* **90**, 899 (1994).
4. Baring, M. G. & Harding, A. K., in *Proc. 23rd ICRC* **1**, 53 (1993).
5. Baring, M. G. & Harding, A. K., *Astrophys. J. (Lett.)*, **481** L85 (1997).
6. Connaughton, V., et al., in *Proc. 24th ICRC (Rome)*, **2** 96 (1995).
7. Fenimore, E. E., Epstein, R. I., & Ho, C., in *Gamma-Ray Bursts*, eds. Paciesas, W. S. and Fishman, G. J., (AIP Conf. Proc. 265, New York) p. 158 (1992).
8. Gould, R. J. & Schreder, G. P., *Phys. Rev.*, **155** 1404 (1967).
9. Hurley, K., et al., *Nature* **372**, 652 (1994).
10. Krolik, J. H. & Pier, E. A., *Astrophys. J.* **373**, 277 (1991).
11. Mannheim, K., Hartmann, D., & Funk, B., *Astrophys. J.* **467**, 532 (1996).
12. Schaefer, B. E., et al., *Astrophys. J. (Lett.)* **393**, L51 (1992).
13. Schmidt, W. K. H., *Nature* **271**, 525 (1978).
14. Sommer, M., et al., *Astrophys. J. (Lett.)* **422**, L63 (1994).
15. Stecker, F. W. & De Jager, O. C., *Space Sci. Rev.* **75**, 401 (1996).

Spectral Variability in Relativistic MHD Winds

Christopher Thompson

Physics and Astronomy, UNC CB3255, Chapel Hill, NC 27599

Abstract. Any cosmological GRB source with a rotation period of ~ 1 msec and the density of nuclear matter plausibly develops a very strong magnetic field $B \sim 10^{15}$ G, and disgorges ordered Poynting flux at the required rate of $\sim 10^{51}$ erg s^{-1} [11,6]. This MHD wind advects outward an intense flux of thermal MeV photons which act as Compton seeds and regulate the thermodynamic state of matter.

Electron-positron pairs created by photon collisions feed back strongly on the emergent spectrum, enhancing the efficiency of energy deposition in the leptonic component, and making regions of the wind with power-law high-energy spectra much brighter than regions with thermal spectra. By contrast, dissipation deep inside the *electron-ion* photosphere plausibly leads to quasi-thermal spectra, and may account for the soft X-ray tails seen by Ginga and soft subpulses seen by BATSE. Explicit solutions to the Kompaneets equation in an expanding wind containing isolated hotspots show that a broken power-law spectrum develops in a pair-dominated atmosphere that covers a very large range ($\sim m_p/m_e$) in radius, and through which the integrated scattering depth significantly exceeds unity. The overall softening trend observed in many bursts may reflect gradual mixing between a high-Γ jet and surrounding lower-Γ material.

We compare double Compton emission and cyclo-synchrotron radiation as sources of Compton seeds. The existence of bursts with soft high-energy cutoffs at *rest frame* energies much less than ~ 1 MeV indicates that quasi-thermal Comptonization is occuring. The γ-ray light-curve may provide interesting information about the central source if the asymptotic Lorentz factor is regulated by neutrino emission, yielding a characteristic luminosity of $L_P \sim 10^{51}$ erg s^{-1}. Off-axis material with Lorentz factor $\Gamma_\infty \sim 1-2$ becomes optically thin to scattering with a delay of ~ 1 day$(E/10^{52}$ erg$)^{1/2}$, and can be a direct source of afterglow radiation.

DISSIPATION IN ULTRACOMPACT MHD WINDS

High energy emission from a γ-ray burst (GRB) source is commonly calculated without regard to the inner boundary conditions on the flow, so that the emergent spectrum develops only at a large distance from the source. In Blazar sources we have little direct knowledge of the physical conditions in the inner jet, but in cosmological GRB models the central compactness is huge, $\ell \sim 10^{15}$, and we can be assured that the flow starts out in a state of local thermodynamic equilibrium.

Thus, a GRB outflow is distinguished from a non-thermal source of low central compactness (such as a pulsar wind) by the presence of an intense bath of blackbody γ-radiation advected outward from large optical depth [7,4]. Ignoring the influence of this radiation is no more sensible than ignoring the effect of the cosmic background radiation on the thermodynamic state of matter in the early universe. Even though its unmodified spectrum bears little resemblence to the bright non-thermal states of classical GRBs, it is an important source of Compton seeds at a large distance from the source [11]. Indeed, the mean energy per photon is $\langle h\nu \rangle \sim T_{\text{eff}} = 0.8(L_\gamma/10^{50} \text{ erg s}^{-1})^{1/4}(P/10^{-3} \text{ s})^{-1/2}$ MeV, if the effective temperature is calculated at the light cylinder radius corresponding to rotation period P. This is remarkably close to the distribution of spectral break energies [10], after allowing for cosmological redshift. A bias against observing GRBs with much higher break energies [9] exists only if the peak flux and break energy of the burst are inversely correlated, in an *opposite* sense to the strong positive correlation between flux and break energy observed during the time evolution of many GRBs. In other words: if hard GRBs are also very energetic (as the total energy budget of $10^{53} - 10^{54}$ erg in most triggering scenarios allows) then the bias against detecting them is small.

VARIABLE Γ_∞, SOFT COMPONENTS, AFTERGLOW

These considerations point to a simple model [11,12] for the soft subcomponents of GRBs seen by Ginga [16] and BATSE [8], with essentially no emission above 300 keV. The radius of the scattering photosphere of the wind is related to the dissipative radius (on an observed timescale Δt) by a very strong function of the asymptotic Lorentz factor $\Gamma_\infty = L_{\text{rel}}/\dot{M}c^2$, namely $R(\tau = 1)/2\Gamma_\infty^2 c\Delta t \sim (m_e/m_p)\Gamma_\infty^{-5}\ell_{\Delta t}$ in the absence of pairs; and $R(\tau = 1)/2\Gamma_\infty^2 c\Delta t \sim \Gamma_\infty^{-5}\ell_{\Delta t}$ when the photon spectrum extends as a powerlaw $\beta \sim -2$ up to a rest frame energy $m_e c^2$. Compton upscattering of the seed photons by *quasi-thermal* electrons in a slow outflow with $\Gamma < \Gamma_{e-p} = (m_e/m_p)^{1/5}\ell_{\Delta t}^{1/5} = 33\, L_{\text{rel},51}^{1/5}(\Delta t/1 \text{ s})^{-1/5}$ will produce a spectrum with a quasi-Wien shape. In the calculation shown in Fig. 1A, adiabatically cooled photons undergo delayed reheating at large scattering depth. Because a high energy cutoff of 100-300 keV lies at an energy much less than $m_e c^2$ in the flow rest frame, it is inconsistent with photon destruction via pair creation. *This points to a dominant quasi-thermal component in the particle distribution function.* Bulk turbulent MHD motions are a promising source of free energy, since magnetic driving of the outflow is favored over neutrino heating due to the stringent limits on baryon contamination [11,6]. Direct Compton drag damps the turbulence effectively, irrespective of the relative inertia carried by baryons and leptons [11,14].

Considerable spectral variation exists between the final pulses of different bursts. The final pulse of GRB870303 (Ginga) was very soft (as indicated by the near absense of emission in the 14-370 keV band [16]), whereas the final three pulses of GRB970228 (BeppoSAX) showed more FRED-like behavior and much flatter spectra [3]. This variation is not unexpected, given the extremely strong dependence of scattering depth on Γ_∞.

Outflows with low Γ_∞ generally carry softer photons. How soft depends on the relative numbers of thermal and non-thermal particles. The simplest case involves continuous heating that balances adiabatic cooling in a thermal plasma. Double Compton emission fails to increase the photon number beyond a radius $R_{\rm d-C}/R_{\rm s} = 1 \times 10^4 \Gamma_\infty^{-7/8} (L_\gamma/10^{51} {\rm\ erg\ s}^{-1})^{3/4} (R_{\rm s}/10^7 {\rm\ cm})^{-1}$, where $R_{\rm s}$ is the source radius. The Lorentz-boosted temperature of the radiation is $\frac{4}{3}\Gamma_\infty T$(rest frame) $= 3\Gamma_\infty^{13/8} (L_\gamma/10^{51} {\rm\ erg\ s}^{-1})^{3/4}$ keV for low Γ_∞. At this radius, the outflow is still very optically thick, the X-ray photons have a blackbody distribution, and cyclo-synchrotron emission by the thermal electrons is unimportant. A large non-thermal electron component reduces the mean photon energy even further. However, if the electron number spectrum is steeper than $p^{-\Gamma} = p^{-2}$, then the advected photons are a more important coolant than is the magnetic field, because synchrotron emission is self-absorbed below an energy $\hbar\omega_{\rm sa}/m_e c^2 \sim 4\,(B/B_{\rm QED})^{(\Gamma+2)/(\Gamma+4)}$.

Afterglow. Low-Γ_∞ components of the outflow can contribute significantly to afterglow radiation. The most probable geometry is a high-Γ_∞ core surrounded by a lower-Γ_∞ sheath. If the magnetic field driving the outflow is amplified in a hot neutron star (or torus), then low-Γ_∞ material will flow out from the source before a high-Γ_∞ flow can develop, and the high-Γ_∞ material must punch through.

Neutrino heating of the (optically thin) surface layers of a centrifugally supported torus *regulates* the total luminosity extracted by magnetic torques to [11] $L_{\rm P} \sim 10^{51}\,(\Gamma_\infty/300)^{-2/3}$ erg s^{-1}. These torques induce a $\nu_e/\bar{\nu}_e$ luminosity $L_\nu \sim L_{\rm P}$ that in turn blows mass off the (optically thin) surface of the torus at a rate $\dot M \sim 3 \times 10^{-7}\,(L_\nu/10^{51} {\rm\ erg\ s}^{-1})^{5/2}$ g s^{-1}. Non-linear damping of MHD waves near the Alfvén/light cylinder radius [14] will raise the photon luminosity to a value comparable to $L_{\rm P}$, and thus greatly increase the entropy per baryon, if the external magnetic field is strongly turbulent. If the MHD wind is powered by the Blandford-Znajek mechanism from a near-extremal central $\sim 2\,M_\odot$ black hole, then the required angular momentum can be provided if the hole forms from a stripped He or CO-core [15] in a tight binary with a neutron star [11].

Material with asymptotic Lorentz factor $\Gamma_\infty = 1-2$ and total energy E becomes optically thin to scattering on a timescale $\sim 10^5\,(E/10^{52} {\rm\ erg})^{1/2}$ sec. This is (perhaps coincidentally) the timescale on which the optical afterglow from GRB970508 showed a sudden rise (IAU circulars 6654-6676). Moreover, the energy density in the ejecta is still many orders of magnitude larger than the energy density in the ambient magnetic field at this time (assuming $B \sim 3 \times 10^{-6}$ G). This favors the ejecta themselves as a direct source of afterglow radiation [13,1]. The preceding calculation suggests that the low-Γ_∞ ejecta may also provide a supplemental source of X-radiation on a $\sim 10^5$ sec timescale, over and above any underlying synchrotron emission. Delayed inverse-Compton emission from a forward shock [5] is guaranteed to lie in the GeV range if the MeV burst photons are themselves the Compton seeds [12]. This requires oblique shocks.

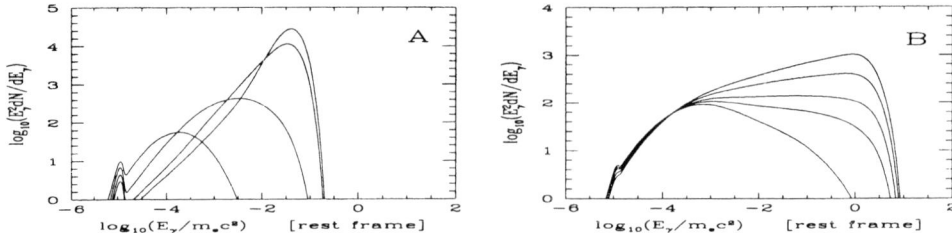

FIGURE 1. A. Continuous Comptonization of cold seed photons in a homogeneous relativistic wind ($T_w/m_ec^2 = .03$) outward from $\tau = 25, 50, 100, 150$ (lower to upper curves). **B**. Heating by hotspots of temperature $T_w/m_ec^2 = 0.4$ and pair density $n_{e\pm} = n_{e^+} + n_{e^-}$ in the two-stage Comptonization model described in the text (for $\int n_{e\pm}\sigma_T c dt = 3, 4, 5, 7, 10$).

PAIRS

Although pair creation has been included for some time in models of BH accretion disk coronae, in that context the formation of a power-law high-energy continuum does not depend essentially on the presence of pairs. The situation is quite different in a relativistic outflow that expands sufficiently rapidly at its base that inhomogeneities fall out of causal contact. Pairs regenerated by $\gamma + \gamma \to e^+ + e^-$ outside the $e - p$ scattering photosphere strongly buffer the decrease of scattering depth with radius in outflows with hard high-energy spectra. One can expect irregularities in the outflow over the entire range of timescales $P < \Delta t < \Delta t_{\max}$, where $\Delta t_{\max}$ is the maximum timescale for (irregular) variability in the burst profile. Dissipation in the pair-loaded region extends over the range of timescales $(\Gamma^{e-p}/\Gamma_\infty)^5 \Delta t_{\max} < \Delta t < \Delta t_{\max}$, as long as the Lorentz factor is less than $\Gamma^{e\pm} = \ell_{\Delta t_{\max}}^{1/5} = 150\, L_{\rm rel,51}^{1/5}(\Delta t_{\max}/1\ {\rm s})^{-1/5}$. When $\Gamma_\infty = \Gamma^{e\pm}$, this atmosphere expands to cover a very wide range of radii $(m_e/m_p)R(\tau = 1) < R < R(\tau = 1)$.

Suppose that the power spectrum of spatial inhomogeneities is flat, so that the photon luminosity grows logarithmically with radius. The Compton parameter accumulated per logarithm of radius, $dy/d\ln R \simeq \frac{1}{3}|d\ln\rho/d\ln R| = \frac{2}{3}$, corresponds to a large scattering depth through the pair atmosphere, $\tau \simeq (d\tau/d\ln R)\ln(m_p/m_e) = 5$. (Given that the turbulence is mildly relativistic with $d\tau/d\ln R \simeq dy/d\ln R$.) Cold pairs created outside hotspots increase τ even further.

The feedback of pair creation on both thermal Comptonization and non-thermal (first-order Fermi) acceleration of leptons is highly non-linear. Fresh pairs are a rich source of suprathermal seeds for Fermi acceleration at a shock, since the gyroperiod of a relativistic electron is orders of magnitude shorter than its cooling time in a Poynting-driven outflow. This forces the minimum Lorentz factor of the non-thermal pairs to unity, $\gamma_{\min} \to 1$, if the outflow is continuously heated. Near the scattering photosphere, the cyclotron energy is $\hbar eB/m_e c = [8\pi\hbar ec^2/\sigma_T(2L_Pc)^{1/2}]\,\Gamma_\infty^3 = 0.04\,(\Gamma_\infty/300)^3\,(L_P/10^{51}\ {\rm erg\ s}^{-1})^{-1/2}$ eV. Thus, the efficiency of electron acceleration is increased at the cost of suppressing the synchrotron energy $\gamma_{\min}^2 e\hbar B/m_e c$ far below the observed range of break energies in GRB spectra. *This indicates that, if the GRB outflow is heated continuously outward from large scattering depth, then the primary emission process is inverse Compton scattering of advected photons.*

The spectrum resulting from delayed Comptonization of adiabatically cooled soft photons by turbulent hotspots (e.g. reconnection sites) with equivalent temperature $T_\mathrm{w} \simeq 0.4 m_e c^2$ is shown in Fig. 1B. This is the rest frame spectrum, which must be multiplied by the bulk Lorentz factor $\Gamma_\infty \sim 300$. Comptonization occurs in two stages. In the first stage, the photons are re-heated close to the total luminosity that they had before adiabatic cooling. The required Compton y-parameter is large, and a simple loss-probability argument [12] indicates a photon index $\alpha = \frac{1}{2} - \sqrt{(9/4) + (4/y)} \simeq -1$ up to a mean energy close to $L_\mathrm{P}/\dot{N}_\gamma \sim 1$ MeV. This reproduces well the low energy spectral indices of GRBs (e.g. [2]).

As dissipation of inhomogeneities with lower wavenumbers continues, Compton drag regulates the spatially averaged y-parameter to a value near unity. Throughout the calculation, photons can leave *and* re-enter hotspots (of radius R). We fix $\tau_\mathrm{T} \equiv (n_{e^+} + n^{e^-})\sigma_\mathrm{T} R$ at 0.3: if τ_T is larger then the Compton cooling time ($R/c\tau_\mathrm{T}$ for $B^2/8\pi \sim U_\gamma \sim \rho c^2$) is less than the residency time ($\frac{1}{3}R/c$) and photons are not upscattered into a non-thermal tail. Calculations including a range of scattering depths are in progress.

This model predicts that the prompt GRB spectrum is cut off above an energy $\sim \Gamma_\infty m_e c^2$, with > GeV photons resulting only from second order Comptonization.

REFERENCES

1. Chiang, J., and Dermer, C.D., preprint (1997).
2. Cohen, E., et al., *Ap. J.* **488**, 330 (1997).
3. Frontera, F., et al., preprint (1997).
4. Goodman, J., *Ap. J.* **308**, L47 (1986).
5. Mészáros, P., and Rees, M.J., *M.N.R.A.S.* **269**, L41 (1994).
6. Mészáros, P., and Rees, M.J., *Ap. J.* **482**, 29 (1997).
7. Paczyński, B., *Ap. J.* **308**, 43L (1986).
8. Pendleton, G.N., et al., *Ap. J.* **489**, 175 (1997).
9. Cohen, E., Piran, T., and Narayan R., preprint (1997).
10. Mallozzi, R.S., et al., *Ap. J.* **454**, 597 (1995).
11. Thompson, C., *M.N.R.A.S.* **270**, 480 (1994).
12. Thompson, C., in *Third Huntsville Symposium on Gamma-Ray Bursts*, ed. C. Kouveliotou, M.S. Briggs and G.J. Fishman, New York: AIP, pp. 802-806 (1996).
13. Thompson, C., in *Relativistic Jets in AGN*, ed. M. Ostrowski, M. Sikora, G. Madejski, and M. Begelman, Springer, in press.
14. Thompson, C., and Blaes O., *Phys. Rev. D*, in press (1997).
15. Woosley, S.E., *Ap. J.* **405**, 273 (1993).
16. Yoshida, A., et al., *P.A.S.J.* **41**, 509 (1989).

Cosmology with GRBs

R. J. Nemiroff[††], G. F. Marani[¶],
J. P. Norris[†], and J. T. Bonnell[*,†]

[††] *Department of Physics, Michigan Technological University, Houghton, MI 49931*
[¶] *CEOSR / George Mason University, Fairfax, VA 22030*
[†] *NASA/Goddard Space Flight Center, Greenbelt, MD 20771*
[*] *Universities Space Research Association*

Abstract. Given the duration and peak flux distribution of GRBs in the BATSE 3B catalog, inherent detection thresholds of BATSE, and the assumption that peak flux is a standard candle irrespective of duration, rates are estimated for GRBs of cosmologically interesting durations and peak fluxes.

We find that current BATSE results do not strongly constrain the rate of very short GRBs. Bursts with durations between 1 and 2 ms have 1 σ rate limits ranging from zero to greater than the measured BATSE burst rate between 8.192 and 16.384 seconds, for peak fluxes in excess of 2 photons per square centimeter per second and canonical assumptions involving GRB spectra, evolution, and cosmology. The burst rate at 1 microseconds is even less well constrained. Bursts this short at redshift unity would be susceptible to detectable gravitational microlensing from a cosmological density of stars and/or MACHOs at the rate of $\Omega_{lens}/8$, given a canonical critical universe. Non-detection of GRB microlensing from future spacecraft could falsify the existence of MACHOs in galactic halos.

Similarly, we find that BATSE results do not strongly constrain the rate of dim, long bursts. Recently Kommers et al. [1] demonstrated that long bursts of low peak flux exist and go undetected by BATSE's onboard trigger criteria but, in some cases, are recoverable by searches on BATSE archival data. Increasingly dim GRBs can be recovered by integrating peak flux over increasingly longer time intervals. Given canonical assumptions of cosmology and the brightness distribution of low peak flux GRBs, we find that recoverable GRBs near ten seconds duration could have a rate in excess of ten times that triggering BATSE. An archival search could sample GRBs at a peak flux level of 0.01 that of BATSE's current completeness limit - corresponding to a redshift of about 20 given standard burst and cosmological assumptions. The log N - log P of GRBs at these early times might probe GRB production mechanisms in the early universe, and from the lack of such sources might be expected to show a turnover.

EXTRAPOLATIONS IN PEAK FLUX AND DURATION

The duration and peak flux distributions of gamma-ray bursts (GRBs) have been measured and discussed continually since discovery [2]. Since 1991 these distributions have been best sampled by the relatively sensitive Burst and Transient Source Detector (BATSE) onboard the Compton Gamma Ray Observatory (CGRO). The intrinsic detection limitations of any detector, including BATSE, place fundamental limits on the duration and peak fluxes of GRBs it can detect. BATSE's onboard trigger criteria make it most sensitive to GRBs with durations of about 1 second and peak fluxes above 1 photon cm^{-2} sec^{-1} [3,4]. Dimmer and/or shorter bursts, however, carry significant cosmological utility.

During BATSE's first three years (i.e. those bursts incorporated in the BATSE 3B catalog [5]), BATSE's onboard trigger criteria were relatively constant. On average, BATSE was trigger sensitive to any point on the sky about 38% of the time, between 50 and 300 KeV on 64 ms, 256 ms, and 1024 ms time scales [4]. Peaks in the time series greater than 5.5 σ over a previous 17 second background triggered the instrument into burst mode, where more detailed time information about the burst was recorded.

GRBs with durations below the BATSE 64 ms trigger threshold are only detected by BATSE when accumulated flux over 64 ms is sufficiently great. Theoretically, a GRB with arbitrarily high peak flux but correspondingly short duration would go undetected by BATSE. Practically, BATSE is increasingly *insensitive* to shorter GRBs, demanding they have increasingly high peak flux for detection [3,6,7].

Additionally, GRBs with durations longer than the 1.024 second trigger accumulation threshold, the longest onboard BATSE, and peak fluxes below 0.25 photons cm^{-2} sec^{-1}, a completeness threshold for BATSE, usually also go undetected [3]. An actual search for long, untriggered GRBs in BATSE data is ongoing by Kommers et al. [1], currently reporting the detection of 91 previously unreported GRB candidates. Isolating cosmic GRBs in archival data is not trivial, however, as triggers related to the Sun [8], electron participation events [9], and background anomalies are quite numerous, and much experience is needed to find a real signal in this noise (Meegan, personal communication, 1996).

To estimate the rates of these undetected GRBs, a Monte Carlo simulation universe was created. In this universe a virtual GRB catalog consistent with the BATSE 3B measured duration histogram and the measured Log N - log P distribution was subjected to various theoretical detection thresholds. The results are shown in Figures 1 and 2.

Figure 2 shows a histogram of durations for BATSE 3B GRBs. Only those GRBs in the BATSE 3B with listed T_{50}, and peak flux on the 64 ms scale above 2 photons cm^{-2} sec^{-1} are plotted in the unfilled histogram region: 295 GRBs in all. BATSE 3B GRBs are expected to be about 98 percent complete to this peak flux level [4]. All rates plotted in Figure 2 are normalized to the BATSE measured rate for GRBs with T_{50} durations between 8.192 and 16.384 seconds and peak fluxes above 2 photons cm^{-2} sec^{-1} – the highest rate recorded by BATSE.

Inspection of the hatched short duration part of the Figure 2 histogram indicates that a significant rate of short GRBs might exist but go undetected by BATSE. In fact, there may be more GRBs between 1 and 2 ms than between 8.192 and 16.384 seconds. Alternatively, bursts this short might not exist at all: both are acceptable to within 1 σ errors. Were these GRBs to exist, however, they would not be recoverable by analysis on existing BATSE archival data because the highest time resolution, continuously recorded BATSE data type has 1.024 sec bins ("DISCLA" data). Similarly, the rate of GRBs below 1 ms, say at 1 μs, is even less well determined, and may exceed the rate at any other time interval, or may be zero.

GRBs vary more rapidly than any other known cosmological phenomena. The lower limits of this variability have not yet been explored. Improvements in detectors would reveal or limit the actual rate of short GRBs. Were microsecond "spike" GRBs to exist and be detectable, they would time-resolve stellar mass objects throughout the universe by their gravitational microlensing effect [10]. The microlensing detection rate for GRBs at redshift unity would be on order $\Omega_{lens}/8$ [11]. *Analyzing the time structure of sufficient numbers of GRB spikes could reveal or limit Ω_{star}, Ω_{MACHO}, and/or Ω_{baryon}* [12].

Inspection of the hatched long duration part of the Figure 2 histogram indicates

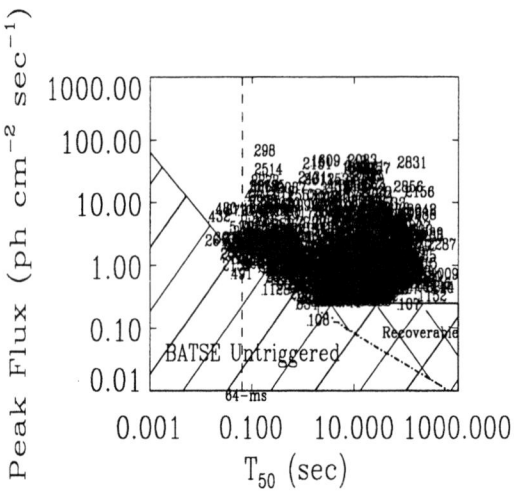

FIGURE 1. A plot of peak flux versus duration for BATSE 3B GRBs. Bursts are represented by their trigger numbers. GRBs occurring in the hatched region with lines running from the lower left to the upper right would not trigger BATSE. GRBs occurring in the hatched region with lines running from the lower right to the upper left may be recoverable from an archival analysis of BATSE data.

that a significant rate of long GRBs might exist, go undetected by BATSE onboard trigger criteria, but be recovered by a search in archival data. In fact, there may be ten times more untriggered GRBs between 8.192 and 16.384 seconds in BATSE's data stream than triggered GRBs. A search for these dim GRBs could falsify extrapolated rate estimates for GRBs, indicate a turn-over in the Log N - Log P, and/or potentially sample GRBs as distant as $z \sim 20$ - more distant than the furthest known QSO.

Because rate estimates like those found here run so high, a significant return could be found from a very detailed search of only *a single day's worth of existing BATSE archival data*. Two major search enhancements might include a background estimation more complex than a local linear fit, and that a bin size be used that is on order the duration of the GRB. The later requirement derives from the idea that every part of a GRB that increases signal to noise should be used for a maximally sensitive detection of that GRB.

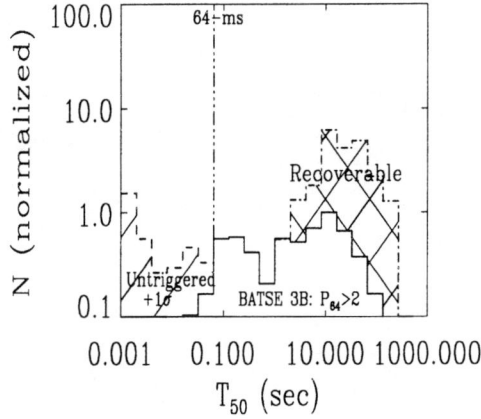

FIGURE 2. A duration histogram. The unhatched region results from GRBs in the BATSE 3B catalog. Burst rate is normalized to the BATSE bin covering durations between 8.192 and 16.384 seconds. Bursts occurring in the hatched region with lines running from the lower left to the upper right are one-sigma upper limits on the rate for GRBs with peak flux greater than 2 photons per square centimeter per second. Bursts occurring in the hatched region with lines running from the lower left to the upper right might be recoverable from an archival analysis of BATSE data.

REFERENCES

1. Kommers, J.M., Lewin, W.H.G., Kouveliotou, C., van Paradijs, J., Pendleton, G.N., Meegan, C.A., Fishman, G.J. *Ap. J.* **491**, 704 (1997).
2. Klebesadel, R., Strong, I. B. & Olson, R. A., *Ap. J.* **182** L85 (1973).
3. Fishman, G. J. et al., *Ap. J. Supp.* **92** 229 (1994).
4. Meegan, C. A. et al., Electronic form only, available from Compton Observatory Science Support Center at
http://www.batse.msfc.nasa.gov/data/grb/catalog/exposure.html (1997).
5. Meegan, C. A. et al., *Ap. J. Suppl.* **106**, 65 (1996).
6. Norris, J.P., Cline, T., Desai, U., & Teegarden, B., *Nature* **308**, 434 (1984).
7. Lee, T.L. & Petrosian, V., *Ap. J.* **470**, 479 (1996).
8. Aschwanden, M.J., Benz, A.O. & Schwartz, R.A., *Ap. J.* **417**, 790 (1993).
9. Datlowe and Imhoff, Lockheed Final Report: 12 Feb. (1994).
10. Paczyński, B., *Ap. J.* **308**, L43 (1986).
11. Nemiroff, R. J., *Ap. J.* **341**, 579 (1989).
12. Nemiroff, R. J., Norris, J. P., Bonnell, J. T., and Marani, G. F., *Ap. J.*, in press (1998).

The Implications of Direct Red-Shift Measurement of γ-Ray Bursts

Ehud Cohen and Tsvi Piran

Racah Institute of Physics, The Hebrew University, Jerusalem, Israel 91904

Abstract. The recent discoveries of X-ray and optical counterparts for GRBs, and a possible discovery of a host galaxy, implies that a direct measurement of the red-shift of some GRBs host galaxies is imminent. We discuss the implications of such measurements. These measurements could enable us to determine the GRBs luminosity distribution, the variation of the rate of GRBs with cosmic time, and even, under favorable circumstances, to estimate Ω. Using GRB970508 alone, assuming standard candles and assuming the GRB source to be at the red-shift of the absorption line observed in the optical transient spectrum, we constrain the intrinsic GRB evolution to $\rho(z) = (1+z)^{-0.5 \pm 0.7}$.

INTRODUCTION

The recent observations of the Italian-Dutch Beppo/SAX satellite of γ-ray bursts [4], with error boxes of a few arc-minutes across, enabled a follow-up by optical and radio observations and the discovery of X-ray, optical and radio counterparts to GRBs (see, e.g., [2,7,9,18,21,23]). The optical observations provided, for the first time, independent estimates of the distances to GRB sources, using absorption lines or association with host galaxies [12], and demonstrated beyond doubt the cosmological origin of GRBs. It is highly possible that in the months to come several GRBs would have independent red-shift estimates. We may use these to obtain estimates of the luminosity diversity and the intrinsic evolution of the GRB rate with cosmic time. Under favorable conditions we might even be able to use GRBs as cosmic probes for estimating the cosmological parameters.

THE LUMINOSITY FUNCTION

A measurement of a red-shift of the optical counterpart of a GRB or the association of a host galaxy with a GRB provides us with the red-shift of the burst, z. Additionally we have the usual peak photon flux parameter, p (ph cm^{-2} s^{-1}), which is transferred to apparent luminosity, l (erg cm^{-2} s^{-1}), using bursts' spectra. The peak photon flux is related to the red-shift and source luminosity as:

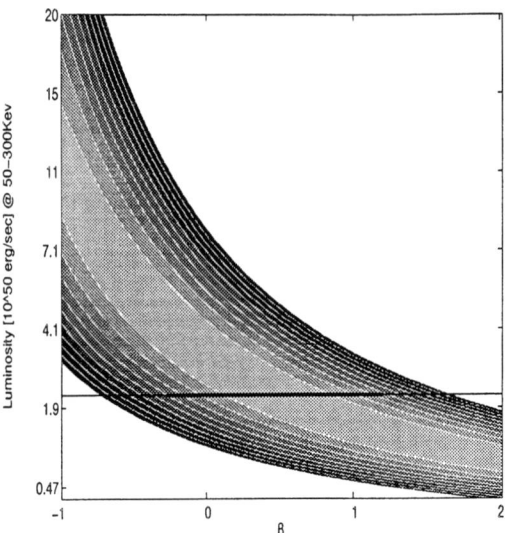

FIGURE 1. The likelihood function (levels 31.6%, 10%, 3.16%, 1%, etc. of the maximum) in (β, L) plane for standard candles, $\alpha = 1.5$, $\Omega = 1$, and evolution given by $\rho(z) = (1+z)^{-\beta}$. Superimposed on this map is the luminosity of GRB970508, with thick line where the likelihood function $> 1\%$. We have used $h_{75} = 1$.

$$p = \frac{N\left[\nu_1(1+z)...\nu_2(1+z)\right]}{4\pi(1+z)(2c/H_0)^2(1-(1+z)^{-1/2})^2}. \quad (1)$$

The detector boundaries ν_1, ν_2 are 150 keV, 300 keV respectively for the BATSE detector. (In order to maintain a uniform catalog, we use the BATSE data for all bursts, including those detected by Beppo/SAX). The Hubble distance is c/H_0, and $N[\nu_1, \nu_2]$ is the number of photons emitted in the range $[\nu_1, \nu_2]$. We have used $\Omega = 1$ and $\Lambda = 0$ in Eq. 1. The effect of Ω and Λ on the luminosity is not large, and we will discuss it later. The luminosity depends on the Hubble constant only via the scale factor $h_{75}^2 = (H_0/(75 \text{ Km s}^{-1} \text{ Mpc}^{-1}))^2$.

When comparing bursts from different red-shifts one must recall that the observed peak-flux is in a fixed energy range, which corresponds to different energy ranges at the sources. In order to discuss a single luminosity that classifies the bursts, we consider $L \equiv \int_{50\text{keV}}^{300\text{keV}} L_\nu \nu d\nu$ at the source. To convert from the observed peak flux to the intrinsic luminosity we assume that the source spectral form is a power law $L_\nu = L(\nu/50\text{keV})^{-\alpha}(2-\alpha)/(6^{2-\alpha}-1)/(50\text{keV})^2$ in the energy range $50\text{keV} < \nu < 300\text{keV}(1+z_{\max})$, so that wherever the source is, the detector sees a power-law spectrum. We use $\alpha = 1.5$ for all bursts. This value is probably a good typical estimate [1], even though the spectra are not the same for all bursts. Using this spectral shape we obtain:

$$L = 7.7 \cdot 10^{50} \text{ erg s}^{-1} \times \left[\frac{l}{10^{-7} \text{erg s}^{-1} \text{ cm}^{-2}} \right] \frac{(1 - (1+z)^{-1/2})^2}{(1+z)^{-\alpha} h_{75}^2}. \tag{2}$$

Using Eq. 2 we determine the luminosity of each burst. Then we estimate the luminosity function using a maximum-likelihood or any other statistical method. To do so we assume a functional shape of the luminosity function, determine its parameters from the data and then estimate the quality of the fit. Using the Cramér-Rao inequality [19] we can estimate the statistical error in this procedure. If the luminosity function has the form of a normal distribution then with twenty bursts we will be able to estimate, with 95% confidence, the standard deviation to ±30% of its true value. For a power-law distribution, twenty bursts will enable us to determine the power-law index to ±0.5, again with 95% confidence. Recall that current data, and in particular the peak-flux statistics of GRBs does not constrain the luminosity distribution of GRBs [10].

The luminosity distribution obtained in this way is for detected bursts only, and therefore it is biased. We define by $V(L)$ the volume from which we can detect a burst of luminosity L. Then the distribution we measure is $\Phi(L)V(L)$ where $\Phi(L)$ is the intrinsic luminosity function. By dividing the measured distribution by $V(L)$, we remove this bias from our estimate.

One of the interesting features that might distinguish between different cosmological GRB models is the rate that GRBs occur per unit time per unit comoving volume: $\rho(z)$. These new measurements could yield a direct estimate of this distribution. Once the GRB luminosity distribution is known we can proceed and compare the theoretical peak-flux statistics (using the observed luminosity distribution) with the observed one. This distribution depends strongly on the intrinsic evolution of GRBs, that is on variation of $\rho(z)$. Following Cohen & Piran [3] we characterize this dependence as $\rho(z) = (1+z)^{-\beta}$. Comparison of the theoretical and observed distribution would limit β.

In fact this comparison can be done even with the current data and assuming a narrow luminosity distribution (standard candles). We can use GRB970508 to constrain the evolution. Using the peak flux $= 1.6 \cdot 10^{-7}$ erg cm^{-2} s^{-1} [8], the red-shift of the absorption lines $z = 0.835$ [12] which sets a lower limit $z > 0.835$ for the burst, and the absence of prominent Lyman-alpha forest in the spectrum which compose an upper limit $z < 2.1$, we obtain $\beta = -0.1 \pm 1.3$ in 99% confidence level. Assuming that the absorption line of GRB970508 corresponds to its own red-shift we estimate $\beta = 0.5 \pm 0.7$ with this confidence level, see Fig. 1. The simplest hypothesis of no evolution $\beta = 0$ is consistent with the observations. A milder assumption of Gaussian luminosity distribution with $\sigma_L = L_{obs}/2$, instead of standard candles, yields a lower limit $\beta > -0.7$ and no upper limit.

Earlier attempts to estimate the bursts evolution without an independent luminosity estimate found only mild limits. Cohen & Piran [3] found no limit on β, Rutledge et al. [20] obtained a limit of $-3 < \beta$ and Loredo & Wasserman [11] obtained a limit of $-2.75 < \beta < 1$. Our preliminary limit which is based on a single burst restricts the limits regarding negative evolution significantly.

ESTIMATES OF COSMOLOGICAL PARAMETERS

Despite numerous attempts to estimate the cosmological closure parameter Ω, its actual value is still unknown and current estimates range from 0.2 to 1 (see eg. [13]). One may wonder whether GRBs would provide a meaningful independent estimate of Ω. Using GRBs peak-flux statistics alone, Ω could not be estimated from the current data [3]. However, given a cosmological distribution of sources with measured red-shifts, we can try to estimate the cosmological closure parameter, Ω, in a similar manner to the attempts to estimate Ω from type I supernovae by [15]. Perhaps GRBs will not be able to contribute meaningfully to the myriad of measurements before other methods become more precise. But it is likely that they provide a useful consistency check, which is based on independent objects. Recall that GRBs are most likely further than the observed type I supernovae.

The observed peak-flux depends on Ω as:

$$l = l(L, \Omega) = L \frac{(H_0/c)^2 \Omega^4 (1+z)^{2-\alpha}}{64\pi (z\Omega/2 + (\Omega/2 - 1)(\sqrt{\Omega z + 1} - 1))^2} \quad (3)$$

Using the known parameters of each burst (peak-flux and red-shift) we obtain for each burst a function $L_i = L_i(\Omega)$. (In the previous sections we have assumed $\Omega = 1$ and obtained $L_i(\Omega = 1)$). For standard candles all L_i must be equal. Given two sources we have two equations $L = L_{1,2}(\Omega)$, with two variable and we should be able to determine Ω.

A luminosity distribution will induce an uncertainly in this estimate that can be approximated by:

$$\sigma_\Omega(z) \approx \left.\frac{d\Omega}{dL}\right|_{\Omega=1,z} \sigma_L = 1/4 \frac{(\sqrt{1+z} - 1)\sqrt{1+z}}{z/2 - 3/2\sqrt{1+z} + 3/2 + z/(4\sqrt{1+z})} \frac{\sigma_L}{L} \quad (4)$$

For $z = 1.5$ we obtain $\sigma_\Omega \approx 2\sigma_L/L$. Thus assuming $\sigma_L/L = 1$, we need 100 bursts with a measured z to estimate Ω with an accuracy of $\sigma_\Omega = 0.2$. Such a goal could be achieved within several years. At present it is not known whether the GRB luminosity distribution is narrow enough and satisfies this condition. However, as we have shown at section 2, the width of the luminosity distribution will be known to 30% when we will have twenty bursts with measured red-shift.

DISCUSSION AND CONCLUSIONS

As expected, the direct red-shift measure of GRB970508 agrees well with estimates made previously using peak-flux count statistics (see e.g., [6,10,20,3]). It is remarkable what could be done with even several additional red-shifts. An estimate of the bursts' luminosity distribution to 30% accuracy can be obtained with twenty bursts.

This luminosity function combined with the observed peak-flux distribution would provide us immediately with an estimate of the cosmological evolution of the rate of GRBs.

It is generally accepted that a fireball (see, e.g., [17]) is inevitable in any cosmological model. Within this model the observed γ-rays are produced during the conversion of a relativistic energy flow to radiation. However, the source itself that produces the flow remains unseen. The limits on cosmological evolution could shed light on the GRB mystery, by distinguishing between different cosmological models. For example the expected rate of merging neutron stars depends on the red-shift [16,22] in a drastically different way than the evolution of AGNs (which seems to decay exponentially at low red-shift) and is even different from the expected rate of supernovae as seen from nucleosynthesis evidence [5].

This research was supported by US-Israel BSF grant 95-328 and by NASA grant NAG5-3516.

REFERENCES

1. Band, D., et al., *ApJ* **413**, 281 (1993).
2. Bond, H.E., IAU Circ. 6654 (1997).
3. Cohen, E. & Piran, T., *ApJL* **444**, L25 (1995).
4. Costa, E., et al., IAU Circ. 6572 (1997).
5. Cowan, J.J., Thielemann, F.-K., Truran, J. W., *Phys. Rep.* **208**, 267 (1991)
6. Fenimore, E.E., et al., *Nature* **366**, 40 (1997).
7. Galama, T.J. et al., *Nature* **387**, 479 (1997).
8. Kouveliotou, C., et al., IAU Circ. 6660 (1997).
9. Heise, J. et al., IAU Circ. 6654 (1997).
10. Loredo, T. J. & Wasserman, I. M., *ApJS* **96**, 261 (1995).
11. Loredo, T. J. & Wasserman, I. M., preprint (1996).
12. Metzger, M., et al., IAU Circ. 6655 (1997).
13. Peebles, P. J. E., in *Some Unsolved Problems in Astrophysics*, Eds. J. N. Bahcall and J. P. Ostriker, Princeton University Press (1996).
14. Narayan, R., Piran, T., Shemi, A., in *Gamma-ray Bursts; Proceedings of the Workshop*, Univ. of Alabama, Huntsville, Oct. 16-18, 1991 (A93-40051 16-93), p. 149-153.
15. Perlmutter, S., et al., astro-ph/9608192 (1996).
16. Piran, T., *ApJ* **389**, L45 (1992).
17. Piran, T., in *Some Unsolved Problems in Astrophysics*, Eds. J. N. Bahcall and J. P. Ostriker, Princeton University Press (1996).
18. Piro, L., et al., IAU Circ. 6656 (1997).
19. Porat, B., *Digital Signal Processing of Random Signals*, Prentice-Hall Press (1993).
20. Rutledge, R.E., Hui, L., Lewin W.H.G., *MNRAS* **276**, 753 (1995).
21. Sahu, K. C., et al., astro-ph/9705184 (1997).
22. Totani, T., astro-ph/9707051 (1997).
23. van Paradijs, J., et al., *Nature* **386**, 686 (1997).

Cooling Synchrotron Spectra and GRB Theory

J. J. Brainerd

Physics Dept., University of Alabama in Huntsville, Huntsville, AL 35899

Abstract. The absence of a cooling synchrotron spectrum places constraints on the physics of the gamma-ray burst emission region. Several theories posit that unattenuated synchrotron emission produces the observed gamma-ray burst spectrum. The characteristic shape of the spectrum is therefore determined by the underlying electron distribution. To produce spectra as hard as those observed, the electrons must be accelerated as they radiate; otherwise, the radiative cooling rate produces a characteristic electron distribution that radiates a softer synchrotron spectrum than is observed. This limits the amount of energy that can be carried by the electrons, which limits the physics of the emission region.

INTRODUCTION

Many theories of gamma-ray bursts assume that synchrotron emission is the radiative process responsible for the gamma-ray flux. In particular, the shock theories of gamma-ray burst emission assume that the synchrotron emission occurs from a region in which the magnetic field and electrons are close to their equipartition energies [2]. For the magnetic field strengths implied under such theories ($B \approx 100\,\text{G}$), the cooling rate is very short compared to the burst duration. One, therefore, does not have total freedom in the choice of the electron distribution; the electron distribution must describe the cooling of electrons to zero energy.

The importance of this is that the electron distribution arising from synchrotron cooling produces a photon spectrum that is no harder than a $\nu^{-3/2}$ power law. In contrast, the photon spectrum for an electron distribution in which the electrons do not cool can be as hard as $\nu^{-2/3}$. Because gamma-ray bursts are often much harder than $\nu^{-3/2}$ [3], the cooling spectrum is not appearing in the gamma-ray energy range for these bursts. On the other hand, the X-ray spectrum of gamma-ray bursts is often above the power-law continuation of the gamma-ray spectrum [4], which may indicate the presence of the cooling component. From these facts, one can place limits on the fraction of emitted energy that is tied up in the electron's kinetic energy.

THE SYNCHROTRON SPECTRUM

The synchrotron emissivity from a single electron is proportional to the function

$$F(x) = x \int_x^\infty K_{\frac{5}{3}}(x') \, dx', \tag{1}$$

where $K_{\frac{5}{3}}(x')$ is a Modified Bessel Function and x is defined as

$$x = \frac{2\epsilon}{3\gamma^2 \bar{B} \sin\phi}, \tag{2}$$

and where ϵ is the photon energy in units of $m_e c^2$, γ is the Lorentz factor of the electron, $\bar{B}$ is the magnetic field strength in units of the critical field $B_c = 4.414 \times 10^{13}$ G, and ϕ is the emission angle relative to the magnetic field.

The model I assume is that electrons move through a region in which some acceleration mechanism counteracts the radiative losses for some time, after which the electron cools radiatively. I refer to this as the "steady" distribution. I assume that the electrons are injected with a single energy. This electron energy distribution is

$$f_s(\gamma) = \frac{\gamma_0}{2\pi} \delta\left(\gamma^2 - \gamma_0^2\right). \tag{3}$$

From this distribution, one derives the energy distribution for electrons cooling through synchrotron emission. This is

$$f_c(\gamma) = \frac{1}{4\pi \alpha_0 \gamma^2 \sin^2\phi}, \tag{4}$$

where

$$\alpha_0 = \frac{2e^4 B^2}{3 m_e^3 c^5}. \tag{5}$$

The photon spectrum is found by convolving equations (3) and (4) with equation (1). If p is the fraction of energy an electron emits while acceleration and radiative cooling are in balance, and $1-p$ is the fraction of energy emitted by an electron in a purely radiative mode, then the synchrotron intensity of an ensemble of electrons injected with the Lorentz factor γ_0 is

$$I \propto x_0^{-\frac{1}{2}} \int_{x_0}^\infty F(x) x^{\frac{1}{2}} \, dx + \frac{p}{1-p} \frac{\int_0^\infty y^{-\frac{1}{2}} \int_y^\infty F(x) x^{\frac{1}{2}} \, dx \, dy}{\int_0^\infty F(x) \, dx} F(x_0), \tag{6}$$

where x_0 is given by equation (2) with γ set to γ_0. The first term on the right is the cooling spectrum and the second term is the steady spectrum. A more general electron distribution—a power-law injection spectrum, for instance—can be incorporated by integrating equation (6) over γ_0.

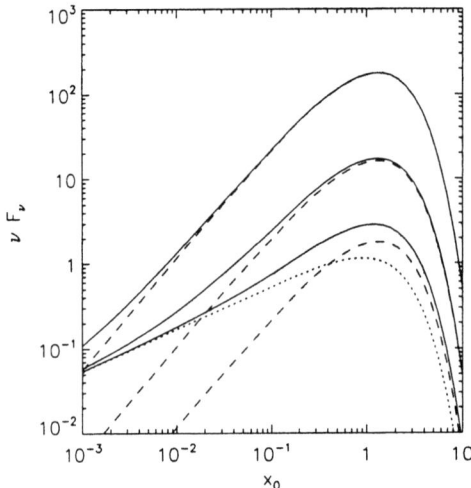

Fig. 1—Synchrotron spectral components. The first term on the right in equation (6), the cooling component, is given as a dotted line, and the second term, the steady component, is given as dashed lines for, from bottom to top, $p = 0.5, 0.1,$ and 0.01. The summed spectra for these values of p are given as solid lines.

The νF_ν spectra corresponding to the energy spectra given by equation (6) are plotted in Figure 1 for three values of p. The peak at $\approx 200\,\text{keV}$ corresponds to $x_0 = 1$, and the emission at $\approx 20\,\text{keV}$ corresponds to $x_0 = 0.1$. What is clear in this figure is that one does not see the $-2/3$ power law at low photon energies in the count spectrum ($4/3$ in the νF_ν spectrum) unless $p > 0.9$. The fraction of the energy emitted by the cooling component as a function of crossover frequency—the frequency at which the cooling component equals the steady component—is given by the solid line in Figure 2. From this figure, one sees that one must have $1 - p \ll 1$ to make the contribution of the synchrotron cooling spectrum at low energies negligible.

DISCUSSION

Gamma-ray burst spectra are hard above 20 keV, and they are often softer below this energy. This softness appears as an excess of X-ray flux compared to a power law extension of the spectrum from gamma-ray energies. The spectra discussed above can be reconciled with these characteristics in one of two ways. First, if the burst spectrum is very hard, with a power law index approaching $-2/3$, then there can be no contribution to the X-ray spectrum from the cooling synchrotron spectrum. Generally, this implies $p > 0.9$ to place the value of x_0 for which less than 10% of the spectrum is from the cooling electrons a factor of ten below the peak of the νF_ν curve. Second, if one desires to explain the X-ray upturn with the cooling spectrum, then $p \approx 0.5$, and the spectrum above $\approx 20\,\text{keV}$ must be relatively soft. This is a prediction that can be tested by seeing if the spectral index at $\approx 40\,\text{keV}$ correlates with the presence of an X-ray excess.

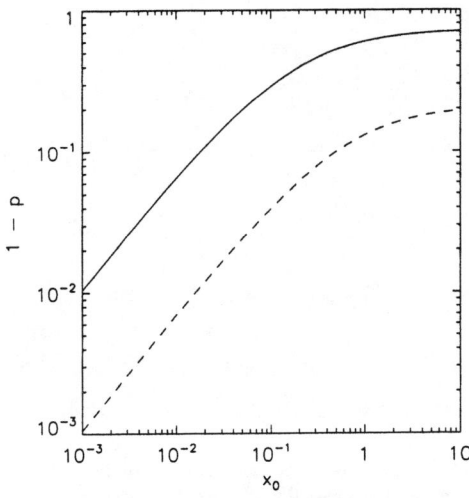

Fig. 2—Energy fraction $1-p$ as function of crossover energy. The crossover energy is defined to be the photon energy at which the two terms in equation (6) are equal, and it is plotted as a solid line. The 10% crossover energy, the energy at which the cooling component is 10% of the steady component, is plotted as a dashed line.

From the standpoint of theory, the one point that must be kept in mind is that γ_0 is smaller than generally assumed in shock models of gamma-ray bursts. For an energy per nucleon E_n released by the shock, the energy residing in an electron is no more than $m_e c^2 \gamma_0 = w E_n$, where $w < p$. This is generally not much of a problem in the shock models, because most observables are such strong functions of the bulk Lorentz factor Γ that a small change in Γ can compensate for a smaller value of γ_0.

Finally, because p must often be much larger than 1/2 to explain the bursts with spectra approaching $-2/3$ power-laws, the acceleration rate must be of order the cooling rate for a large fraction of gamma-ray bursts. This places limits on the acceptable acceleration mechanisms in the synchrotron emission model.

REFERENCES

1. Brainerd, J. J., & Lamb, D. Q., *ApJ* **313**, 231 (1987).
2. Mészáros, P., Rees, M. J., & Papathanassiou, H., *ApJ* **432**, 181 (1994).
3. Preece, R. D., Briggs, M. S., Mallozzi, R. S., et al., these proceedings.
4. Preece, R. D., Briggs, M. S., Pendleton, G. N., et al., *ApJ* **473**, 310 (1996).

Emission and Cooling Processes in a Hybrid Thermal-Nonthermal Plasma

D. Lin and E. P. Liang

Department of Space Physics and Astronomy, Rice University, Houston, TX 77005-1892

Abstract. In a hybrid thermal-nonthermal plasma, we find that the dominant emission and absorption mechanisms are synchrotron by nonthermal electrons and bremsstrahlung by thermal electrons. These two processes significantly change the spectrum from inverse Compton scatterings at low energies. We also find that Coulomb collisions are effective in cooling down the lower energy electrons but do not significantly alter the emission pattern. Compton cooling is more effective in changing emission and absorption coefficients when the photon energy density is high.

INTRODUCTION

The accurate positioning of GRBs by BeppoSAX has allowed observation of gamma-ray bursts over a wide range of wavelengths. Spectra from these observations provide crucial tests of different emission models, one of which is the inverse Compton scattering (ICS) model [1]. The ICS model explains from first principles most of the known spectral evolution properties of GRBs [1,2]. Under the ICS model, GRB spectra suggest that the sources are thermal-nonthermal hybrid plasma [1,2] whose electron distribution is described as

$$N(\gamma) = \begin{cases} f_{\rm th}(\gamma) = A\gamma\sqrt{\gamma^2-1}\, e^{-\frac{(\gamma-1)}{T_e}} & \gamma \leq \gamma_{\rm th} \\ f_{\rm nth}(\gamma) = B\gamma^{-p} & \gamma \geq \gamma_{\rm th}, \end{cases} \quad (1)$$

where A and B are determined by normalization and the continuity condition at $\gamma_{\rm th}$. This hybrid particle distribution has its own distinctive emission and absorption properties, which certainly affect the emerging spectra from the plasma. In this article, we will calculate the emission and absorption coefficients and incorporate them into the ICS model to make the model applicable to the multiwavelength spectra.

The significant roles of nonthermal electrons in emission and Compton scattering make it important to study the cooling processes of these electrons. Coulomb collisions and Compton scattering are the two processes we study. Aiming at modeling GRBs, we focus our calculation on the sources [1,2] with magnetic fields between

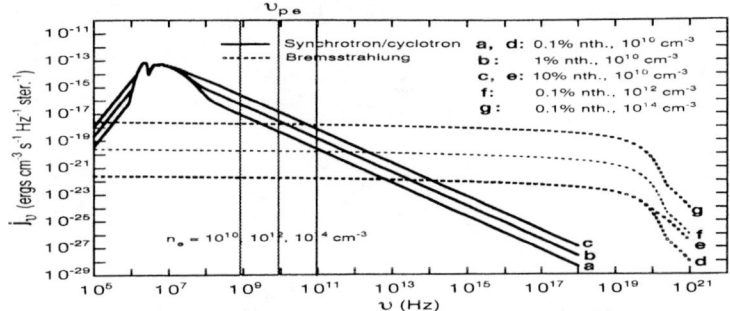

FIGURE 1. Emission coefficients of synchrotron/cyclotron and bremsstrahlung emissions, where ν_{pe} is the plasma frequency. The emission coefficients of cases f and g have been reduced by a factor of 10^2 and 10^4 respectively, so that they can be compared with cases a, b and c. For this calculation, B = 1 Gauss and T_e = 50 keV.

0.1-10 Gauss, T_e less than 100 keV, electron densities from 10^{10}-10^{14} cm^{-3}, and nonthermal fraction less than 10%.

EMISSION AND ABSORPTION COEFFICIENTS

Three major emission mechanisms in the hybrid source are cyclotron, synchrotron and bremsstrahlung emissions of electrons. For an arbitrary normalized particle distribution function $f(\gamma)$, the emission coefficient of cyclotron/synchroton emission is [3]

$$j_\nu^{cyc} = \frac{e^2 w n_e}{2c} \sum_{n=1}^{\infty} \int_{\beta_{min}}^{\beta_{max}} \frac{f(\gamma)}{\gamma\sqrt{\gamma^2-1}}[(1-\frac{\gamma_\parallel^2}{\gamma^2})\frac{J_n'(\xi)}{\gamma_\parallel} + \frac{(\cos\phi - \beta_\parallel)^2}{\sin^2\phi}J_n(\xi)]d\beta_\parallel, \quad (2)$$

where $\gamma = \frac{\gamma_n}{1-\beta_\parallel \cos\phi}$, $\gamma_n = \frac{n\nu_c}{\nu}$, ν_c is the cyclotron frequency, and $\xi = \frac{\nu}{\nu_c}\sin\phi\sqrt{\gamma(1-\beta_\parallel^2)-1}$. The integration limits in equation (1) are set to make ξ a real number.

The emission coefficient of bremsstrahlung emission is [4]

$$j_\nu^{ff} = \frac{4e^6 n_e^2}{3\pi m_e^2 c^4} \int_1^\infty \frac{\gamma f(\gamma)}{\sqrt{\gamma^2-1}}[ln(\frac{2mc^2(\gamma^2-1)}{h\nu}) - \frac{\gamma^2-1}{2\gamma^2}]d\gamma. \quad (3)$$

The absorption coefficients of both processes are calculated by $\alpha_\nu = \frac{c^2}{8\pi h\nu^3}j_\nu'$, where j_ν' is obtained from eqn. (2) or (3) by replacing $f(\gamma)$ with $f'(\gamma)$ [5]. $f'(\gamma)$ is:

$$f'(\gamma) = (\frac{f(\gamma^*)}{\gamma^*\sqrt{\gamma^{*2}-1}} - \frac{f(\gamma)}{\gamma\sqrt{\gamma^2-1}})\gamma\sqrt{\gamma^2-1}, \quad (4)$$

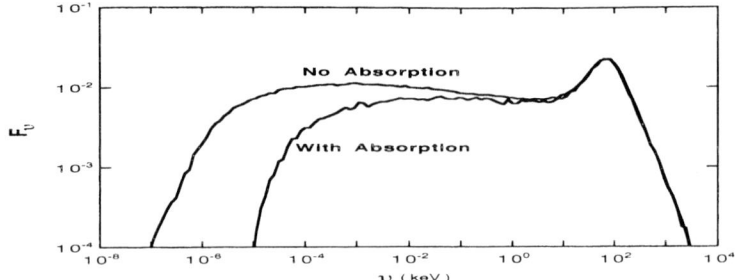

FIGURE 2. Spectra from ICS with and without absorption. The soft photon source with $T_r = 5.11 \times 10^{-7}$ keV sits in the center of a spherical plasma. For this calculation, B = 10 Gauss, $\tau_T = 10$, $T_e = 20$ keV, $n_e = 10^{10}$ cm^{-3}, and nonthermal fraction = 10%

where $\gamma^* = \gamma - \frac{h\nu}{mc^2}$. Calculation results (see Fig. 1) show that cyclotron emission can not propagate in the plasma, because its frequency is well below the plasma frequency. Nonthermal bremsstrahlung emission is also negligible. Nonthermal synchrotron and thermal bremsstrahlung are the dominant emission mechanisms. The same conclusion can be drawn for the absorption processes. Fig. 2 shows how the absorption and emission processes change the spectra from ICS. The spectrum is significantly cut-off at lower energies. This may be the reason why not many radio counterparts are detected [6].

COOLING OF NONTHERMAL ELECTRONS

As shown above, emission spectra are largely dependent on the particle distribution. The cooling processes of nonthermal particles affects not only the soft photon emission but also the inverse Compton scattering. Coulomb and Compton coolings are the two dominant processes. Synchrotron cooling is neglected because the magnetic field energy density is much less than electron's kinetic energy density. The cooling processes are governed by $\frac{d\gamma}{dt} = -\frac{\gamma}{\tau_c}$, where τ_c is the characteristic cooling time of both processes [5,7].

Calculation results (see Fig. 3) show that Coulomb thermalization quickly cools down low energy electrons. However, these less energetic electrons contribute little to synchrotron emission. The emission coefficient above the plasma frequency does not significantly change as thermalization occurs. Compton cooling, on the other hand, cools down high-energy electrons first. Loss of the high-energy electrons significantly reduces synchrotron emission when the photon energy density is high.

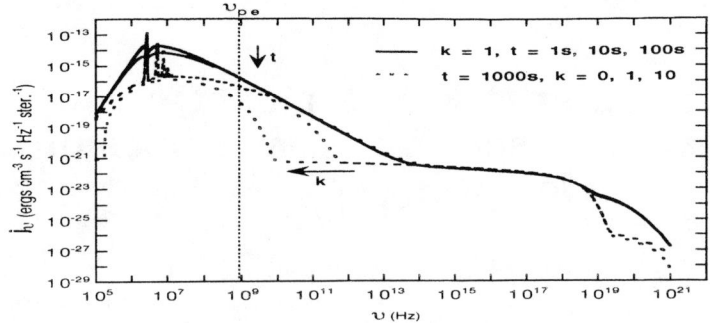

FIGURE 3. Effects of cooling processes on emission. Solid lines show the Coulomb collision effects, and dotted lines show Compton scattering effects. In this calculation, B = 1 Gauss, $n_e = 10^{10}$ cm^{-3}, T_e =10 keV, $\gamma_{th} = 1.1$ and $k = \frac{U_{ph}}{n_e kT}$, where U_{ph} is the photon energy density.

SUMMARY

We have calculated the emission and absorption coefficients for a hybrid thermal-nonthermal plasma and find that nonthermal synchrotron and thermal bremsstrahlung are the dominant mechanisms. These two mechanisms have been incorporated into our inverse Compton scattering model so that it may be compared with radio observations of GRBs. Future multi-wavelength observations will be critical tests of this model. Our calculations also show that Coulomb thermalization does not significantly change the emission pattern but Compton cooling does.

ACKNOWLEDGEMENTS

This work was partially supported by NASA grant NAG 5-3824.

REFERENCES

1. Liang, E. P., et al., *Ap. J. Lett.* **479**, L35-38 (1997).
2. Liang, E. P., *Ap. J. Lett.* **491**, in press (1997).
3. James, J., *PhD Thesis*, Harvard University (1985).
4. Zheleznyakov, V. V., *Radiation in Astrophysical Plasma*, Netherland: Kluwer, 1997.
5. Rybicki, G. & Lightman, A., *Radiative Processes in Astrophysics*, New York: Wiley, 1979.
6. Smith, I. A., et al., *Ap. J. Lett.* **487**, L5-L7 (1997).
7. Ginzburg, V. L., *Applications of Electrodynamics in Theoretical Physics and Astrophysics*, New York: Gordon and Breach, 1989.

Constraining the Intergalactic Magnetic Field with Cascading TeV Emission from Cosmological GRBs

Bruce Roscherr and Paolo S. Coppi

Physics Department, Yale University, PO Box 208101, New Haven, CT 06520-8101

Abstract. The high energy (> 100 GeV) emission of cosmological GRBs will be attenuated by photon-photon pair production on infrared and optical background light. The energy of these "absorbed" photons is reprocessed by cascading to lower energies. The reprocessing induces a delay in the arrival of cascade X-rays and results in a finite angular size for the source. If an intergalactic magnetic field is present then the X-ray flux as a function of both time and angle from the source change in a way that is characteristic of the field strength. We here calculate the cascade X-ray flux expected from a bright GRB and show that fields as low as 10^{-22} G have a discernable effect on the flux. The flux level, however, is too low to currently be detected, but a similar analysis might be more fruitfully applied to blazar AGNs.

INTRODUCTION

A number of mechanisms have been suggested for producing an intergalactic magnetic field (IGMF). These range from the esoteric option of quantum fluctuations locked in during the inflationary era [1,2] to the more mundane fields that might be generated by the injection of plasma into the intergalactic medium by active galactic nuclei [3]. We have currently no knowledge of the actual structure of the IGMF. We do though have a structure-dependent upper limit to the field strength from Faraday rotation measurements. For the case of a globally constant field the constraint is $B < 10^{-10}$ G and for a randomly oriented field of constant strength with coherence length of 1 Mpc, $B < 10^{-9}$ G [4]. We here explore a method orginally suggested by Plaga [5] which may strengthen these constraints by a number of orders of magnitude.

Typical GRB spectra are roughly flat in νF_ν over the whole observable range of five orders of magnitude up to GeVs. If their emission extends to more than a few hundred GeV then it becomes increasingly likely that these high energy photons will pair produce on the diffuse infrared and optical background light. The resulting electrons and positrons lose energy by Compton upscattering background photons. A fraction of these upscattered photons themselves have enough energy to pair produce and a cascade results. Eventually pair production ceases and the electrons and positrons slowly cool. An electron with Lorentz factor γ upscatters photons to an energy of $k'_B = \frac{4}{3}\gamma^2 k_B$ where k_B is the initial photon energy. The cosmic microwave background (CMB) dominates the electron cooling. CMB photons are upscattered to X-ray energies by electrons with $\gamma_x \approx 10^3$. It takes a TeV electron on the order of $N = 10^6$ scatterings to cool to this level. Each scattering deflects the electron by an angle of only $\frac{k_B}{\gamma} \approx 10^{-13}$ rad. To a good approximation then, we can take it that the electron travels in a straight line. If the electron travels a distance λ_c before it generates an X-ray photon then this X-ray photon will be delayed by $t_d = \frac{\lambda_c}{2\bar{\gamma}^2 c}$ with respect to the gamma-ray pulse. $\bar{\gamma}$ is the average Lorentz factor for the electron over the cooling length. Electrons with energy of a TeV or lower are in the Thompson regime for Compton scattering off the CMB. Hence the mean free path for Compton scattering is $\lambda_{cs} \approx 1$ kpc. The typical delay time thus is on the order of 10^9 s.

An upscattered photon emerges from the Compton scattering interaction within a cone of opening angle $\frac{1}{\gamma}$ from the electron direction of motion. If the source radiates into a cone wider than $\alpha = \frac{1}{\gamma_x}(1 - \frac{\lambda_c}{D})$ then we will detect X-rays generated by electrons that were emitted with angles less than α to our line of sight, D being the source distance. These photons have travelled a longer path to reach us and so will have a slightly longer delay, $t_d = \frac{\lambda_c}{2\bar{\gamma}^2 c} + \frac{\lambda_c}{2c}\frac{1}{\gamma_x^2}(1 - \frac{\lambda_c}{D})$. A further consequence is that the source will appear to have a finite angular size on the sky. The typical angular size will be $\frac{\lambda_c}{D}\frac{1}{\gamma_x}$.

If a magnetic field is present then the electron is deflected by an angle $\phi = \frac{\lambda_{cs}}{r_L}$ between interactions, where r_L is the electron's Lamor radius. The curvature in the electron's path introduces an additional time delay. For a globally constant field $t_{d,B} \simeq \frac{\lambda_c^3 q^2}{24\bar{\gamma}^2 m^2 c^5} B^2$. The effects of the magnetic field become noticable when $t_{d,B}$ is on the order of the delay caused by the subluminal electron velocity, i.e. when $B_c \simeq \sqrt{\frac{12 m^2 c^4}{\lambda_c^2 q^2}} \simeq 10^{-23}$ G. If the field is randomly fluctuating with a coherence length of 1 Mpc then the critical field value rises to 5×10^{-23} G. This is still a very low field strength. The magnetic field changes the X-ray flux profile as a function of time and angle and so X-ray observations of cascade radiation provide a highly sensitive probe of the IGMF strength.

We have written a Monte Carlo code that follows a cascade in detail. Position and velocity information is kept for all particles at all times. We can calculate the expected X-ray flux for a cosmolgical source as a function of time and angle from the source position.

RESULTS

We have calculated the expected flux of 1-10 keV X-rays as a function of both time and angle for a GRB situated at $z = 0.1$ ($H_0 = 75$ km s^{-1} Mpc^{-1}) with $E^2 dN/dE = 0.1$ MeV cm^{-2} s^{-1}, typical of a bright burst. Photons of 1-10 TeV were injected from an E^{-2} distribution. The IR/optical background used is flat in $E^2 dN/dE$ from 0.02 to 13.6 eV (the Tyson background, [6]) and is normalized to 2×10^{-3} eV/cm^3. The magnetic field was chosen to have a constant magnitude and a coherence length of 1 Mpc with the orientation changing randomly from one 1 Mpc cell to the next.

Figure 1 shows the cascade spectrum generated from a sample of 500 input photons. The bulk of the power is received at around 10 GeV, with only a very small fraction emerging in the 1-10 keV window.

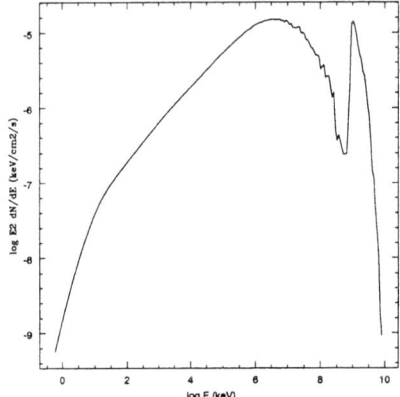

FIGURE 1. Cascade spectrum generated by a sample of 500 input photons drawn from an E^{-2} distribution between 1 and 10 TeV.

FIGURE 2. 1-10 keV flux as a function of time after the burst for various magnetic field strengths. The curves from top to bottom correspond to strengths of $10^{-23}, 10^{-22}, 5 \times 10^{-22}, 10^{-21}, 5 \times 10^{-21}$ and 10^{-20} G respectively.

Figure 2 shows the X-ray flux received as a function of time for various magnetic field strengths. In all cases the flux peaks between 10^8 and 10^9 seconds after the burst. We see that as the field strength increases from 10^{-23} G to 10^{-20} G that there is a substantial change in the flux.

The X-ray flux is also angle dependent. Figure 3 shows the total number of 1-10 keV photons received as a function of the apparent angle to the source for the same range of field strengths depicted in Figure 2. For fields of 10^{-23} G or less the source has an apparent angular size of some 20 arcsecs. As the field grows, the angular size grows, already reaching 200 arcsecs for a field of 10^{-20} G. The angular size is, of course, also time dependent. Figure 4 shows how the flux evolves as a function of time for each of three angular bins for three different field strengths. In all cases, the angular size grows with time. At any given time, the ratio of the flux in each of the three angular bins is characteristic of the field strength.

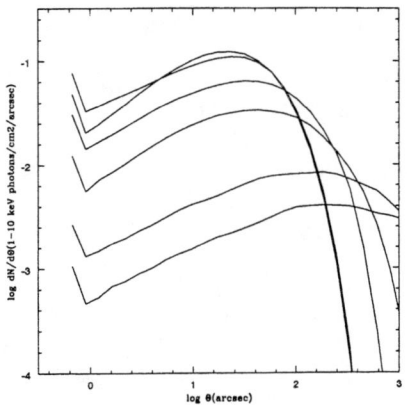 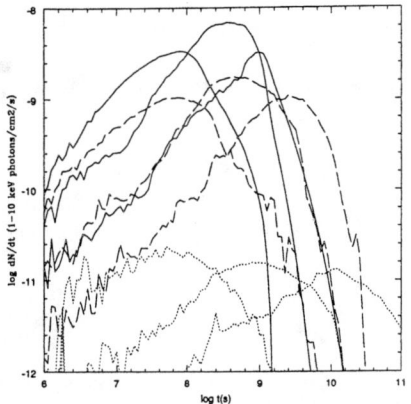

FIGURE 3. Total number of 1-10 keV photons received over all time as a function of the apparent angle from the source for various field strengths. All flux received within 0.5 arcsec of the source is placed in the first bin. The curves from top to bottom correspond to the same range of strengths as in Figure 2.

FIGURE 4. 1-10 keV flux received in each of 3 angular bins as a function of time after the burst for 3 different field strengths. The solid curves are for $B=10^{-23}$G, the dashed curves for $B=5 \times 10^{-22}$ G, and the dotted curves for $B=10^{-20}$ G. The curves that peak first correspond to the flux received between 0 and 10 arcsecs, followed by the flux between 10 and 60 arcsecs, and the flux between 60 and 250 arcsecs.

DISCUSSION

Our results show that magnetic fields as low as 10^{-22} G have a discernable effect on the X-ray part of the cascade spectrum. The X-ray flux as a function of both time and angle evolve in a way that is characteristic of the field strength. We thus have a very sensitive potential means of constraining the IGMF. The expected flux, however, is very low. The maximum flux of 7×10^{-9} photons cm^{-2} s^{-1} occurs for fields weaker than 10^{-23} G. If we compare this to the background rate for XTE of 3.2×10^{-3} photons cm^{-2} s^{-1} [7], we see that the cascade X-rays are undetectable. The source size could easily be resolved by AXAF, but again the flux is too low by at least two orders of magnitude to allow a detection in any reasonable integration time. This same analysis might be more fruitfully applied to blazar AGNs. If these sources continuously emit TeV radiation then we get a substantial boost in the expected flux. The flux may be at a detectable level in this case, but the background halo produced by the scattering of prompt X-rays off dust in our galaxy would have to be carefully calculated [8,9].

REFERENCES

1. Ratra, B., *Ap. J.* **391**, L1-L4 (1992).
2. Turner, M.S. & Widrow, L.M., *Phys. Rev. D* **37**, 2743-2754 (1988).
3. Thomson, R.C. & Nelson, A.H., *M.N.R.A.S.* **201**, 365-383 (1982).
4. Kronberg, P.P., *Space Science Rev.* **75**, 387-399 (1996).
5. Plaga, R., *Nature* **374**, 430-432 (1995).
6. Tyson, J.A., in *Galactic and Extragalactic Background Radiation, IAU Symposium* **139** ed. Bowyer, S. and Leinert, C., (Dordrecht: Kluwer Academic), 245.
7. Bradt, H.V. et.al., in *Observatories in Earth Orbit and Beyond* ed. Kondo, Y. (Dordrecht: Kluwer Academic), 89-110 (1990).
8. Mathis, J.S. & Lee, C.W., *Ap.J.* **376**, 490 (1991).
9. Klose, S., *A&A* **248**, 624 (1991).

Variability in Shell Models of GRBs

M. C. Sumner and E. E. Fenimore

NIS-2, MS D436, Los Alamos National Laboratory, Los Alamos, NM 87545

Abstract. Many cosmological models of gamma-ray bursts (GRBs) assume that a single relativistic shell carries kinetic energy away from the source and later converts it into gamma rays, perhaps by interactions with the interstellar medium or by shocks within the shell. Although such models are able to reproduce general trends in GRB time histories, it is difficult to reproduce the high degree of variability often seen in GRBs. We investigate methods of achieving this variability using a simplified external shock model. Since our model emphasizes geometric and statistical considerations, rather than the detailed physics of the shell, it is applicable to any theory that relies on relativistic shells. We find that the variability in GRBs gives strong clues to the efficiency with which the shell converts its kinetic energy into gamma rays.

The "external shock" models of gamma-ray bursts (GRBs) assume that the gamma rays are emitted from a single shell of material traveling at highly relativistic speeds (for example, [1] and [2]). Fenimore, Madras, and Nayakshin [3] have found that the envelope of emission for a single relativistic shell fits the overall shape of a few GRBs, but it does not account for the wide variability found in GRBs. In this paper, we investigate methods of achieving the observed degree of variability in GRB time histories using randomly placed active regions on the shell.

DESCRIPTION OF THE MODEL

The "time" that is measured in GRB time histories is the time at which photons arrive at the detector (denoted by T), and is not the time of emission as would be measured in the rest frame of the detector (equivalent to the rest frame of the shell's expansion, denoted by t). We set $t = 0$ to be the time at which the central explosion occurs, and $T = 0$ to be the arrival time of a photon emitted at $t = 0$; therefore

$$T = (1 - \beta \cos \theta)t = \Lambda \gamma^{-1} t \quad , \tag{1}$$

where Λ is defined as $\Lambda = \gamma(1 - \beta\mu)$, and $\mu = \cos\theta$. By setting Equation 1 equal to a constant, one finds that the surface of constant arrival time is an ellipsoid

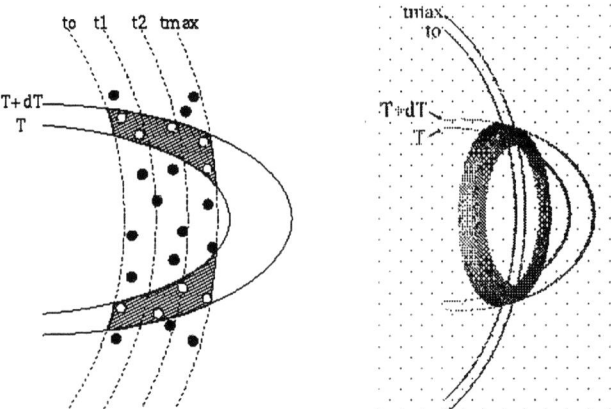

FIGURE 1. *a. (left)* The position of the spherical shell at four different times as it expands from t_o to t_{max}. The two ellipses represent surfaces of constant arrival time. Photons emitted from any point on the inner ellipsoid will arrive at the detector simultaneously at time T, and photons emitted from any point on the outer ellipsoid will arrive at the detector at time $T + dT$. The cross-hatched region represents the volume that will be "seen" by the detector between these two times. Each small circle represents the volume swept out by an entity during its emission lifetime of ΔT_p. Only the white entities will be seen between T and $T + \Delta T$. *b. (right)* A three-dimensional view of the cross-hatched area shown at left. The volume of this annulus is given by $\Upsilon(T)dT$.

[4]. All of the photons emitted from such an ellipsoid will arrive at the detector simultaneously.

Our model considers a thin shell (as required by the observations [3]) that expands outward from a central point so that $R = \beta ct$. The shell expands to a radius of $R_o = \beta ct_o$, and then emits gamma rays until it reaches a radius of $R_{max} = \beta ct_{max}$. The gamma-ray emission must occur from small, independent patches on the shell, which we call "entities" [3]. Each entity begins as a small perturbation, perhaps caused by the shell's interactions with the interstellar medium or by instabilities within the shell. The entity grows at the sound speed $c_s \sim c$ until it reaches a maximum size of $\Delta R_\perp = \Gamma c \Delta T_p$ (see Table 2 in [3]). Thus, each entity represents a causally connected region, and many entities can fit on the surface of the shell. We assume that each entity emits isotropically with a power-law photon spectrum in its own reference frame, $\Phi'(E') = E'^{-\alpha}$ photons (entity $dt'\ dE'\ d\Omega'$)$^{-1}$ (primed quantities are measured in the rest frame of the entity).

Figure 1 gives a pictorial representation of our model. It is important to emphasize that we are considering a *thin* shell. Thus, the four dashed lines in Figure 1a represent the shell at four different times. The area between t_o and t_{max} represents the volume *swept up* by the shell over its emission lifetime. Likewise, the entities are small *areas* on the shell. The small spheres in Figure 1a represent the volumes swept out by the entities as the shell expands.

A more quantitative description can be made by calculating the volume enclosed in the cross-hatched region, which we denote as $\Upsilon(T)dT$ (Figure 1b):

$$\Upsilon(T)dT = 2\pi \int_{\min(R_o,R_{\text{ell,in}})}^{\min(R_{\max},R_{\text{ell,out}})} \int_{\mu(T)}^{\mu(T+dT)} r^2 d\mu \, dr \quad , \tag{2}$$

where $R_{\text{ell,out}} = \beta c(T+dT)/(1-\beta)$, and $R_{\text{ell,in}} = \beta c T/(1+\beta)$. Equation 1 gives $\mu(T) = [\beta^{-1} - Tc/r]$, and $\mu(T+dT) = [\beta^{-1} - (T+\Delta T)c/r]$. Evaluating the integral yields the volume seen per dT:

$$\begin{aligned}
\Upsilon(T) &= 0 & &\text{if } T < T_o \\
&= \frac{\pi c(\beta c)^2}{(1-\beta)^2}\left(T^2 - T_o^2\right) & &\text{if } T_o < T < T_{\max} \\
&= \frac{\pi c(\beta c)^2}{(1-\beta)^2}\left(T_{\max}^2 - T_o^2\right) & &\text{if } T_{\max} < T < \Gamma^2(1+\beta)^2 T_o \quad (3) \\
&= \frac{\pi c(\beta c)^2}{(1-\beta)^2}\left(T_{\max}^2 - \frac{T^2}{\Gamma^4(1+\beta)^4}\right) & &\text{if } \Gamma^2(1+\beta)^2 T_o < T < \Gamma^2(1+\beta)^2 T_{\max} \\
&= 0 & &\text{if } T > \Gamma^2(1+\beta)^2 T_{\max} \;,
\end{aligned}$$

where $T_o = t_o/(1-\beta)$, and $T_{\max} = t_{\max}/(1-\beta)$. Note the rather surprising result that the volume "seen" by the detector is constant for $T_{\max} < T < \Gamma^2(1+\beta)^2 T_o$ for the relativistically expanding shell. During these times, the number of entities in the cross-hatched region in Figure 1a is a constant with respect to T. In comparison, in the non-relativistic case, the volume seen, and therefore the number of entities seen, would increase as the area of the shell increased.

The time history from such a shell is given by

$$V(T)dT = 2\pi \int_{\min(R_o,R_{\text{ell,in}})}^{\min(R_{\max},R_{\text{ell,out}})} \int_{\mu(T)}^{\mu(T+dT)} \rho C' \Lambda^{-(\alpha+1)} r^2 d\mu \, dr \quad , \tag{4}$$

where $V(T)$ is the expected time history in photons $(dT \, dA_{\text{det}})^{-1}$, C' is the photon flux as observed in the rest frame of the entity, and $\Lambda^{-(\alpha+1)}$ incorporates the relativistic effects. We define ρ as the "density" of entities so that $\rho \, dV$ gives the number of entities within the cross-hatched region in Figure 1a. (The density ρ is proportional to the fraction of the shell's surface that emits gamma rays during the shell's evolution, and therefore gives some idea of how efficiently the shell converts its kinetic energy into gamma rays.) The envelope is [5]

$$\begin{aligned}
V(T) &= 0 & &\text{if } T < T_o \\
&= \psi\frac{T^{\alpha+3} - T_o^{\alpha+3}}{T^{\alpha+1}} & &\text{if } T_o < T < T_{\max} \quad (5) \\
&= \psi\frac{T_{\max}^{\alpha+3} - T_o^{\alpha+3}}{T^{\alpha+1}} & &\text{if } T_{\max} < T < \Gamma^2(1+\beta)^2 T_o
\end{aligned}$$

767

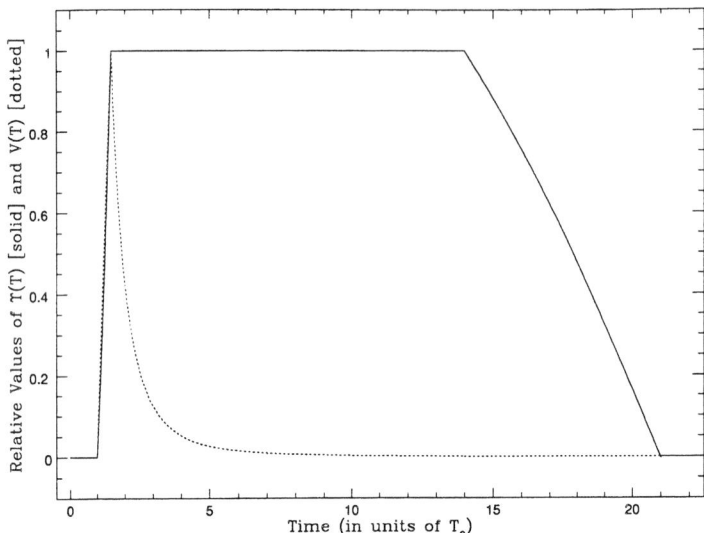

FIGURE 2. The volume seen by the detector per ΔT is shown as the solid line, and the resulting photon flux at the detector is shown as the dashed line. The volume, $\Upsilon(T)$, and the detector signal, $V(T)$, both increase as T^2 for $T_o < T < T_{max}$; this is identical to the results for a non-relativistic shell. However, in the case of a relativistic shell for $T > T_{max}$, the volume remains constant over a large range of T, and the signal shows a power-law decay. Since the detector sees a constant volume, and thus a constant number of entities, per ΔT for $T > T_{max}$, the decrease in the photon flux after T_{max} is entirely determined by relativistic effects.

where ψ is a constant (see Figure 2).

Equation 5 gives the *average* signal that would be expected from a collection of entities scattered over the surface of a relativistic shell. However, the number of entities seen in any volume $\Upsilon(T)dT$ is a random quantity; since the entities are independent and discrete, the randomness follows simple Poisson statistics. Therefore, this model predicts time histories with a mean given by Equation 5 and Poisson variations about that mean, which would look very similar to the "peaks" usually observed in GRB time histories. A low value for ρ gives a low number of entities and correspondingly large Poisson fluctuations, leading to a "spiky" time history with many peaks. Conversely, a high efficiency leads to a smooth time history. To illustrate these points, we present two simulated time histories in Figures 3a and 4a with characteristics selected to match particular BATSE bursts. We estimate that the typical fraction of a single relativistic shell's surface that emits gamma rays during its lifetime is $\sim 10^{-3}$ [6].

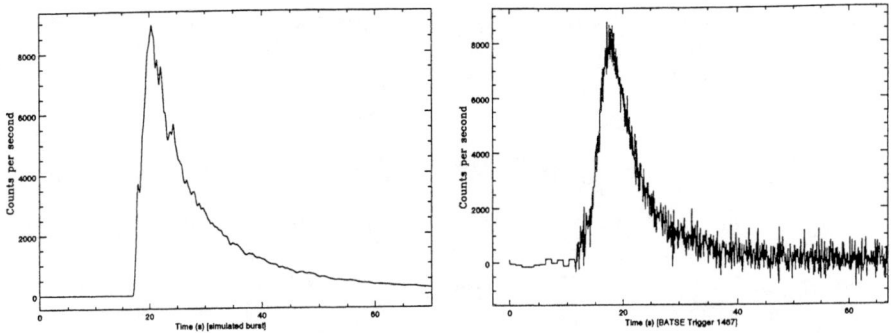

FIGURE 3. *a. (left)* Simulated time history using a high density of entities. Nearly 100% of the shell's surface emitted gamma rays in this simulation. Note that the high efficiency gives a smooth profile. *b. (right)* BATSE trigger 1467.

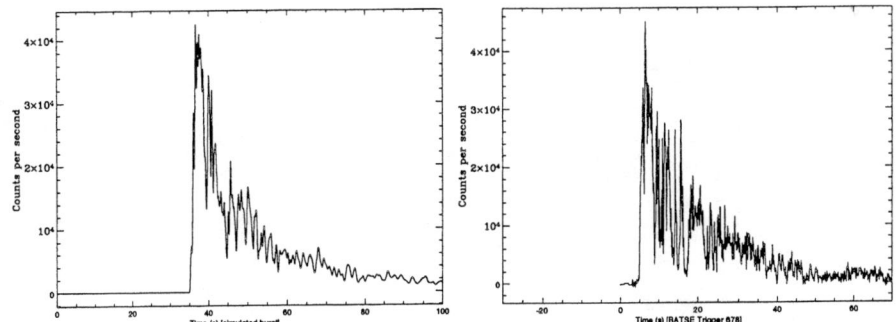

FIGURE 4. *a. (left)* Simulated time history assuming a low density of entities. Only 1% of the shell's surface emitted gamma rays in this simulation. Note that this low efficiency gives a spiky profile, with many "peaks". *b. (right)* BATSE trigger 678.

CONCLUSIONS

In order to achieve the observed variability in GRBs using a single-shell model, we have found that the gamma-ray emission must occur from small patches on the shell (entities). We have derived two significant characteristics of the time histories expected from such a model. First, the average number of entities contributing to the signal remains constant throughout much of the time history, although the overall photon flux decreases due to relativistic effects. Second, the "peaks" in a time history can be ascribed to Poisson variations in the actual number of entities contributing to the signal at any given time. Taken together, these properties imply that the relative variations in a GRB (i.e. the heights of the peaks relative to the average envelope of the signal) should remain constant throughout a time history.

Qualitatively, this result is consistent with visual inspections of GRBs. In general, bursts are equally "spiky" during the first half of the time history as the second half.

REFERENCES

1. Mészáros, P., and Rees, M. J., *ApJ* **405**, 278 (1993).
2. Piran, T., Shemi, A., and Narayan, R., *MNRAS* **263**, 861 (1994).
3. Fenimore, E. E., Madras, C. D., and Nayakshin, S., *ApJ* **473**, 998 (1996); astro-ph/9607163.
4. Rees, M. J., *Nature* **211**, 468 (1966).
5. Fenimore, E. E., and Sumner, M. C., *All-Sky X-Ray Observations in the Next Decade*, Eds. M. Matsuoka and N. Kawai, in press (1997); astro-ph/9705052.
6. Fenimore, E. E., Ramirez, E., and Sumner, M. C., these proceedings.

Radiative Efficiency in Gamma-Ray Bursts and Afterglows

A. Panaitescu and P. Mészáros

Department of Astronomy & Astrophysics
Pennsylvania State University, University Park, PA 16802

Abstract. We present numerical simulations of gamma-ray bursts arising from external shocks in the impulsive and wind models, including a weak or a strong coupling between electrons and protons + magnetic fields, and analyze their features in each scenario. The dynamics of the fireball-external medium is followed until later times in order to simulate the hydrodynamics of the remnant and the temporal and spectral evolution of the afterglow.

INTRODUCTION

Gamma-ray bursts (GRBs) are simulated in the framework of the "external shock model" [2], thus the bursts obtained numerically resemble only those that exhibit a single hump light-curve or a low ($\lesssim 10$) number of pulses. Even if the GRB originates in "internal shocks", its lower energy emission ("afterglow") must be released during the interaction of the relativistic wind with the external medium. The details of the energy release mechanisms are presented in Panaitescu & Mészáros [5,6]. Here we mention only the most important ones:
1. Electrons are accelerated by one of the two shocks ("reverse" or "forward") that sweep up the relativistic ejecta or the external medium (respectively), and lose energy through synchrotron emission in the presence of a turbulent magnetic field, and through inverse Compton scattering of the synchrotron photons. The initial distribution of electrons is a power law of index p.
2. The full shape of the synchrotron spectrum is used to calculate the synchrotron emission from the two shocks. However, when the inverse Compton spectrum is calculated, the spectrum of the synchrotron radiation emitted by each shock is approximated as monochromatic, at an intensity-averaged frequency. Furthermore, the inverse Compton spectrum itself is approximated as monochromatic, at its peak frequency.

The interaction between the expanding shell and the external matter is simulated using a 1D hydrodynamical code suitable for relativistic flows involving shocks.

The temporal (i.e. bolometric and band light-curves) and spectral features (a set of instantaneous spectra and the averaged spectrum) of the burst are calculated by integration over lab-frame time t, volume of the shocked fluid and electron distribution.

GAMMA-RAY BURSTS

The peak of the synchrotron emission from the forward shock, which dominates the overall emission of the main burst and of the afterglow, is at

$$h\nu_p = 2 \times 10^{-3} (1+z)^{-1} \varepsilon_{el}^2 (\varepsilon_{mag} n_0)^{1/2} \Gamma^4 \text{ eV} , \tag{1}$$

where n_0 is the external medium particle density in cm^{-3}, z is the burst redshift and Γ is the flow Lorentz factor. In deriving equation (1) it was assumed that $h\nu_p$ is determined by the least energetic electrons. The energy release parameters ε_{el} and ε_{mag} represent the fraction of the total internal energy of the shocked gas that is stored in electrons and magnetic field, respectively. For a fireball initial Lorentz factor Γ_0, the average flow Lorentz factor during the main burst is $\simeq 1/2\Gamma_0$. In all simulations discussed below, $\Gamma_0 = 500$, $p = 2.5$, $n_0 = 1$, $\varepsilon_{el} = \varepsilon_{mag} = 1/3$. Then equation (1) gives $h\nu_p \sim 250$ keV for $z = 1$.

If after shock acceleration electrons do not exchange energy with protons and the magnetic field (**weak coupling model**), most internal energy of the shocked fluid remains locked up in protons. If electrons are re-accelerated behind the forward shock (**strong coupling model**), then the internal energy is depleted very fast and the shocked structure is in a radiative regime. Figure 1 shows the burst spectrum for an impulsive fireball with weak and strong coupling in the shocked ejecta. Each spectrum has six components: one synchrotron and two inverse Compton from each shock, but not all of them can be distinguished (for more details see Panaitescu & Mészáros [6]). The most intense component is the blast wave synchrotron emission. Longward of it is the synchrotron emission from the reverse shock.

In the **impulsive model**, the co-moving frame density of the fireball at $t \lesssim t_{dec}$ is $\sim \Gamma_0^2$ times larger than that of the external medium, leading to a quasi-newtonian reverse shock ($\Gamma_R \simeq 1.1$) that is inefficient in converting the ejecta's kinetic energy into heat. A more relativistic reverse shock can be obtained if the fireball is less dense. This can be achieved if the fireball results from an energy release that lasts more than few seconds (**wind model**). For a wind of duration $t_{wind} = 33$ s, Γ_R increases to $\simeq 1.5$, while the Lorentz factor of blast wave decreases by a factor 5/3. A more relativistic reverse shock radiates more efficiently, while a less relativistic forward shock yields a softer and weaker GRB, as shown in Figure 1 (long dashed curve).

In Figure 2 we compare the light-curves and spectral evolution of the bursts obtained in the three models discussed so far. Note that the reverse shock is contributing more to the observed burst if there is a strong coupling. A strong coupling also leads to light-curves that have a temporal asymmetry closer to that

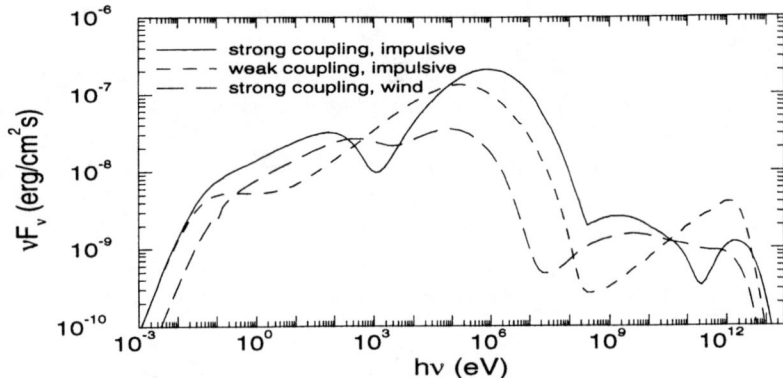

FIGURE 1. Comparison between averaged spectra obtained in three models. The initial fireball kinetic energy is $E = 10^{52}$ erg sr^{-1}. Note the harder and more intense burst resulting from an impulsive fireball and strong coupling. An extended energy release at the place where the fireball originates results in a softer burst, in which the two shocks radiate comparable amounts of energy.

observed in real bursts [4]. For the same set of parameters, the wind model produces a soft burst with the slowest softening rate, while the impulsive model with strong coupling gives the hardest burst.

AFTERGLOWS

The spectral evolution of the afterglow is mainly determined by the Lorentz factor of the shocked fluid and we assume all other parameters (such as ε_{el} and

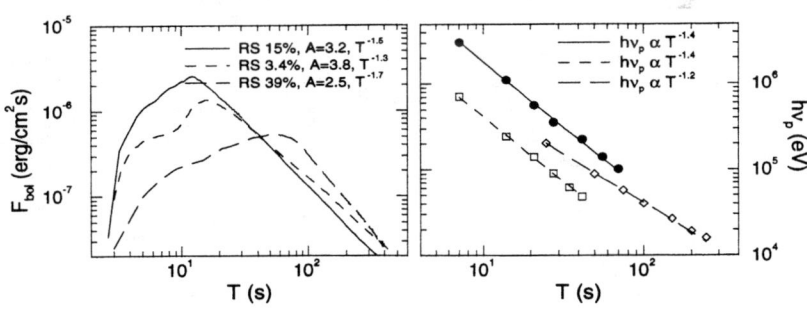

FIGURE 2. Temporal and spectral evolution of the bursts whose spectra are shown in Figure 1. The legend of the left graph gives the efficiency of the reverse shock, the temporal asymmetry of the light-curve (fall time to rise time ratio), and the burst $T^{-\alpha}$ fall. The right graph shows the evolution of the peak of νF_ν.

$\varepsilon_{\mathrm{mag}}$) to be constant. The expected behavior of Γ is $\propto t^{-3}$ if the remnant is radiative and $\propto t^{-1.5}$ if it is adiabatic [1]. In order to have a radiative remnant, electrons must be rather strongly coupled with protons and the magnetic field and must be themselves radiative (for a complete discussion of all regimes see Mészáros, Rees & Wijers 1997): $t_{\mathrm{cool}} <$ few $\times t_{\mathrm{dec}}$, where t_{cool} is the lab-frame synchrotron cooling timescale. For the bursts shown in Figure 1, $t_{\mathrm{cool}} \sim 4 \times 10^{-4} t_{\mathrm{dec}}$ and using $t_{\mathrm{cool}} \propto \Gamma^{-2}$, it results that electrons become adiabatic when Γ drops below ~ 5; after that the remnant is also adiabatic no matter how strong the coupling is.

It can be shown [6] that the evolution of the Lorentz factor of an adiabatic remnant is independent of the fireball initial Lorentz factor

$$\Gamma = 6.2 \, (E_{52}/n_0)^{1/8}(1+z)^{3/8}(T/1\,\mathrm{day})^{-3/8} \,, \tag{2}$$

where $E = 10^{52} E_{52}$ erg sr^{-1}, which implies that electrons become adiabatic at $T \simeq 4.0$ days. Equations (1) and (2) gives the detector time when the peak of the synchrotron spectrum of the forward shock emission is at detector frame energy $h\nu_p$:

$$T_p = 2.1 \, (1+z)^{1/3} \varepsilon_{\mathrm{el}}^{4/3} (\varepsilon_{\mathrm{mag}} E_{52})^{1/3}(h\nu_p/1\,\mathrm{eV})^{-2/3} \text{ days} \,. \tag{3}$$

At equipartition, $h\nu_p = 1$ eV for $T_p = 0.42$ days, consistent with Figure 3. Equation (2) gives $\Gamma_{1\,\mathrm{eV}} = 11$, thus electrons are radiative during the optical afterglow. In the case of a continuous post-shock re-acceleration, the electrons in the high energy part of the power-law distribution contribute more to the synchrotron emission than those in the low energy part, and equation (1) under-estimates the true $h\nu_p$ of the synchrotron spectrum. For the same energy release parameters, the spectra of a strong coupling afterglow is harder than that of weak coupling one.

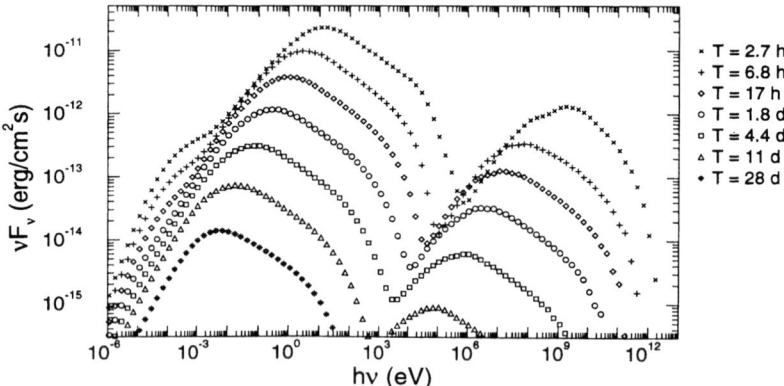

FIGURE 3. Spectral evolution of the afterglow in the weak coupling model. The legend indicates the observer time for each spectrum. The hydrodynamic and energy release parameters are those used for Figures 1 and 2.

CONCLUSION

Summarizing, the features of bursts and afterglows arising from impulsive/wind fireballs with strong/weak coupling of electrons with baryons and magnetic fields are:
1. Wind fireballs produce softer GRBs than impulsive ones for the same set of hydrodynamical and energy release parameters, and increase the efficiency of the reverse shock, producing a brighter optical and UV counterpart. An appropriate change in model parameters (particularly Γ_0) could shift the softer spectrum produced by wind fireballs back into the γ-ray domain, but the fact that the GRB efficiency is reduced makes this possibility less likely to be the real scenario.
2. Strong coupling yields more time-symmetric peaks than weak coupling does, in better agreement with the observed pulse time-asymmetry. It also leads to harder spectra and allows the radiative phase of the afterglow to extend to later times than in the weak coupling model. It can also explain the X-ray paucity observed in many GRBs in that it maintains the synchrotron emission from the blast wave in the γ-ray range.
3. The energy release parameters (i.e. those for the magnetic field intensity and for electron energy) should not be too much below equipartition, otherwise the main burst will have a peak below 10 keV. As a consequence, all remnants start their evolution with a radiative phase. The adiabatic phase begins earlier for a weak coupling remnant than for one with strong coupling. Electrons are radiative in the early afterglow (T less than few days) and adiabatic for the rest of the afterglow.

ACKNOWLEDGEMENTS

This research has been supported by NASA NAG5-2857 and NAG5-2362.

REFERENCES

1. Blandford, R. D. & McKee, C. F., *Phys. Fluids* **19**, 1130 (1976).
2. Mészáros, P. & Rees, M. J., *ApJ* **415**, 181 (1993).
3. Mészáros, P., Rees, M. J., Wijers, R., *ApJ*, submitted (astro-ph/9709273) (1997).
4. Mitrofanov, I. G.. et al., *ApJ* **459**, 570 (1996).
5. Panaitescu, A. & Mészáros, P., *ApJ* **492**, in press (astro-ph/9703187) (1998).
6. Panaitescu, A. & Mészáros, P., in preparation.

Cosmic Rays and Neutrinos from Gamma-Ray Bursts

Jörg P. Rachen and P. Mészáros

Pennsylvania State University, University Park, PA 16802
jrachen@astro.psu.edu, pmeszaros@astro.psu.edu

Abstract. We review the hypothesis that the acceleration of protons at internal shocks in gamma-ray bursts (GRB) could be the origin of the ultra high-energy cosmic rays (UHECR) observed at earth, $E_{\max} \gtrsim 10^{19}$ eV. We find that, even though protons may be accelerated to such energies, their ejection into the interstellar/intergalactic medium is problematic because it is likely to be accompanied by considerable adiabatic losses in the expanding shell. The problem is circumvented by neutrons produced in photohadronic interactions, which are not magnetically bound and thus effectively ejected in the moment of their production. They can be both produced in sufficient number and be able to leave the emission region if the optical depth of the emission region to photohadronic interactions is of order 1. We show that this requirement can be fulfilled under the same conditions which allow acceleration of protons to the highest energies. The production of neutrinos in this process correlates the fluxes of cosmic rays and neutrinos, and makes the hypothesis of UHECR origin in GRBs testable.

ACCELERATION OF PROTONS IN GRBS

The hypothesis that gamma-ray bursts (GRB) might be responsible for the origin of ultra high-energy cosmic rays (UHECR) has been proposed by Waxman [1], who assumed cosmic ray acceleration by momentum diffusion (2nd order Fermi acceleration), or at internal shock waves (1st order Fermi acceleration) within the expanding wind. We investigate here in some more detail the problems and consequences of this idea, in particular in reference to the internal shock scenario [2], where the observed variability in GRBs is explained by the presence of shocks in an unsteady outflow which occur when expanding subshells of different velocity catch up and merge.

The energy of protons accelerated at internal shocks in GRBs is essentially constrained by two conditions: (1) confinement in the acceleration region, $r'_g = E'_p/eB' < R' \sim cT_{\text{var}}\Gamma$; (2) balance of acceleration and synchrotron cooling, $t'_{\text{acc}} \sim 2\pi r'_g/c < t'_{\text{syn}}$. Here, T_{var} is the observed variability time scale, and Γ is the bulk Lorentz factor of the wind; primed quantities refer to the comoving frame. It can be shown that energy losses due to photohadronic interactions and other cooling processes are less relevant [3]. Expressing the magnetic field relative to its equipartition value with the radiation, $U'_B = \xi_{B\gamma} U'_\gamma$, the maximum proton energy has to satisfy the conditions

$$E_{p,20} \lesssim 3(L_{51}\xi_{B\gamma})^{1/2}\Gamma_2^{-1} \tag{1a}$$

$$E_{p,20} \lesssim \tfrac{1}{3}(L_{51}\xi_{B\gamma})^{-1/4}T_{-1}^{1/2}\Gamma_2^{5/2} . \tag{1b}$$

Quantities are expressed in canonical units: $E_{p,20} = E_p/10^{20}\,\text{eV}$, $L_{51} = L/10^{51}\,\text{erg sec}^{-1}$, $T_{-1} = T/0.1\,\text{sec}$, and $\Gamma_2 = \Gamma/100$. These equations can be rewritten to express the minimum requirements on the physical conditions in the GRB wind in order to produce the highest energy cosmic rays:

$$\Gamma_2 \gtrsim E_{p,20}^{3/4} T_{-1}^{-1/4} \tag{2a}$$

$$\xi_{B\gamma} \gtrsim 0.2\, E_{p,20}^{7/2} T_{-1}^{-1/2} L_{51}^{-1} . \tag{2b}$$

Since the highest energy cosmic rays are observed with energies up to 3×10^{20} eV [4], GRB winds must have bulk Lorentz factors $\Gamma \gtrsim 100$ and require $U'_B \gtrsim U'_\gamma$. The first condition is remarkably close to the canonical value assumed for GRBs in the internal shock scenario [2]. The latter condition is reasonable if the cosmic ray energy density dominates about that of relativistic electrons, $U'_p \gtrsim U'_e \sim U'_\gamma$, because the magnetic field could still be in equipartition with the total energy density in relativistic particles, $U'_B \sim U'_p + U'_e$.

THE EJECTION PROBLEM FOR UHE PROTONS

For cosmic ray acceleration at internal shocks, the acceleration site is placed within a relativistically expanding wind. The acceleration takes place typically at radii $r_i \sim 10^{14}$ cm, while the expansion continues until the wind hits the decelerated material behind the external shock at radii $r_e \sim 10^{16}$ cm. In a Poynting dominated flux, the magnetic field decreases as $B' \propto r^{-1}$ and is largely transversal, because the longitudinal component decays with r^{-2}. Magnetic reconnection can, however, maintain isotropy of the magnetic field, leading to $B' \propto r^{-2}$ [5], consistent with the generic approach of a matter dominated flow where some mechanism keeps the magnetic field in equipartition with the thermal gas. In a decreasing magnetic field, charged particles suffer adiabatic energy loss due to the constancy of the adiabatic invariant $B'r'^2_g$, which leads to an energy evolution of the nonthermal particle component with $E' \propto B'^{1/2}$. The condition for adiabaticity is that the time scale of particle gyration is much shorter than the expansion time scale, $r'_g/c \ll r/c\Gamma$, which is equivalent to the confinement condition during acceleration for protons sufficiently below the maximum energy defined by $r'_g \sim R' = r/\Gamma$. During expansion, $r'_g/R' \propto r^{-1/2}$ if $B \propto r^{-1}$, and confinement during acceleration implies that even the most energetic protons remain confined in the expanding shell and cool adiabatically. For $B' \propto r^{-2}$, r'_g/R' remains constant and adiabaticity applies only for protons with $E'_p \ll E'_{p,\text{max}}$, but some cooling should be also expected for $E'_p \sim E'_{p,\text{max}}$ (this requires further calculations).

When the material hits the outer shell at $r = r_e$, its bulk Lorentz factor, Γ, drops as a power law to values close to unity [6]. In this deceleration phase, the magnetic field confinement may break up and the energetic particles can be released. Their energy in the comoving field is then $E'_{\text{ej}} \lesssim (r_i/r_e)^{\alpha/2} E'_{\text{acc}}$, if $B' \propto r^{-\alpha}$, and $\Gamma_{\text{ej}} \ll \Gamma_{\text{acc}}$ would additionally reduce the energy in the observers frame. Hence, we expect a reduction of the energy of most protons *at ejection* by some orders of magnitude compared to their observer frame energy immediately after acceleration. This would essentially rule out a dominant contribution of GRBs to the UHECR spectrum above 10^{19} eV.

NEUTRON AND NEUTRINO PRODUCTION

The problem of adiabatic losses can be circumvented if the ejection of neutral particles is considered, because they are not coupled to the magnetic field. The obvious candidates are here neutrons, which are produced by protons in charged current photomeson-production, e.g. $p\gamma \to n\pi^+$. This is the same process which is also responsible for the production of neutrinos as a result of the pion decay, $\pi^+ \to \mu^+ \bar{\nu}_\mu$, $\mu^+ \to e^+ \nu_\mu \bar{\nu}_e$. Neutrinos are produced with an energy $E_\nu \lesssim 0.05 E_n$; the neutrino energy can be considerably below this limit, if energy losses of pions and muons are relevant, which is the case in GRBs [3]. The neutrino flux produced by this mechanism was recently proposed to reach observable levels above 100 TeV [7]. One can show that the conditions for the acceleration of protons to $\gtrsim 10^{20}$ eV implies that neutrinos up to 10^{16} eV must be produced [3].

The neutrons are left with about 80% of the proton energy and therefore carry cosmic-ray energy efficiently. The production spectrum of neutrons depends on both the proton spectrum and the spectrum of background photons. The relevant photon energies for the reaction are $\epsilon' \gtrsim 150\,\text{MeV}\gamma'^{-1}_p$; for $\gamma'_p \gtrsim 10^5$ this is below the break energy of GRB spectra, so that the integrated number of photons above the reaction threshold rises only slowly with Lorentz factor. The neutron spectrum would than be expected to follow the proton spectrum, which is canonically assumed as $N'_p \propto E'^{-2}_p$; the same is the case for the accompanying neutrino spectrum (for details see Ref. [3]).

In order to escape from the GRB shell, neutrons must fulfill two conditions: Their decay time in the lab frame, $\tau_n \gamma_n$, must be considerably larger than the total time of the burst, and the probability of reabsorption by a $n\gamma \to p\pi^-$ reaction must be small. It is easy to see that the first condition is satisfied in GRBs, since $c\tau_n \gamma_n > r_e$ for $\gamma_n > 10^3$; for $\gamma_n \gtrsim 10^{10}$, it may even leave any possible stronger magnetized environment of the GRB before undergoing β-decay. Less trivial is the second condition: The probability of a neutron to leave the GRB shell, which has a thickness $R' = r/\Gamma$, is

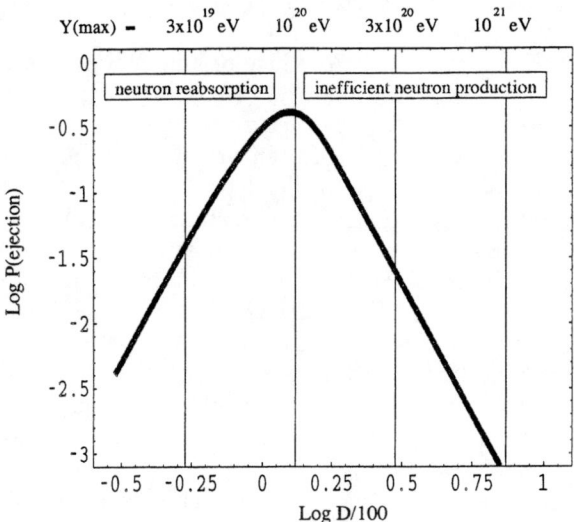

FIGURE 1. The ejection probability of neutrons from GRBs vs. $D = \Gamma T_{-1}^{1/4} L_{51}^{-1/4}$, and the correlated maximum proton energy scaled with luminosity, $Y_{\max} = E_{p,\max} L_{51}^{-1/3}$.

$$P_{\text{esc}} \sim \int_0^{R'/c} \exp\left(-\frac{t'}{t'_{p\gamma \to n}}\right) dt' , \quad (3)$$

where the integral covers the range of distances of the point of the production of the neutron to the border of the shell. This probability is directly related to the probability of the neutron to be produced, $P_{\text{prod}} \approx 1 - \exp(-t'_{\text{ad}}/t'_{p\gamma \to n})$, where t_{ad} is the time scale for adiabatic cooling of the protons. In the relativistically expanding GRB shell we have $t'_{\text{ad}} \approx R'/c$. Consequently we can write the probability P_{ej} for a UHECR proton to be ejected from the GRB as a neutron as

$$P_{\text{ej}} \sim \frac{ct'_{p\gamma \to n}}{R'} \left[1 - \exp\left(-\frac{R'}{ct'_{p\gamma \to n}}\right)\right]^2 . \quad (4)$$

The characteristic ratio involved in this expression can be expressed by canonical GRB parameters, $R'/ct'_{p\gamma \to n} \sim \frac{1}{2}\Gamma_2^4 T_{-1} L_{51}^{-1}$. The probability P_{ej} as a function of $D \equiv \Gamma_2 T_{-1}^{1/4} L_{51}^{-1/4}$ is shown in Fig. 1, together with the lower limits on D to produce UHECR protons. We see that the same conditions which make GRBs perfect proton accelerators, makes them also to almost perfect "neutron bombs" with ejection efficiencies of order 1–30%.

CONSEQUENCES

Under the assumption that GRBs produce protons with an energy density comparable to the radiation, and that each proton produces about 1 pion during its lifetime, i.e. transferring about 20% of the cosmic ray energy to neutrinos, Waxman and Bahcall [7] have shown that GRBs can produce a diffuse background flux of neutrinos above 100 TeV, which would lead to about 10 to 100 GRB correlated events per year in a km^3 underground neutrino detector. This flux should be easily detectable above the background due to the possibility of a correlation in both direction and time with the GRB. The same conditions would suffice to contribute a large fraction of the observed UHECR flux [1], but the connection of cosmic ray and neutrino ejection efficiencies depends in the GRB parameters.

Here, we have argued that energetic protons cannot be emitted directly from a GRB without losing most of their energy in adiabatic expansion, but that neutrons produced in charged current photohadronic interactions can escape the GRB and contribute to the cosmic ray proton spectrum after β-decay, provided that every proton produces on average one pion or less. This "one-pion-requirement" constrains the physical parameters of GRBs, but allows the acceleration of $\sim 10^{20}$ eV protons, however, with a strongly decreasing ejection efficiency for larger energies. One conclusion might be that only the most luminous GRBs can produce cosmic rays of the highest energies, $E \gtrsim 3 \times 10^{20}$ eV. The neutrino flux is then one-to-one correlated to the emitted cosmic ray flux, thus VHE neutrino observations could test the hypothesis of UHECR origin from GRBs; non-observation of a GRB correlated neutrino flux at the level predicted by Waxman and Bahcall [7] would rule out this hypothesis for standard assumptions of UHECR propagation.

ACKNOWLEDGEMENTS

This work was supported in part by the NASA under grant NASA5-2857.

REFERENCES

1. Waxman, E., *Phys. Rev. Lett.* **75**, 386 (1995).
2. Rees, M.J., and Mészáros, P., *Astrophys. J.* **430**, L93 (1994).
3. Rachen, J.P., and Mészáros, P., to be submitted to *Phys. Rev. D*.
4. See references in, e.g., Burdman *et al.*, hep-ph/9709399.
5. Thompson, C., *MNRAS* **270**, 480 (1994).
6. Mészáros, P., and Rees, M.J., *Astrophys. J.* **476**, 232 (1997).
7. Waxman, E., and Bahcall, J., *Phys. Rev. Lett.* **78**, 2292 (1997).

MODELS

Gamma-Ray Bursts as Hypernovae

Bohdan Paczyński

Princeton University Observatory, Princeton, NJ 08544-1001, USA
e-mail: bp@astro.princeton.edu

Abstract. A standard fireball/afterglow model of a gamma-ray burst (GRB) relates the event to a merging neutron star binary, or a neutron star - black hole binary, which places the events far away from star-forming regions, and is thought to have an energy of $\sim 10^{51}$ erg. A hypernova, the death of a massive and rapidly spinning star, may release $\sim 10^{54}$ erg of kinetic energy by tapping the rotational energy of a Kerr black hole formed in the core collapse. Only a small fraction of all energy is in the debris ejected with the largest Lorentz factors, those giving rise to the GRB itself, but all energy is available to power the afterglow for a long time. In this scenario GRBs should be found in star-forming regions, the optical afterglows may be obscured by dust, and the early thermal emission of the massive ejecta may give rise to X-ray precursors, as observed by Ginga.

The optical and X-ray afterglows of GRB970228, GRB97508, and GRB97828 provide some evidence that these bursts were located in galaxies, most likely in dwarf galaxies, in or near star-forming regions.

INTRODUCTION

The discovery of the two "gold plated" gamma-ray burst afterglows, GRB970228 and GRB970508, made a major breakthrough in the field, providing the first direct evidence for the interaction between the explosive event and its environment, and in providing the first ever direct distance estimate for one of them, GRB970508 (see [4,25,19,18,3,12,12,6], and many presentations at this meeting). It seems that a now standard fireball/afterglow model agrees with the observations reasonably well (cf. [28,27] and references therein), though it has a number of explicit as well as hidden adjustable parameters (cf. [17] and references therein). The most popular ultimate energy source is usually thought to be related to a merging/colliding pair of neutron stars or a neutron star and a stellar mass black hole, i.e. an old product of the evolution of a massive binary star. An alternative are the "failed SN Ib" [29] and hypernova [17] scenarios, in which a young massive and rapidly spinning star, following the core collapse, forms a Kerr black hole with a dense torus rotating around it. Two different modes of energy transfer from the ultimate energy source (almost always gravitational) to a fireball were proposed: neutrino - antineutrino annihilations and superstrong magnetic fields. The kinetic energy injection is considered to be $\sim 10^{51}$ erg s^{-1} in a "classical" fireball model, and $\sim 10^{54}$ erg s^{-1} in the hypernova scenario.

The afterglow is produced when the ultra-relativistic GRB ejecta interact with the ambient medium, be it circumstellar, interstellar, or intergalactic. To a good approximation the afterglow does not depend on the nature of the GRB, just on the amount of kinetic energy released in the explosion. In an essential way, this is analogous to the formation of supernovae remnants in interstellar space, and giant radio lobes in intergalactic space, which are the products of supernovae and galactic nuclear explosions, respectively.

A HYPERNOVA SCENARIO

A massive star, at the end of its nuclear evolution, creates an iron core which is too massive to make a neutron star. Following the gravitational collapse a black hole forms. There are about a dozen binary stars known with black hole components of $\sim 10~M_\odot$ (cf. [21, p. 615]). If the star is spinning rapidly then its angular momentum prevents all matter from going down the drain, and a rotating, very dense torus forms [29] around the rapidly spinning Kerr black hole. The largest energy reservoir, which may be accessed with a super-strong magnetic field (cf. [2]), is the rotational energy of the black hole:

$$E_{\rm rot,max} \approx 5 \times 10^{54}~[{\rm erg}] \left(\frac{M_{\rm BH}}{10~M_\odot}\right). \tag{1}$$

The maximum rate of energy extraction by the field was estimated by Macdonald et al. [10, Eq. 4.50] to be

$$L_{\rm B,max} \approx 10^{51}~[{\rm erg~s}^{-1}] \left(\frac{B}{10^{15}~G}\right)^2 \left(\frac{M_{\rm BH}}{10~M_\odot}\right)^2. \tag{2}$$

It is not clear how a superstrong field is generated, even though it has become popular in theoretical papers over the last few years [5,15,23,16,29,30,26,11,17]. The following is a possible scenario. A rapidly rotating massive star, just prior to its core collapse, has a convective shell [31]. According to [1] a large scale magnetic field may be generated in the shell, and it may reach equipartition with the convective kinetic energy density. Following the collapse the polar caps of the shell end up in the black hole, while the equatorial belt becomes part of the torus. At least two topologies are possible. In one, the magnetic field lines link the torus to the black hole, while in the other the field connection is severed. In both cases the collapse increases the field strength while the magnetic flux is conserved, and a substantial radial component can lead to rapid field increase driven by differential rotation. In the second case the magnetic field assists torus accretion into the black hole by redistributing its angular momentum, and helps the release of its gravitational binding energy in the process. In the first case, a much larger amount of rotational energy of the black hole can be extracted by the Blandford & Znajek [2] mechanism.

It is far from clear if this scenario can work, and if yes, then how often does it happen? For the rate to be relevant to GRBs it has to be $\sim 10^4$ times less common than ordinary supernovae. It may take a very long time to find the answer with model computations, as two decades of massive effort have not provided a clear picture of the "core bounce" that is needed for Type II supernovae. It may be more productive to seek observational evidence for or against this scenario.

A pre-hypernova must be a member of a short period massive binary, so that tidal interaction can keep the star in a synchronous, i.e. rapid rotation. Examples of such systems are some Wolf-Rayet binaries, and in particular Cyg X-3, with its ~ 5 hour orbit.

The death of a massive star cannot be more than a few million years away from its birth time, and therefore it explodes within its star-forming region, or very close to it. This makes it distinct from a popular merging neutron star model: as the merger follows a long gravitational radiation driven orbital evolution, so it is likely to be $\sim 10^9$ years after the original binary had formed. During this time the system traveled tens of kiloparsecs, having acquired a high velocity during the two supernovae explosions [22].

The star forming site for the GRBs in the hypernova scenario implies that on many occasions the optical afterglow may be heavily obscured by the dust commonly present in such regions [9]. Gradual emergence of the fireball out of the circumstellar dust shell may affect the early afterglow, possibly accounting for the early rise in the GRB970508 optical light curve.

In a hypernova scenario a small fraction of all kinetic energy is in the debris ejected with the largest Lorentz factor required to generate gamma-ray emission, but the bulk of the ejecta is less relativistic, or even sub-relativistic. Note, that if $\sim 10\ M_\odot$ is given $\sim 10^{54}$ erg then a typical velocity is $\sim 10^{10}$ cm s$^{-1} \approx c/3$. When the fireball is gradually slowed down by the ambient medium the slower moving ejecta gradually catch up, and they provide a long lasting energy supply to the afterglow, much larger than the original one related to the GRB shell. Therefore, the afterglow may persist for much longer than predicted by the standard fireball model. In case of the nearest GRBs, even thermal X-ray emission from the hypernova remnant may be detectable, acting as a calorimeter, and providing a fairly direct measurement of the total energy release.

GRB ENVIRONMENT

The afterglows of three GRBs (GRB970228, GRB970508, and GRB970828) provide some information about their environment.

GRB 970228. The fading optical afterglow is located near the edge of a fuzzy ~ 25 mag object, most likely a dwarf galaxy at a moderate redshift (see [8], and references therein).

GRB 970508. The absorption [12] and emission line [13] redshifts are identical: $z = 0.835$, but the emission line region is not resolved by the HST [7]. The compactness of the emission line region makes it very likely that it is related to the GRB. Its small size indicates it is a dwarf galaxy, possibly a star-forming region.

GRB 970828. No optical afterglow was detected, and the upper limits are very stringent [24]. The absence of optical emission may be explained by dust extinction [9]. Indeed, the X-ray spectrum indicates the presence of absorbing gas, with $N_H = 4 \times 10^{21}$ cm^{-2}, assuming no redshift [14]. The source is at a high galactic latitude, hence the absorption is likely to be at a large redshift, possibly near the source. If the redshift $z \approx 1$ is adopted then the absorbing gas column density is higher by a factor ~ 10, and the observed R-band is in the near UV at the source. Adopting a standard dust to gas ratio provides more than enough extinction to make the optical afterglow undetectable. If the absorbing gas is near the GRB then this GRB was located in a distant galaxy, possibly in a star forming region.

While the evidence for the GRBs to be located near star forming regions is marginal, it will become robust, one way or the other, when a few dozen of new afterglows are detected. The answer to the question: "Are GRBs related to star forming regions?" will be answered with future observations, independently of speculative theoretical models.

REFERENCES

1. Balbus, S., private communication (1997).
2. Blandford, R. D., & Znajek, R. L., *MNRAS* **179**, 433 (1977).
3. Bond, H., et al., *IAU Circular* 6654 (1997).
4. Costa, C., et al., preprint: astro-ph/9706065 (1997).
5. Duncan, R. C., & Thompson, C., *ApJ* **392**, L9 (1992).
6. Frail, D., et al., *IAU Circular* 6662 (1997).
7. Fruchter, A., Bergeron, L., & Pian, R., *IAU Circular* 6674 (1997).
8. Fruchter, A., et al., these proceedings.
9. Jenkins, E. B., private communication (1997).
10. Macdonald, D. A., Thorne, K. S., Price, R. H., & Zhang, X-H., in *Black Holes, the Membrane Paradigm*, eds. Thorne, K. S., Price, R. H., & MacDonald, D. A., Yale Univ. Press, 1986.
11. Mészáros, P., & Rees, M. J., *ApJ* **482**, L29 (1997).
12. Metzger, M. R., et al., *IAU Circular* 6655 (1997).
13. Metzger, M. R., et al., *IAU Circular* 6676 (1997).
14. Murakami, T., these proceedings.
15. Narayan, R., Paczyński, B., & Piran, T., *ApJ* **395**, L83 (1992).
16. Paczyński, B., in *Texas/PASCOS '92: Relativistic Astrophysics and Particle Cosmology*, eds. C. W. Akerlof & M. A. Srednicki, Ann. of the NY Acad. Sci. **688**, 1993, p. 321.
17. Paczyński, B., astro-ph/9706232 (1997).
18. Piro, L., et al., *IAU Circular 6617*, 6656 (1997).

19. Sahu, K. C., et al., *Nature* **387**, 476 (1997).
20. Schaefer, B., these proceedings.
21. Tanaka, Y., & Shibazaki, N., *ARA&A* **34**, 607 (1997).
22. Tutukov, A. V., & Yungelson, L. R., *MNRAS* **268**, 871 (1994).
23. Usov, V. V., *Nature* **357**, 472 (1992).
24. van Paradijs, J., these proceedings.
25. van Paradijs, J., et al., *Nature* **386**, 686 (1997).
26. Vietri, M., *ApJ* **471**, L91 (1996).
27. Waxman, E., preprint: astro-ph/9705229 (1997).
28. Wijers, R. A. M. J., Rees, M., & Mészáros, P., preprint: astro-ph/9704153 (1997).
29. Woosley, S. E., *ApJ* **405**, 273 (1993).
30. Woosley, S. E., in *Proceedings of 3rd Huntsville Conference on GRBs*, AIP Conf. Proc. 384, eds. C. Kouveliotou, M. S. Briggs, & G. J. Fishman, 1996, p. 709.
31. Woosley, S. E., private communication (1997).

A Binary Neutron Star GRB Model

J. R. Wilson[*†], J. D. Salmonson[*], & G. J. Mathews[†]

[*]*University of California, Lawrence Livermore National Laboratory, Livermore, California 94550*
[†]*University of Notre Dame, Department of Physics, Notre Dame, Indiana 46556*

Abstract. In this paper we present the preliminary results of a model for the production of gamma-ray bursts (GRBs) through the compressional heating of binary neutron stars near their last stable orbit prior to merger.

Recent numerical studies of the general relativistic (GR) hydrodynamics in three spatial dimensions of close neutron star binaries (NSBs) have uncovered evidence for the compression and heating of the individual neutron stars (NSs) prior to merger [1,2]. This effect will have significant effect on the production of gravitational waves, neutrinos and, ultimately, energetic photons. The study of the production of these photons in close NSBs and, in particular, its correspondence to observed GRBs is the subject of this paper.

The gamma-rays arise as follows. Compressional heating causes the neutron stars to emit neutrino pairs which, in turn, annihilate to produce a hot electron-positron pair plasma. This pair-photon plasma expands rapidly until it becomes optically thin, at which point the photons are released. We show that this process can indeed satisfy three basic requirements of a model for cosmological gamma-ray bursts: 1) sufficient gamma-ray energy release ($> 10^{51}$ ergs) to produce observed fluxes, 2) a time-scale of the primary burst duration consistent with that of a "classical" GRB (~ 10 seconds), and 3) the peak of the photon number spectrum matches that of "classical" GRB (~ 300 keV).

NEUTRON STAR HEATING

The method for solving the general relativistic field equations in three spatial dimensions has been discussed in [1,2]. At each time slice we obtain an exact (to numerical accuracy) instantaneous solution to the GR field equations. The hydrodynamic equations are then evolved for the moving matter against these GR fields. This method ignores gravitational waves, however in [1] it was shown that the effect is very small; $\dot{J}/\omega J \sim 10^{-4}$, where J is the angular momentum and ω is the angular frequency of the NSB.

The computational evolution calculation of NSBs and their GR fields begins by generating two identical non-spinning neutron stars with an initial mass and an equation of state (EOS). The stars are allowed to evolve until a stationary orbit is

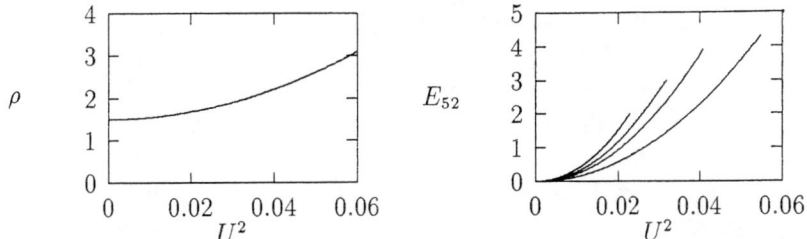

FIGURE 1. Plots of central density ρ ($\times 10^{15}$gm/cc) and released gravitational energy E_{52} ($\times 10^{52}$ergs) as a function of U^2. The E_{52} plot shows a range of star masses: (from left to right) 1.50, 1.45, 1.40, 1.35 $M_\odot$. The EOS used for these examples has a critical mass $M_{\text{crit}} = 1.64 M_\odot$.

achieved with a prescribed initial angular momentum. A variety of systems have been studied over a range of star masses, equations of state and initial angular momentums. The key result to report here is that the proper baryonic density of the stars was observed to increase prior to the stars reaching their last stable orbit. This compression and heating can be parameterized in terms of U^2, the squared amplitude of the spatial components of the 4-velocity (Figure 1). See [2–4] for discussion of why the compression is not adiabatic.

In [2] it is argued that the gravitational binding energy will be converted into thermal energy. This thermal energy will be radiated via neutrino luminosity. The time scale for the energy emitted up to collapse ($t < 0$) can be estimated from the gravitation wave emission

$$E(t) = \frac{E_{\text{tot}}}{[1 - (64/5)m^{5/3}\omega_0^{8/3}t]^{1/2}}, \tag{1}$$

where ω_0 is the final angular orbital velocity.

Some of the $\nu\bar{\nu}$ pairs emanating from the surface of a hot neutron star will annihilate to create e^+e^- pairs. In order to calculate an estimate of the efficiency of this process, the numerical supernova model of Mayle & Wilson [5,2] was used. It was found that the $\nu + \bar{\nu} \to e^+ + e^-$ reaction is 3% efficient at the end of a standard supernova calculation when the neutron star at the center is at its hottest and most compact. This simulation includes all GR effects, except the neutrinos are assumed to travel in straight lines. In the strong field environment around a NSB the neutrino trajectories will be significantly bent, thus increasing the chances of $\nu\bar{\nu}$ annihilation. New estimates show that this bending will augment $\nu\bar{\nu}$ annihilation by a factor of ~ 3. So the efficiency of $\nu + \bar{\nu} \to e^+ + e^-$ is estimated to be $\sim 10\%$. It is also found that the entropy per baryon of the pair plasma is greater than 10^{10} so relatively very few baryons are liberated from the surface of the star.

From Figure 1 we see that a typical star emanates $\sim 3 \times 10^{52}$ ergs of gravitational energy as $\nu\bar{\nu}$ pairs. Thus we have $\sim 6 \times 10^{52}$ ergs in $\nu\bar{\nu}$ pairs from both stars. A 10% efficiency in $\nu + \bar{\nu} \to e^+ + e^-$ gives us $\sim 6 \times 10^{51}$ ergs of energy in the form of an e^+e^- pair plasma-photon gas. The temperature of this gas near the NS is several MeV.

THE GAMMA-RAY BURST

Having roughly defined the initial parameters of the hot e^+e^- pair wind blowing off of a NS, we wish to follow its evolution and characterize the observable gamma-ray emission. It is important to note that there are no free parameters in this model, barring uncertainties in understanding and correctly calculating the physics; our signature either corresponds to an experimentally observed phenomenon (i.e. GRBs) or it does not.

The expanding e^+e^- pair plasma emanating from a NS is modeled as a spherically symmetric special relativistic fluid by the following hydrodynamic equations:

$$\frac{\partial D}{\partial t} = -\frac{1}{r^2}\frac{\partial}{\partial r}(r^2 DV), \qquad (2)$$

$$\frac{\partial E}{\partial t} = -\frac{1}{r^2}\frac{\partial}{\partial r}(r^2 EV) - P\left[\frac{\partial \gamma}{\partial t} + \frac{1}{r^2}\frac{\partial}{\partial r}(r^2 \gamma V)\right], \qquad (3)$$

$$\frac{\partial S}{\partial t} = -\frac{1}{r^2}\frac{\partial}{\partial r}(r^2 SV) - \frac{\partial P}{\partial r}, \qquad (4)$$

where the radial 4-velocity U, Lorentz factor γ and coordinate velocity V are defined as

$$U \equiv \frac{S}{D + \Gamma E}, \quad \gamma \equiv \sqrt{1 + U^2}, \quad V \equiv \frac{U}{\gamma}. \qquad (5)$$

The coordinate densities of baryons and thermal energy (e^+e^- and photons) are given by D and E, respectively.

The total energy equation, including photons and e^+e^- pairs, is

$$E_{\text{tot}} = aT^4 + E_{\text{pairs}}(T). \qquad (6)$$

To track the e^+e^- pairs we define a pair equation

$$\frac{\partial N_{\text{pairs}}}{\partial t} = -\frac{1}{r^2}\frac{\partial}{\partial r}(r^2 N_{\text{pairs}} V) + \overline{\sigma v} N_{\text{pairs}}(N^0_{\text{pairs}}(T) - N_{\text{pairs}})/\gamma^2, \qquad (7)$$

where the coordinate pair number density is N_{pair}, $\overline{\sigma v}$ gives the mean pair annihilation rate, and $N^0_{\text{pairs}}(T)$ is given by a Fermi integral with a chemical potential of zero.

To model the material blown off the surface of a NS we inject baryon and pair-photon energy densities into the innermost zone at a rate determined by the time derivative of the heating energy given in Equation 1.

The hydrodynamic equations are evolved, allowing the plasma to expand. Once the system becomes transparent to Compton scattering, assuming no further scattering, the calculation is stopped and the photon signal is analyzed. In the results presented here we have set $E_{tot} = 10^{51}$ ergs. Since the entropy per baryon of the wind is quite high we define the rate of injection of baryons as $\dot{D} = 10^{-10}\dot{E}$.

Since the photons and e^+e^- pairs appear to decouple at virtually the same time throughout the entire fireball (radius $\sim 10^{12}$ cm), we take this event to be instantaneous and to occur when the cloud becomes optically thin to Compton scattering. We then look at two observables, the time integrated number spectrum $N(\epsilon)$ and the total energy received as a function of time $E(t)$.

To get the spectrum, as mentioned above, we assume that the e^+e^- pairs and photons are in thermodynamic equilibrium when they decouple. Thus the photons in the fluid frame (denoted with a prime) make up a Plank distribution. We calculate the observed number spectrum, per photon energy ϵ, per steradian, of a relativistically expanding spherical shell with radius R, thickness dR in cm, velocity v where $c = 1$, Lorentz factor γ and fluid-frame temperature T' to be

$$N(\epsilon) = 4\pi R^2 \, dR \, \frac{\epsilon T'}{v\gamma} \log\left[\frac{1 - exp[-\gamma\epsilon(1+v)/T']}{1 - exp[-\gamma\epsilon(1-v)/T']}\right], \tag{8}$$

which has a maximum at $\epsilon_{max} \cong 1.39\gamma T'$ eV for $\gamma \gg 1$. We may then sum this spectrum over all shells of our fireball to get the total spectrum shown in Figure 2 (top). Since we *a priori* assume the photons are thermal, our spectrum has a high-frequency exponential tail.

A key feature of this spectrum is that its peak agrees with observation. It is interesting to note that the bulk of the photons have a fluid frame temperature of only $\epsilon \sim 5 - 15$ eV, but are Lorentz boosted by $\gamma \sim 10^4 - 10^5$. Thus our spectrum derives from a more relativistic fluid than other models. The photons to be observed at early times have about twice the energy of the later photons.

To acquire the observed light curve $E(t)$ we consider two effects. First is the relative arrival time of the first light from each shell. Second is the shape of the light curve from a single shell [6]. We find that, for our Plank distribution of photons, a relativistically expanding shell of radius R will have a time profile $E(\tau > R/c) \sim (R/c\tau)^4$.

In Figure 2 (bottom) we see an example of $E(t)$ for a NSB of equal star mass. Variation in the ratio of star mass in the NSB effects the relative compression and heating rate of each star, thus allowing a variety of GRB durations. We estimate a range of burst durations from several seconds to a few 10s of seconds.

ACKNOWLEDGEMENTS

Grant support is NSF PHY-97-22086, PHY-9401636 and DOE W-7405-ENG-48.

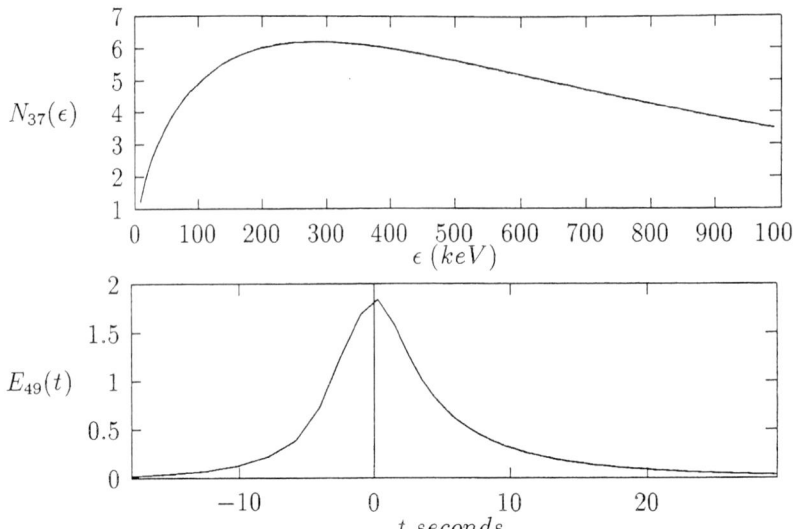

FIGURE 2. (Top) Photon number spectrum $N_{37}(\epsilon)$ ($\times 10^{37}$ photons/keV/4π) from e^+e^- pair plasma fireball. (Bottom) Light curve $E_{49}(t)$ ($\times 10^{49}$ ergs/second/4π)

REFERENCES

1. Wilson, J. R., Mathews, G. J. & Marronetti, P., *Phys. Rev. D.* **54**, 1317 (1996).
2. Mathews, G. J., & Wilson, J. R., *Astrophysical J.* **482**, 929 (1997).
3. Wilson, J. R., & Mathews, G. J., *Phys. Rev. D*, submitted (1997).
4. Mathews, G. J., *Phys. Rev. Lett.*, submitted (1997).
5. Wilson, J. R., & Mayle, R. W., *Phys. Rep.* **227**, 97 (1993).
6. Fenimore, E. E., et al., *Astrophysical J.* **473**, 998 (1996).

Coalescing Neutron Stars as Possible Gamma-Ray Burst Sources

Maximilian Ruffert[1,2] and Hans-Thomas Janka[2]

[1] Inst. of Astronomy, Madingley Road, Cambridge CB3 0HA, U.K.
[2] MPI für Astrophysik, Postfach 1523, D-85740 Garching, Germany

Abstract. We have simulated a variety of models, in order to shed light on the accessible physical conditions during the mergers, on the amount of matter dynamically ejected during the merging, on the timescales of mass accretion by the (forming) black hole, on the conversion of energy into neutrino emission, on the amount of energy deposited by $\nu\bar{\nu}$-annihilation, and on the baryon loading of the created pair-plasma fireball. To this end, we varied the masses and mass ratios as well as the initial spins of the neutron stars, changed the impact parameter to consider spiral-in orbits and direct collisions, included a black hole (vacuum sphere) in our simulations, and studied the dynamical evolution of the accretion torus around the black hole formed after the neutron star merging until a (nearly) stationary state was reached. While the neutrino emission during the dynamical phases of the mergings is definitely too small to power gamma-ray bursts (GRBs), we find that the masses, lifetimes, and neutrino luminosities of the accretion tori have values that might explain short ($\mathcal{O}(0.1\text{--}1\,\text{s})$) and not too powerful ($\sim 10^{51}/(4\pi)\,\text{erg}/(\text{s}\cdot\text{sterad})$) gamma-ray bursts.

SUMMARY OF NUMERICAL PROCEDURES AND INITIAL CONDITIONS

The hydrodynamical simulations were done with a code based on the Piecewise Parabolic Method (PPM) developed by Colella & Woodward [1]. The code is basically Newtonian, but contains the terms necessary to describe gravitational wave emission and the corresponding back-reaction on the hydrodynamical flow (Blanchet et al. [2]). The terms corresponding to the gravitational potential are implemented as source terms in the PPM algorithm. In order to describe the thermodynamics of the neutron star matter, we use the equation of state (EOS) of Lattimer & Swesty [3]. Energy loss and changes of the electron abundance due to the emission of neutrinos and antineutrinos are taken into account by an elaborate "neutrino leakage scheme" [4]. The energy source terms contain the production of all types of neutrino pairs by thermal processes and of electron neutrinos and antineutrinos also by lepton captures onto baryons. Matter is rendered optically thick to neutrinos due to the main opacity-producing reactions which are neutrino-nucleon scattering and absorption of neutrinos onto baryons. More detailed information about the employed numerical procedures can be found in Ruffert et al. [4]. The following modifications, compared to the previously published simulations ([4] and [5]), were also made: (a) The models were computed on multiply-nested and refined grids to increase locally the spatial resolution while at the same time computing a larger total volume. Our grid handling is described in detail in Section 4 of Ruffert [7]. (b) An entropy equation, instead of the equation for the total specific energy (specific internal energy plus specific kinetic energy), was used to calculate the temperature of the gas in order to be able to determine low temperatures more accurately. (c) The equation of state table was extended to $100\,\mathrm{MeV} < T < 0.01\,\mathrm{MeV}$ and $2.9 \times 10^{15}\,\mathrm{g/cm^3} > \rho > 5 \times 10^7\,\mathrm{g/cm^3}$ because in extreme cases very high temperatures can be reached and the density of ejected gas decreases to low values at large distances.

We started our simulations with two cool (temperatures initially a few 0.01 MeV to a few MeV) Newtonian neutron stars with baryonic masses between 1.2 $M_\odot$ and 1.8 $M_\odot$ (depending on the particular model), a radius of approximately 15 km, and an initial center-to-center distance of 42–46 km. The distributions of density ρ and electron fraction Y_e were taken from one-dimensional models of cold, deleptonized neutron stars in hydrostatic equilibrium. We prescribed the orbital velocities of the coalescing neutron stars according to the motions of point masses, as computed from the quadrupole formula. The tangential components of the velocities of the neutron star centers correspond to Kepler velocities on circular orbits, while the radial velocity components result from the emission of gravitational waves that lead to the inspiral of the orbiting bodies. For an observer not corotating with the system, the neutron stars were given either no spins or the two stars were assumed to be tidally locked or to have spins opposite to the direction of the orbital angular momentum [4].

SOME RESULTS FOR MERGING NEUTRON STARS

From the long list of our models, we present here in detail some results for only one exemplary case: the coalescence of a 1.8 $M_\odot$ primary with a 1.2 $M_\odot$ secondary (baryonic masses), both initially corotating (solid-body like). The results of this model were rather typical and not the most extreme in any direction.

The neutrino luminosities during the neutron star merger are shown in Fig. 1. During the phase of dynamical merging, a constantly rising neutrino luminosity is produced. It saturates after about 7 ms, when the secondary has effectively been tidally shredded and wrapped up along the surface of the primary. This evolution of the neutrino luminosity is very similar to what has been published for equal-mass neutron stars (Ruffert et al. [5]), although somewhat higher luminosities of $2\text{--}5\times10^{53}$ erg/s were found in the recent simulations, because the use of the larger grid (out to about 200 km distance from the center of mass) kept track of the mass flung to very large radii and then falling back towards the center and contributing

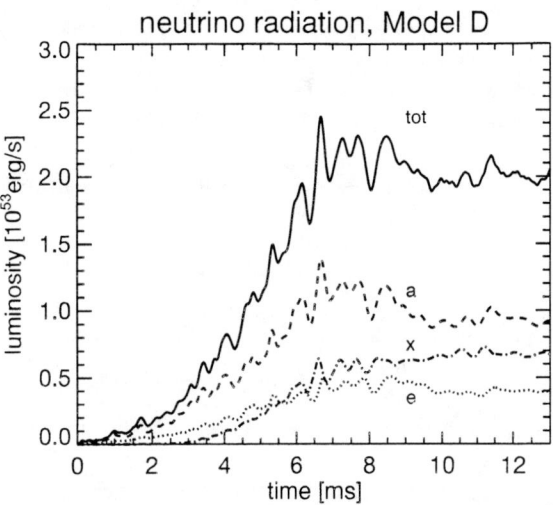

FIGURE 1. Luminosities of individual neutrino types (ν_e, $\bar{\nu}_e$, and the sum of all ν_x) and of the total of all neutrinos as functions of time for the merging of a 1.2 $M_\odot$ and a 1.8 $M_\odot$ neutron star.

to the neutrino emission. Since the geometry and neutrino luminosities are similar for equal-mass and nonequal-mass neutron star mergers, the energy deposition rate by $\nu\bar{\nu}$-annihilation is of the same order. We calculated values of several 10^{51} erg/s, about a factor of 10 larger than the ones given in [5], mainly because of the larger neutrino luminosities of the more recent models. However, the dynamical phase of the merging lasts only a few milliseconds before one must expect the collapse to a black hole of the merged object with significantly more than 2 $M_\odot$. Therefore the energy pumped into a $e^\pm\gamma$ fireball during this phase hardly exceeds a few 10^{49} erg. Even worse, by far most of this energy is deposited in the surface-near regions of the merged stars and will drive a mass loss ("neutrino wind") which will pollute the fireball with an unacceptably large baryon load.

The neutrino emission from the coalescence of two neutron stars is very different from the case of head-on collisions in which neutrinos are emitted in two very short (about 1 millisecond) and extremely luminous bursts reaching peak values of up to 4×10^{54} erg/s [6]! This gigantic neutrino luminosity leads to an energy deposition of about 10^{50} erg by $\nu\bar{\nu}$-annihilation within only a few milliseconds. However, the surroundings of the collision site of the two neutron stars are filled with more than 10^{-2} $M_\odot$ of ejected matter and the maximum values of the Lorentz-factor Γ are only about 10^{-3}, five orders of magnitude lower than required for relativistic fireball expansion that could produce a GRB.

The merging of the 1.2 $M_\odot$–1.8 $M_\odot$ binary proceeds in the following way: Initially (0 ms $< t <$ 3 ms) the secondary is tidally elongated by the primary and a mass transfer is initiated. The flow of gas is concentrated through the L1-point onto the surface of the primary. As more and more matter is taken away from the secondary it becomes ever more elongated ($t \approx 5$ ms). Most of the matter of the secondary finally ends up forming a rapidly rotating surface layer of the primary, while a smaller part concentrates in an additional, extended thick disk around the primary ($t \approx 10$ ms). This matter has enough angular momentum to stay in an accretion torus even after the massive central body has most likely collapsed to a black hole.

NEUTRON TORI AROUND BLACK HOLES AND GRBS

If the central, massive body did not collapse into a black hole, it would continue to radiate neutrinos with high luminosities, like a massive, hot proto-neutron star in a supernova. But instead of producing a relativistically expanding pair-plasma fireball, these neutrinos will deposit their energy in the low-density matter of the surface and thus will cause a mass flow known as neutrino-driven wind (e.g., [8]). Most of this energy is consumed lifting baryons in the strong gravitational potential and the expansion is nonrelativistic.

This unfavorable situation is avoided if the central object collapses into a black hole on a timescale of several milliseconds after the merging of the neutron stars. We simulated the subsequent evolution by replacing the matter inside a certain radius by a vacuum sphere (black hole) of the same mass. The region along the system axis was found to be evacuated on the free-fall timescale of a few milliseconds as the black hole sucks the baryons. Thus an essentially baryon-free funnel is produced where further baryon contamination is prevented by centrifugal forces. This provides good conditions for the creation of a clean $e^{\pm}\gamma$ fireball by $\nu\bar{\nu}$-annihilation. Also the thick disk closer to the equatorial plane loses matter into the black hole until only gas with a specific angular momentum larger than $j^* \cong 3R_s v_k(3R_s) = \sqrt{6}GM/c$ ($R_s = 2GM/c^2$ is the Schwarzschild radius of the black hole with mass M, $v_k(3R_s)$ the Kepler velocity at the innermost stable circular orbit at $3R_s$) is left on orbits around the black hole. We find torus masses of up to $M_t \approx 0.2$–$0.3\,M_\odot$ at a time when a quasi-stationary state is reached. The temperatures in the tori are 3–10 MeV, maximum densities a few 10^{12} g/cm^3. Typical neutrino luminosities during this phase are of the order of $L_\nu \approx 10^{53}$ erg/s (60% $\bar{\nu}_e$, 35% ν_e). With a maximum radiation efficiency of $\varepsilon_r \approx 0.057$ for relativistic disk accretion onto a nonrotating black hole we calculate an accretion rate of $\dot{M}_t = L_\nu/(c^2\varepsilon_r) \sim 1\,M_\odot$/s and an accretion timescale of $t_{acc} = M_t/\dot{M}_t \sim 0.2$–$0.3$ s. This is in very good agreement with the analytical estimates of Ruffert et al. [5].

From our torus models we find that the energy deposition rate by $\nu\bar{\nu}$-annihilation near the evacuated system axis is about $\dot{E}_{\nu\bar{\nu}} \approx 5 \times 10^{50}$ erg/s. The corresponding annihilation efficiency ε_a is of the order of $\dot{E}_{\nu\bar{\nu}}/L_\nu \approx 5 \times 10^{-3}$, which is rather large because of the large ν_e and $\bar{\nu}_e$ luminosities and because the torus geometry allows for a high reaction probability of neutrinos with antineutrinos. In order to account for an observed GRB luminosity of $L_\gamma \sim 10^{51}$ erg/s the required focussing of the $\nu\bar{\nu}$-annihilation energy is moderate, $\delta\Omega/(4\pi) = \dot{E}_{\nu\bar{\nu}}/(2L_\gamma) \sim 1/4$, corresponding to an opening angle of about 60 degrees. This value does not seem implausible for the geometry of a thick accretion torus. Taking into account general relativistic effects in the treatment of $\nu\bar{\nu}$-annihilation reduces the above estimates only by about 10–50%.

In summary, our numerical simulations show that stellar mass black holes with accretion tori that form after the merging of neutron star binaries have masses, lifetimes, and neutrino luminosities that might provide enough energy by $\nu\bar{\nu}$-annihilation to account for short and not too powerful GRBs. The longer bursts and very energetic events, however, would require an alternative explanation, e.g., failed supernovae (or "collapsars") [9] where a stellar mass black hole could have a greater than 10 times more massive accretion torus than in the binary neutron star scenario.

ACKNOWLEDGMENTS

The calculations were performed at the Rechenzentrum Garching on an IBM SP2.

REFERENCES

1. Colella, P., Woodward, P.R., *JCP* **54**, 174 (1984).
2. Blanchet, L., Damour, T., Schäfer, G., *MNRAS* **242**, 289 (1990).
3. Lattimer, J.M., Swesty, F.D., *Nucl. Phys.* **A535**, 331 (1991).
4. Ruffert, M., Janka, H.-Th., Schäfer, G., *A&A* **311**, 532 (1996).
5. Ruffert, M., Janka, H.-Th., Takahashi, K., Schäfer, G., *A&A* **319**, 122 (1997).
6. Ruffert, M., Janka, H.-Th., in *The Eighth Marcel Grossmann Meeting on General Relativity*, eds. T.Piran & A. Dar, World Scientific Press, 1997.
7. Ruffert, M., *A&A* **265**, 82 (1992).
8. Woosley, S.E., Baron, E., *ApJ* **391**, 228 (1992).
9. Woosley, S.E., *ApJ* **405**, 273 (1993).

Black Hole–Neutron Star Coalescence as a Source of Gamma–Ray Bursts

William H. Lee and Włodzimierz Kluźniak[†]

Physics Department, University of Wisconsin
Madison, WI, 53706
[†]*Also Copernicus Astronomical Center, ul. Bartycka 18, 00-716 Warszawa, Poland*

Abstract. We present the results of hydrodynamic (SPH) simulations showing the coalescence of a black hole with a neutron star to be a promising theoretical source of short-duration gamma-ray bursts. The favorable features of the process include rapid onset, millisecond variability, a duration much longer than the dynamical timescale, and a range of outcomes sufficient to allow variety in the properties of individual gamma-ray bursts. Interestingly, the process of coalescence differs rather markedly from past predictions.

INTRODUCTION

The coalescence of a tight binary composed of two extremely compact objects (two neutron stars or one neutron star and a black hole) has been suggested as a possible source of cosmological GRBs [9]. In the black hole–neutron star scenario, the neutron star was expected to be tidally disrupted and to form a long–lived accretion torus around the black hole, thus powering a relativistic fireball that produces the observed GRB. Our aim is to investigate the outcome of the coalescence from the standpoint of the hydrodynamics, and explore how the initial mass ratio and the degree of tidal locking in the binary affect the final outcome. In particular, it is of great interest to determine if the neutron star is completely disrupted in a single encounter and if a baryon–free axis persists throughout the process.

NUMERICAL METHOD AND ASSUMPTIONS

For the calculations presented here we have used a three dimensional Newtonian smooth particle hydrodynamics [8] code [5]. The neutron star is modeled via a stiff polytropic equation of state, i.e $P = K\rho^\Gamma$, with $\Gamma = 3$. The unperturbed radius of the spherical neutron star is $R_{\rm NS} = 13.4$ km and its mass is $M_{\rm NS} = 1.4\ M_\odot$. The black hole is modeled as a Newtonian point mass with an absorbing boundary at the Schwarzschild radius $r = 2GM_{\rm BH}/c^2$. Any particle in the simulation that crosses this boundary is absorbed by the black hole and its mass is added to that of the black hole. For the different simulations we vary the initial mass ratio of the binary $q = M_{NS}/M_{BH}$ by adjusting the mass of the black hole only.

RESULTS

Tidally Locked Binaries

For any value of the mass ratio one can construct tidally locked initial configurations that are in equilibrium [10], provided the binary separation is large enough so that no mass transfer occurs. If the neutron star fills its Roche lobe, any further decrease in separation will produce mass transfer, which can be stable or unstable, depending on the initial mass ratio. We study the coalescence of the binary by performing dynamical runs starting with initial configurations that are on the verge of initiating mass transfer.

High Mass Ratios

For an initial mass ratio of unity, the neutron star overflows its Roche lobe at a separation $r = 2.78\ R_{NS}$ (the orbital period at this separation is 2.3 ms). A very fast episode of mass transfer ensues, in which 0.9 $M_\odot$ are accreted by the black hole (Figure 1). Note that we have not included a gravitational backreaction force in our

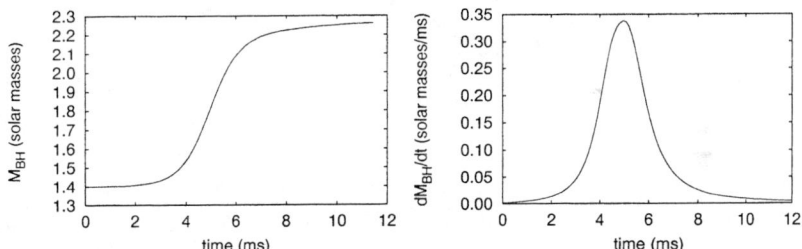

FIGURE 1. Black hole mass (left) and accretion rate onto the black hole (right) for an initial mass ratio of $q = 1$ in a tidally locked binary.

calculations, but the orbital decay is driven by hydrodynamical effects which are comparable in magnitude to angular momentum losses to gravitational radiation. We observe the formation of a transient accretion torus (Figure 2) containing about 0.1 solar masses around the black hole, lasting for several orbits. To the limit of our resolution ($10^{-4}\ M_\odot$), there is a baryon-free axis along the rotation axis of the binary throughout the simulation. We are unable at present to follow the further evolution of the matter spread around the black hole for more than 11 ms. Note that the neutron star is not completely disrupted by the initial encounter, but that

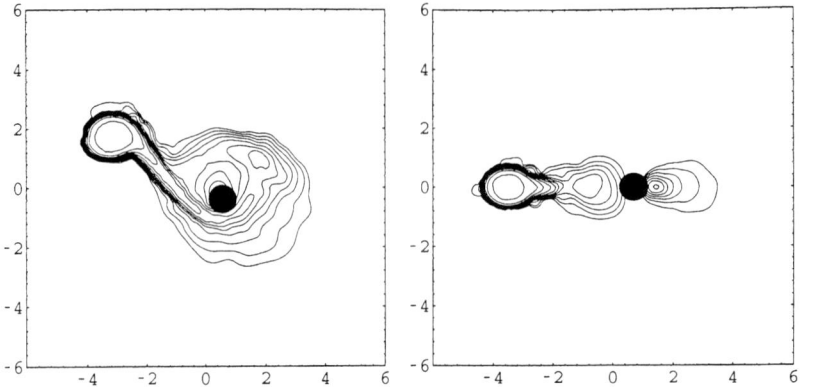

FIGURE 2. Density contours for the initially tidally locked binary with $q = 1$ in the orbital plane (left) and a meridional plane (right). Orbital rotation is counterclockwise. There are eleven evenly spaced logarithmic contours between 5×10^{14} kg m^{-3} and 5×10^{17} kg m^{-3}. The axes are labeled in units of $R_{\rm NS}$.

a low–mass cores survives and is transferred to a higher orbit by conservation of angular momentum. This core contains approximately 0.43 $M_\odot$, and thus the final mass ratio in the binary is $q = 0.19$, while the separation has increased to about 47 km.

Lower Mass Ratios

We have also investigated the outcome of a coalescence for lower mass ratios, i.e. with a more massive black hole. For an initial mass ratio of $q = 0.31$, corresponding to a black hole mass of 4.5 $M_\odot$, the evolution of the binary is quite different than that presented above. The separation corresponding to the onset of mass transfer is $r = 50.4$ km, but for this case the neutron star is not violently disrupted. Instead, mass transfer through Roche–lobe overflow occurs, with about 1% of the mass of the neutron star being transferred to the black hole. The binary separation increases slightly (Figure 3), and the whole episode conserves total orbital angular momentum. All the matter stripped from the neutron star is accreted by the black hole, hence no torus forms, and a baryon–free axis is also present. However, the mass transfer episode lasts about 11 ms, and angular momentum losses to gravitational radiation cannot be ignored. Also, the assumption of tidal locking is not thought to be realistic [1].

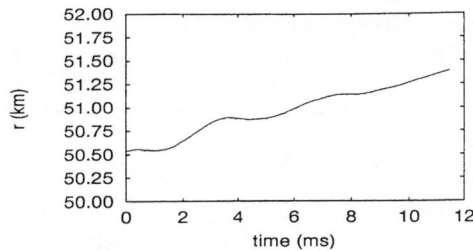

FIGURE 3. Binary separation r as a function of time for an initially tidally locked binary with an initial mass ratio of $q = 0.31$.

Binaries which are not Tidally Locked

Suppose we now remove the restriction of tidal locking. If the neutron star is initially spherical and non-rotating, the presence of the black hole will create a tidal bulge on the neutron star. The neutron star will then be spun-up to some degree, and this spin angular momentum will be extracted from the orbital component. The orbit will thus decay on a shorter time scale. We have performed just such a calculation, starting with an initial mass ratio of $q = 0.31$ as above, and with a neutron star that is initially spherical and not spinning. The orbital evolution is now much more rapid, with an episode of mass transfer lasting less than 4 ms (Figure 4). The total mass transferred is 0.6 $M_\odot$, and again the core of the neutron star (0.8

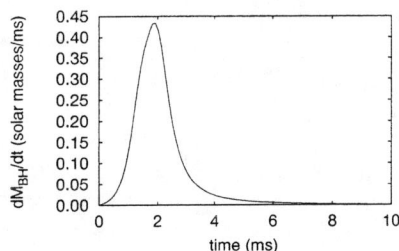

FIGURE 4. Mass accretion rate onto the black hole for a binary with an initial mass ratio $q = 0.31$ that is not tidally locked (the neutron star is initally not spinning).

$M_\odot$) is not disrupted, but transferred to a higher orbit. The final separation is approximately 70 km.

DISCUSSION

Our simulations [3,6] lead us to believe that for the equation of state used here the survival of the neutron star core is a robust result. The timescale for the entire coalescence is thus lengthened from a few milliseconds to at least several tens of milliseconds, since a binary with a lower mass ratio and greater separation (the result of the initial episode of mass transfer) will take a longer time to decay via angular momentum losses to gravitational radiation. Furthermore, to the limit of our resolution, there is a baryon–free axis along the rotation axis of the binary present throughout the simulation in every case. Calculations with improved resolution are required to determine if the baryon loading is low enough to accomodate the requirements of the blast–wave model for GRBs [7].

We note that without the formation of a torus, it is difficult to extend the time scale of the coalescence to many seconds, but we nevertheless believe that the coalesence of a neutron star with a black hole is a promising candidate source for the central engine of the shorter gamma ray bursts in the bimodal distribution [4].

These results also suggest that black hole–neutron star binaries might well be the production sites for low–mass neutron stars unstable to explosion. The details of how the remnant core would react to the violent episode of mass loss depend on the equation of state, but it has been shown [2,11] that if the mass were to drop below the minimum required for stability, a violent explosion would ensue. The timescale for this event is not certain, but estimates range from a few milliseconds to several tens of seconds.

ACKNOWLEDGEMENTS

This work was supported in part by Poland's Committee for Scientific Research under grant KBN 2P03D01311 and by DGAPA–UNAM.

REFERENCES

1. Bildsten, L., Cutler, C., *Astrophys. J.* **400**, 175 (1992).
2. Colpi, M., Shapiro, S.L., Teukolsky, S.A., *Astrophys. J.* **369**, 422 (1991).
3. Kluźniak, W., Lee, W.H., *Astrophys. J. Lett.*, submitted (1997)
4. Kouveliotou, C., et al., in *Gamma-Ray Bursts*, eds. C. Kouveliotou, M.S. Briggs, G.J. Fishman, AIP: New York, 1995, p. 42.
5. Lee, W.H., Kluźniak, W., *Acta Astron.* **45**, 705 (1995).
6. Lee, W.H., Kluźniak, W., in preparation (1997).
7. Mészáros, P., Rees, M.J., *Astrophys. J.* **405**, 278 (1993).
8. Monaghan, J.J., *Ann. Rev. Ast. & Astrophys.* **30**, 543 (1992).
9. Paczyński, B., *Acta Astron.* **41**, 257 (1991).
10. Rasio, F., Shapiro, S.L., *Astrophys. J.* **432**, 242 (1994).
11. Sumiyoshi, K., Yamada, S., Suzuki, H., Hilldebrandt, W., preprint: astro-ph/9707230 (1997).

Gamma-Ray Bursts Near the Horizon

Ralph A.M.J. Wijers

Inst. of Astronomy, Univ. of Cambridge, Madingley Road, CB3 0HA Cambridge, UK
ramjw@ast.cam.ac.uk

Abstract. Gamma-ray bursts are much brighter than supernovae, and could therefore possibly probe the Universe to high redshift. Since most proposed mechanisms for GRBs link them closely to deaths of massive stars, it is a reasonable ansatz to assume that their rate density in the past was proportional to the star formation rate. Work by Wijers et al. [16] does indeed show that the GRB flux distribution calculated from this assumption agrees well with the data. However, the implied luminosity of GRBs is 20 times higher, and the density 150 times lower, than for conventional no-evolution fits to a cosmological distribution. I briefly discuss some implications of this finding for GRBs and cosmology.

THE MODEL

A gamma-ray burst (GRB) releases an amount of energy similar to that of a supernova explosion, which combined with its rapid variability suggests an origin related to neutron stars or black holes. In all such models (except possibly mergers [15]) the delay between a typical event and the birth of the original stars is small, and one can neglect the delay relative to the age of the Universe even at fairly high redshift. Therefore the GRB rate should trace the formation rate of (massive) stars.

Further, we assume that GRBs are standard candles in the restframe 30–2000 keV band, following Fenimore and Bloom [5]: since the spectra of bursts vary, the flux in the BATSE 50–300 keV band as a function of redshift is not the same for each burst. We therefore calculate the redshift-flux relation as the median of relations for each of 50 spectra from a bright sample by Band et al. [1].

The star formation rate as a function of redshift has recently been studied extensively, and is determined observationally with some confidence (see Madau [8]): the luminosity density in the rest frame U and B band is used to deduce the star formation rate.

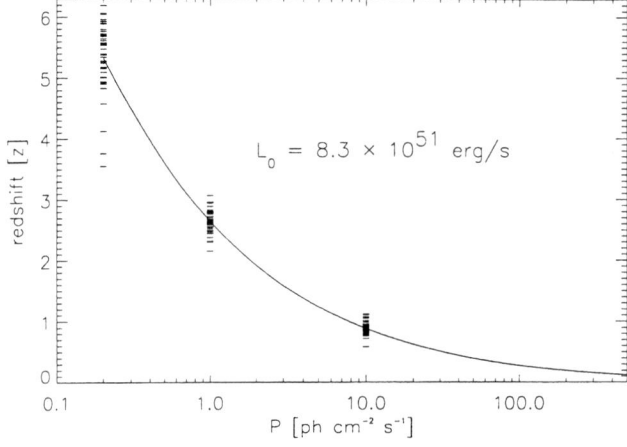

FIGURE 1. The redshift as a function of flux for the evolving model fit. The solid curve gives the average over spectral shapes. For three flux values, the individual redshifts for each of 48 measured spectra [1] are shown to indicate the considerable variation due to spectral shape.

RESULTS

All results quoted are for a cosmology with $\Omega_0 = 1$, $\Lambda = 0$, and $H_0 = 70\,\mathrm{km\,s^{-1}\,Mpc^{-1}}$. The fit of the model to the data is only for two normalizations, namely the local GRB rate density ρ_0 and the standard-candle 30–2000 keV luminosity L_0. The fit is done by χ^2 minimization for the same 11 flux bins of combined PVO and BATSE data used by Fenimore and Bloom [5]. It gives $L_0 = 8.3^{+0.9}_{-1.5} \times 10^{51}\,\mathrm{erg\,s^{-1}}$ and $\rho(z=0) \equiv \rho_0 = 0.14 \pm 0.02\,\mathrm{Gpc^{-3}\,yr^{-1}}$. This translates into a rate of 0.025 GEM (Galactic Events per Myr). The median redshift of bursts with $P = 1\,\mathrm{photon\,cm^{-2}\,s^{-1}}$ is 2.6. The best-fit redshift-flux relation is seen to extend to above $z = 6$ for bursts at the BATSE threshold of $P = 0.2\,\mathrm{photons\,cm^{-2}\,s^{-1}}$ (Fig. 1).

For comparison, we also fitted the same data with a non-evolving rate density. In that case, we recover the previously known result that $L_0 = 0.44 \times 10^{51}\,\mathrm{erg\,s^{-1}}$ and $\rho_0 = 3.7\,\mathrm{GEM}$ [5]. The redshift at $P = 1\,\mathrm{photon\,cm^{-2}\,s^{-1}}$ is then 0.68. This means that our assumption about the evolution of the GRB rate has quite drastic consequences: it increases the GRB luminosity by a factor 19, and the local rate is decreased by a factor 150.

IMPLICATIONS

The twenty-fold increase in luminosity of the bursts has important implications. First, the total energy released in gamma rays in a 10 s burst goes up to 10^{53} erg, which only a "hypernova" [13] can provide if the emission is isotropic. Of course, gamma-ray beams of bursts may illuminate no more than a few percent of the sky, in which case most gamma-ray bursters escape detection in gamma rays. Since all the models of interest entail the collapse of an already-rotating system and a non-vanishing angular momentum implies cylindrical symmetry, such beaming is quite plausible in the context of these models.

The increased distance scale also removes the "no-host problem" for GRBs, because greater distances imply that a given magnitude limit of potential host galaxies translates to a weaker limit on their luminosities. Deep searches of GRB error boxes (e.g. [14]) have been used to set limits on the absolute brightness of host galaxies of 3.5 to 5.5 magnitudes fainter than $L_\star$ [14], suggesting that GRBs do not come from galaxies. But since these estimates depend on L_0 and now have to be adjusted by 3.2 magnitudes just from the increased distance, and by another 1–2 magnitudes, depending on galaxy type, due to increased K corrections [7]. This changes the limits on host galaxy luminosities to between 1.5 magnitudes above and 1.5 magnitudes below $L_\star$, so they are no longer inconsistent with the assumption that GRBs are in normal galaxies.

An important difference between the various compact stellar remnant models is the distance that a gamma-ray burster travels between where it was born as a massive (binary) star and where it produces the burst. A direct supernova origin or hypernova would occur in short-lived objects with low space velocity, which would therefore still be in the star-forming regions where they were born. In this case, an optical counterpart to a GRB should always be embedded in a galaxy or star-forming region. A neutron star-neutron star or neutron star-black hole merger occurs in a system that has obtained a moderate to high $(100–300\,\mathrm{km\,s^{-1}})$ space velocity from the two supernova explosions that have taken place in the binary some 100 Myr before the merger. This means that such GRBs can easily occur 30 kpc away from any star-forming region (corresponding to $6''$ at $z = 3$) depending on whether the host galaxy has a deep enough potential to hold it. The optical counterpart to GRB970228 is embedded in an extended object. That of GRB970508 is at least 25 kpc away from any host so far detected [12], but the [OII] emission line seen in its spectrum [9] suggests that it lies in a star-forming region (see [2]). The recent detection [11] of a large absorption column in the X-ray spectrum of GRB970828 suggests that it, too, may lie close to a star-forming region. There may thus be some tentative evidence favouring progenitors with low space velocities and very short delays between formation and burst.

If they are in star-forming regions, the hypernova model gains credibility, given how much brighter in optical light GRBs are than supernovae. Using a flat spectrum in optical at peak [6] for GRB970508 to do K corrections and $m_{V,\mathrm{peak}} = 20$, the range of absolute magnitudes at peak is -25.5 to -29.8 for the allowed redshift range 0.835–2.3 [9]. Compared with a typical peak magnitude -19 for type Ia supernovae, it may truly be called a hypernova, whatever its underlying mechanism. Nonetheless, the energy generation rate in the Universe is still dominated by ordinary supernovae, as can be deduced from the above numbers: take $L_0 \times 10\,\mathrm{s}$ to be the energy generated, then combining the inferred density with energy per event, we find an energy generation rate at the present time ($z = 0$) of $4 \times 10^{44}\,\mathrm{erg\,Gal^{-1}\,yr^{-1}}$ for a non-evolving GRB population, and $0.6 \times 10^{44}\,\mathrm{erg\,Gal^{-1}\,yr^{-1}}$ for an evolving one. (These numbers do not depend on beaming because if bursts are beamed the lower inferred energy per burst is exactly offset by the consequent higher space density.) Taking a type II supernova rate of $0.02\,\mathrm{Gal}^{-1}\,\mathrm{yr}^{-1}$ and an energy of 10^{51} erg per event, it follows that they generate $2 \times 10^{49}\,\mathrm{erg}\,\mathrm{Gal}^{-1}\,\mathrm{yr}^{-1}$, some 5 orders of magnitude more than GRBs. For similar reasons, unless very different types of nuclear reactions could be identified as taking place in GRBs, it is unlikely that they would leave a chemical signature that is measurable.

Also, if stars formed with similar mass functions and binary properties in the early Universe as now, one predicts that there should be a wave of gamma-ray bursts from the first episode of star formation which made the metals seen in the most distant quasars. This wave may have occurred as early as $z = 10$ [10]. It is worth exploring the possibility of searching for these, since they may well be the only directly observable phenomenon from such high redshift. ROSAT and AXAF are easily sensitive enough to detect such bursts, but their fields of view are too small, whereas the SAX WFCs have the opposite problem.

CONCLUSION

The consequence of assuming that GRBs are related to remnants of the most massive stars in almost all of the ways hitherto proposed is that the GRB rate is proportional to the formation rate of massive stars in the Universe. This assumption is consistent with the GRB flux distribution. Compared to previous, non-evolving, models of cosmological bursts we find a twenty-fold increase of the required GRB luminosity, which suggests that the gamma-ray emission is significantly beamed in order that the emitted energy can be supplied by merger/collapse models. The redshifts of GRBs are also greatly increased, and the very dimmest known ones are at $z \gtrsim 6$, beyond the farthest known quasars. This makes them the most distant known objects, and their optical counterparts very valuable probes of the early evolution of stars and interstellar gas.

REFERENCES

1. Band, D., et al., *ApJ* **413**, 281 (1993).
2. Bloom, J. S., et al., these proceedings.
3. Brainerd, J. J., *ApJ* **487**, 96 (1997).
4. Dezalay, J.-P., et al., these proceedings.
5. Fenimore, E. E., and Bloom, J. S., *ApJ* **453**, 25 (1995).
6. Groot, P., et al., these proceedings.
7. Lilly, S. J., Tresse, L., Hammer, F., Crampton, D., and Le Fèvre, O., *ApJ* **455**, 108 (1995).
8. Madau, P., in *Star Formation Near and Far*, AIP Conf. Proc., AIP: New York, 1996 (also available as `astro-ph/9612157`).
9. Metzger, M. R., Djorgovski, S. G., Steidel, C. C., Kulkarni, S. R., Adelberger, K. L., and Frail, D. A., *IAUC* 6655 (1997).

10. Miralda-Escudé, J., and Rees, M. J., *ApJ* **478**, L57 (1997).
11. Murakami, T., these proceedings.
12. Natarajan, P., et al., *New Astronomy*, in press (1997).
13. Paczyński, B., *ApJ Letters*, submitted (1997) (also available as `astro-ph/9706232`).
14. Schaefer, B. E., Cline, T. L., Hurley, K. C., and Laros, J. C., *ApJ*, in press (1997) (also available as `astro-ph/9704278`).
15. Totani, T., *ApJ* **486**, L71 (1997) (also available as `astro-ph/9707051`).
16. Wijers, R. A. M. J., Bloom, J. S., Bagla, J. S., Natarajan, P., *MNRAS*, in press (1998).

Gamma-Ray Bursts from Closed and Filled Neutron Star Magnetospheres

M. Böttcher[1], B. Eastlund[2] and B. Miller[3]

[1] *Rice University, Space Physics & Astron. Dept., 6100 Main St., Houston, TX 77005*
[2] *Eastlund Scientific Enterprises Corporation, 6615 Chancellor Drive, Spring, TX 77379*
[3] *Miller Energy Corporation, 7805 Fox Gate Court, Bethesda, MD 20817*

Abstract. The beaming of curvature radiation from closed and filled neutron star magnetospheres can overcome a number of well-known problems associated with older γ-ray burst models invoking dead pulsars. Our calculations naturally explain the spectral evolution of γ-ray bursts, while sufficiently reducing the energy required to produce them as compared to isotropic emission, making it possible to observe these events at cosmological distances.

INTRODUCTION

After the discovery of γ-ray bursts, galactic neutron stars were considered the most favorable candidates for their origin. Among the proposed mechanisms were crust quakes [7], core quakes [6], and the infall of a solid body onto the neutron star surface (e.g., [1]). These models, in their original versions, mostly considered shock waves propagating through the neutron star magnetospheres and accelerating particles, eventually initiating pair cascades. The resulting radiation, dominated by synchrotron and thermal bremsstrahlung, did not exhibit a very strong angle-dependence.

However, when compared to more recent results of the BATSE instrument, these neutron star models have difficulty explaining a number of observational results, such as the complicated pulse shapes, the γ-ray burst spectra and, most importantly, the large amount of energy needed to explain the observed luminosity at the large distances implied by the apparent isotropy of γ-ray burst locations on the sky.

In this paper, we show that many of the problems with models invoking dead pulsars as the sources for γ-ray bursts can be overcome if the particles which are responsible for the observed emission are emitting predominantly curvature radiation from closed and filled magnetospheric shells. Our calculations on this idea are based on the results of [4] and [5], where this model was successfully applied to a

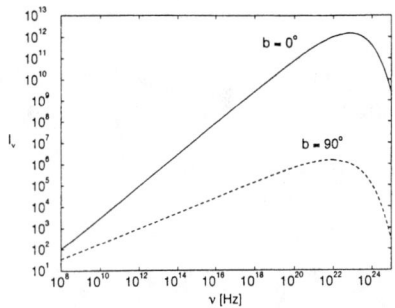

FIGURE 1. Angle-dependence of the observable curvature radiation from particles at $r = 0$. $A = 0$, $\gamma = 2.56 \cdot 10^5$.

number of pulsars, but in particular to the Crab, fitting its broadband spectrum as well as its frequency-dependent pulse shapes.

EMISSION FROM CLOSED AND FILLED MAGNETOSPHERIC SHELLS

Following [9] and [5] exactly, we consider a neutron star whose magnetic axis is inclined at an angle A with respect to the rotational axis. The angle between the line of sight to the observer and the rotational axis is called b. We assume that close to the surface of the pulsar, relativistic particles in the magnetosphere are moving exactly along the magnetic field lines. Near the light cylinder the particles can have a non-negligible component of velocity perpendicular to the magnetic field lines described by the parameter r, defined as

$$r = \frac{\beta_\perp}{\beta}, \qquad (1)$$

where $\beta_\perp$ is the component of the particle velocity orthogonal to the plane of curvature of the field line. The total emission observed from all particles moving on trajectories parametrized by r at curvature radiation mode number m is then given by

$$F_m = N \int_1^\infty d\gamma\, f(\gamma) \int_0^{2\pi} d\delta\, (C_m + 1)\, P_m(r, \theta)\, \sin\theta, \qquad (2)$$

where

$$P_m(r, \theta) = \frac{m^2 \omega_0^2}{(1 - r\beta \cos\theta)^3} \left| \left(\frac{\cos\theta - r\beta}{\sin\theta} \right)^2 J_m^2(x) + \beta_\perp^2 {J'_m}^2(x) \right|. \qquad (3)$$

Here, N is the total number of emitting particles, C_m is the coherence coefficient (which becomes negligible in the case of very high mode number), δ is the azimuthal angle around the magnetic axis, θ is the angle between the line of sight and the normal to the plane of the magnetic field line defined by δ, $x = m\beta_\perp \sin\theta/(1 - r\beta\cos\theta)$, $f(\gamma)$ is the particle distribution for electrostatic mirroring and $\omega_0 = c/R_c$ is the fundamental frequency, where R_c denotes the radius of curvature of the field line.

THE BEAMING EFFECT

Cones and beams of radiation are predicted from closed and filled magnetospheric shells of an aligned rotator by [4]. A well-known feature of the curvature radiation of a single particle is that it is strongly beamed along its trajectory. This implies that the emission from all particles moving exactly along a given magnetic field line ($r = 0$) is very strongly restricted to the plane of curvature of this field line. Therefore, an observer looking along the magnetic axis of the pulsar will see a contribution to the observed radiation from all field lines, whereas an observer looking at a large angle to the magnetic axis only sees the radiation from particles moving along one single field line. For this reason the radiation pattern described by Eqs. (2) and (3) is strongly beamed along the magnetic axis of the pulsar. This beaming effect is illustrated in Fig. 1.

For particles on a magnetic field line close to a typical pulsar, the fundamental frequency is expected to be $\omega_0 \sim 10^4$ s^{-1}. Therefore, in order for the curvature radiation spectrum to peak at hard X-rays or soft γ-rays, Lorentz factors of $\gamma \sim 10^5$ are required. Under this assumption, the beaming enhancement along the magnetic axis as compared to the magnetic equator is of order 10^7. The width of the resulting beam is $\Delta\theta \approx 6''.5$.

In the case of particles forced to move at a finite angle with respect to the magnetic field line ($r > 0$), the beaming pattern opens up to a cone inside which no emission can be observed. The enhancement of radiation along directions on the cone is slightly weaker than for the single beam ($r = 0$), but still considerable.

SPECTRAL EVOLUTION AND LIGHT CURVES

There is now strong evidence that most γ-ray bursts show spectral evolution, generally described by a softening of the X-ray to γ-ray spectrum with a decrease of the peak frequency. A statistical analysis of the spectral evolution of a small set of γ-ray bursts can be found in [3], and the left panel of Fig. 2 shows an example of such evolution. The right panel of Fig. 2 illustrates that in the framework of curvature emission in neutron star magnetospheres, such kind of evolution can result from a gradual increase in the parameter r. If this interpretation is correct, this suggests that the acceleration site (e.g. a shock wave) moves outward to regions where the shock acceleration results in particle trajectories at $r > 0$. Because the

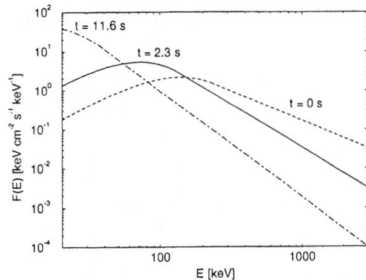 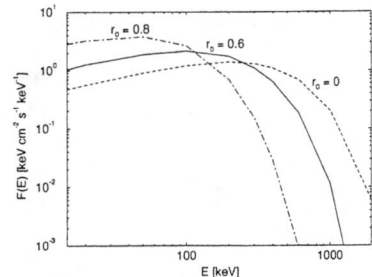

FIGURE 2. Spectral evolution of GRB910927 (left) compared to curvature radiation along the beaming direction for different values of r; $A = 82.9°$, $\gamma = 2.56 \cdot 10^5$ (right).

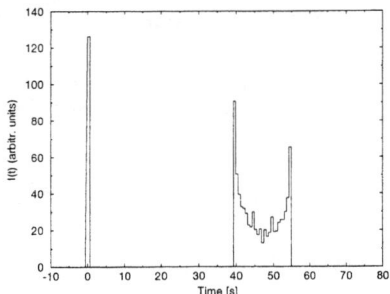 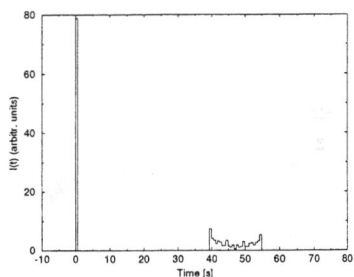

FIGURE 3. Example of γ-ray burst light curves at two different mode numbers (photon energies). Parameters: $A = 0.042°$, $b = 45°$, $\gamma = 2.56 \cdot 10^5$, $P_{ns} = 100$ s, left panel: $m = 4 \cdot 10^{16}$; right panel: $m = 2.2 \cdot 10^{16}$.

energy loss time scale of ultrarelativistic electrons in the magnetosphere is short compared to the duration of the burst, the instantaneous γ-ray burst spectrum is expected to be governed by the properties (and location) of the acceleration site.

The observed spectral evolution tells us how the parameter r evolves with time. With this knowledge we can predict typical γ-ray burst light curves. Fig. 3 shows an example of such a light curve at two different photon energies where we assumed a linear decrease in the number of relativistic particles with time and a linear increase in the r parameter up to a saturation value of 0.7. The resulting light curves look very similar to the ones observed in GRB970228 [2].

STATISTICAL CONSIDERATIONS

An important aspect of the idea outlined above is that in order for a γ-ray burst from ultrarelativistic particles in a neutron star magnetosphere to be detectable, the rotation of the neutron star must be very slow, because otherwise the beam would pass the detector too quickly. Our analysis indicates that rotational periods of at least several seconds are required. Even though most of the observed pulsars have periods below ~ 1 s, the total number of slower neutron stars is expected to be much greater than the one following from counts of the observed pulsars because slow pulsars radiate inefficiently and are difficult to detect. However, even very slow pulsars continue to decelerate their rotation. Therefore, most of the existing neutron stars are expected to have rotational periods $P \gg 1$ s, and the total number of neutron stars in our galaxy is estimated to be $N_{ns} \sim 10^8$ [8].

Assuming that γ-ray bursts originate in an extended halo around our galaxy, the energy required to produce a typical burst is around $E_{GRB} \sim 10^{38}$ erg. Given that due to the very narrow beaming pattern we only expect to detect 1 out of $\sim 10^7$ events, this requires a typical event rate of 10 events per year per neutron star.

The extreme beaming provided by the mechanism proposed here even allows us to consider a cosmological scenario for these events. In this case, a total energy of $E_{GRB} \sim 10^{44}$ erg is required, which is of the order of the energy supplied by the infall of a solid body of mass $m \sim 10^{23-24}$ g or by a strong crust quake. This would have to happen once every 10^7 years per neutron star.

ACKNOWLEDGMENTS

M. Böttcher acknowledges support by the NASA grant NAG 5-4055

REFERENCES

1. Colgate, S. A., & Petschek, A. G., *ApJ* **248**, 771 (1981).
2. Costa, E., Frontera. F., Heise, J., et al., *Nature* **387**, 783 (1997).
3. Crider, A., Liang, E. P., Smith, I. A., et al., *ApJ* **479**, L39 (1997).
4. Eastlund, B. J., & Miller, B., *Nature*, submitted (1997).
5. Eastlund, B. J., Miller, B., & Michel, F. C., *ApJ* **483**, 857 (1997).
6. Ellison, D. C., & Kazanas, D., *A&A* **128**, 102 (1983).
7. Fabian, A. C., Icke, V., & Pringle, J. E., *Astroph. Space Sci.* **42**, 77 (1976).
8. Michel, F. C., private communication (1997).
9. Miller, B., & Eastlund, B. J., *ApJ* **464**, 359 (1996).

Saturated Compton Cooling Model of Cosmological Gamma-Ray Bursts

Edison P. Liang

Rice University, Houston, Texas, 77005-1892

Abstract. We generalize the saturated Compton cooling model of gamma-ray burst spectral evolution to the cosmological context. In order for this model to be self-consistent, the Compton cooling plasma must be a relativistically expanding thin shell of bulk Lorentz factor $> 10^2$, consisting of mostly thermal leptons of keV comoving temperature and less than a few percent nonthermal population.

INTRODUCTION

Recent discovery of the redshifted absorption lines with $z = 0.835$ [1] in the optical afterglow of GRB970508 strongly suggests that at least some gamma-ray bursts (GRBs) are cosmological. Hence it is necessary to reconsider the saturated Compton cooling model [2], which was motivated by the spectral evolution of bright long BATSE bursts, in the context of cosmological distances. It turns out that the saturated Compton cooling model can indeed be consistent with cosmological distances, but only for a restricted emission parameter space. Here we briefly summarize the main constraints on such a model. Details can be found in Liang [3].

THE SATURATED COMPTON COOLING MODEL

The saturated Compton cooling model [2] was motivated by: (a) the exponential decay of the spectral break energy E_{pk} with photon fluence [4]; (b) the evolution of the Band et al. [5] α spectral index; (c) the high initial value of α, often exceeding $+1$ in many hard-to-soft pulses [6]; and (d) the clustering of E_{pk} around a few hundred keV [7]. However, in a static configuration this model leads to a maximum distance of only 100 kpc due to causality constraints. To apply this model to cosmological GRBs we must relax the static assumption. It turns out that a relativistically expanding thin shell that is cooling via saturated Comptonization can indeed be self-consistent provided that the emission parameters satisfy certain constraints.

RELATIVISTIC COMPTON COOLING SHELLS

Here we summarize the parameter space of such shells that are consistent with causality and observed spectral evolution properties of bright BATSE bursts. For details see Liang [3].

Unlike the standard relativistic blast wave scenarios [8,9] which are extremely Thomson-thin, adiabatically cooling, and radiating via optically thin synchrotron, here we are talking about dense thin shells of Thomson depth $\gg 1$ being relativistically ejected from a central engine. Each GRB pulse corresponds to the energization and cooling episode of a single shell. Hence multipulse bursts would consist of multiple shell ejections. These shells are dominated by radiative rather than adiabatic cooling. The high Thomson depth leads to saturation of Compton cooling, so that the spectral break corresponds to the average kinetic energy of the leptons, Lorentz boosted by the bulk motion. We also assume that the bulk Lorentz factor of the shell changes little during the GRB phase.

For a pulse rise time of the order of one second and burster distance of Gpc, the typical shell parameters are constrained to: bulk Lorentz factor $> 10^2$; shell radius 10^{15} cm; shell thickness $< 10^{11} - 10^{12}$ cm; comoving lepton density $> 10^{13}$ cm^{-3}; comoving thermal temperature around a few keV; nonthermal lepton fraction less than a few percent; Thomson depth from a few to 100; effective absorption depth $< 10^{-4}$; mean magnetic field around 0.1 G; total lepton number $> 10^{57}$; total internal energy in the comoving frame $> 10^{48}$ ergs; total leptonic bulk kinetic energy $> 10^{53}$ ergs. Hence only a tiny fraction of the bulk motion is converted to gamma-ray radiation. This model predicts that during the GRB phase the low-energy source spectrum should extend down to 10^{12} Hz before it is cutoff by (synchrotron and free-free) self-absorption. It also predicts that at late times the pulse should decay as a power law in time with index not far from unity.

Acknowledgements

The work is partially supported by NASA grant NAG5-3824

REFERENCES

1. Metzger, M., et al., *Nature* **387**, 879 (1997).
2. Liang, E., et al., *Ap. J.* **479**, L35 (1997).
3. Liang, E., *Ap. J.*, in press (1997).
4. Liang, E., and Kargatis, V., *Nature* **381**, 49 (1996).
5. Band, D., et al., *Ap. J.* **413**, 281 (1993)
6. Crider, A., et al., *Ap. J.* **479**, L39 (1997).
7. Mallozzi, R., et al., *Ap. J.* **471**, 636 (1996).
8. Meszaros, P., and Rees, M., *Ap. J.* **418**, L59 (1993).
9. Fenimore, E., et al., *Ap. J.* **473**, 998 (1996).

Single Black Holes as Parent Bodies of Galactic and Extragalactic Gamma–Ray Bursts

G.M. Beskin[1], A. Shearer[2], M. Redfern[2], A. Golden[2], R. Butler[2], C. Bartolini[3], A. Guarnieri[3], N. Masetti[3] & A. Piccioni[3]

[1] *Special Astrophysical Observatory of RAS, Nizhnij Arkhyz, Russia*
[2] *University College of Galway, Ireland*
[3] *Dipartimento di Astronomia, Università di Bologna, Italy*

Abstract. There are several pieces of evidence that the sample of gamma-ray bursts (GRBs) is inhomogeneous and that these objects can be situated in the Galaxy as well as at cosmological distances. We consider single black holes (BHs) of $10 - 100$ $M_\odot$ as parent objects for GRBs of both types. The origin of GRBs is connected to the release of energy generated by magnetic field inhomogeneities together with interstellar plasma accretion onto a regular magnetosphere around a BH. The percolation nature of the energy release for this kind of instability yields a universal power law for the GRB energy distribution, that is $E^{-\beta}$ (with $\beta = 1.7 - 1.8$), which is the same as that for the flares observed in the Sun, in flare stars and in X-ray sources. The number of GRBs per galaxy is about 10^3 for both galactic (with luminosities up to 10^{38} erg s^{-1}) and cosmological (with luminosities up to 10^{52} erg s^{-1}) populations. We also discuss the possibility of detecting galactic BHs as parent bodies of GRBs in quiescent phase, as stars of magnitude 20 to 29 with continuum spectra and fast variability.

INTRODUCTION

The detection of gamma-ray bursts (GRBs) in the X-ray, optical and radio ranges seems to have indicated the cosmological origin for at least some GRBs. Nevertheless, from our point of view, it is perhaps too early to consider that the hypothesis of their cosmological origin comes from only one source. While the currently preferred model of associating GRBs with star formation at $z \approx 1.25$ (with consequent production of neutron star binary systems) is consistent with most observations, it is still important to consider other hypotheses for a cosmological population of GRBs. There is some evidence (e.g. from the bimodal duration distribution) that the GRB sample consists of two populations caused by different physical processes [3,13,14,16]. In this paper we consider a model based upon flare phenomena around $10 - 100$ $M_\odot$ black holes (BHs). We initially consider the rate of GRBs for a contemporary galactic population of BHs and then scale it to a cosmological population.

We consider BHs of $10 - 100$ $M_\odot$ to be the parent bodies of GRBs. Local inhomogeneities of the magnetic field in the magnetospheres around BHs are the energy source (of fast particles) for these GRBs. Transformation of these inhomogeneities into electron and proton motion is caused by spontaneous development of magnetohydrodynamic instabilities of different types. This approach allows us to give a quantitative interpretation to the phenomenon of GRBs, both of galactic and cosmological origin.

HYPOTHESES

Burst Mechanism

Here we formulate the set of premises which give the possibility of powerful energy release near $10 - 100$ $M_\odot$ BHs.

- During the accretion of interstellar plasma onto a BH, magnetic fields strengthen up to $10^5 - 10^6$ Gauss and probably form a regular magnetosphere [4,8,20]. The formation of this magnetosphere is caused by the presence of plasma with angular momentum and by the BH rotation (under the Kerr metric).

- There are local inhomogeneities in the magnetosphere which are similar to those of the Sun and of flare stars [6,17].

- The energy of these inhomogeneities as the result of spontaneous development of plasma instabilities produces fast particles and anomalous heating. The GRB is the burst of photons that have been generated by such a process. We think this is an analogue of solar flares, flashes of UV Ceti type stars, and of activity evident in X-ray binaries with compact components [2,5,12,15,19].

- The energy distribution of flares has a universal character – $F(E)dE \propto E^{-\beta}$, where $\beta = 1.7 - 1.8$. Such a universality is caused by the percolation nature of the break of stability in an active region [9,18,23].

As there is no self-consistent theory of these processes, we will estimate the parameters of this distribution on the basis of observational features. First of all we point out that we will consider a quite natural ergodic hypothesis – that the distribution of bursts for one object coincides with one for the whole set. To simplify we will consider that the time of magnetosphere formation is less than the mean interval between bursts.

Main Features of BH Sample

The objects under consideration may belong to the disc population as well as to population II. In the first case their masses are greater than 10 $M_\odot$, while in the second they are greater than 10^3 $M_\odot$. These limits are derived from dynamical, photometric and chemical data [1,7]. Their density in the solar neighbourhood is not to exceed $0.1 M_\odot$ pc^{-3}, i.e. the density of local dark matter ρ_{dark} [1].

If we consider the disc population and restrict to the luminosity of bursts with Eddington limit L_{Edd} for the cutoff value of flux $\sim 10^{-6}$ erg s^{-1}cm^{-2}, we will obtain a limit distance to the observed events $R \sim 5$ kpc (in the galactic plane). Accepting the value $r \sim 200$ pc for the disc half-thickness, we estimate the number of BHs where GRBs can be observed as

$$N_{\text{BH}} \sim 2\pi R^2 \cdot r \cdot \rho_{\text{dark}} \frac{1}{M} \sim 3.14 \cdot 10^9 \frac{1}{M},$$

where M is the BH mass in solar masses; when $M \geq 10$ $M_\odot$, then $N_{\text{BH}} \leq 3 \cdot 10^8$. Taking the number of GRBs observed during one year as $N_\gamma < 10^3$ / galaxy, we obtain their mean recurrent time

$$\tau = \frac{1}{P} \sim \frac{N_{\text{BH}}}{N_\gamma} > 3 \cdot 10^5 \text{ years},$$

where P is the mean probability or frequency of a burst.

Now it is easy to obtain parameters of the GRB energy distribution $dN = nE^{-\beta}dE$ (here we consider that all bursts have duration of ~ 1 s). From the normalization condition we have

$$nE_0^{-\beta+1} = \beta - 1,$$

where E_0 is the minimal energy of the distribution; on the other hand, for the mean probability of burst detection we have

$$\langle P \rangle = \frac{n^2}{(\beta-1)^2} E_0^{-\beta+1} \cdot E_1^{-\beta+1},$$

where $E_1 \sim L_{\text{min}}$ (taking a duration ~ 1 s) is the minimal luminosity of the observed GRBs. But $E_1 \sim L_{\text{min}} \sim 10^{33}$ erg s^{-1} (with a threshold flux value of 10^{-6} erg cm^{-2} s^{-1} and a minimal distance of 3 pc); hence

$$E_0 = E_1 \langle P \rangle^{\frac{1}{\beta-1}},$$

and, for $\langle P \rangle \sim 3.3 \cdot 10^{-6}$, we get $E_0 \sim 1.5 \cdot 10^{25}$ erg and

$$n = (\beta-1)E_0^{\beta-1} \sim 3 \cdot 10^{17}.$$

Critical Experiment

As it was shown by Shvartsman [20], haloes of accreted ISM plasma onto BHs with masses >10 $M_\odot$ emit in the optical and infrared ranges; moreover, their spectra have no lines. Of course, the luminosity depends on the ISM density and on the speed and mass of the BH according to the formula [10,11,20]

$$L \sim 4.5 \left(\frac{M}{M_\odot}\right)^3 \left(\frac{\rho_\infty}{10^{-24} \text{ g} \cdot \text{cm}^{-3}}\right)^2 \left[\left(\frac{v}{16.6 \text{ km} \cdot \text{s}^{-1}}\right)^2 + \frac{T_\infty}{10^4 \text{ K}}\right]^{-3}.$$

Hence, it follows that we can hope to detect the closest BHs in quiescent state; such observations of a local population would give strength to the possibility that a cosmological distribution of BHs could exist. Because the local ISM has low density, they will be observed as faint stars of magnitude 20 to 29 with high proper motion ($10^{-2} - 1$ arcsec yr^{-1}) and maximum emission in red and infrared ranges. The brightness of these objects is to be variable on time scales $\tau_{\text{var}} \sim 10^{-5} - 10^{-2}$ s [20]. This feature is a critical test when searching for isolated BHs. The search for BHs was carried out at the 6-meter SAO/RAS telescope in the framework of the MANIA experiment. Limits on the BH density relative to bright haloes (slow and massive BHs) on the level of $5 \cdot 10^{-4}$ pc^{-3} were obtained from the local number of stars [21,22].

GRBS OF COSMOLOGICAL LOCALIZATION

Using the parameters of the GRB energy distribution obtained above we can estimate the frequency of GRBs in far galaxies. In this case $E_0 \sim 10^{52}$ erg (once again we suppose a burst duration ~ 1 s). The probability of the burst with energy exceeding this level in the Galaxy is then

$$P_{\text{GAL}} \sim \left(1 - \frac{n}{\beta - 1} E_0^{1-\beta} + \frac{n}{\beta - 1} E^{1-\beta}\right) \sim 1.7 \cdot 10^{-19}.$$

In 10^{11} galaxies containing $2 \cdot 10^{11}$ BHs each, we would then see $N_\gamma \sim 3.5 \cdot 10^3$ per year. Taking into account all our assumptions, the estimate is quite good.

The detection and characterization of galactic BHs (via Shvartsman radiation) would give an important constraint on the degree at which this mechanism is tenable for some GRBs. At cosmological distances the number density of BHs would be expected to be radically different to our local population.

ACKNOWLEDGEMENTS

We are thankful to the Russian Fund of Fundamental Researches, Science–Education Centre "Cosmion" (grant 95-02-03691) and to the Italian Ministry of Foreign Affairs for financial support. We also thank V. Komarova for assistance in preparing the paper.

REFERENCES

1. Bahcall, J., *ApJ* **219**, 1008 (1984).
2. Bartolini, C., et al., *ApJS* **92**, 455 (1994).
3. Belli, B.M., *ApJ* **479**, L31 (1997).
4. Beskin, V.S., *Uspechi Fiz. Nauk* **167**, 689 (1997).
5. Beskin, G.M., et al., in *Flares and Flashes*, eds. Greiner, J., Duerbeck, H.W., Gershberg, R.E., Springer: Berlin, 1994, p. 330.
6. Byrne, P.B., in *Physics of Solar & Stellar Coronae*, eds. Linsky, J.L., Serio, S., Kluwer: Dordrecht, 1992, p. 489.
7. Carr, B., *ARAA* **32**, 531 (1994).
8. Chakrabarti, S.K., *Phys. Rep.* **266**, 229 (1996).
9. Feder, J., *Fractals*, Plenum Press: New York, 1998.
10. Heckler, A.F., Kolb, E.W., *ApJ* **472**, L85 (1996).
11. Ipser, J.R., Price, R.H., *ApJ* **255**, 654 (1982).
12. Katsova, M.M., Livshits, M.A., *SvA* **68**, 131 (1991).
13. Katz, J.I., Canel, L.M., *ApJ* **471**, 915 (1996).
14. Link, B., Epstein, R.I., *ApJ* **466**, 764 (1997).
15. Meekins, J.F., *ApJ* **278**, 288 (1992).
16. Pendleton, et al., *ApJ* **489**, 175 (1997).
17. Priest, E.R., in *Flares and Flashes*, eds. Greiner, J., Duerbeck, H.W., Gershberg, R.E., Springer: Berlin, 1994, p. 3
18. Pustil'nik, L.A., in press (1997).
19. Rothschild, R.E., et al., *ApJ* **189**, L13 (1974).
20. Shvartsman, V.F., *SvA* **15**, 377 (1971).
21. Shvartsman, V.F., et al., *SvA Lett* **15**, 337 (1989).
22. Shvartsman, V.F., et al., *Astrofizika* **31**, 457 (1989).
23. Wentzel, D.G., Seiden, P.E., *ApJ* **390**, 280 (1992).

The Galactic Model of GRBs

Stirling A. Colgate and Hui Li

*Theoretical Astrophysics, LANL,
Los Alamos, NM 87545; colgate@lanl.gov; hli@lanl.gov*

Abstract. The galactic model of gamma-ray bursts (GRBs) is based upon the observed production of soft gamma-ray repeaters (SGRs) in our galaxy and the consequences of a reasonable model to explain them. In this view GRBs are the long term result of the burn-out conditions of the SGRs in this and in other galaxies. A delay of ~ 30 million years before GRBs are being actively produced can be understood as the time required for the ejected matter during the SGR phase to cool, condense, and form planetesimals that are eventually captured by the central neutron star. The amount of disk matter and the interaction between each GRB and the disk determine the rate of burst production and turn-off time of GRBs. The X-ray afterglow as well as optical emission is derived from X-ray fluorescence and ionization of previously ablated matter.

INTRODUCTION

The primary motivation for considering GRBs to originate at cosmological distances has been the extraordinary isotropy measured by BATSE of all bursts. This has then been reasonably interpreted by the "cosmologically inclined community" as "only the universe could be that isotropic!". However, this simplistic view is not the case in the event that GRBs originate from fast objects ejected from our galaxy (and from all other galaxies as well) [1,2]. It is rather surprising to realize that the acceptable parameter space delineated by Bulik, Coppi, and Lamb from fitting the BATSE data is so large and overlaps the conditions predicted by the present model, based upon fast neutron stars and planetesimal accretion [3,4]. This parameter space is bounded by three conditions: 1) the ejection velocity of neutron stars must be $\gtrsim 800$ km/s and $\lesssim 1500$ km/s in order to diminish galactic gravitational distortion and not approach M31 too closely; 2) the turn-on time should be within $3 - 5 \times 10^7$ y, long enough so that the first bursts are far enough away from both us and the center of the galaxy; and 3) the turnoff time $\sim 5 \times 10^8$ y, permitting averaging over a galactic rotation period and excluding M31.

The starting point of the model is that GRBs are the result of the expected evolution of the burn-out conditions of SGRs (see also [5]). During the SGR phase, the thickness of the captured disk of matter decreases due to both accretion and ablation (from SGRs) until the disk is too thin to support the alpha viscosity accretion mechanism. The resulting quiescent disk then evolves to planetesimals, which are then scattered into neutron stars producing GRBs. The formation of planetesimals accounts for the delayed turn-on of GRBs, exactly as in planet formation within the solar system.

SGRS

A sizable amount of matter ($\sim 1-10\%\ M_\odot$) will be captured by the high velocity neutron star in a near-miss collision with its companion [4]. This matter forms a thick accretion disk which subsequently evolves both radially outward and inward, accreting a fraction onto the central neutron star, i.e., the SGR phase. The total energy emitted during the SGR phase is $\sim 10^{47}$ ergs (the total mass required is $\sim 10^{27}$ g), mostly from steady, soft X-ray emission at $\sim 3 \times 10^{35}$ ergs/s and a fraction from the SGRs themselves. The steady accretion is at the Eddington limit corresponding to the enhanced iron-like element opacity ($\sim 10^3 \times$ Compton opacity). SGRs are from the episodic high state accretion of ionized plasma that is so thick, $\gtrsim 10^4$ g cm^{-2} as to be self-shielded against radiation pressure, hence, leading to episodes of much larger luminosities than the above Eddington limit.

The X-ray flux of an SGR event terminates the inner disk mass inflow by ablation so that a subsequent period of mass replenishment from outer radii requires ~ 10 revolutions at a fraction of an AU, i.e., a year or so between bursts. Each SGR event will ablate roughly 10 g cm^{-2} of matter out to 1 AU, $\sim 10^{27}$ g per year or 10^{31} g total. This large mass in Keplerian orbit is thick enough to sustain $\sim 10^4$ events until the residual thickness is $\lesssim 100$ g cm^{-2}. This is the critical thickness necessary to contain heat (and entropy) within the disk for ~ 10 turns. Below this density the enhanced alpha viscosity no longer can transmit torque, and the SGR phase terminates. A quiescent, and therefore cooling disk of matter, roughly 100 g cm^{-2} thick, $\sim 10^{30}$ g in total mass, extending out to several AU, circulating around a neutron star, is then the starting point for planetesimal and then planet formation.

THE FORMATION OF PLANETESIMALS

The only difference between the protoplanetary disk of the sun and that of the depleted SGR is that the former is formed by accretion of a solar mass from the outside and the SGR disk is formed from the inside by the same alpha viscosity but acting within a smaller mass, $\sim 0.1-0.01\ M_\odot$, initially captured by the neutron star from its companion at formation. Thus the disk around the neutron star extends only to 3 − 10 AU. Thus the inner disks of the solar and neutron star should be similar, but with likely enrichment in condensible solids of the captured disk. Thus the evolution time to solid planetesimal sizes should be similar. We next summarize a simplified theory of planet formation from reviews by Pollack [6], by Lissauer [7] and by Woolfson [8].

Both geophysical evidence and theoretical modeling lead to times for the formation of the Earth of a few $\times 10^7$ years. We consider next the sequence of growth by accretion (or sticking) in the disk of bodies whose geometric cross section (σ_{geo}) is less than their gravitational scattering cross section (σ_{grav}), or bolloids. We reserve the word planetesimals to describe bolloids large enough such that $\sigma_{\text{geo}} > \sigma_{\text{grav}}$.

The Growth of Molecules. As the now stable accretion disk matter cools, molecules of metal oxides and silicates bind first, leading to a high temperature gas of these condensible solids. Further cooling and the first molecules collide and initiate crystal or grain growth. One can estimate a characteristic time for this process from the number density n, collision cross section σ and RMS velocity v_{RMS}, $\tau_{\text{collision}} = 1/(n\sigma v_{\text{RMS}})$. We can further divide the cross section into a geometric part (σ_{geo}) and a sticking probability s, where $s \sim 0.01$ for both crystal growth as well as for very much larger bodies like rocks and boulders, provided in both cases, the velocity is not so great as to destroy the molecule, grain, or bolloid. We further note that $n \propto 1/H$ where H is the height of the disk, and $H = R(v_{\text{RMS}}/v_{\text{K}})$ where R is the orbit radius with Keplerian velocity v_{K}. If we consider a molecular weight of $A \sim 100$ for the condensibles and $\sigma_{\text{geo}} = \pi r_{\text{molecule}}^2$ where $r_{\text{molecule}} \sim 2 \times 10^{-8}$ cm. Then the density will be 2.5 g cm^{-3}, like rock, and the geometric collision time leads to $\tau_{\text{molecularcollision}} \sim 5 \times 10^{-3}(R/R_{\text{AU}})^{3/2}$ seconds, which becomes ~ 0.5 s with a small sticking probability of 1%.

The Growth of Bolloids and the Formation of Planetesimals. We now extend this growth rate to larger bodies noting that the geometric cross section scales from molecular sizes to bolloid sizes. As the particles grow in size, they decrease in number for a fixed total mass in orbit so that $\tau_{\text{growth}} \propto m_{\text{particle}}/r^2 \propto r$. This implies that all the time of accretion is spent at the largest bolloid size. This scaling breaks down when a bolloid reaches a critical mass (i.e., planetesimal) such that the gravitational potential at its surface exceeds the RMS kinetic energy of its "thermal" distribution. Note that v_{RMS} is both heated by the Keplerian velocity shear and cooled by disintegrating collisions. A natural velocity limit is when the dynamic pressure of impact of bolloids is within the strength of the bolloids or simply that of rock. (For the outer planets where ice forms the principle solid, the critical velocity will be considerably less.) This dynamic pressure, P_{dynamic} is ~ 100 atmospheres, so $m_{\text{planetesimal}} \propto P_{\text{dynamic}}^{3/2} \sim 10^{22}$ g at a radius of $r = 10^7$ cm. A statistical wide range of $m_{\text{planetesimal}}$ is likely. Thus at a critical mass of a planetesimal of $m_p \sim 10^{22}$ g, $r_p \sim 10^7$ cm, and velocity $v_p \sim \sqrt{Gm_p/r_p} \sim 10^4$ cm/s, the growth time is $\tau_{\text{growth}} = \tau_{\text{moleculargrowth}}(r_{\text{planetesimal}}/r_{\text{molecule}}) = 1.5 \times 10^7$ y at an orbit of an AU and ~ 50 million years at 3 AU. This thus spans the desired turn on time of the GRBs.

The Formation of a Planetoid. Once the planetesimals reach this critical mass/size, runaway accretion takes place due to the gravitationally-enhanced collision cross section. The importance of gravitational scattering in the evolution of the planetesimal mass distribution has been calculated and emphasized by many authors, particularly by analogy to plasma physics by Safronov [9] and later by Goldreich, by Tremaine and by Ward [10,11], and numerically confirmed by Aarseth, Lin, and Palmer [12,13]. The run-away accretion is so fast that nearly all the evolution time is spent reaching the critical size, $m_\text{planetesimal}$ and little time afterwards. Part way through the run-away accretion process to the planet size, an intermediate condition of a "proto planet" of perhaps $m_\text{protoplanet} \sim 10^4 \times m_\text{planetesimal}$ will form. The strong large-angle scatterings between proto planet and planetesimals lead to heating rather than cooling of v_RMS. So a fraction of the planetesimals will diffuse in orbit exactly the same as we even now observe comets scattered by Jupiter and Saturn into the inner solar system. This is the initiation of the GRB phase.

THE FORMATION OF THE GRB

Once scattered inside a radius of $\sim 10^{11}$ cm from the neutron star in an elliptical orbit, the planetesimal will break-up due to tidal forces and the limited strength of rock. The friction between fragments at periastron will ensure evolution to a circular orbit and further evolution by friction describes an accretion disk of rock fragments around the neutron star. The pressure within the disk becomes $P = (\Sigma H)(M_\text{NS}G/R^2)(H/R)$, which should be supported by degeneracy pressure, $P_\text{degen} \gtrsim 10^{13}$ dynes cm^{-2} at a radius of $\sim 3 \times 10^8$ cm. This degenerate, fluid disk evolves in density and energy density reaching the Alfven radius where $P_\text{degen} = B_\text{dipole}^2/8\pi$ or at $R_\text{Alven} \sim 10^7$ cm. This is then a thin conducting disk rotating in a strong dipole field, or a classical unipolar generator. The electrical break-down of this generator is ensured because of the high potential, between dipole lines of force threading both the neutron star and the inner most and outer radii of the disk. This voltage is $V_\text{unipolar} = \int B \times (v_\text{keplerian}/c) dR = 5 \times 10^{19} B_{12} R_\text{NS}((R_1/R_\text{NS})^{-5/2} - (R_2/R_\text{NS})^{-5/2})$ volts or 1.5×10^{17} volts at R_Alfven. The strong torque due to the current I_unipolar crossing the field lines in the disk allows rapid accretion onto the neutron star, while the current outside of the disk flows parallel to field lines and we assume that it may excite a disruption instability. As in the classical tokamak disruption (possibly due to lower hybrid waves), all the current is carried by relativistic runaway electrons, which will radiate synchrotron emissions in equilibrium with their acceleration. Then the GRB luminosity is the electrical power, or $L_\text{GRB} = I_\text{unipolar} \times V_\text{unipolar} = 10^{41}$ ergs/s and thus $I_\text{unipolar} = 2 \times 10^{14}$ amperes at R_NS. Balancing the synchrotron cooling with acceleration, runaway electrons can achieve maximum γ from $I_\text{unipolar} \times V_\text{unipolar} = (4/3)(I_\text{unipolar}R_\text{NS}/e)\gamma^2(B_\text{NS}^2/8\pi)\sigma_\text{Compton}$. Then $\gamma^2 B_{12} = 2.3 \times 10^3$ and the photon emission, $\gamma^2 h\nu_\text{cyclotron}$ would be at ~ 30 MeV. But these high-energy photons will immediately cascade through pair creation in the magnetic field creating a pair plasma further limiting the particle γ. The emission from the creation and annihilation of this pair plasma is the GRB.

When a GRB with the emission of $\sim 10^{41-42}$ ergs occurs at the center of the planetesimal distribution, the gamma-ray flux penetrates the surface of all bolloids as well as the planetesimals a distance determined by the gamma-ray mean free path. The specific heat of the matter corresponding to this depth as well the flux determines the rate of heating and the blow-off velocity. Provided the surface is heated to the vaporization temperature, then for a heat flux of $\sim 10^{14-15}$ ergs cm^{-2} s^{-1}, a mean gamma-ray energy of $\sim 1/4$ MeV, and delivered in bursts over a period of seconds, a mass of $\sim 10^{3-4}$ g cm^{-2} will be ablated at a velocity of $\sim 3 \times 10^5$ to 10^6 cm/s. As a consequence a planetesimal will recoil with a positive, radial velocity impulse of $\Delta v_{\rm RMS} \sim 0.01 - 0.03 \times v_{\rm RMS}$. The progressive heating of $v_{\rm RMS}$ by a sequence of ~ 100 bursts will prevent further accretion of planetesimals until cooling has again taken place. This cooling should take roughly 100 orbits or 100 years and so determines the frequency of GRBs. This also determines the depletion time or turn-off time of the GRB phase, namely 10^6 planetesimals in 10^8 years requiring 10^{28} g or 1% of the orbiting mass after the SGR phase. The X-ray afterglow is created from the X-ray fluorescence of the GRB flux as it traverses the ablated matter.

It will be ironic that the delay time, the size of bursts, and the frequency, or equivalently the turn-off time will depend upon something as rudimentary, prosaic, and mundane as the strength of rock.

REFERENCES

1. Li, H., and Dermer, C.D., *Nature* **359**, 514 (1992).
2. Bulik, T., Coppi, P.S., and Lamb, D.Q., *Ap.J.*, submitted (1997).
3. Colgate, S.A., and Leonard, P.J.T., in *High Velocity Neutron Stars and Gamma-Ray Bursts*, AIP Conf. Proc., eds. R.E. Rothschild & R. Lingenfelter, 1996, p. 269.
4. Colgate, S.A., & Li, H., in *Gamma-Ray Bursts: Huntsville*, eds. C. Kouveliotou, M. Briggs, and G. Fishman, 1996, p. 734.
5. Duncan, R.C., and Li, H., *Ap.J.* **484**, 720 (1997).
6. Pollack, J.B., *Ann. Rev. Astron. Astrophy* **22**, 389 (1984).
7. Lissauer, J.L., *Ann. Rev. Astron. Astrophy* **31**, 129 (1993).
8. Woolfson, M.M., *Q.J.R. Astr. Soc.* **34**, 1 (1993).
9. Safranov, V.S., in *Evolution of the Protoplanetary Cloud and the Formation of the Earth and Planets*, Moscow: Nauka Press, 1969.
10. Goldreich, P., and Tremaine, S., *Icarus* **34**, 227 (1978).
11. Goldreich, P., and Ward, G.R., *Ap.J.* **183**, 1051 (1973).
12. Aarseth, S.J., Lin, D.N.C., and Palmer, P.L., *Ap.J.* **403**, 35 (1993).
13. Palmer, P.L., Lin, D.N.C., and Aarseth, S.J., *Ap.J.* **403**, 336 (1993).

Gamma-Ray Bursts from Electrical Discharges

Tipei Li

High Energy Astrophysics Lab., Inst. of High Energy Physics, Beijing, China

Abstract. Typical spectra, energy dependence of time profiles and spectral evolution observed in GRBs can be produced by explosive discharges in plasmas.

MODEL

In an isotropic thermal emission with mean photon energy $\overline{h\nu}$, the mean energy of a scattered photon from an electron with an energy ϵ is $\overline{E} = \frac{4}{3}(\epsilon/mc^2)^2 \overline{h\nu}$. For thermal emission of temperature $T = 10^6$ K and optical depth $\tau_0 = 0.1$ at $h\nu = 1$ kT the mean energy of a scattered photon from an electron with $\epsilon = 100$ MeV can be calculated as $\sim 3.7 \times 10^3$ keV. Therefore, a large voltage and high temperature discharge column is an effective radiator of hard X-rays and soft γ-rays.

A simple discharge model is used in this paper; an initial discharge voltage V_0 is produced at time $t = 0$, the discharging current has a Gaussian rise, $i(t) \propto \exp(-(\frac{t_m-t}{a_1})^2)$ to its maximum at $t = t_m$ and then falls exponentially with a decay constant a, $i(t) \propto \exp(-\frac{t-t_m}{a})$. The potential difference falls proportionally to the total quantity of charge transferred by the current, $V(t) = V_0 - B \int_0^t i(t) dt$. For example, let $V_0 = 200$ MV, $a_1/a = 0.4$, $a = 10$ s and the voltage at the end of discharge $V_D = 20$ MV, the corresponding time histories of discharge voltage and current are shown in Fig. 1(a).

TIME-AVERAGED SPECTRA

The GRB spectra accumulated by the Spectroscopy Detectors (SD) of BATSE are described well by the following empirical Band et al. model [1]: $E^\alpha \exp(-E/E_0)$ for $(\alpha-\beta)E_0 \geq E$ and $[(\alpha-\beta)E_0]^{\alpha-\beta} \exp(\beta-\alpha)E^\beta$ for $(\alpha-\beta)E_0 \leq E$. The peak power energy $E_{\text{pk}} = (2+\alpha)E_0$. Our calculations show that γ-ray flashes produced by large voltage and high temperature discharge plasma columns can also be described well by this model [2]. As an example, the crosses in Fig. 1(b) show the expected time-averaged spectrum produced by the electrons in the discharge process (Fig. 1(a))

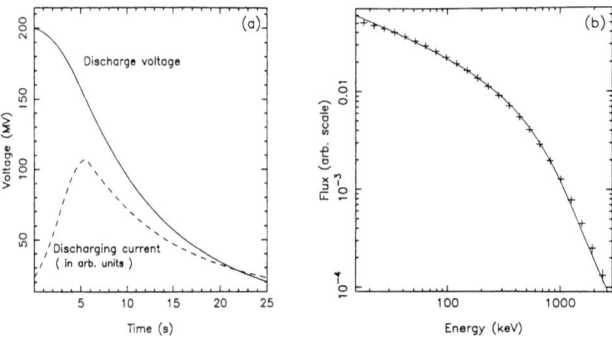

FIGURE 1. An example of the discharge model. (a) Time histories of discharge voltage (solid line) and current (dashed line). (b) The calculated time-averaged spectrum (crosses) from a discharge column described in the text with time histories of voltage and current shown in (a).

passing through an isotropic thermal emission with $T = 10^5$ K and $\tau_0 = 0.1$, and observed by SD of BATSE for scattered photons with scattering angle 0.25°. The error bars were estimated by assuming the total number of recorded photons of $E \geq 20$ keV being 2×10^4, a E^{-2} background spectrum with a flux of 1×10^{-2} cm^{-2}s^{-1}keV^{-1} at 100 keV and in considering the SD full-energy peak efficiency. Fitting the Band et al. model to the simulated spectrum in the region of $E \geq 20$ keV we got the solid line in Fig. 1(b) with $\alpha = -0.48$, $\beta = -2.8$, $E_{\rm pk} = 803.1$ keV and the reduced $\chi^2 = 1.0$, showing the discharge model can produce a typical GRB spectrum.

More calculations with different values of the discharge model parameters show that producing Band et al. spectra is a general property of the discharge mechanism and that a summation of several spectra from different discharge columns with different parameters can also be well described by the Band et al. model.

TIME PROFILES

For the discharge column shown in Fig. 1 we calculated the light curves in four energy channels: $20 - 50$, $50 - 100$, $100 - 300$, and > 300 keV, and their autocorrelation functions respectively. The calculated autocorrelation function decreases more rapidly with time lag in the higher energy channels compared to the lowest one, as shown in Fig. 2(a), which is consistent with that observed in GRBs [3]. Cross-correlation functions of two time profiles in different energy channels were also calculated and shown in Fig. 2(b). A time delay between soft and hard emissions can be seen in Fig. 2(b), which is also observed in GRBs [4–6].

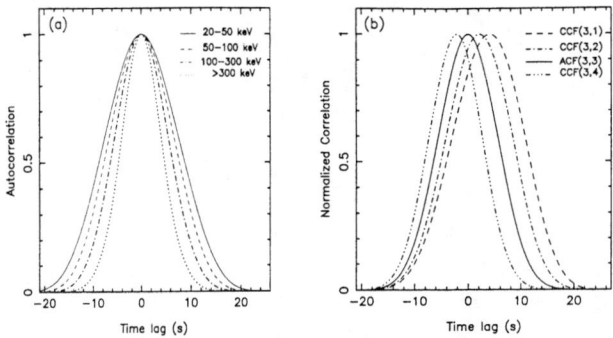

FIGURE 2. Correlation functions of time profiles in four energy channels expected by the discharge model of Fig. 1. (a) Auto-correlation functions. (b) Cross-correlation functions CCF(i,j) between energy channel i and j, ACF(3,3)=CCF(3,3).

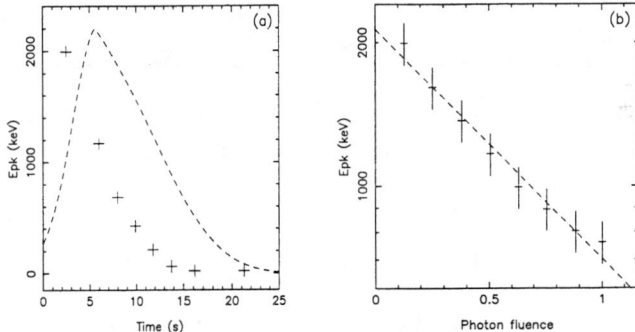

FIGURE 3. Spectral evolution expected for the discharge model described by Fig. 1. (a) Time histories of E_{pk} (keV) (crosses) and photon flux (dashed line, in arbitrary units). (b) E_{pk} vs. photon fluence (in units of total fluence).

PEAK ENERGY EVOLUTION

A hard-to-soft spectral evolution has been observed for many bright GRBs [7]: the peak energy, E_{pk}, softens over the burst. Liang & Kargatis [8] further found that for a sample of bursts consisting of well-resolved, isolated pulses E_{pk} decreases exponentially with photon fluence (running time integral of the flux). Time-resolved spectra for the discharge model described by Fig. 1 were calculated, and the obtained time history of E_{pk}, Fig. 3(a), shows a typical hard-to-soft evolution. In Fig. 3(b) E_{pk} is plotted against photon fluence (the relative errors of E_{pk} are simply taken as 10%, a typical value for bright GRBs of BATSE), showing an exponential decay. The softening trend in E_{pk} is not a universal property for the discharge model. For the case of both the initial discharge voltage and temperature being high, the cross section of Compton scattering falling with the increase of electron energy causes E_{pk} to not considerably change or even increase with time. Ford et al. [7] have already observed such evolution in a few bursts.

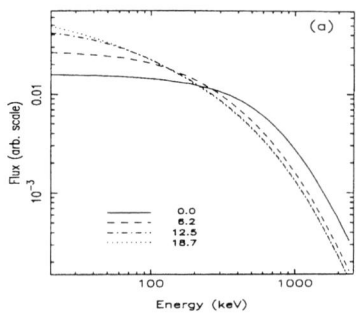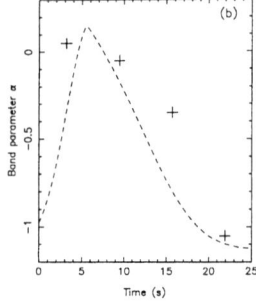

FIGURE 4. Evolution of the Band et al. spectral function for a burst from the discharge described by Fig. 1. (a) Time-resolved spectra. Each kind of line is marked with the time corresponding to the beginning of the time bin. (b) Evolution of α with respect to time. The dashed line shows photon flux (in arbitrary units).

LOW-ENERGY PHOTON SPECTRA

Crider et al. [9] reported evidence that the low-energy power law slope α of GRB spectra evolves with time and a positive correlation exists between the time-resolved spectral break energy E_{pk} and α. These patterns can also be produced by the discharge model. Fig. 4 shows the expected evolution of time-resolved spectra of the discharge model Fig. 1, which is similar to that observed in GRBs [9].

DISCUSSION

McBreen et al. [10] noted that there is similarity in the statistical properties of the light curves and peak flux distributions of GRBs and terrestrial lightning. Fishman et al. [11] discovered terrestrial gamma-ray flashes (TGFs). An evident correlation of TGFs with storm system indicates that these events are caused by electrical discharges to the stratosphere or ionosphere of the Earth, thus TGFs give observational evidence that the discharge process can produce high-energy bursts like GRBs.

Alfvén [12] has stressed the importance of studying electric current circuits in plasmas to understand phenomena occurring on dimensions from the magnetosphere to galaxies, and held that many of the explosive events observed in cosmic physics are produced by exploding electric double layers. The energy released by a GRB is dependent on its distance. The long-term variability obviously observed in durations and hardness ratios of long bursts [13] and more significantly in their correlation coefficients of short bursts [14,15] impel us to seriously consider the possibility of GRBs originating within the heliosphere. The typical energy released by a GRB produced at a distance of 100 AU is about $10^{26} f$ ergs, where f is the beaming solid angle fraction. If emission is beamed into an angular region of ~ 1 deg^2, the typical energy of GRBs will be only $\sim 2 \times 10^{21}$ ergs, equivalent just an average thunderstorm. A burst with total released energy 2×10^{21} erg emits $\sim 2 \times 10^{28}$ photons (60 keV/ph assumed). If each discharge electron produces one burst photon through the inverse Compton scattering with thermal photons, a discharging current of $\sim 2 \times 10^8$ A can produce the burst emission with a 10 s duration. A current flowing in the heliospheric plasma may contract by the magnetic confinement and form a plasma cable with much larger density than the surroundings [12]. In the disruptive discharge the cable is further pinched into a very narrow column (Z pinch). The interaction length of the Compton scattering between energetic electrons and thermal photons in a plasma column with diameter $d = 1$ m and temperature 10^6 K, which is pinched from a initial current cable with diameter $d_0 = 0.1$ AU$= 1.5 \times 10^{12}$ cm and density $N_0 = 1$ cm^{-3}, is only ~ 10 km. The density of the pinch column can be estimated as $N \sim 2 \times 10^{20}$ cm^{-3} and optical depth $\tau_0 \sim 0.1$.

REFERENCES

1. Band, D.L., et al., ApJ **413**, 281 (1993).
2. Li, T.P., & Wu, M., Chin. Phys. Lett. **14**, 557 (1997).
3. Link, B., Epstein, R.I., & Priedhorsky, W.C., ApJ **408**, L81 (1993).
4. Chipman, E., in Gamma-Ray Bursts, AIP Conf. Proc. 307, AIP: New York, 1995, p. 202.
5. Cheng, L.X., et al., A&A **300**, 746 (1995).
6. Band, D.L., ApJ **486**, 928 (1997).
7. Ford, L.A., et al., ApJ **439**, 307 (1995).
8. Liang, E., & Kargatis, V., Nature **381**, 49 (1996).
9. Crider, A., et al., ApJ **479**, L39 (1997).
10. McBreen, B., et al., MNRAS **271**, 662 (1994).
11. Fishman, G.F., et al., Science **264**, 1313 (1994).
12. Alfvén, H., Cosmic Plasma, D.Reidel Publishing Company, 1981.
13. Li, T.P., Chin. Phys. Lett. **13**, 637 (1996).
14. Li, T.P., Acta Astrophys. Sinica **17**, 256 (1997).
15. Li, T.P., Ap&SS, submitted (also available as astro-ph/9704264).

The Physical Source of Gamma-Ray Bursts

W. Kluźniak‡ and M. Ruderman*

‡*University of Wisconsin, Madison, WI 53706*
‡*Copernicus Astronomical Center, PL 00-716 Warszawa*
**Columbia University, New York, NY 10027*

Abstract. Gamma-ray bursts are thought to come from a relativistic blast-wave emitting gamma-rays at a distance of 10 to 1000 AU from the central engine which releases the explosive energy. To account for the observed duration and variability of GRBs, the central engine powering the γ-ray emission must be active from several to many seconds and yet must strongly fluctuate in its output on much shorter timescales. A differentially rotating neutron star of millisecond period and canonical magnetization (DROMP) will wind up a toroidal magnetic field of $> 10^{16}$ G and then will release its magnetic energy, $E \sim 10^{51}$ erg, in a rapid sub-burst. This process will repeat on a timescale of $t \sim 1$ to 10^3 s (depending on the initial conditions) until the kinetic energy of differential rotation is exhausted after several (N) sub-bursts. The calculated values E, t, N are in agreement with observations. The baryon loading in each sub-burst was found to be also in agreement with theoretical requirements for cosmological GRBs. DROMPs would be expected to be created in several kinds of astrophysical events, at least two of which would be expected to occur at the apparent GRB rate of $\sim 10^{-6}$ y^{-1} per galaxy. Some DROMPs could be created close to, while others far from, galaxies.

INTRODUCTION

One of the most remarkable features of gamma-ray bursts–and one rarely addressed in theoretical models–is the diversity of the observed light curves. This diversity in itself suggests that the source of emission is a system with many internal degrees of freedom. Even more remarkable is the coexistence within a single light curve of many disparate time-scales: individual sub-bursts, each perhaps exhibiting millisecond rise times and/or spikes, may be separated in the longer bursts by a quiescent interval of many seconds or even minutes. In view of the extreme energy requirements ($\sim 10^{51}$ erg in γ-rays at 3 Gpc) most models of GRBs involve a single catastrophic energy release–the challenge in such models is to explain how the gamma-ray emission may last for tens of seconds or minutes, let alone how to obtain millisecond variability.

The now standard relativistic blast-wave model [5] successfully predicted the (occasionally) observed radio [8], optical [6] and X-ray afterglows [12]. At the same time, the detection of these afterglows, with their characteristic decay timescale of weeks, underscores the necessity of explaining the GRB durations of seconds or minutes, which (on the one hand) are very short compared to the time scale of evolution of the blast wave in the observer's rest frame, and (on the other) are very long compared to the dynamical timescale of the physical objects most likely involved, i.e., stellar mass black holes or neutron stars. Calculations of binary mergers [4,9,13] indicate that at most a subclass of GRBs can arise in coalescence events.

Below we suggest a particular source of energy (the central engine) eventually powering the γ-ray release at a distance of $10^{2\pm1}$ AU from the engine (and a few Gpc from the observer). We imagine, but do not show, that the γ-rays will be released in internal shocks, such as the ones discussed by Sari and Piran [10] and Mochkowitch & Daigne [7]. These authors have shown how a central engine of requisite properties may drive internal shocks in the relativistic blast-wave to produce the observed light curves. Our task is to supply the central engine.

THE CENTRAL ENGINE OF GRBS

We find that differentially rotating millisecond "pulsars" (DROMPs) are excellent candidates for the central engine of GRBs. (Here, "pulsar" is shorthand for "a magnetized neutron star." We do not predict millisecond pulsations in the γ-ray

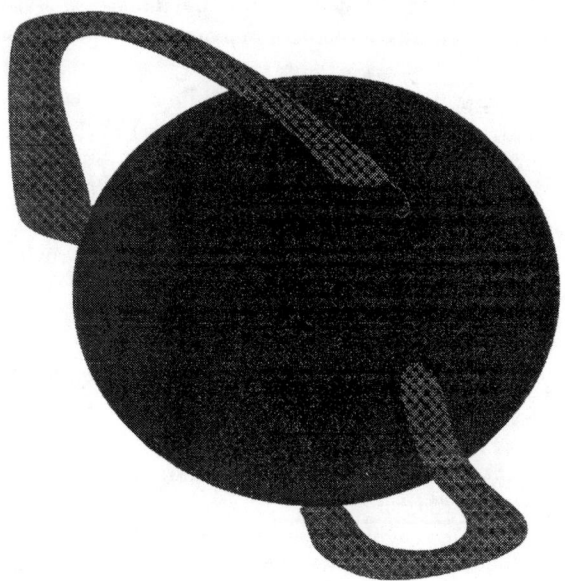

FIGURE 1. The magnetic torus emerges from the neutron star.

light curve.) It is crucial that the neutron star does not rotate as a solid body and that the kinetic energy in internal motions (over and above any net total angular momentum of the star) be $\sim 10^{52}$ erg. When these conditions are met, a magnetic field of enormous strength will be wound up in the interior and will form a buoyant flux tube (torus). After the torus floats up to the surface, its energy will be released. The resulting huge Poynting flux will be capable of powering internal shocks in a relativistic blast-wave, and even the blast-wave itself. Each time the magnetic energy is released (as the magnetic field is repeatedly amplified by the internal motions of the star) a new sub-burst occurs. These repeating episodes thus eventually power successive peaks of the γ-ray light curve.

As the neutron star exists for the whole duration of the GRB and may even survive after the GRB activity is over, ours is not, strictly speaking, a catastrophic model of GRBs. Nonetheless, a neutron star may live only twice as a DROMP, once when it is born and once when it dies [1]. The ordinary radio pulsars are not sources of GRBs, of course. However, some of the radio-pulsars may be long-lived remnants and/or progenitors of GRB sources.

A DROMP may be formed in the accretion induced collapse of a rapidly spinning white dwarf (as in Usov's model [11] of GRBs), in the collapse to a black hole of a millisecond pulsar on the supramassive sequence (i.e., one initially, for $\sim 10^{10}$ y, stabilized against collapse by rapid rotation), in the binary merger of two neutron stars (if the equation of state of dense matter is stiff enough [3], or if the masses of the coalescing neutron stars are low enough), or as a result of catastrophic spin-up of the neutron star in a collision with another compact stellar remnant. Some of these processes are more likely to occur in globular clusters. Note that two of the four processes mentioned would be expected to occur at the rate of $\sim 10^{-6}$ y^{-1} per galaxy.

Fuller details of the energy release process will be given elsewhere. Here we note that the magnetic field value (and its associated energy) as the torus breaks through the surface is calculable, as is the interval between sub-bursts. The computed values are given in the abstract. The total duration, Nt, of the GRB would be the time necessary to (repeatedly) convert the kinetic energy of the internal motions to magnetic energy. We find that Nt scales inversely with the initial strength of the magnetic field in the neutron star and also inversely with the square root of the internal kinetic energy, a "typical" value is $Nt \sim 100$ s.

End Notes

After presenting this contribution we learned from Dr. J.R. Wilson that, according to relativistic simulations, the final stages of binary evolution of two neutron stars lead to internal motions which are sufficient to build up a magnetic field of 10^{17} G (see [13] and references therein). This raises the interesting possibility that the neutron stars in coalescing neutron-star or black-hole binaries are individually GRB sources through the mechanism discussed in the previous section, above. For such a system, there is a continual supply of orbital energy to the internal motions, this process would terminate as the stars individually collapse to black holes (*op. cit.*) or coalesce with the companion.

We also note that several authors have suggested that initial orbital motion of a binary system may lead to differential rotation in the possibly resulting disk-like configuration with attendant build-up of the magnetic field and the eventual release of energy in a Poynting flux. We are now aware of seminal contributions by Drs. J. Katz, B. Paczyński, T. Piran, P. Mészarós and M. Rees. We would especially like to acknowledge conversations with Jonathan Katz and to point the interested reader to his paper [2].

ACKNOWLEDGEMENTS

This work supported in part by Komitet Badań Naukowych through grant 2P03D01311.

REFERENCES

1. Fleming, I., in *You Only Live Twice* (1965).
2. Katz, J., *ApJ in press*, also available as astro-ph/9701116 (1997).
3. Kluźniak, W., *Bulletin of the AAS* **29**, 793 (1997).
4. Lee, W.H., & Kluźniak, W., these proceedings, also available as astro-ph/9711172 (1997).
5. Mészarós, P., & Rees, M., *ApJ* **405**, 278 (1993).
6. Mészarós, P., & Rees, M, *ApJ* **476**, 232 (1997).
7. Mochkovitch, R., & Daigne, F., these proceedings.
8. Paczyński, B., & Rhoads, J.E., *ApJ* **418**, L5 (1993).
9. Ruffert, M., & Janka, H.T., these proceedings.
10. Sari, R., & Piran, T., *ApJ* **485**, 270 (1997).
11. Usov, V.V., *Nature* **357**, 472 (1992).
12. Vietri, M., *ApJ* **478**, L9 (1997).
13. Wilson, J.R., Salmonson, J.D., & Mathews, G.J., these proceedings, also available as astro-ph/9711307 (1997).

INSTRUMENTATION
AND TECHNIQUES

First Year Results from LOTIS

G. G. Williams[1], H. S. Park[2], E. Ables[2], D. L. Band[3],
S. D. Barthelmy[4,5], R. Bionta[2], P. S. Butterworth[4], T. L. Cline[4],
D. H. Ferguson[6], G. J. Fishman[7], N. Gehrels[4], D. H. Hartmann[1],
K. Hurley[8], C. Kouveliotou[7], C. A. Meegan[7], L. Ott[2], E. Parker[2],
and R. Wurtz[2]

[1] *Dept. of Physics and Astronomy, Clemson University, Clemson, SC 29634-1911*
[2] *Lawrence Livermore National Laboratory, Livermore, CA 94550*
[3] *CASS 0424, University of California, San Diego, La Jolla, CA 92093*
[4] *NASA/Goddard Space Flight Center, Greenbelt, MD 20771*
[5] *Universities Space Research Association*
[6] *Dept. of Physics, California State University at Hayward, Hayward, CA 94542*
[7] *NASA/Marshall Space Flight Center, Huntsville, AL 35812*
[8] *Space Sciences Laboratory, University of California, Berkeley, CA 94720-7450*

Abstract. LOTIS (Livermore Optical Transient Imaging System) is a gamma-ray burst optical counterpart search experiment located near Lawrence Livermore National Laboratory in California. The system is linked to the GCN (GRB Coordinates Network) real-time coordinate distribution network and can respond to a burst trigger in 6-15 seconds. LOTIS has a total field-of-view of $17.4° \times 17.4°$ with a completeness sensitivity of $m_V \sim 11$ for a 10 second integration time. Since operations began in October 1996, LOTIS has responded to over 30 GCN/BATSE GRB triggers. Seven of these triggers are considered *good* events subject to the criteria of clear weather conditions, < 60 s response time, and $> 50\%$ coverage of the final BATSE 3σ error circle. We discuss results from the first year of LOTIS operations with an emphasis on the observations and analysis of GRB971006 (BATSE trigger 6414).

INTRODUCTION

Our knowledge of the nature of gamma-ray bursts (GRBs) has greatly increased as a result of recent detections of X-ray, optical, and radio counterparts. X-ray observations of several GRBs including but not limited to GRB970228 and GRB970508 by BeppoSax [1,2] and GRB970828 by XTE/ASCA [3] have provided precise localizations which have allowed for deep optical follow-up searches. These searches have resulted in the identification of two GRB optical counterparts, namely GRB970228 [4] and GRB970508 [5] and a single radio counterpart, GRB970508 [7]. Despite the wealth of information that has been obtained from these discoveries, the physical mechanisms which cause a gamma-ray burst remain a mystery. The lack of a bright host galaxy for either optical counterpart has further confused matters. Although the identification of the "Hurley 100" [8] may pin down the nature of the afterglows, multiwavelength observations of GRBs *simultaneous* with the gamma-ray emission may be a more direct method of probing their origin. If the physical processes that produce the prompt gamma-ray emission and the lower energy afterglow differ, as Katz and Piran [6] have suggested, a broad-band spectrum revealing the nature of the source environment can only be produced from simultaneous observations. Small, wide field-of-view telescopes, such as LOTIS, which were originally designed to provide more precise burst locations by detecting the simultaneous optical emission may assist in producing or constraining this broad-band spectrum.

OBSERVATIONS

LOTIS is a second generation simultaneous optical counterpart search experiment. The precursor experiment, called Gamma-Ray Optical Counterpart Search Experiment (GROCSE), found no evidence of simultaneous optical activity brighter than $m_V = 7.5$ [10].

TABLE 1. LOTIS GRB Triggers

Trig	UTC Date	Fluence/10^{-6} (erg cm^{-2})	Stat Error (deg)	Hunt-GCN Error (deg)	t_{res} (s)	Duration (s)
5634	961017	0.51	2.9	2.7	11	3
5719	961220	1.8	1.5	3.6	9	15
6100	970223	48.0	0.73	2.0	11	30
6117	970308	0.81	5.8	13.6	14	2
6307	970714	1.7	2.8	7.1	14	1
6388	970919	2.3	3.0	5.1	12	20
6414	971006	9.3	0.6	6.8	17	150

The LOTIS telescope, located 25 miles East of Livermore, CA, consists of four individual cameras arranged in a 2 × 2 array. Each camera has a field-of-view of 8.8° × 8.8° which yields a total field-of-view of 17.4° × 17.4° allowing for a 0.2° overlap. A detailed description of the system is provided in Park *et al.* [11] as well as at http://hubcap.clemson.edu/~ggwilli/LOTIS/.

From the start of routine operation in early October 1996 through early October 1997 LOTIS has responded to 36 GCN/BATSE GRB triggers [9]. Six of these triggers were a result of particle events occurring in the BATSE detectors. Two triggers were caused by known soft gamma-ray repeaters (SGRs). Of the remaining 28 triggers 26 were unique GRBs while two were refined coordinates of previously triggered GRBs (GCN LOCBURST). Of the 26 unique triggers 7 were considered *good* events subject to the criteria of clear weather conditions, < 60 s response time, and > 50% coverage of the final BATSE 3σ error circle. By far the hardest

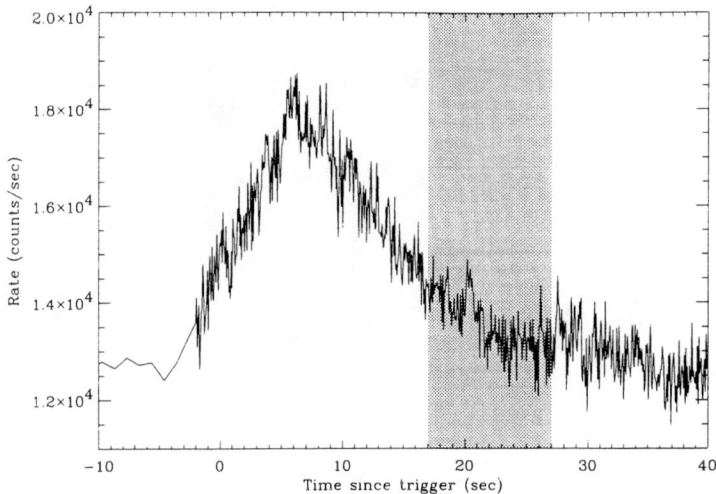

FIGURE 1. GRB971006 gamma-ray light curve. The shaded area represents the integration time of the first LOTIS image

criterion to meet was the coverage criterion owing to the difficulty in determining accurate GRB locations from the first few seconds of gamma-ray emission. Based on these statistics LOTIS responds to ~ 1 good GRB event every 52 days.

Data from the seven good events which LOTIS responded to is given in Table 1. The fluence values were determined by summing the fluence in the four energy channels given in the current BATSE GRB catalog with the exception of triggers 6100 and 6414 which are discussed below. The fourth column gives the statistical error in the final BATSE postion while the fifth column gives the angular difference between the initial and final BATSE coordinates. The LOTIS response time and the total duration of the burst are given in the last two columns. In four cases LOTIS began imaging while gamma-rays were still being emitted making the observations truly simultaneous.

The event with the largest gamma-ray fluence was GRB970223 which was among the top 3% of all BATSE GRBs. Although no optical transients were identified for this burst LOTIS placed an upper limit on the ratio of optical flux at 700 nm to gamma-ray flux at 100 keV of $R_{F_{\mathrm{simultaneous}}}(t = 11 - 21 \mathrm{s}) = F_{\mathrm{opt:700nm}}/F_{\gamma:100\mathrm{keV}} < 475$ and on the ratio of optical to gamma-ray fluence of $R_L = L_{\mathrm{opt:500-850nm}}/L_{\gamma:20-2000\mathrm{keV}} < 1.1 \times 10^{-4}$. The full analysis of this event is given in Park et al. [11].

The longest burst which LOTIS responded to was GRB971006. This burst had a main pulse duration of ~ 30 s but exhibited weak pre- and post-burst emission resulting in a total duration of ~ 150 s. The light curve of GRB971006 is shown in Figure 1. LOTIS began imaging the field centered on the initial GCN coordinates

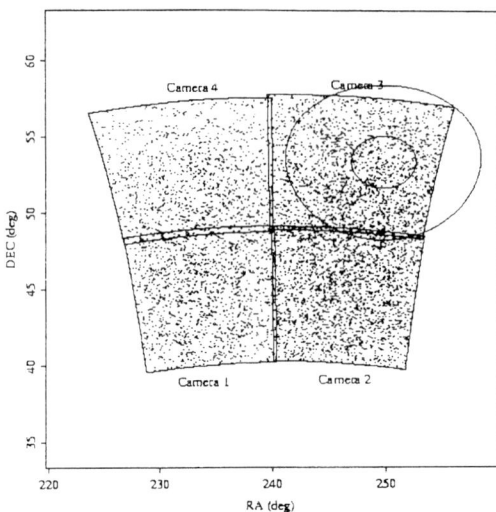

FIGURE 2. LOTIS coverage of GRB971006. Each of the individual points represent a stellar object above the 4σ threshold. The small and large ellipses represents the BATSE 1σ and 3σ error circles respectively. LOTIS obtained approximately 75% coverage of the BATSE 3σ error circle.

(RA = 241.1, Dec = 49.2 J2000) ~ 17 s after the start of the burst. The shaded region of Figure 1 represents the 10 s integration time of the first LOTIS image. The final BATSE coordinates of GRB971006 (RA = 249.8, Dec = 53.3) were well within the LOTIS field-of-view. Figure 2 shows the LOTIS coverage for this burst. The small and large ellipses represent the BATSE 1σ and 3σ error circles (including the 1.6° systematic error) respectively. There was no Interplanetary Network (IPN) [12] localization available for this burst and therefore it was necessary to search the entire 3σ error circle for transient objects. This search found no transients with a point spread function (psf) consistent with the stellar psf.

From a histogram plot of stellar magnitudes in camera 3 we determined the completeness magnitude (the faintest magnitude for which 100% of the stars were detected) for this event to be $m_V \sim 11.0$. Following the analysis in Park et al. [11] this yields an upper limit to the flux density at 700 nm of $F_\nu(700\text{nm}) < 2.7 \times 10^{-24}$ erg cm^{-2} s^{-1} Hz^{-1}. The BATSE flux density at 100 keV was found by fitting the spectrum from LAD3 during the integration time of the first LOTIS image to the Band GRB functional form [13] which yielded a value of $F_\nu(100 \text{ keV}) = 1.7 \times 10^{-27}$ erg cm^{-2} s^{-1} Hz^{-1}. The resulting upper limit of the optical to gamma-ray flux for this event is $R_{F_{\text{simultaneous}}}(t = 17-27s) = F_{\text{opt:700nm}}/F_{\gamma:100\text{keV}} < 1600$.

The total gamma-ray fluence was determined by integrating the Band GRB functional form for the entire burst from 20 keV to 2000 keV. The total gamma-ray fluence was $L_{\gamma:20-2000\text{keV}} = 9.3 \times 10^{-6}$ erg cm^{-2} while the upper limit to the GRB's optical fluence, again following Park *et al.* [11], is $L_{\text{opt}:500-850\text{nm}} < 5.4 \times 10^{-9}$ erg cm^{-2}. The resulting upper limit for the optical to gamma-ray fluence ratio is $R_L = L_{\text{opt}:500-850\text{nm}}/L_{\gamma:20-200\text{keV}} < 5.8 \times 10^{-4}$.

DISCUSSION

Although LOTIS has already placed upper limits on the simultaneous optical to gamma-ray flux for specific events, we hope to further constrain the ratio with an upgrade to thermo-electric cooled CCDs in January 1998. In the future we plan to investigate GRB spectral evolution focusing on how the low-energy power-law index, α, of the Band GRB functional form [13] effects optical constraints. We also plan to implement Super-LOTIS [14], a dedicated 0.6 m reflector with a design sensitivity of $m_V \sim 19$ (10 s integration time) early next year.

REFERENCES

1. E. Costa *et al.*, *Nature* **387**, 783 (1997).
2. L. Piro *et al.*, *A & A*, in press (1997).
3. T. Murakami *et al.*, IAUC 6729 (1997).
4. J. van Paradijs *et al.*, *Nature* **386**, 686 (1997).
5. S. G. Djorgovski *et al.*, *Nature* **387**, 876 (1997).
6. J. I. Katz & T. Piran, *ApJ* **490**, 772 (1997).
7. D. A. Frail *et al.*, *Nature* **389**, 261 (1997).
8. K. Hurley, these proceedings.
9. S. D. Barthelmy, these proceedings.
10. H. S. Park *et al.*, *ApJ* **490**, 99 (1997a).
11. H. S. Park *et al.*, *ApJL* **490**, L21 (1997b).
12. K. Hurley *et al.*, in AIP Conf. Proc. **307**, *Gamma-Ray Bursts, 2nd Huntsville Symposium*, ed. G. J. Fishman, J. J. Brainerd, & K. Hurley, 27 (1994).
13. D. L. Band *et al.*, *ApJ* **413**, 281 (1993).
14. H. S. Park *et al.*, these proceedings.

Super-LOTIS
A High-Sensitive Optical Counterpart Search Experiment

H. S. Park[1], E. Ables[1], D. L. Band[5], S. D. Barthelmy[3], R. M. Bionta[1], P. S. Butterworth[3], T. L. Cline[3], D. H. Ferguson[6], G. J. Fishman[4], N. Gehrels[3], D. Hartmann[2], K. Hurley[7], C. Kouveliotou[4], C. A. Meegan[4], L. Ott[1], E. Parker[1], G. G. Williams[2]

[1] *Lawrence Livermore National Laboratory, Livermore, CA 94550*
[2] *Dept. of Physics and Astronomy, Clemson University, Clemson, SC 29634-1911*
[3] *NASA/Goddard Space Flight Center, Greenbelt, MD 20771*
[4] *NASA/Marshall Space Flight Center, Huntsville, AL 35812*
[5] *CASS 0424, University of California, San Diego, La Jolla, CA 92093*
[6] *Dept. of Physics, California State University at Hayward, Hayward, CA 94542*
[7] *Space Sciences Laboratory, University of California, Berkeley, CA 94720-7450*

Abstract. We are constructing a 0.6 meter telescope system to search for early time gamma-ray burst (GRB) optical counterparts. Super-LOTIS (Super-Livermore Optical Transient Imaging System) is an automated telescope system that has a $0.8° \times 0.8°$ field-of-view, is sensitive to Mv $\sim$ 19 and responds to a burst trigger within 5 min. This telescope will record images of the gamma-ray burst coordinates that are given by the GCN (GRB Coordinate Network). A measurement of GRB light curves at early times will greatly enhance our understanding of GRB physics.

INTRODUCTION

The origin and nature of gamma-ray bursts (GRBs) remains an important unresolved problem in astrophysics. GRBs are brief bursts (< 100 sec duration) of high-energy radiation that appear at random in the sky. Much of the difficulty in studying gamma-ray bursts results from the poor directional precision ($\sim$1-10°, 1σ statistical error) available from current gamma-ray detection experiments and their short duration ($\sim$ 1-100 s). Even though recent fading X-ray, optical and radio counterpart observations by the Italian-Dutch satellite (BeppoSAX) [1–3] provided information on their distance scale [4], these observations were made many hours later than the GRB. These afterglows may be due to different process from the GRB production mechanism. An observation of optical activity simultaneous with the GRB may provide clues to understanding this process.

In an attempt to search for simultaneous optical counterparts of GRBs, we initially utilized an existing wide-field-of-view telescope at Lawrence Livermore National Laboratory (LLNL) to rapidly image GRB coordinates distributed by the BATSE real-time coordinate distribution network [5]. This first experiment, the Gamma Ray Optical Counterpart Search Experiment (GROCSE), did not find optical optical counterparts at the Mv $\sim$ 7.5 sensitivity level [6].

Subsequently, we constructed the Livermore Optical Transient Imaging System (LOTIS) which has a 17.4° $\times$ 17.4° field of view to image the entire error circle of the rapid GCN notice (BATSE-Original trigger coordinates arrive in 5 s but have $\sim$15° errors). LOTIS has been operating since October 1996 and we have not yet observed any simultaneous optical activity at Mv $\sim$ 11 level [7,8].

In order to provide better GRB coordinates to search for counterparts, the GCN has installed new triggers called LOCBURST and RXTE which utilize the best analysis performed by the BATSE team (involving an operator's interaction) and the RXTE satellite which uses the hard X-ray afterglow to calculate a better position. The LOCBURST trigger error is $\sim$0.2–2°; and the RXTE trigger error is $\sim$6–40 arcmin depending on the statistics. Although delayed ($\sim$15–35 min. for LOCBURST; $\sim$ 3–5 hr. for RXTE triggers), the smaller error box enables a more conventional, deeper telescope (larger aperture but smaller field of view than LOTIS) to follow up on the GRBs. Since the bursts are random, this telescope will need to be dedicated and automated to always be ready for new triggers.

In an attempt to make such observations, we are constructing a large aperture telescope system dedicated to this search.

SUPER-LOTIS

The telescope is a Boller and Chivens 0.6 meter reflective telescope with f/3.5. Figure 1 shows the telescope. It has superb optical quality and mechanical structure; however, it is not equipped with computer controllable drives nor an electro-optical imaging sensor. We are converting this telescope to Super-LOTIS by refurbishing the motor drive, installing a CCD camera, and placing it at a remote site for dedicated observation. As for the sensor, we are installing a LOTIS CCD camera which utilizes a Loral 442A 2048 $\times$ 2048 CCD (15 $\times$ 15 μm pixels) with LLNL built readout electronics. The CCD is cooled by thermoelectric cooling (to -30°C) which minimizes dark current and readout noise.

The Super-LOTIS will have 0.84 $\times$ 0.84° field-of-view (1.5 arcsec/pixel), which is sufficient for BATSE/RXTE trigger types distributed by the GCN. We plan to dither around the GCN "Original" trigger coordinates which has only a 5 s delay, but a large 15° error box. When we receive refined positions, i.e. LOCBURST or RXTE triggers, we will scan the region and stay at that location the rest of the night. Our scanning strategy and automation will allow us to record GRB optical activity as early as a few minutes.

The basic on-line software has already been written and has been operating on LOTIS. We will need very minimal modification to the existing software for the entire data acquisition control.

We have estimated the sensitivity of the Super-LOTIS system. The calculation includes the measured camera dark current at -30°C, readout noise, typical sky background at an observing site and shot noise. Figure 2 shows the resulting signal to noise ratio with 10 and 60 s integration time vs. the visual magnitude. The calculation indicates that the Super-LOTIS will see Mv $\sim$ 19 stars at a signal to noise ratio of 10 with 10 s and Mv $\sim$ 21 with 60 s integration times.

Utilizing our successful experience in construction, operation, data handling, and data analysis of the GROCSE and LOTIS systems we expect Super-LOTIS to be constructed and operational within a year. Detection of optical emission at early times (or placing stringent constraints) would provide a crucial link between the multiwavelengths properties of the burst and its afterglow. Burst and afterglow emission are likely to probe different aspects of the GRB model (e.g. internal

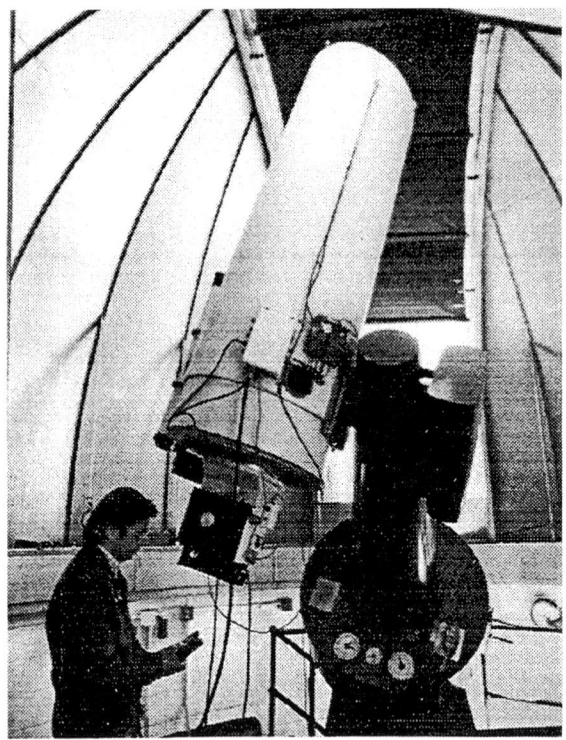

FIGURE 1. Super-LOTIS Boller and Chivens 0.6 meter reflective telescope. After refurbishing by adding computer controlled motors and installing a CCD camera, this automated system will be dedicated to the GRB optical counterpart search.

vs. external shocks.) So far only two afterglows have been detected and no optical detection simultaneous with or shortly after the gamma-ray burst has been made. The 90 minute delayed emission of high energy photons from GRB940217 [9] suggests that similar emissions in the optical wavelength could accompany some bursts. Super-LOTIS would detect such a new spectral component of the bursts to a magnitude level of $> Mv \sim 19$. While upper limits will be useful for constraining the models, Super-LOTIS will establish the GRB light curves at early times which will provide a crucial step toward understanding GRB phenomenon.

REFERENCES

1. Costa, E., et al., *IAU Circ.*, 6572 (1997).
2. Costa, E., et al., *IAU Circ.*, 6649 (1997).
3. Heise, J., et al., *IAU Circ.*, 6654 (1997).
4. Metzger, M., et al., *IAU Circ.*, 6655 (1997).
5. Barthelmy, A., et al., these proceedings.
6. Park, H., et al., *AstroPhys. J.* **490**, 99 (1997).
7. Park, H., et al., *AstroPhys. J. Lett.* **490**, L21-L24 (1997).
8. Williams, G., these proceedings.
9. Hurley, K., et al., *Nature* **372**, 652 (1994).

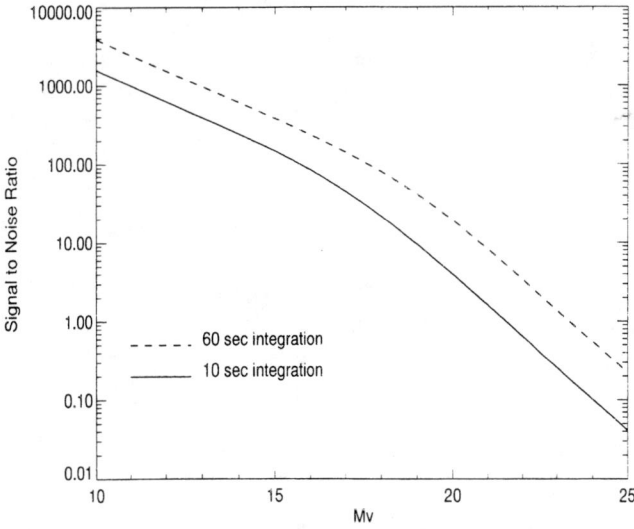

FIGURE 2. Super-LOTIS sensitivity: Predicted signal to noise ratio vs. visual magnitude. Super-LOTIS will be able to detect $Mv \sim 19$ objects with 10 s and $Mv \sim 21$ with 60 s integration times.

TAROT: A Status Report

Michel Boër, J.L. Atteia, M. Bringer, A. Klotz, C. Peignot*,
R. Malina, P. Sanchez[†],
H. Pedersen[‡]
G. Calvet, J. Eysseric, A. Leroy, M. Meissonier[||]
C. Pollas, J. de Freitas Pacheco[¶]

*Centre d'Etude Spatiale des Rayonnements (CNRS)
BP 4346, F 31028 Toulouse Cedex 4, France
[†]Laboratoire d'Astronomie Spatiale (CNRS), Marseille, France
[‡]Copenhagen University Observatory, Denmark
[||]INSU-CNRS, Division Technique, Paris, France
[¶]Observatoire de la Côte d'Azur, Nice, France

Abstract. TAROT-1 is an automatic, autonomous ground based observatory whose primary goal is the rapid detection of the optical counterparts of cosmic gamma-ray burst sources. It will be able to begin imaging any GRB localization 8 seconds after receipt of an alert from CGRO/BATSE or HETE-2. TAROT-1 will reach the 17th V magnitude in 10 seconds, at a 10σ confidence level. TAROT will be able to observe GRB positions given by Beppo-SAX or RXTE, EUV transients from ALEXIS alerts, etc. TAROT will also study a wide range of secondary objectives and will feature a complete automatic data analysis system, and powerful scheduling software. TAROT will be installed this fall on the Plateau du Calern, 1200m above sea level. We report on the status of the project.

SCIENTIFIC GOALS

Cosmic Gamma-Ray Bursts

The Télescope à Action Rapide pour les Objets Transitoires (Rapid Action Telescope for Transient Objects) has as primary objective the detection of optical transients associated with gamma-ray bursts. Its construction was decided in 1995, at a time when the optical emission of GRBs was unknown, but predicted by a few models [1]. At that time we thought that the best chance was to observe the GRB source in optical during its gamma-ray activity, meaning a fast-moving wide-field telescope. We also decided, given the various constraints, to reach a sensitivity level somewhat below the theoretical constraints, i.e. $V_{min} \geq 16$ or 17. The recent detection of the afterglow from GRB970228 and GRB970508 [2,3] at optical and X-ray wavelengths demonstrates that this objective is achievable, since given the time delays involved, the light emitted by GRB970508 would have been easily detectable by TAROT-1. Popular models [4] invoke the shock of a relativistic fireball with the interstellar medium. TAROT will be able to detect it quite early, but also, it will detect the emission due to internal shocks within the fireball itself [5]. In this latter case, the optical emission is predicted to be simultaneous with the gamma-ray emission, and in the range of V magnitude between 14 and 16 [1]. With a maximum delay of 8 seconds from the burst onset to the TAROT observation, we will be able to catch 70% of the sources while they are still active, provided that the error box is small enough (4 square degrees). In the case of BATSE we hope to detect one or two bursts per year within this delay.

Detection of this emission by TAROT or by a similar experiment would be an important objective, since it may lead to the confirmation of the model, and give data on the physical source conditions. An early detection of a fireball at a relatively high level (e.g. magnitude 16) would trigger observations at larger telescopes, in order to take a spectrum of the source itself, as well as of the host, and to detail the light curve of the optical transient from the beginning of the event, and to see the transition between the internal shock regime and the afterglow.

Moreover, since TAROT has a large field of view (4 square degrees) and will operate continuously, it will be able to detect optical transients which may be related to undetected GRBs if the emission is beamed [6]. Large areas of the sky will be surveyed both for secondary objectives and to establish a reference catalogue for later detection of GRB optical counterparts. This has two advantages: 1) In the case we detect an object within the error box of a GRB, from the inspection of our catalogue we can determine its nature, real new object or variable or flaring object active at the GRB time; 2) During this survey we will detect a large number of variable or new objects, since a substantial fraction of the sky will be observed every night. How we will be able to separate GRB afterglows from variable or flare stars is another problem which is currently under study, but the information will be in our data.

Secondary Objectives

An automated, versatile telescope like TAROT has a wide range of possible applications. Objects may be observed upon alert or in a systematic mode. In addition, the wide field of view will result in a lot of serendipitous detections.

In alert mode we shall try to identify EUVE transients detected by the ALEXIS satellite and so far of unknown nature, as well as X-ray transients upon alert from SAX, RXTE, BATSE, HETE-2, INTEGRAL, MOXE, etc.

In the routine mode we plan to observe systematically several late type flare stars in order to test our ability to detect optical transients. Our programme includes also the detection of supernovae, symbiotic stars, asteroids and comets.

TABLE 1. Summary of TAROT technical data

Aperture	25cm, f/3.3
Field of view	2 x 2 degrees
Optical resolution	20 microns
Maximum time to slew to target	3 seconds (180 degrees)
Maximum slew speed	120 deg/s
Tracking speed	Adjustable α and δ, range from 0 to 60 deg/s
Maximum acceptable wind speed	80 km/h
Mount type	Equatorial
CCD size	2048 x 2048, 3 x 3 cm
CCD pixel size	15 x 15 microns
CCD readout noise	≤ 10 e$^-$
CCD readout time (imaging mode)	10 seconds
Filter wheel	6 positions (B, V, R, I, B+V, R+I, Blank)
Limiting V magnitude	17 @ 10σ in 10s, 19 in 1 min.
Typical integration time (alert)	20 seconds
Single exposure maximum integration time	5 minutes

The detailed program of TAROT is currently being elaborated and will be made available through our server and in later publications.

TECHNICAL DESIGN

The actual design of TAROT is summarized in Table 1. As can be inferred from the table, the goal of observing GRBs while they are active has driven the design. We list below some technical features.

- Mechanics: The mechanical design has been extensively studied in order to ensure the stability and the reliability of the telescope. The requirement was that TAROT should be able to track (without a guiding star) an object for at least 5 minutes, without any noticeable displacement on the CCD camera. Also, TAROT will accommodate wind speeds as high as 80 km/h (50 mph). The behaviour of the mount has been simulated to ensure that no vibrations are generated during the acceleration/deceleration phase.

- Motors and controls: The drives have been chosen in order to accomodate the large accelerations needed by TAROT. They will be able to make a move to any point in less than 3 seconds, meaning a maximum speed of 120 degrees/second., and accelerations as large as 100 deg/s^2. For simplicity, we decided to use the same motors for the declination and right ascension axis. All drives, including the focus and filter wheel mechanisms, are controlled from the telescope control software via a single PC card. Extensive protection of all electric and electronic parts is used against lightning.

- Filter wheel: We use a custom designed filter wheel with 6 positions. In addition to a transparent position, a set of standard Cousins B,V, I filters will be used, and two wide band filters, of transmission approximating the overall band pass of the B+V and R+I filters.

SOFTWARE

The software is one of the most sensitive parts of TAROT, since it should run in complete autonomy. The interfaces for the alerts and routine observations will use the Web, e-mail, and socket processes (for GCN/BACODINE). In addition, a local interface will be available, mainly for testing and debugging purposes. Our objective is that the telescope operates unmanned for periods as long as 3 months. Hence the control program will be responsible for night operations, day/night transition, calibrations, focusing, etc. This software will take into account the data from the environmental sensors to decide what operation to perform, and will run the telescope accordingly. Routine operations can be interrupted at any time to process an alert. In addition the control software will perform general tasks such as housekeeping, logging, etc.

Routine observations, and follow-up alert observations will be scheduled through particular software called *Majordome*. This software implements several algorithms in order to ensure a maximum efficiency of the observations. Objects should be observed at minimal airmass (unless there are other constraints), and the number of possible observations should be maximized, according to various parameters such as the Moon, user constraints, observation types (periodic, repeated, time tagged, etc.) and priorities. If an alerts occurs, the routine program is interrupted, and the alert processed according to a predefined sequence. The alert modifies in turn the input of *Majordome* in order to introduce follow-up observations.

We began to design a module to process automatically the data taken by TAROT. Our decision was based on the fact that TAROT will produce an average of 3 Gb per night, and on the necessity to react quickly after an alert. This software will produce a list of sources detected in the image, together with their characteristics (photometry, spatial extension, apparent motion, etc.), will compare each object with the TAROT database (whenever possible), and with other available catalogues to search for a possible variability, or change in properties, or to detect candidate new objects. In order to ensure their nature, each observation will be done twice.

In addition to the above mentioned routine and alert modes, TAROT will be able to scan the sky according to two modes: in imaging mode we take a normal 2K x 2K image, and in scanning mode the telescope scans a wide area, while the CCD is read-out continuously. This later mode will be mainly used for BACODINE/BATSE alerts. In this mode a typical error box is scanned in less than 5 minutes. The scanning mode may be used also to build quickly a first database of TAROT objects, to a limiting magnitude of 17 (V).

CURRENT SCHEDULE

The mechanics and the optics of TAROT have been delivered and integrated in September 1997 together with the drives. The software is currently being integrated and tested in the lab, and the optics will soon be submitted to interferometric measurements.

This fall (1997), the telescope will be moved to its final location, the "Plateau du Calern", 1200 m above sea level and the French Riviera. It will be installed in a building with a fully retractable roof, which has been recently refurbished in order to ensure maximum sky coverage.

After that the telescope will enter in an extensive test period (mechanics, software, security checks, optics, scientific validation, etc.). During this period we hope to be able to receive alerts at least through the GCN network. Routine scientific observations and automatic image processing should start running during the second semester of 1998.

CONCLUSION

Though its dimensions are rather modest, TAROT will be a very efficient instrument, optimized for its prime objective, the detection of high energy transients. Given that 5 seconds are needed to obtain the coordinate information from BATSE/BACODINE or HETE-2, TAROT will be able to get data from the source less than 8 seconds after the burst onset, while most sources are still active, and to eventually detect the internal shock from the GRB fireball. TAROT will be able also to estimate the background of transient events over the sky, to detect putative "optical GRBs", and to address a wide range of secondary objectives. Its schedule is well in accordance with BATSE, SAX, RXTE and HETE-2 satellites.

ACKNOWLEDGEMENTS

The TAROT project is funded by the Centre National de la Recherche Scientifique (CNRS / INSU) in France, and by the Carlsberg Fondation in Denmark.

REFERENCES

1. Mészáros, P., and Rees, M., *ApJ* **432**, 181 (1994).
2. Costa, E., et al., *Nature* **387**, 783 (1997).
3. Piro, L., et al., IAUC 6656 (1997).
4. Wijers, R.A.M.J., Rees, M.J., and Mészáros, P., *MNRAS* **288**, L51 (1997).
5. Mészáros, P., and Rees, M.J., *ApJ* **476**, 232 (1997).
6. Rhoads, J.E., *ApJ* **487**, L1 (1997).

EN: Real-Time Optical Data for GRBs

René Hudec, Zdeněk Ceplecha, Pavel Spurný, Jan Florián, Aleš Kovář, Jaroslav Boček and Jiří Borovička

Astronomical Institute of Czech Academy of Science, Observatory Ondřejov, 251 65 Ondřejov, Czech Republic

Abstract. Despite many efforts in rapid follow-up optical observations of GRBs, optical real-time data as well as pre-burst data are still rarely available. We report on a system of 11 Czech Stations of the EN (European Fireball Network) providing extended sky monitoring. The stations are equipped with fish-eye lenses imaging the whole visible sky hemisphere on photographic emulsion. This allows a large fraction of GRBs to be observed (so far more than 90) with limiting magnitudes between 7 and 11 at times before, during, and after gamma–ray triggers.

INTRODUCTION

Despite many efforts in rapid follow-up optical observations of GRBs, real-time data as well as pre-burst data are still rarely available. The most rapid detailed optical observations of GRBs are still delayed by 3 hrs or more. On the other hand, there are hypotheses that the optical emission of GRBs may peak within 3 hrs after a GRB, or even that there may be optical emission preceding GRBs. Providing of real-time and pre-burst data (sky monitoring) is hence extremely important.

We report on a system of 11 Czech stations of the EN (European Fireball Network) providing extended sky monitoring. The stations are equipped with fish-eye lenses imaging the whole visible sky hemisphere on photographic emulsion. This allows a large fraction of GRBs to be observed (so far more than 90) with limiting magnitudes between 7 and 11 at times before, during and after the GRB triggers. This still represents the faintest real-time and pre-burst limits of GRB optical emission.

THE NETWORK

All the 11 stations in the Czech Republic are equipped with the all–sky detectors with parameters listed below. The system is operated every cloudless night.

Parameters

- Optics: Fish-Eye Objective F-Distagon 3.5/30.

- Detector: Planfilm FOMAPAN 400 ASA or 100 ASA (panchromatic emulsion) 90 x 120 mm, sky diameter 80 mm.

- Typical exposure time: 3 hrs for guided cameras, whole night for fixed cameras.

- 2 stations equipped with guided and fixed cameras.

- 9 stations equipped with fixed cameras.

- Sensitivity for brief 1 sec triggers 2-3 mag, for stars up to mag 11.

- Response limited to the red light above 400 nm.

Preferences

The large number of stations, the large sky coverage as well as the prolongated exposure times represent a unique system for obtaining real–time optical data for GRBs with:

- Large sky coverage (full visible hemisphere).

- Large fraction of observation time: 2400 to 6000 sr-h for one station/year.

- Multiplicity of data to eliminate background triggers easily.

- Classification of detected triggers by parallax.

- Simultaneous and pre-burst optical data (limits) for GRBs.

RESULTS

1. No optical emission above mag 5 (1 sec duration assumed) or mag 13 (full exposure time) or $L_g/L_o \leq$ 100–300 has been detected for a few GRBs. The faintest limit (320) exists for GRB830313 [2].

2. No optical emission above magnitudes 0–3 (1 s duration assumed) or 4-11 (full exposure time) or $L_g/L_o \leq$ 0.1–10 has been detected for many (90) GRBs.

3. Optical brightening has been found $\sim$7 h after GRB790929 inside its error box, V=8.8 mag [1] coinciding with star HDE249119.

SAX Related Results

There are no direct simultaneously taken plates for the precisely localized GRBs detected by the BeppoSAX satellite, but there are plates taken just a few hours before and/or after. In the following we briefly list the obtained results.

GRB960720 11:37 UT

This GRB occurred at daytime in Central Europe; however, there are plates available for nights before and after:
960719/20 7 plates 20:40 - 01:27 UT lim mag 4-11 for stars,
960720/21 8 plates 20:45 - 01:30 UT lim mag 4-11 for stars.
Results: No optical activity 10 h before the GRB with limiting magnitude 11, nothing 9 h after the GRB with a limit of 11.

GRB970111 09:44 UT

The GRB occurred at daytime; plates are available for the following night.
970111/12 1 plate 16:45-20:15 UT lim mag 4.
Results: nothing 7 h after the GRB with a limit of 4.

GRB970228 02:28 UT

This GRB occured at night time but below the local horizon at the time of the event. Plates are available for the same night.
970227/28 2 plates 17:56-21:58 UT lim mag 4,
970227/28 9 plates 00:50-04:32 UT lim mag 4-10 but GRB below horizon,
970228/0301 23 plates 17:56-01:46 lim mag 4-11.
Results: Nothing 4.5 h before the GRB with lim mag 4, nothing 13.5 h after the GRB with lim mag 10.

GRB970508 21:25 UT

This burst occurred at nightime on the visible sky. Unfortunately, no plates are available due to bad weather.

CONCLUSIONS

The photographic data provided by the EN network still represent the only optical pre–burst data as well as simultaneous data for GRBs. Also for the time immediately following the GRBs, the EN network with limiting magnitudes of up to 11 still are among the most sensitive systems available.

The role of all-sky monitors remains important in optical analyses of GRBs. There are numerous new rapidly–moving systems for optical follow–up observations of GRBs in development, but none of them will be able to provide data before or during the beginning of the bursts.

No simultaneously taken plates are available for the recent GRBs detected by BeppoSAX, due to day time of most of the bursts as well as observing conditions. On the plates taken before and after the gamma–ray events, no optical activity exceeding mag 11 has been revealed.

For other bursts, limits of up to mag 5 (1 s duration assumed) or mag 13 (full exposure time) or $L_g/L_o \leq 100$–300 have been obtained. For majority of the GRBs with simultaneously taken EN plates (around 90), these limits are magnitudes 0–3 (1 s duration assumed) or 4-11 (full exposure time) corresponding to $L_g/L_o \leq 0.1$–10. Optical brightening has been found ~ 7 h after GRB790929 inside its error box.

ACKNOWLEDGEMENTS

The investigations of GRBs on EN patrol plates are supported by The Ministry for Education and Youth of the Czech Republic, Project No. ES002/1996 (CEI/KONTAKT).

REFERENCES

1. Borovička, J., Hudec, R. and Dědoch, *A&A* **258**, 379 (1992).
2. Hudec, R., *Astroph. Letters and Communications* **28**, 359 (1993).

BART:
Burst Alert Robotic Telescope

Jan Soldán[1], René Hudec[1], Miloš Němček[2], Tomáš Rezek[1]

[1] *Astronomical Institute, Academy of Science of the Czech Republic*
251 65 Ondřejov, Czech Republic,
[2] *Technical University Ostrava,*
třída 28. listopadu, Czech Republic

Abstract. The design and current progress in development of the Ondřejov Burst Alert Robotic Telescope (**BART**) are discussed. The telescope system will provide rapid (< 1 min) optical follow–up observations of GRB positions provided by satellites down to magnitude 13 (over a FOV of 20 degr.) and/or 19 (over FOV of 0.3 degr.).

The telescope is able to receive e–mail messages about GRB events, provided by the Compton Gamma Ray Observatory, project BACODINE), analyse them and automatically investigate their positions at selected optical wavelengths. We propose to adapt such a system also for the International Gamma Ray Astrophysics Laboratory (INTEGRAL) in the future.

INSTRUMENT DESCRIPTION

- Schmidt–Cassegrain Telescope (SCT).
- NF CCD Camera in primary focus of the SCT.
- WF CCD Camera on the top of the main telescope.
- Motorized Color Filter Wheel for NF CCD Camera.
- Pentium PC with Windows NT operation system.

Figure 1 shows the main parts of the system.

SOFTWARE SOLUTION

- Computer language: Microsoft Visual C++, version 5.0.
- Technology: OOP – Object Oriented Programming technology.

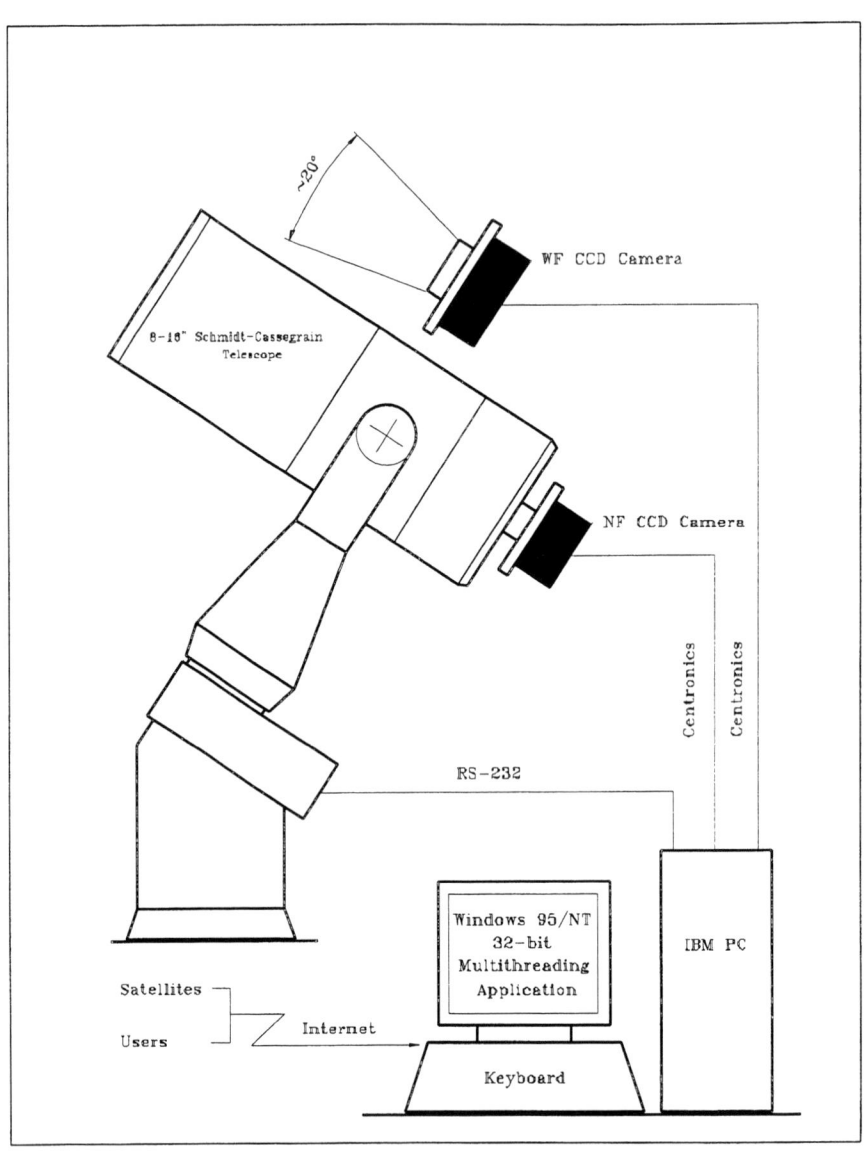

FIGURE 1. The Burst Alert Robotic telescope

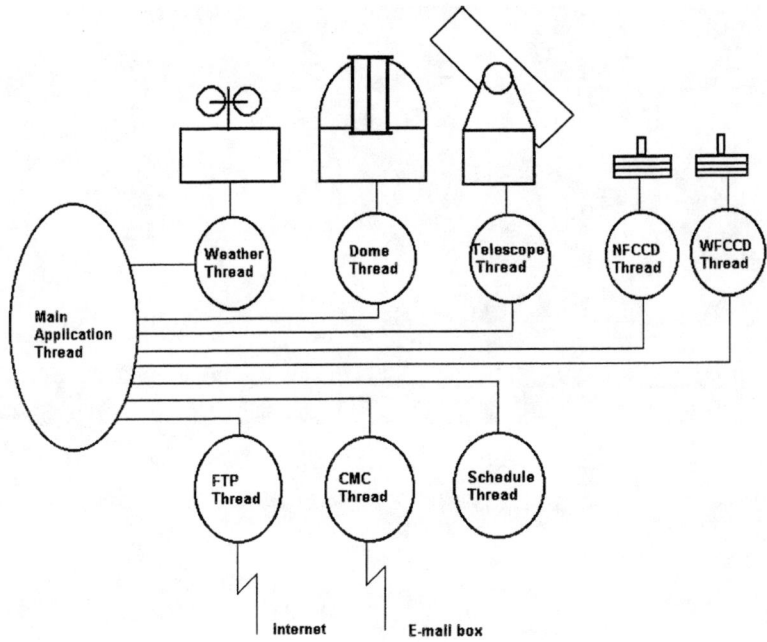

FIGURE 2. Structure of the threads

- Application: 32-bit multithreaded application running under NT workstation, version 4.0.

- NT kernel driver for controlling both CCD cameras.

The software is based on the several *user–interface threads*. Figure 2 shows the design of the cooperative threads.

CURRENT STATE OF THE PROJECT

1. *Mechanical parts – telescope and CCD cameras.*
   The telescope system was assembled and equipped with necessary accessories. The CCD cameras and control computer were also selected.

2. *Software*
   The structure of the software was proposed and its first prototype was implemented. The 32-bit multithreading approach, object-oriented environment and C++ was chosen.

   The authors have implemented several threads, namely, *Main application thread, Internet CMC thread, Telescope thread and WFCCD and NFCCD*

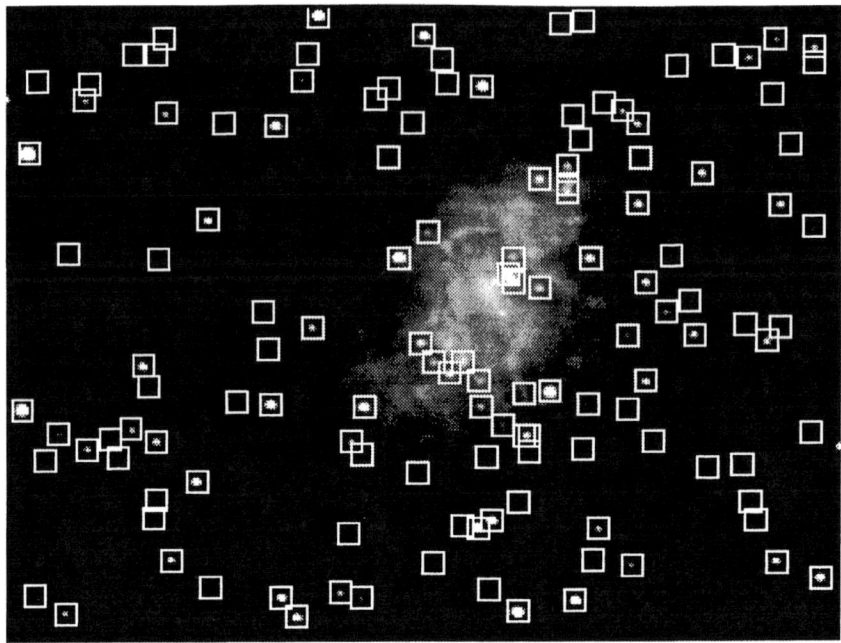

FIGURE 3. Crab nebula processed with our adaptive search algorithm

threads. The other threads *Weather thread, Dome thread, Schedule thread and FTP thread* are under development.

The algorithms for the image data preprocessing was also developed (dark frame subtraction, flat-field correction, artifact removing and sky background estimation). The adaptive algorithm which searches for star objects was developed and tested. Figure 3 shows results of the star search algorithm.

ACKNOWLEDGEMENTS

The project is supported by the Ministry of Education and Youth of the Czech Republic, projects ES036/1996 and ES02/1996.

OMC Camera Experiment for INTEGRAL and Search for Compton GRO BATSE LOCBURST Optical Transients

Tomáš Rezek[1], René Hudec[1], Filip Hroch[2], Jan Soldán[1], Miguel Mas–Hesse[3] and Alvaro Giménez[3]

[1] *Astronomical Institute of Czech Academy of Science, Observatory Ondřejov, 251 65 Ondřejov, Czech Republic,*
[2] *Department of Theoretical Physics and Astrophysics, Faculty of Science, Masaryk University, 611 57 Brno, Czech Republic,*
[3] *Laboratorio de Astrofísica Espacial y Física Fundamental, Villafranca del Castillo, P.O. BOX 50.727, E-28080, Madrid, Spain*

Abstract. The test camera of the Optical Monitoring Camera (OMC) experiment for INTEGRAL spacecraft achieving an angular pixel size of 18 arcsec and a field of view $7.5° \times 5.1°$ has been succesfully developed and tested at the Astronomical Institute Ondřejov.

The test camera is able to provide imaging down to 15 mag over the whole field of view within one exposure of 300 seconds. Although developed primarily to test the OMC performance and help with software developement, this device is ideally suited for use as a ground–based camera for sites where Compton Gamma Ray Observatory BATSE Locburst triggers are followed-up in the optical waveband and also for wide-field sky monitoring in general. The low cost of this camera makes it possible to duplicate the system at a number of observing sites.

A chart and a corresponding CCD–image for the BACODINE Locburst Position 6368 taken with the OMC test camera at Ondřejov observatory are also presented. The image taken 18 hours after the trigger was computer–blinked with the frame taken 30 days later. No optical activity has been found down to 13.5 mag.

OMC TEST CAMERA

Optical Monitoring Camera [1] is one of the four scientific instruments of the INTEGRAL (International Gamma-Ray Astrophysics Laboratory) satellite, selected by ESA as the medium-size mission to be launched in 2001.

The OMC Test Camera, developed in 1997 at the Astronomical Institute Ondřejov to test its performance, imaging capabilities and the software

TABLE 1. OMC Test Camera parameters

CCD Chip	Kodak KAF1600, 1530×1020 pixels
	pixel size 9 × 9 μm
CCD Camera ST8	A/D resolution 16 bits
	Read-Out Noise 15 e$^-$ rms
	Full Well Capacity 80 ke$^-$
Wide Field Lens	Focal Length 109 mm
	Aperture 82.5 mm
	f/1.32
Field Of View	7.6° × 5.1°
Real Pixel Size	17.9 arcsec
Detection Limit	15.1 mag in V filter (one exposure of 300 sec)

for image processing, is designed to have similar parameters as the original OMC instrument. The most important parameters are the angular pixel size, the number of pixels in a CCD image and the aperture of the optics. The parameters of the OMC test camera are specified in Table 1.

Components necessary to complete the design described here are commercially available: $7000 for the ST8 CCD camera [2], $300 for the wide field lens made by Meopta Přerov. This cost makes it possible to duplicate the system at a number of observing sites.

BATSE LOCBURST POSITIONS

GCN (The GRB Coordinates Network — formerly BACODINE) [3] was modified in May 1997 to incorporate the distribution of BATSE GRB locations using the Huntsville LOCBURST Processing. These locations are derived for bursts brighter than 1000 ± 20 counts, which yelds approximately one LOCBURST Location Notice per week. The time delay, including 5 minutes of post-trigger data collection and processing in Huntsville, is 15 – 35 minutes. The error boxes are usually less than 2° statistical with a 2° systematic contribution. Latter we present one of typical GRB–related observations as an example.

POST–TRIGGER OPTICAL OBSERVATION OF BATSE TRIGGER #6368

The reduced LOCBURST Location Notice for Trigger #6368 with the information about the time and position of the burst can be found in Table 2.

The optical follow-up observations of Trigger #6368 were performed with the OMC test camera during the night of 07 Sep 1997, approximately 18 hours after the BATSE trigger. The limiting magnitude of the image, taken with a 180 seconds exposure, is about 13.6 mag in V filter.

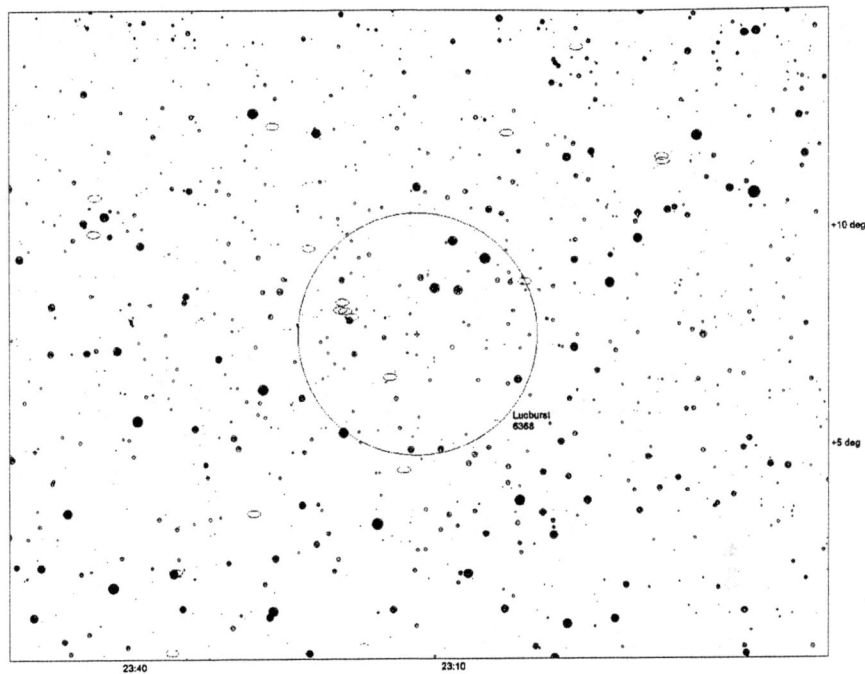

FIGURE 1. Chart for the position of BATSE/LOCBURST Trigger #6368. The error box is marked with the circle.

We present both the chart (Figure 1) for the position of the GRB and the real image (Figure 2) taken with OMC test camera. The large circle is the BATSE/LOCBURST error box.

The image has been computer–blinked with the frame obtained by the same instrument 30 days after the trigger. No optical activity has been found down to 13.5 mag in V band.

CONCLUSIONS

If we compare the size of the LOCBURST error box with the field of view of the OMC Test Camera we find that the OMC Test Camera is an ideal instrument for ground based optical follow-up monitoring of GRBs, especially LOCBURST locations. The camera will now be used at AIO to collect post-trigger and in the near future (see [4]) also real-time optical data for GRBs and it will be duplicated at other stations worldwide. Any kind of wide-field sky monitoring would be possible with this instrument as well.

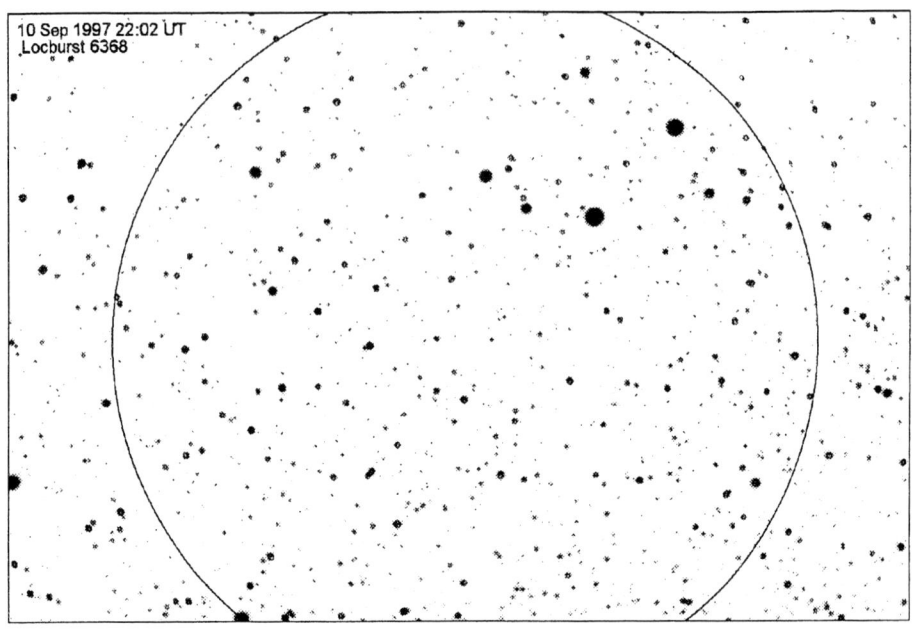

FIGURE 2. Image of the position of BATSE/LOCBURST Trigger #6368, taken with OMC Test camera. Exposure time 180 seconds. The error box is marked with the circle.

TABLE 2. LOCBURST Notice for Trigger #6368

TITLE:	LOCBURST BURST POSITION
NOTICE_DATE:	Sun 07 Sep 97 04:07:03 UT
NOTICE_TYPE:	Hunts_Locburst
TRIGGER_NUM:	6368
GRB_RA:	348.36d {+23h 13m 26s} (J2000),
	348.33d {+23h 13m 19s} (current),
GRB_DEC:	+7.67d {+07d 40' 21"} (J2000),
	+7.66d {+07d 39' 36"} (current),
GRB_ERROR:	2.8 [deg radius, statistical only]
GRB_INTEN:	2293 [cnts] Peak=2293 [cnts/sec]
GRB_TIME:	11738.91 SOD {03:15:38.91} UT
GRB_DATE:	10698 TJD; 250 DOY; 97/09/07
SUN_DIST:	166.02 [deg]
COMMENTS:	Locburst Coordinates.
COMMENTS:	Definite GRB.

ACKNOWLEDGEMENTS

This work is supported by the Ministry of Education and Youth of the Czech Republic, project ES0036/1996 (ESA/KONTAKT).

REFERENCES

1. Hesse, M. M., *Optical Monitoring Camera (Science Performance Report)*, LAEFF-INTA (1996).
2. Santa Barbara Instrument Group, *CCD Camera Operating Manual for Model ST7 and ST8*, SBIG, Santa Barbara (1996).
3. Barthelmy, S., *The GRB Coordinates Network (CGN): A Status Report*, these proceedings.
4. Soldán, J., et al., *BART – Burst Alert Robotic Telescope*, these proceedings.

Optical Transient Monitor

Martin Bernas[1], Petr Páta[1], René Hudec[2], Jan Soldán[2],
Tomáš Rezek[2], Alberto J. Castro-Tirado[3]

[1] *Faculty of Electrical Engineering, Department of Radioelectronics*
Czech Technical University Prague, Czech Republic,
e-mail: bernas@feld.cvut.cz, pata@feld.cvut.cz
[2] *Astronomical Institute of Czech Academy of Science, Ondřejov, Czech Republic,*
e-mail: rhudec@asu.cas.cz, jsoldan@asu.cas.cz, rezek@asu.cas.cz
[3] *Laboratorio de Astrofisica Espacial y Fisica Fundamental, Madrid, Spain,*
e-mail: ajct@laeff.esa.es

Abstract. Although there are several optical GRB follow-up systems in operation and/or in development, some of them with a very short response time, they will never be able to provide true simultaneous (no delay) and pre-burst optical data for GRBs. We report on the development and tests of a monitoring experiment expected to be put into test operation in 1998. The system should detect Optical Transients down to mag 6–7 (few seconds duration assumed) over a wide field of view.

The system is based on the double CCD wide-field cameras ST8. For the real time evaluation of the signal from both cameras, two TMS 320C40 processors are used. Using two channels differing in spectral sensitivity and processing of temporal sequence of images allows us to eliminate man-made objects and defects of the CCD electronics. The system is controlled by a standard PC computer.

INTRODUCTION

Probably the first visible counterpart of a GRB was detected on 1997 February 28 less than 21 hours after the γ–burst [4]. A week after that discovery astronomers at New Technology Telescope and the Keck Telescope identified an extended source at the location of the suspected GRB. The fading of an optical counterpart was tracked by the Hubble Space Telescope too. Consequently, this speeded up the development of ground-based systems capable of regular monitoring and checking suspicious sky fields during the year. In 1995 we started building a system OTM (Optical Transient Monitor) which should be able to detect transient objects.

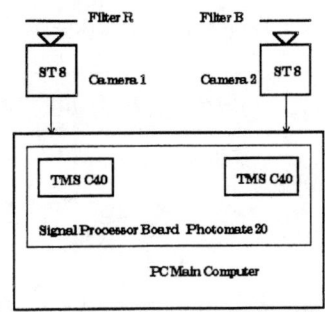

FIGURE 1. The hardware of the system.

TABLE 1. The size of field of view of CCD cameras at focal lengths $F = 50$ mm and $F = 25$ mm for used resolution. Bining mode $1:1 \times 1:1$.

Size of CCD	Resolution	Field of view	
		$F = 50$ mm	$F = 25$ mm
Full	1530×1020	$15.7 \times 10.5°$	$34.6 \times 24.7°$
1/4	765×510	$7.8 \times 5.2°$	$19.0 \times 12.9°$
1/9	510×340	$5.2 \times 3.5°$	$13.0 \times 8.7°$
ST6 mode	375×242	$3.8 \times 2.5°$	$9.6 \times 6.2°$

BASIC IDEA OF OTM SYSTEM

The basic idea of OTM is permanent watching of a wide field of the sky and looking for new objects with parameters expected for optical transients of GRBs.

We assume these basic parameters are: a brief change of the brightness (up to 7–6 mag on few minutes only), a stable position on the sky, and a defined spectral radiation in the optical band (different from spectral radiation of satelites and other false objects). The real time evaluation of all these parameters allows us to eliminate most false triggers and consequently to work on the automatic station, reducing manpower to a minimum.

DESCRIPTION OF THE HARDWARE

The block diagram of the hardware is in Figure 1. The hardware of the system consists of two CCD cameras type ST8 (Santa Barbara Instruments Group, USA), the Signal Processor Board Photomate 20 (Daimler Benz Aerospace, FRG), which is based on two fast processors type TMS 320C40 and a PC compatible main controller.

Optical lenses are wide-field Pentacon lenses 50/1.8 or shorter (25mm). Photometrical filters correspond to the Johnson and Morgan UBVRI extended system.

TABLE 2. Transfer time for images from camera ST8 for different binning modes.

Bining	Resolution	Transfer time
$1:1 \times 1:1$	1530×1020	58 s
$2:1 \times 2:1$	765×510	19 s
$3:1 \times 3:1$	510×340	10 s
ST6 mode	375×242	7 s

The two CCD cameras ST8 are mounted on the base plate which allows orienting both cameras so that their optical axes are parallel. The imaging chip is KAF 1600 (non–ABG) with the following basic parameters: resolution 1530×1020 pixels, active area of $9 \times 9\,\mu m$ and an image can be electronically binned in modes 2:1, 3:1. Both detectors are cooled.

The camera ST8 produces 3.12 MB of data per image in the full resolution mode. For the real time evaluation of the temporal image sequence, a fast signal processor is therefore necessary. We decided to use the Signal Processor Board Photomate 20 utilizing two TMS 320C40 signal processors. Each processor evaluates images from one camera with the possibility of cross-correlating each other. The detailed description of the architecture of the board Photomate 20 and processor TMS 320C40 can be found in [3,6]. The basic parameters of the system hardware are given in Table 1 and Table 2.

SYSTEM SOFTWARE

The software of the system consists of the software running on the main PC computer, which is responsible for control of the whole system, and the image processing software running on both processors of the Photomate board. Each processor on the Photomate board evaluates image sequences from one camera.

Figure 2 presents the schema of the working cycle, which consists of a parallel exposure on both cameras, a transfer of the images to the PC, and transfer of the images to both CPUs on the Photomate board. Then the image processing starts on both CPUs along with the start of a new exposure.

The time duration of each section is presented in Table 3. Because the time duration of image processing is actually shorter then the transfer from both cameras and exposure, the duration of the whole working cycle can be expressed as (see Figure 2) $T_\text{w} = T_\text{exp} + T_{t1} + T_{t2}$. Consequently, the time is about 142 seconds (see Table 3). This time bounds the class of detected objects, because there is little possibility to detect faster variable sources.

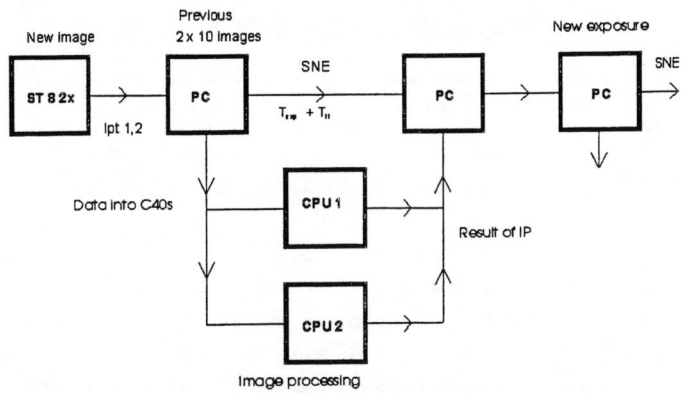

FIGURE 2. Scheme of working cycle. SNE — Start of New Exposure.

Image processing step by step

The image processing consists of the following routines:
1) Compensation of the image using dark image and flat field.
2) Counting of the background value using histogram and iteration method.
3) Identification and classification of objects.
4) Image photometry (using aperture photometry [1]). Compiling of photometric table of found stars and others objects.
5) Comparison of the table with previous tables from the same camera. Construction of light curves for all objects in the image.
a) Pseudocorrelation method – finding of transformation matrix, translation vector.
b) Differential photometry – step with automatic selection of standard stars.
c) Searching for new objects based on change of brightness.
6) Comparison of the actual table with the actual table from the second camera which differs in spectral sensitivity, elimination of background (false) images such as objects with suspicious color index and/or detected by one camera only.

CONCLUSION

We have briefly introduced our system for real-time detection and evaluation of optical transients. The whole working cycle and proper image processing was presented with characteristic duration times. These values restrict detected objects. Our system isn't able to discover optical transients with duration shorter than $T < 142$ sec. Short exposition time restricts the brightness of detected optical transients. The system should detect objects down to mag 6–7 over a wide field of view. For prolonged exposures, the limiting magnitude should reach mag 13 with decreased time resolution.

TABLE 3. Proper duration of every section of the working cycle.

Name of time	Used symbol	Time
Exposure	T_{exp}	10 − −20 s
Transfer from cameras	T_{t1}	116 s (both images)
Transfer to Photomate 20	T_{t2}	16 s (both images)
Image processing	T_{ip}	40 s

Technical properties of the OTM allow using the system for the detection of new variable stars and transient (flaring) objects. The OTM will be placed also at two observatories in Spain as part of the BOOTES system [2]. Both experiments (OTM and BOOTES) will also test the procedures and performance of the INTEGRAL onboard experiment OMC as well as provide rates and parameters for objects expected to be seen by the OMC.

ACKNOWLEDGEMENTS

The project is supported by the Ministry of Education and Youth of the Czech Republic, projects ES002/1996 (CEI/KONTAKT) and ES0036/1996 (ESA/KONTAKT).

REFERENCES

1. Buil, Ch., *CCD Astronomy*, Wilhman - Bell, Inc., Richmond Virginia (1991).
2. Castro-Tirado, A., et al., *4th Huntsville Gamma-Ray Burst Symposium*, **I-P12** (1997).
3. DASA, *Handbuch Photomate 20* (1994).
4. van Paradijs, J. et al., *Nature*, **386**, 686 (1997).
5. Páta, P., Bernas, M. *Workshop 97, CTU Prague*, part III, 941 (1997).
6. Texas Instruments Inc., *TMS320C4x Technical Brief, Digital Signal Processing Products* (1991).

Optical Imaging of Gamma-Ray Bursts with the LONEOS Telescope

R. M. Wagner[1], E. Bowell[2], K. H. Cook[3], S. B. Howell[4], B. W. Koehn[2], C. R. Shrader[5], S. G. Starrfield[6], and C. W. Stubbs[7]

[1] *Department of Astronomy, Ohio State University; mailing address: Lowell Observatory, 1400 West Mars Hill Road, Flagstaff, AZ 86001*
[2] *Lowell Observatory, 1400 West Mars Hill Road, Flagstaff, AZ 86001*
[3] *Lawrence Livermore National Laboratory, V Division, MS L-401, PO Box 808, Livermore, CA 94550*
[4] *Department of Physics and Astronomy, University of Wyoming, PO Box 3905, University Station, Laramie, WY 82071*
[5] *NASA/GSFC, Laboratory for High Energy Astrophysics, Code 660.1, Greenbelt, MD 20771*
[6] *Department of Physics and Astronomy, Arizona State University, Tempe, AZ 85287-1504*
[7] *Department of Astronomy, University of Washington, Seattle, WA 98195-1580*

Abstract. The optical identification of gamma-ray bursts discovered and localized by BACODINE/LOCBURST using the Lowell Observatory Near-Earth Object Search (LONEOS) 58-cm Schmidt-type telescope and mosaic CCD camera is described. In its final form, LONEOS images 10 square degrees of the sky ($3.2° \times 3.2°$) to ~22nd mag (2σ) in a 5 minute integration. Identification of optical transients will be based on variability by comparison with subsequent images or previous scans of the region. To date, optical images have been obtained of three BATSE triggers processed by LOCBURST for development and evaluation purposes.

INTRODUCTION

The recent identification of two gamma-ray bursts (GRBs) with optical transients based on the initial discovery and subsequent localization by the BeppoSAX experiment have revolutionized the field [1-3]. These observations have established, at least in the case of GRB970508, that some of the GRBs are cosmological in nature [3]. They have also shown that the optical transients (OTs) associated with GRBs are unlikely to be ~8th-12th mag events lasting a few seconds, but ~20th mag events lasting a few days to weeks. While the success of the BeppoSAX experiment has been nothing short of phenomenal in this regard, it is only capable of discovering a few bursts per year. A full description of the types and physical nature of the bursts is a statistical question requiring many counterparts studied in detail at optical, radio, and X-ray wavelengths. Only the BATSE experiment has been successful in identifying a statistically large enough sample of bursts.

We have initiated a program to identify OTs of GRBs discovered by BATSE and reported by BACODINE using the Lowell Observatory Near-Earth Object Search (LONEOS) Schmidt telescope and mosaic CCD camera. LONEOS is a NASA funded project (E. Bowell, PI) to undertake a systematic large-scale observational search to increase the known population of near-Earth asteroids and comets that might represent a hazard to Earth. The field-of-view (FoV) of the LONEOS system is well matched with respect to BACODINE/LOCBURST positions and typical error boxes. One of the major scientific objectives of our program is the measurement of the optical light curve of an event early in its evolution. The large FoV of the LONEOS telescope, its deep limiting magnitude, a dark-sky site, and the availability of considerable observing time make it an ideal platform to search for either the optical counterparts of GRBs or other optical transients.

LONEOS INSTRUMENTATION

The LONEOS telescope consists of a completely refurbished 58-cm Schmidt telescope originally owned by the Ohio Wesleyan and Ohio State Universities and relocated to the Lowell Observatory Anderson Mesa dark-sky site near Flagstaff. The diameter of the corrector plate is 15 in., but it will be replaced in early 1998 with a 24 in. plate resulting in a deeper limiting magnitude for the same exposure time. The telescope can slew at a rate of $> 3°$ per second and point with an accuracy of better than an arcminute. Currently, the detector consists of a thermoelectrically cooled Loral $4k \times 2k$ pixel front-illuminated CCD with proceeding reimaging optics controlled by a Sun Ultra 1/140e computer. The field of view is 5.0 deg^2 ($3.2° \times 1.6°$) at an image scale of 2.8 arcsec/pixel.

All components of the telescope and facility are in place, including a new observing floor, computerized control, automatic dome rotation control, a temperature-regulated computer room, a control room for operation, and a T1 Internet link to the outside world for rapid communication. Images are subsequently transferred to various analysis computers, which emphasize different processing steps and individual science programs. First light for the LONEOS telescope was achieved in late July 1997. The telescope and camera system are currently undergoing testing and characterization, but routine stare-mode science observations are in progress, including imaging of GRBs and other potential optical transients. In early 1998, a second Loral CCD will join the current CCD. The effective array will consist of $4k \times 4k$ pixels covering 10 deg^2 ($3.2° \times 3.2°$) on the sky at the same scale. The FoV and limiting magnitude of LONEOS (see below) are ideally suited to imaging BATSE GRB triggers which have been subsequently processed by LOCBURST.

PERFORMANCE

Given the parameters of the telescope with the 24 cm corrector plate, camera system, and sky, we can calculate the performance of the LONEOS telescope as a function of apparent magnitude. Images of GRBs with LONEOS will be obtained in the R-band to be consistent with other groups in the construction of light curves and to provide better contrast in bright time. We assume a dark sky with an average surface brightness in the R-band of 20.9 mag/arcsec^2. The algorithm used to calculate the performance of LONEOS has been used to successfully predict the performance of a wide range of direct imaging CCD systems and telescopes at both private and national facilities.

We find that the 2σ detection limit of our system occurs at $\sim$22nd mag, the 3σ detection limit is near 21.5 mag, and the 10σ detection limit is at 20th mag assuming a total integration time of 300 s. A comparison of these results with the observed optical light curves of GRB970228 and GRB970508 indicates that we could easily have imaged both of these OTs with the LONEOS system. In addition, recent tests based on images of stellar fields with the 15-in corrector indicate that we will be close to realizing the predicted performance after installation of the 24 in. corrector.

IMAGING GRBS WITH LONEOS

Upon notification of a GRB by BACODINE, we first ascertain its viewing constraints from Flagstaff. In general, we confine our observations to the dark of the Moon, declinations north of -25°, and galactic latitudes $|b| \geq 15°$. We will image events at the first opportunity if they satisfy our selection criteria and are available during forthcoming night. An event received during routine LONEOS nighttime operations will be imaged immediately, if it is accessible, or at the first opportunity.

Identification of OTs of GRBs will be based on variability to distinguish these OTs from other potential optical candidates such as variable stars and asteroids. This approach will require us to obtain frames of the same error box 24 to 48 hours after the acquisition of our first frame shortly after the discovery of a GRB. While the details of the identification algorithm are under development, several detection methods are applicable. First, we can perform simple blinking of the pairs of images and examine the results by eye in an effort to find a relatively bright counterpart. Second, either image can be compared against the Digital Sky Survey, the USNO-A1.0 catalog, or previous scans of the region in an effort to discover a counterpart. Third, the large FoV and depth of LONEOS images can be used in an archival fashion to identify an optical counterpart based on a subsequent GRB localization obtained by the RXTE/PCA. Finally, we can digitally identify all stellar objects in each pair of images down to the detection limit and perform digital aperture photometry or PSF fitting of each object to generate a catalog of positions and magnitudes. Identification of a potential optical counterpart will then depend upon an object at the same position exhibiting a statistically significant change in brightness with respect to a defined threshold.

Since first light we have imaged the fields of three GRBs for testing and evaluation purposes. The field surrounding GRB970828 and images of a subsequent weak burst discovered by BATSE, GRB970907, were both imaged on 1997 September 10 UT. These images covered a 3° × 3° field centered on the LOCBURST position. Unfortunately, an improperly figured corrector plate, an undersized shutter, and an electronic problem hindered both images. If LONEOS were in routine operation, GRB970907 would have been imaged within 2 minutes of our notification by BACODINE since it was in a dark sky at the time of the event.

After fixing these problems and improving the performance of the telescope over the following month, we obtained images (Fig. 1) of the field surrounding GRB971029 (trigger 6453). This burst met the preliminary criteria for an RXTE/PCA follow-up observation, but the viewing constraints for RXTE were not optimal and no X-ray observations were performed. Our first set of LONEOS images (Fig. 1, left panel) were obtained on 1997 October 29.277 and only 5.2 hours after the burst which occurred at 1997 October 29.06155. Our second set of images were obtained one night later, on 1997 October 30.356 (Fig. 1, right panel). Both epochs simultaneously cover a 3.1° × 1.2° region of the GRB error box centered on the LOCBURST position. The galactic latitude of this burst was only 7.4° and well below our imaging constraint, but in spite of crowding the frames have been useful for developing identification algorithms and astrometry techniques. Several potential optical counterparts of GRB971029 with $R \sim 20$ mag are present on these images, which include only part of the GRB error box.

FUTURE PLANS

In early 1998, the 15 in. corrector plate will be replaced with a new and properly figured 24 in. corrector plate. At that time, we expect to realize the calculated performance. Considerable software development is currently underway to develop algorithms to aid in the identification. It is anticipated that a fast and dedicated Sun Ultra 2 or Ultra 30 computer will be utilizing the software to process LONEOS optical images of GRB error boxes in an attempt to identify possible optical counterparts. We expect that LONEOS will be in full science operation by spring 1998. Until then, we will image GRB triggers that are available to us in an effort to adjust our identification techniques and discover additional optical counterparts of GRBs.

REFERENCES

1. van Paradijs, J., et al., *Nature* **386**, 686 (1997).
2. Djorgovski, S. G., et al., *Nature* **387**, 876 (1997).
3. Metzger, M. R., et al., *Nature* **387**, 879 (1997).

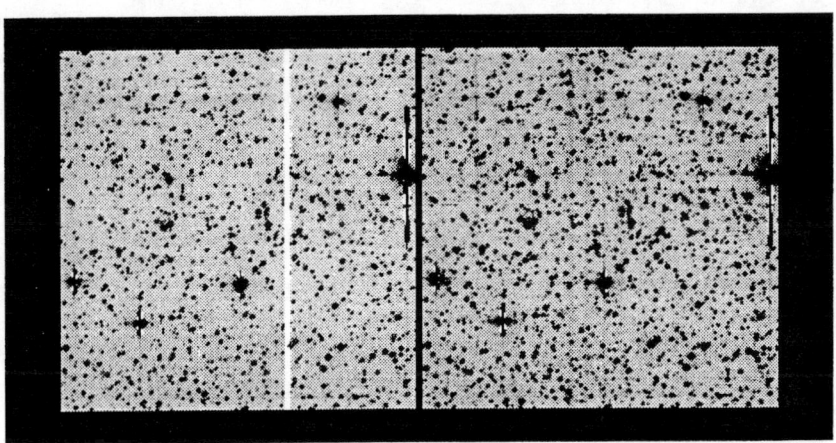

FIGURE 1. Images centered on the LOCBURST error box for GRB971029 (1997 October 29.06155) obtained on 1997 October 29.3 (left) and October 30.4 (right) with LONEOS. The October 29.3 image was obtained only 5.2 hours after the GRB. These images only show the inner square 24 arcmin of the original images which cover a FoV of $3.1° \times 1.2°$. The October 29th image was constructed from three disregistered 60 s exposures while the October 30th image was formed from five disregistered 60 s exposures. The faintest stars have $R \simeq 20.5$ mag.

The Status of the Burst Observer and Optical Transient Exploring System (BOOTES)

Alberto J. Castro-Tirado[1], Javier Gorosabel[1], René Hudec[2], Jan Soldán[2], Martin Bernas[3], Petr Pata[3] and T. Rezek[2]

[1] *Laboratorio de Astrofísica Espacial*
y Física Fundamental (LAEFF-INTA)
P.O. Box 50727, E-28080, Madrid, Spain
[2] *Astronomical Institute, Academy of Sciences of Czech Republic*
251 65 Ondrejov, Czech Republic
[3] *Prague Technical University, Department of Radioelectronics*
16627 Prague 6, Czech Republic

Abstract. The Burst Observer and Optical Transient Exploring System (BOOTES) is considered as a part of the preparations for the ESA's satellite project INTEGRAL, and is currently being developed in Spain, in collaboration with two Czech institutions. It will make use of two sets of wide-field cameras 240 km apart, and two robotic 30 cm telescopes. All the instruments will be placed in Southern Spain. It is expected that BOOTES will provide rapid follow-up observations of GRBs detected by BATSE, BeppoSAX, RossiXTE and future experiments, as well as clarify whether optical transients, related or not to GRBs, are indeed of cosmic origin.

BOOTES: THE CONCEPT

The *Burst Observer and Optical Transient Exploring System* (BOOTES) is expected to contribute significantly to the GRB and OT fields. The system consists of two parts: (i) two pairs of wide field cameras and (ii) two 0.30 m robotic telescopes. See Fig. 1.

The Wide Field Cameras

Commercial wide-field lenses (Nikkor, Japan) will be used, attached at two ST8 CCD cameras (produced by Santa Barbara Instruments Group, USA), similarly to the *Optical Transient Monitor* (OTM) that is already in operation at the Astronomical Institute in Ondrejov.

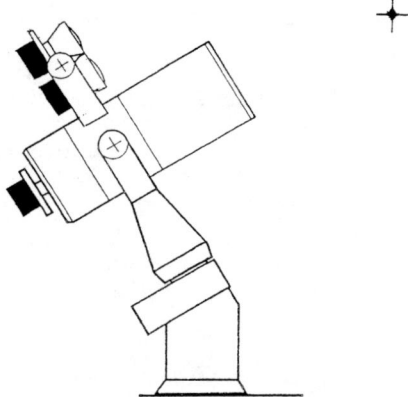

FIGURE 1. A sketch of the BOOTES system, showing the robotic telescope and the three CCD cameras (*filled black*): two ST-8 atop the telescope and one ST-7 at the Cassegrain focus.

TABLE 1. The BOOTES Wide Field Cameras.

Elements	Configuration 1	Configuration 2
Lenses	f/1.6 @ f = 6 mm	f/1.2 @ f = 50 mm
CCD	1534 x 1020 pixels	1534 x 1520 pixels
FOV	88° x 66°	16° x 11°
angular resolution	3.44'	0.63'
limiting magnitude[a]	10	13

[a] Estimated for an integration time of 10 s.

Each pair of the BOOTES wide-field cameras is mounted atop the 0.3-m LX 200 Meade telescope (Meade, USA), allowing long integrations of previously selected high galactic latitude regions. The four cameras will monitor the same region of the sky, both in the V and R-bands. At the proposed locations, it is expected to reach a limiting magnitude V $\sim$ 13 for an integration time of 10 s, and V $\sim$ 14.5 for 300 s (see Table 1).

It is very likely that one pair of cameras will be placed at El Arenosillo, a station owned by the Instituto Nacional de Técnica Aeroespacial (INTA), whereas the other set would be placed at the Estación Experimental de La Mayora, 240 km apart. The latter is run by the Consejo Superior de Investigaciones Científicas (CSIC).

The system is based on the digital signal processor board Photomate 20 Signal Processor Board (Daimler Benz Aerospace, Germany) which consists of two TMS320C40 processors [1]. The system, to be operated by commercial PCs, is planned to work most of the time as discussed in [2]. The frames taken at each station at the very beginning of the night will be loaded into memory (the *primary* frames). During the rest of the night, sucessive frames (*secondary* frames) will be compared with the primary frame. Hence bad pixels and plane flashes can be ruled

TABLE 2. The BOOTES Robotic Telescope.

Elements	Configuration
Telescope	D = 0.30 m @ f/6.3
CCD	762 x 510 pixels
FOV	18' × 12'
angular resolution	1.5"
limiting magnitude[a]	18.5

[a] Estimated for an integration time of 300 s.

out.

If a real flash is detected in the secondary frames, the coordinates of the flashing object and the images themselves will be transferred to LAEFF-INTA. If the flash is indeed of cosmic origin, an object at an identical position should have also been recorded in the second twin station. Hence such a configuration will allow distinguishing flashing objects closer than 1 million km, therefore ruling out satellite glints and so on.

When information on a GRB position is obtained on the basis of an already established collaboration with the Huntsville (BATSE) and Roma (BeppoSAX) groups, the cameras will provide images of the corresponding error boxes. In fact, such a capability has proven to be very useful on the basis of the large size of GRB error boxes monitored in Ondrejov. These are currently being provided by the BACODINE and RRC Network [3–5].

The Robotic Telescope

This is based on the *Burst Alert Robotic Telescope* (BART) [6], that uses commercially available hardware components. The telescope has been succesfully tested in Ondrejov [7]. It is expected that positional information of GRBs will be obtained by means of the Internet network, and therefore BOOTES will perform rapid follow-up observations of events detected by BATSE, BeppoSAX, RossiXTE and future instruments.

In relation to the BOOTES Wide Field Cameras, the telescope should be able to react inmediately and take deep frames at the position of suspected OTs detected by the cameras.

In the meantime, selected objects (variable stars, nearby galaxies, bright QSOs, etc.) will be regularly monitored, searching for flaring behaviour.

Two telescope units will be placed at the proposed stations, under enclosures that will be opened automatically, according to weather conditions. The main characteristics are shown in Table 2.

SUMMARY OF SCIENTIFIC OBJECTIVES

- **The observation of GRB error boxes simultaneously with GRB occurrences** ($\sim$ 6 GRBs per year). Although the two first detected optical counteparts were not brighter than 20 mag a few hours after the burst [8,9], nobody has yet detected transient optical emission *simultaneous* with the event, which is expected to lie in the range V $\sim$ 10-15. The faint transient emission that has been detected a few hours after the event is a consequence of the expanding remnant that the GRB leaves behind it [10]. This provides information about the surrounding medium, but not about the burster itself. Many theories predict that there should be simultanoeus X-ray/optical emission. An extrapolation of the γ–ray power-law spectrum indicates that the simultanoeus optical flash should lie in the range 10-15 mag, depending on the burst intensity [11].

- **The detection of optical flashes (OTs) of** *cosmic* **origin**, that could be unrelated to GRBs and constitute a new type of different astrophysical phenomenon (perhaps associated to QSOs/AGNs). If some of them are related to GRBs, the most recent GRB models predict that there should be a large number of bursting sources in which only transient X–ray/optical emission should be observed, but no γ-ray emission. The latter would be confined in a jet-like structure and pointing towards us only in a few cases [12,13].

- **The observation of the sky in the V and R filters, as a part of preparations for ESA's satellite project INTEGRAL**, in which Spain has, for the first time, the leadership in one of the instruments, the Optical Transient Camera (OMC). The preparation includes tests of technologies, methods, data processing, ground-based observational network, etc.

- **The monitoring** of several objects (bright AGNs/QSOs, old GRB positions, etc.), looking for *recurrent* optical transient optical emission arising from these sources. There are hints that sudden and rapid flares occur, though of smaller amplitude. This will be achieved by means of the 0.3-m BOOTES telescope.

CONCLUSIONS

As we have previously discussed, the main objectives of BOOTES, as a part of the preparatives for OMC/INTEGRAL are: (i) the detection of simultaneous transient optical emission associated with cosmic γ-ray bursts previously localized by space instrumentation (BATSE, BeppoSAX, RossiXTE, etc.) and (ii) the capability to distinguish whether OTs are indeed of cosmic origin or not. The precision in the localization of a flaring optical counterpart by the Wide Field Cameras (0.8') and eventually by the Robotic Telescope (1.5") would lead in a short time interval

(≤ 1 d) to geting several CCD frames and spectra at major astronomical observatories. These detections would be a major step in order to clarify the GRB and OT mysteries.

ACKNOWLEDGEMENTS

We are very grateful for the support given by the Space Sciences Division (DCE) at INTA, through the project IGE 4900506. We thanks J. Maiz, M. Mas-Hesse, M. Santos-Lleó, and J. Torres for very fruitful conversations. The Czech contribution is supported by the Grant Agency of the Czech Republic, grant No. 205/93/0890.

REFERENCES

1. Bernas, M., Abraham, M., Zajízek, K., Páta, P. & Hudec, R. *Proc. of the Radiolektronika 96 Conference*, Brno, Czech Rep. (1996).
2. Castro-Tirado, A. J., Hudec, R., & Soldan, J., AIP Conf. Proc. **384**, 814 (1996).
3. Barthelmy, S., et al., AIP. Conf. Proc. **307**, 643 (1984).
4. Hudec, R., *A&SS* **231**(1/2), 239 (1995).
5. Kouveliotou, C., et al., talk presented at the Elba Workshop, on *Latest Results on GRBs* (1997).
6. Hudec, R. & Soldán, J., *A&SS* **231**, 311 (1995).
7. Soldán, J. Hudec, R. & Nemcek, M., AIP Conf. Proc. **384**, 643 (1996).
8. Castro-Tirado, A. J., Gorosabel, J., Benítez, N., et al., *Science*, in press (1998).
9. Guarnieri, A., et al., *A&A* **328**, L13 (1997).
10. Meszaros, P. & Rees, M. J., *ApJ* **476**, 232 (1997).
11. Band, D. & Ford, L., *A &SS* **231** (1-2), 247 (1996).
12. Dar, A., astro-ph/9704187 (1997).
13. Katz, J. I. & Piran, T., astro-ph/9706141 (1997).

Observing GRBs with INTEGRAL

Christoph Winkler[1]

Astrophysics Division, Space Science Department, ESA-ESTEC
2200 AG Noordwijk, The Netherlands

Abstract. The International Gamma-Ray Astrophysics Laboratory INTEGRAL is dedicated to the fine spectroscopy and fine imaging of celestial gamma-ray sources in the energy range 15 keV to 10 MeV. This paper summarizes the GRB detection and observation capabilities of the INTEGRAL instruments and describes the dissemination of INTEGRAL GRB alerts to the science community to enable rapid follow-up observations of GRB error boxes for counterpart searches and subsequent investigations.

INTRODUCTION

INTEGRAL [16] is ESA's next major 15 keV – 10 MeV gamma-ray mission in collaboration with Russia (PROTON launcher) and NASA (ground stations). The observatory is scheduled for launch in 2001 with a lifetime of two years to up to five years. The scientific objectives focus around high-energy resolution spectroscopy and fine imaging/location of gamma-ray sources and include: study of compact objects (black holes, neutron stars); explosive and hydrostatic nucleosynthesis (supernovae and novae, WR stars); high-energy transients and GRBs; mapping of the Galactic structure (Galactic Centre) and ISM; normal galaxies and clusters; AGN, Blazars and Seyferts; cosmic diffuse background and identification of high-energy sources. The payload consists of a Germanium spectrometer (SPI) optimised to perform high-resolution spectroscopy of gamma-ray lines and observations of the large scale diffuse emission, a CdTe/CsI imager (IBIS) optimised to provide fine point source images with accurate locations and sensitive studies of broad lines, and two monitors in the X-ray (JEM-X) and optical energy range (OMC) which identify and monitor high-energy sources at the low end of the INTEGRAL energy band. All instruments onboard INTEGRAL are co-aligned with overlapping FOVs and they operate simultaneously. The three high energy instruments utilize coded aperture masks. A particle radiation monitor complements the payload. The key

---

[1] This paper is written on behalf of the INTEGRAL Science Working Team consisting of: T. Courvoisier, N. Gehrels, A. Gimenez, S. Grebenev, W. Hermsen, F. Lebrun, N. Lund, G. Palumbo, J. Paul, R. Sunyaev, B. Teegarden, P. Ubertini, V. Schönfelder, G. Vedrenne, C. Winkler

features of the instruments are shown in Table 1. As a baseline, INTEGRAL[2] will be launched by a Russian PROTON launcher into a High Earth Orbit (48 h period, 46000 km perigee, 51.6° inclination) allowing the science instruments to operate entirely outside the Earth's trapped proton belt, and largely outside the electron belts. The entire orbit can be used for real-time scientific observations (downlink data rate = 86 kbps).

GRB DETECTIONS AND OBSERVATIONS

From the scientific objectives and the instrument characteristics in Table 1 it follows that GRBs are one important scientific objective for INTEGRAL, but the payload is not specifically designed to support primarily a GRB mission. However, INTEGRAL is well suited to provide important new observational data on the GRB phenomenon. In particular, INTEGRAL will provide GRB locations to arcminute accuracy or better, spectral coverage from 3 keV (JEM-X) up to 10 MeV (IBIS, SPI) with concurrent optical monitoring and high resolution spectroscopy (3 keV FWHM @ 1.33 MeV). Recent discoveries of optical, X-ray and radio counterparts, as discussed during this Symposium, show, as expected, that major progress in understanding the GRB phenomenon comes from accurate locations, broad-band coverage and rapid response. SPI [7], IBIS [14] and JEM-X [15] would detect GRBs in their nominal operational modes (on-ground, photon-by-photon events). In addition it its foreseen that IBIS will provide an on-board trigger to OMC [4] which subsequently switches into a burst trigger mode. The implementation of that trigger mechanism and the impact of false trigger events is currently under investigation by the instrument teams. The expected number of GRBs per year detected within the fully coded field of view has been estimated to 13 (SPI), 6 to 24 (IBIS) – where the lower and upper limits are given by the BATSE detection rate and extrapolation of the BATSE lgN–lgP distribution [13] – and 2 to 6 (JEM-X). IBIS will be able to locate GRBs with its upper CdTe detector area (2621 cm^2) to < 30″ (Table 1). A simulation of an IBIS GRB detection [12] is shown in Figures 1 and 2. Upon receipt of an IBIS trigger, the OMC [8] would accumulate a sequence of binned CCD images and compare them to the reference image accumlated at trigger time. The image is searched and areas of significant flux variations are identified and transmitted to ground.

TABLE 1. Key performance parameters of the INTEGRAL payload.

	SPI	IBIS	JEM-X	OMC
Detection area	500 cm^2	2621 cm^2 (CdTe)	1000 cm^2	1k × 1k CCD
Energy range	20 keV – 8 MeV	15 keV – 10 MeV	3 – 35 keV	550 nm (V)
ΔE (FWHM)	3 keV	7 keV	1.5 keV	n/a

[2]) Further details on INTEGRAL can be found in [16] and references therein as well as on the WWW using URL http://astro.estec.esa.nl/SA-general/Projects/Integral/integral.html

TABLE 1. Continued.

	SPI	IBIS	JEM-X	OMC
	@ 1.33 MeV	@ 100 keV	@ 10 keV	
FCFOV[a]	16°	9×9°	4.8°	5×5°
PCFOV[b]	35°	29×29°	13.2°	n/a
Angular resol.	2°	12′	3′	17.6″/pixel
10σ source location	< 15′	< 30″	< 20″	8″
S_c[c]	7×10^{-8}	4×10^{-7}	9×10^{-6}	19.7^{m_v} (10^3 s)
S_l[d]	5×10^{-6}	1×10^{-5}	1.7×10^{-5}	-
Timing	100 μs	100 μs – 1000 s	100 μs	> 1 s

[a] Fully Coded Field of View
[b] Partially Coded Field of View (Zero Response)
[c] Cont. sensitivity (3σ, 10^6 s), at 1 MeV (SPI), 100 keV (IBIS), 6 keV (JEM-X), units = ph cm^{-2} s^{-1} keV^{-1}
[d] Line sensitivity (3σ, 10^6 s), at 1 MeV (SPI), 100 keV (IBIS), 6 keV (JEM-X), units = ph cm^{-2} s^{-1}

Routine observations (pointings of up to typically 14 days) allow in principle the observation of pre- and post-GRB emission in γ, X-ray and optical domains. Due to specific requirements by the SPI instrument to improve its coded mask imaging, the spacecraft will perform, during many pointings, a number (7 – 25) of small (2°) dithering steps around the nominal target position along a hexagonal or rectangular pattern. This increases the effective sky area viewed by the instruments considerably. Quiescent or afterglow counterparts can principally be captured during nominal target pointings. Alternatively the spacecraft can react to GRBs outside its FOV through the standard Target of Opportunity mechanism which allows a re-scheduling of the observing programme and re-pointing of the

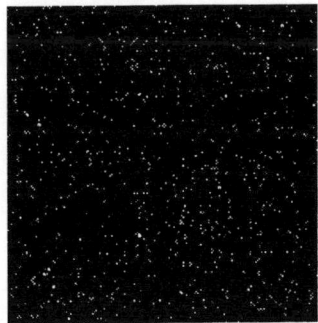

FIGURE 1. Simulation of 10σ GRB (600 background events, 250 source events) within IBIS 9×9° detection plane.

FIGURE 2. Deconvoled image (pixel size = 5′). The GRB is in the lower right corner.

spacecraft on average within ∼30 hours. It is noted that INTEGRAL will perform, as part of its Core Programme [5], weekly scans of a b = ± 20° wide strip along the entire Galactic plane to detect high-energy transients (e.g. GRS 1915+105 etc.), and GRBs would be covered due to their spatial and temporal serendipity. We have estimated whether INTEGRAL would have detected the afterglow of GRB970228, which may or may not be "typical" for GRB afterglow emission processes. The 55 s burst itself – as observed by SAX – can be fit at high energies (13 keV - 700 keV) by a -1.95 power law [3]. An observation of 5515 s duration by SAX/LECS of the fading counterpart (1.5 – 10 keV), 8 hours after the event, resulted in a -1.9 power law spectrum [11]. During this integration time, JEM-X would have possibly detected at 3σ the counterpart at < 10 keV and marginally at 30 keV. At higher energies the afterglow is too soft to be detected. On shorter timescales, the prospects for observing the fading counterpart are more promising: X-ray afterglow data for 4 GRB counterparts in 1997 indicate a generic $t^{-1.1}$ decay curve up to 10^6 s after the event. IBIS and JEM-X would have sufficient sensitivity to discover afterglow emission on ∼ mCrab level at around one hour after the event.

INTEGRAL GRB ALERTS AND THE SCIENTIFIC COMMUNITY

Following the successful utilisation of the BACODINE/GCN network [1] and the BATSE/COMPTEL/NMSU rapid response system [9], it is planned to install a similar GRB alert system for INTEGRAL [12]. Its purpose is to faciliate rapid follow-up observations. Based on INTEGRAL detections, a largely automatic system at the INTEGRAL Science Data Centre (ISDC) in Geneva monitors the real–time telemetry in order to detect, localize and validate an event before an alert is distributed via TCP/IP sockets or e-mail to the community. Once the spacecraft has achieved stable pointing after a slew, the predicted attitude is known to a few arcminutes accuracy. A snapshot of the actual (refined) spacecraft attitude is available 5 – 10 minutes into stable pointing (absolute pointing error < 1') and will be sent from ESA to the ISDC in real-time. Various steps from detection, via ground station links to on-ground processing steps like deconvolution, validation, localization and alert release will result in a delay (i.e. time difference between dissemination of GRB alert message and GRB trigger time) of about 14 s (best case) to 344 s (worst case) [12], where the main delay (of the order of 300 s worst case) at present is associated with the ground station performances (delay of science telemetry at NASA ground stations). For the best case scenario it is noted that a significant number of GRBs, as deduced from the BATSE 3B distribution of durations [10], have durations longer than 14 s.

INTEGRAL, THE 4TH IPN AND BATSE

Following [6], the 4th Interplanetary Network (IPN) could consist of 3 interplanetary missions (Mars surveyor, Pluto express and Ulysses) while INTEGRAL, and possibly CGRO would provide large detectors in near Earth orbits. It is planned to use parts of the SPI anticoincidence subsystem to provide simple time histories (50 ms binning) for ~ 300 GRBs/year above 4×10^{-7} erg/cm^2. A GRB event of moderate strength would produce a typical annulus width of $\sim 7''$, assuming Ulysses at a distance of 4.3 AU. After the succesful reboost of the CGRO in 1997 into a higher Low Earth Orbit, the prospects of operating CGRO during the next decade are high. The BATSE instrument onboard CGRO could be of great value to support, as an all-sky monitor, the INTEGRAL mission. In particular the following GRB related areas would be of importance [2]: (i) During the Core Programme, INTEGRAL will perfom weekly scans of the Galactic plane ($b \sim \pm 20°$) while BATSE would cover the high-latitude range, i.e. $|b| > 20°$; (ii) BATSE would provide ancillary time history and spectral evolution data for those GRBs that are inside the INTEGRAL FOV.

REFERENCES

1. Barthelmy, S., et al., these proceedings.
2. Fishman, G., et al., *Proc. 2nd INTEGRAL Workshop*, ESA-SP **382**, 537 (1997).
3. Frontera, F., et al., these proceedings.
4. Gimenez, A., et al., *Proc. 2nd INTEGRAL Workshop*, ESA-SP **382**, 613 (1997).
5. Gehrels, N., et al., *Proc. 2nd INTEGRAL Workshop*, ESA-SP **382**, 587 (1997).
6. Hurley, K., *Proc. 2nd INTEGRAL Workshop*, ESA-SP **382**, 491 (1997).
7. Mandrou, P., et al., *Proc. 2nd INTEGRAL Workshop*, ESA-SP **382**, 591 (1997).
8. Mas – Hesse, M., et al., *OMC Science Performance Report*, Issue 2, 1997.
9. McNamara, B., et al., *ApJSS* **103**, 173 (1996).
10. Meegan, C., et al., *ApJSS* **106**, 65 (1996).
11. Owens, A., et al., *A&A*, preprint (1997).
12. Pedersen, H., et al., *Proc. 2nd INTEGRAL Workshop*, ESA-SP **382**, 433 (1997).
13. Skinner, G., et al., *Proc. 2nd INTEGRAL Workshop*, ESA-SP **382**, 487 (1997).
14. Ubertini, P., et al., *Proc. 2nd INTEGRAL Workshop*, ESA-SP **382**, 599 (1997).
15. Westergaard, N., et al., *Proc. 2nd INTEGRAL Workshop*, ESA-SP **382**, 605 (1997).
16. Winkler, C., *Proc. 2nd INTEGRAL Workshop*, ESA-SP **382**, 573 (1997).

Simulated Observations of Gamma-Ray Bursts With GLAST

J. T. Bonnell[1], J. P. Norris[1]
B. L. Dingus[2], J. D. Scargle[3]

[1] *Code 661, NASA Goddard Space Flight Center, Greenbelt, MD 20771*
[2] *Department of Physics, University of Utah, Salt Lake City, UT 84112*
[3] *NASA Ames Research Center, MS 245-3, Moffett Field, CA 94035*

Abstract. The Gamma-ray Large Area Space Telescope (GLAST) will incorporate high sensitivity, large field of view, and precision tracker technology, providing arcminute localizations of gamma-ray bursts (GRBs) and exploring GRB physics up to ~ 100 GeV. We have simulated the response of GLAST to GRBs with power-law spectra extending to GeV energies to determine the detailed burst localization capability. The simulated properties of GRBs are based on the BATSE peak flux and duration distributions. GLAST's hodoscopic calorimeter design has sufficiently good angular resolution and discrimination power against cosmic rays that > 1 GeV gammas which are only detected in the calorimeter may be utilized for bright sources such as GRBs. Our results indicate that GLAST will detect and image ~ 230 GRBs yr^{-1} with sensitivity to ~ 100 GeV for ~ 20 bursts yr^{-1}. Many bright burst localizations will be comparable in size to the current InterPlanetary Network error boxes thus probing the cosmological burst parameter space at nearby redshifts and enabling counterpart searches at all lower energies.

INTRODUCTION

The Gamma-ray Large Area Space Telescope (GLAST) mission is designed to observe celestial gamma-ray sources in the 0.01 to 100 GeV band. In particular, the instrument incorporates features resulting in high sensitivity and a large field of view. GLAST has an exciting potential for exploring gamma-ray burst (GRB) physics at high energies. Here we describe detailed simulations of power-law spectrum GRBs and the burst localization capability of GLAST for one year of observations.

INSTRUMENT SIMULATION

The Gismo simulation package [1] was used to generate the response of the baseline GLAST instrument to an incident power-law spectrum of photons. The instrument configuration simulated was 5 x 5 towers of 17 Si/Pb tracker planes above a calorimeter composed of 6 horizontal layers of 3 cm square cross section CsI logs. The total geometric area of the 25 towers was 26500 cm^2.

Separate instrument simulation runs were made adopting input gamma spectra with power-law indices of α = 1.6, 1.8, 2.0, 2.2, and 2.4. For each power-law spectrum, an n-tuple file containing the response to 50,000 gammas was produced for a minimum energy of 100 MeV and an on-axis beam. These n-tuples form the parent distribution from which gammas were drawn in the simulations and represent the GRB photon spectrum well in the absence of unknown spectral features or breaks [6].

The GLAST point spread function (PSF) off-axis was approximated by the parametric form $\text{PSF}_{\text{on-axis}}/\cos(\theta)$, fitted empirically by sets of simulation runs at zenith angles θ = 0°, 15°, 30°, 45°, 60°, and 75°. Its energy dependence was characterized as Gaussian. Simulation runs at 0.1, 0.3, 1, 5, and 30 GeV indicate that a Gaussian with $\sigma = C_1 * E^{-a} + C_2$ yields an adequate representation between E = 0.3 and 30 GeV for a = 0.85, C_1 = 0.30, and C_2 = 0.025.

The incident photon directions were determined by the standard tracker reconstruction and by a moments analysis reconstruction in the calorimeter. The relative tracker and calorimeter efficiencies were calculated from these simulations.

GAMMA-RAY BURST SIMULATION

During a one-year period of simulated GRB observations, GLAST was assumed to operate in a zenith-pointed mode with an acceptance angle of 75° from normal incidence. Reasoning from the observed BATSE log N – log P and isotropic sky distribution yields 300 bursts in the GLAST field of view.

The fluence distribution of power-law spectrum GRBs was assumed to be proportional to the observed BATSE peak-flux distribution (log N – log P) and to a brightness-independent duration distribution [3], spanning 0.1 s to 500 s. Thus, in the simulations, a bright, long BATSE burst produces more high-energy gammas than a dim short burst. This distribution was normalized to the average 1 MeV fluxes of 15 bright BATSE bursts as follows.

The 15 brightest bursts (BATSE 64 ms peak fluxes > 30 ph cm^{-2} s^{-1}, through trigger 5621), which had suitable data, were used. Power-law spectra were fitted to BATSE spectroscopy detector data averaged over the time interval determined by each burst's measured T_{90} duration [3] and an energy range of 500 – 1500 keV. The average of this distribution of the ratio of burst flux (ph cm^{-2} s^{-1} MeV^{-1}) at 1 MeV to the peak flux was adopted for the entire range of BATSE log N – log P.

The distribution of spectral indices determined compares favorably with the results of [2,6,5]. The dominant peak in this distribution contains $\sim 90\%$ of the bursts and has $\alpha_{mean} = 2.0 \pm 0.2$. A Gaussian with a corresponding mean and standard deviation was used to represent this distribution in the burst simulations.

The burst simulations were realized by scaling the BATSE log N - log P distribution to yield a flux (ph cm^{-2} s^{-1} MeV^{-1}) at 1 MeV, then choosing α and T$_{90}$ and integrating above 100 MeV to yield the fluence or N$_{tot}$ γs > 100 MeV cm^{-2}.

GRB LOCALIZATIONS

Within the GLAST field of view, an isotropic distribution of 270 GRBs with fluences determined as above was generated. This number results from reducing the previously stated $\sim$300 per year in GLAST's field of view based on the estimate that 10% of GRBs have steeper αs (softer spectra) than accounted for in our assumed spectral index distribution. The N$_{tot}$ photons incident on GLAST for a given burst were then selected from the parent power-law distribution n-tuple data based on simulated tracker and calorimeter efficiencies. Low energy thresholds of 100 MeV in the tracker and 1 GeV in the calorimeter were adopted.

The GLAST PSF for photons above 100 MeV is sufficiently narrow, $\sim 2.25°$ radius corresponds to 68% containment on-axis in the tracker, but quickly broadens at lower energies. We therefore chose this energy as a threshold for localization purposes. Photons detected in the calorimeter number $\sim$2.5 times those detected by the tracker, but only photons above $\sim$1 GeV from the calorimeter-only mode may contribute to the burst (due to trigger restrictions understood so far). Considering a power-law index of 2, the > 1 GeV calorimeter-only flux is $\sim$2-3% of the tracker flux > 100 MeV. While these few photons are worth including to characterize the higher energy spectra, they contribute only a small improvement to the GRB localization.

The localization uncertainty was computed by first summing the normalized PSFs of the individual tracker photons for a burst, offset in X and Y coordinates by the difference between reconstructed and input Monte Carlo vectors, then convolving the summed probability distribution with a master template probability distribution of detected tracker photons whose only parameter was the power-law index (no offsets from bullseye center), and finally choosing the (X,Y) peak of the convolution as the GRB localization position. This approach yields results ranging from $\sim$1 to 10 times better than simply using the highest energy photon. Algorithms which somehow make superior use of the available information, for instance weighting by energy (E) as E^x, might be envisioned. We have tried just a few such possibilities, but have found none that afforded an improvement over the current results. Note that using the 2-D PSF of each photon already weights the estimate by E^{-1}.

Ten realizations per burst were computed (ten different sets of N$_{tot}$ photons were chosen from the parent distribution) and the 7th ranked error box size was chosen as the 68% containment radius. A scatter plot of a year's worth of GRB localizations vs. number of tracker detected photons is illustrated in Figure 1.

RESULTS AND DISCUSSION

Previous work [4] has demonstrated that GLAST GRB localization errors are strongly dependent on the assumed power-law spectral index α. For example, it was found that if $\alpha = 2.0$ for *all* simulated bursts then ~95% of GRBs in the GLAST field of view would produce at least 1 photon above 100 MeV in the tracker. This fraction falls to ~76% at $\alpha = 2.5$ and ~13% at $\alpha = 3.0$.

The major improvement in these GRB simulations is the adoption of an empirically determined distribution of high-energy spectral indices and an improved treatment of the bright burst flux normalization at 1 MeV. The new results (Figure 1) indicate that ~230 GRBs per year will produce at least 1 tracker detected photon above 1 MeV. This corresponds to 76% of the GRBs in the GLAST field of view. Approximately 120 of these bursts would have localizations with 68% containment radius better than 10 arcminutes. Special on-board trigger and processing modes under consideration would make these error regions rapidly available, enabling follow-up observations at all lower energies.

These simulations of high-energy properties of GRBs are necessarily scalings and extrapolations of the lower-energy distributions. Whenever possible we have

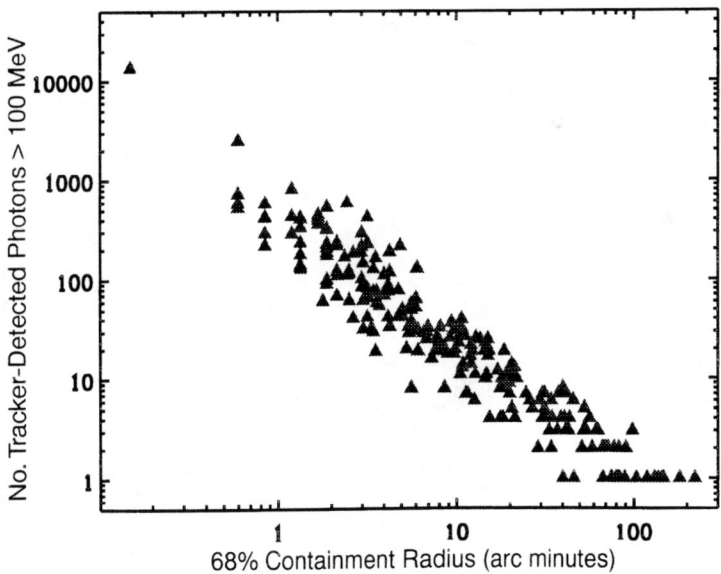

FIGURE 1. The GLAST 68% localization radii for GRBs with a power-law index distribution of 2.0 ± 0.2. The majority of localizations are determined to better than 10 arcminutes.

adopted established results and conservative assumptions. As described in the previous section we have ignored the small fraction of bursts with soft spectra ($\alpha > 2.5$) which would tend to produce few high-energy photons and less accurate localizations. We have also ignored the sub-population of harder, dim bursts, mostly unprobed by BATSE, whose existence is only just being investigated [7].

Our simulations also indicate (see Figure 2) that, for $\alpha = 2.0 \pm 0.2$, if GRB spectra continue up to 100 GeV, then $\sim$5 bursts yr^{-1} will be detected at or above that energy by GLAST, thus defining the spectrum to the regime covered by TeV ground-based observatories.

REFERENCES

1. Atwood, W. B. et al., *NIM, A,* **342**, 302 (1994).
2. Band, D. et al., *ApJ* **413**, 281 (1993).
3. Bonnell, J. T. et al., *ApJ* **490**, 79 (1997).
4. Bonnell, J. T. et al., *25th ICRC Proceedings*, eds. Potgieter, M. S., Raubenheimer, B. C., van der Wait, D. J., **3**, 69 (1997).
5. Catelli, J., private communication (1997).
6. Dingus, B. L., *Ap Space Sci* **231**, 187 (1995).
7. Harris, M., these proceedings.

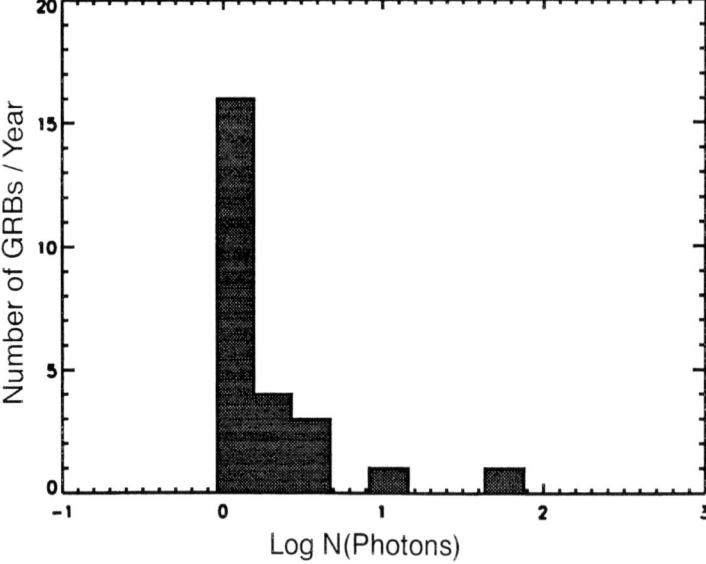

FIGURE 2. The number of GRBs per year producing N(photons) > 100 GeV which could be detected by GLAST. Power-law spectra with a distribution approximating $\alpha_{\mathrm{mean}} = 2.0 \pm 0.2$ were used. No spectral cut-offs or features were considered.

Development of a Hard X-Ray Polarimeter for Gamma-Ray Bursts

M.L. McConnell, D.J. Forrest, J. Macri,
J.M. Ryan and W.T. Vestrand

Space Science Center, University of New Hampshire, Durham, NH 03824

Abstract. We describe recent work on the development of a Compton scatter polarimeter for measuring the polarization of hard X-rays (100–300 keV) from astrophysical sources. Results from measurements with a laboratory prototype are summarized, along with comparisons to Monte Carlo simulations. We also present a new design concept that envisions a complete polarimeter module on the front end of a 5-inch position-sensitive PMT. Although the emphasis of our effort is measuring hard X-ray polarization in solar flares, our design has the advantage that it is sensitive over a rather large FoV (> 1 sr), a feature that makes the design especially attractive for γ-ray burst studies.

INTRODUCTION

The measurement of hard X-ray polarization in γ-ray bursts would add yet another piece of information in our effort to resolve the true nature of these enigmatic objects. Here we report on the development of a hard X-ray polarimeter for solar flares that, because of its relatively large FoV, may be useful in studies of γ-ray bursts.

The basic physical process used to measure linear polarization of hard X-rays (100–300 keV) is Compton scattering. The measurement is based on the fact that the incident photons tend to be scattered at right angles to the incident electric field vector. A Compton scatter polarimeter consists of two detectors that are used to determine the energies of both the scattered photon and the scattered electron. One detector (the *scattering detector*) provides the medium for the Compton interaction to take place. This detector must be designed to maximize the probability of a single Compton interaction with a subsequent escape of the scattered photon. The primary purpose of the second detector (the *calorimeter*) is to absorb the full energy of the scattered photon. To be recorded as a polarimeter event, an incident photon Compton scatters from one (and only one) of the scattering detectors into the central calorimeter. The incident photon energy can be determined from the sum of the energy losses in both detectors and the scattering angle

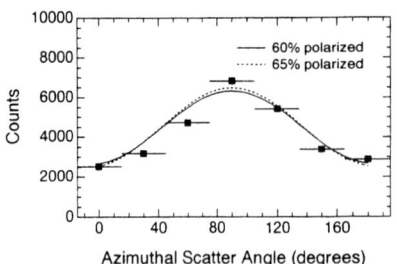
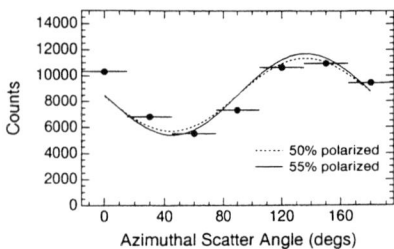

FIGURE 1. The prototype response to an on-axis polarized beam. The smooth curves represent simulation results.

FIGURE 2. The prototype response with the polarization vector rotated 45° with respect to that in Figure 1.

can be determined by the azimuthal angle of the associated scattering detector. When the polarimeter is arranged so that the incident flux is parallel to the symmetry axis, unpolarized radiation will produce an axially symmetric coincidence rate. If the incident radiation is linearly polarized, then the coincidence rate will show an azimuthal asymmetry whose phase depends on the position angle of the incident radiation's electric vector and whose magnitude depends on the degree of polarization.

LABORATORY PROTOTYPE

In an earlier paper, we discussed a polarimeter design consisting of a ring of twelve individual scattering detectors (composed of low-Z plastic scintillator) surrounding a single NaI calorimeter [1]. The characteristics of this design were investigated using a series of Monte Carlo simulations (based on a modified version of GEANT). We have recently prototyped this design in the laboratory to validate our Monte Carlo code. For prototype testing, we set up a semicircular array around a central NaI detector, eliminating the redundancy and simplifying the hardware and associated electronics. Seven plastic scintillators (each 5.5 cm × 5.5 cm × 7.0 cm in size) were positioned at a radius of 15 cm from a 7.6 cm diameter × 7.6 cm high cylindrical NaI(Tl) detector.

Polarized photons were generated by Compton scattering photons from a radioactive source [2]. The exact level of polarization is dependent on both the initial photon energy and the photon scatter angle. The use of plastic scintillators as a scattering block permits the electronic tagging of the scattered (polarized) photons. This is used to provide a coincidence signal to the polarimeter. For our laboratory measurements we used a ^{137}Cs source to generate a beam of polarized 288 keV photons.

The laboratory data (Figure 1) led to a measured polarization value of 64.0%(±3.0%), in good agreement with the estimated value of 50-60% based on analytical estimates [3]. This result demonstrates: a) the ability of a simple Comp-

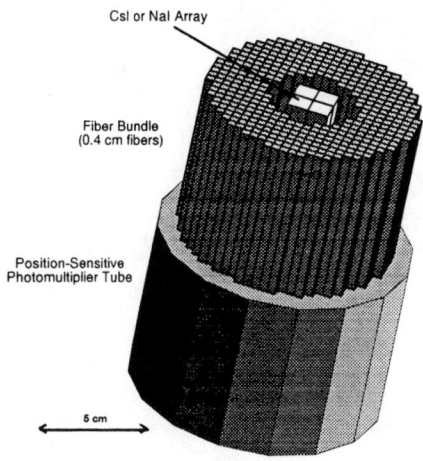

FIGURE 3. Schematic diagram of a polarimeter module.

ton scatter polarimeter to measure hard X-ray polarization; b) the ability of our Monte Carlo code to predict the polarimeter response; and c) the ability to generate a source of polarized photons using a simple scattering technique. In another laboratory measurement (Figure 2), the plane of polarization of the incident beam was rotated ~ 45° with respect to that used in the first set of data. The measured shift of 50.4° in the polarization vector is consistent with the uncertainties in our experimental setup.

A NEW DESIGN CONCEPT

There are at least two possible means of improving the polarimeter performance: 1) by more precisely measuring the scattering geometry of each event; and 2) by rejecting those events that undergo multiple Compton scattering within the scattering elements. (Our simulations indicate that roughly 30-40% of the events recorded in the prototype polarimeter as valid events involved multiple scattering within a single scatter element.) Improvements in either area will lead directly to a more clearly defined modulation and, therefore, a better polarization sensitivity.

We have developed a new design that places an entire device on the front end of a single 5-inch diameter position-sensitive PMT (PSPMT) [4]. A bundle of scintillation fibers (each with a cross section of 4 mm × 4 mm) provides the improved spatial resolution in the scattering elements. The bundle is in the form of an annulus with an outside diameter of 10 cm and an inside diameter of 4 cm. A 2 × 2 array of 1 cm inorganic scintillators is positioned within the annulus, each scintillator being coupled to its own independent PMT for light collection and signal timing. Figure 3 shows a schematic view of such an assembly.

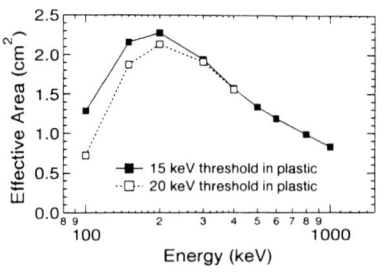

FIGURE 4. Effective area versus energy.

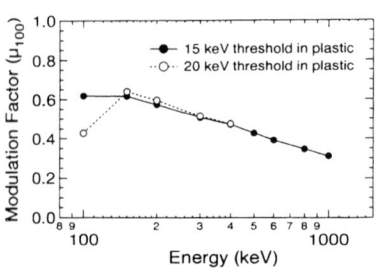

FIGURE 5. Modulation factor versus energy.

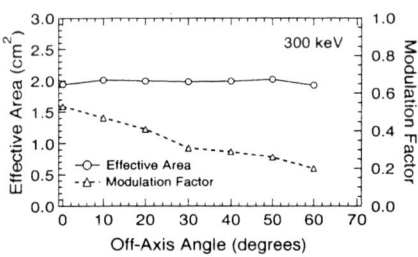

FIGURE 6. The modulation factor and effective area as a function of incidence angle for a photon energy of 300 keV.

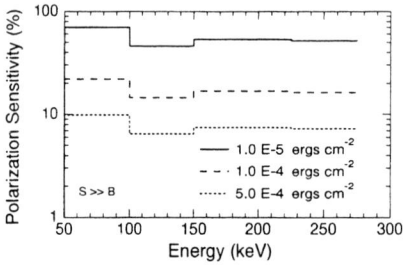

FIGURE 7. The 3σ sensitivity level of a 16-module array to γ-ray bursts. Fluence levels are for the 50–300 keV range.

Monte Carlo simulations have been used to determine the characteristics of this design. Figures 4 and 5 show the modulation factor and the effective area, respectively, as a function of energy. The low energy response is very sensitive to the energy threshold in the fiber array. Figure 6 shows the off-axis response of the design, which suggests a useful FoV of at least one steradian.

SUMMARY

We anticipate that this design would be used in the context of a (not necessarily contiguous) array of polarimeter modules. Figure 7 shows the sensitivity of an array of 16 modules, suggesting that a sensitivity level of about 15% may be achieved for the strongest γ-ray bursts.

The use of polarimetry in X-ray and γ-ray astronomy has so far been largely limited to energies below 100 keV [5-7,2,8], with an emphasis on the study of non-transient sources. Several higher energy experiments offer polarimetry as a secondary capability [9,10]. Although designs similar to that proposed here have been discussed in the literature [6,11], we are unaware of any other *active* effort to specifically measure polarization in γ-ray bursts at energies above 100 keV.

ACKOWLEDGEMENT

This work has been supported by NASA grant NAGW-5704.

REFERENCES

1. M. McConnell, D. Forrest, K. Levenson, and W.T. Vestrand, "The design of a gamma-ray burst polarimeter", in AIP Conf. Proc. 280, *Compton Gamma-Ray Observatory*, M. Friedlander, N. Gehrels and D.J. Macomb, Eds. New York: AIP, 1993, pp. 1142-1146.
2. H. Sakurai, M. Noma, and H. Niizeki, "A hard x-ray polarimeter utilizing Compton scattering", in *SPIE Conf. Proc.* **1343**, pp.512-518 (1990).
3. W.H. McMaster, "Matrix representation of polarization", *Reviews of Mod. Phys.* **33**, no. 1, pp. 8-28 (1961).
4. M.L. McConnell, et al., "Development of a hard X-ray polarimeter for solar flares and gamma-ray bursts", submitted to *IEE Trans. Nucl. Sci.* (1998).
5. R. Novick, "Stellar and solar X-ray polarimetry", *Space Science Reviews* **18**, pp. 389-408 (1975).
6. G. Chanan, A.G. Emslie, and R. Novick, "Prospects for solar flare X-ray polarimetry", *Solar Physics* **118**, pp. 309-319 (1988).
7. P. Kaaret, et al., "The Stellar X-ray Polarimeter - a focal plane polarimeter for the Spectrum X-Gamma mission", *Optical Engineering* **29**, pp. 773-780 (1990).
8. E. Costa, M.N. Cinti, M. Feroci, G. Matt, and M. Rapisarda, "Scattering polarimetry for X-ray astronomy by means of scintillating fibers", *SPIE Conf. Proc* **2010**, pp. 45-56 (1993).
9. E. Aprile, A. Bolotnikov, D. Chen, R. Mukherjee and F. Xu, "The polarization sensitivity of the liquid xenon imaging telescope", *ApJ Supp* **92**, pp. 689-692 (1994).
10. T.J. O'Neill, et al., "Tracking, imaging and polarimeter properties of the TIGRE instrument", *Astron. Astrophys. Suppl. Ser.* **120**, pp. C661-C664 (1996).
11. T.L. Cline, et al., "A Gamma-Ray Burst Polarimeter Study", in *Proceedings of the 25th Internat. Cosmic Ray Conf.* **5**, pp. 25-28 (1997).

The Konus-Wind and Konus-A Instrument Response Functions and the Spectral Deconvolution Procedure

M. M. Terekhov, R. L. Aptekar, D. D. Frederiks, S. V. Golenetskii, V. N. Il'inskii, E. P. Mazets

Ioffe Physico-Technical Institute, St. Petersburg, 194021 Russia

Abstract.

The Konus-Wind and Konus-A detector response function to γ-rays is calculated using the Monte Carlo method. The laboratory energy calibration in the energy range 6.4 keV to 2614 keV is described and the relative light output for monoenergetic electrons in the detector scintillator is determined. The process of γ-ray photon spectral deconvolution from the count rate spectra, using Monte Carlo response functions, and the results of the laboratory calibration are also discussed.

INSTRUMENTATION

The Konus-Wind [1] and Konus-A instruments are γ-ray burst detectors designed to obtain detailed timing and spectral measurements in the energy range 10 keV – 10 MeV for about four minutes whenever a γ-ray burst is detected by a rapid count rate increase over the current background in the range 50-200 keV. During this time 64 spectra are measured with accumulation times adjusted according to the count rate. At other times, the instruments are operating in the background mode, recording only limited data about the radiation environment.

The Konus-Wind instrument is installed on board the GGS Wind spacecraft, in accordance with an agreement between NASA and the Russian Space Agency. The Wind spacecraft is in a complex orbit extended toward the Sun, spending most of the time far outside the Earth's magnetosphere. The spacecraft is stabilized by its rotation around an axis perpendicular to the ecliptic plane. The spectrometric detectors S1 and S2 are placed on opposite faces of the spacecraft, viewing the south and north hemispheres of the sky, respectively, with their axes along the rotation axis.

Cosmos-2326, containing the Konus-A instrument, is in a circular orbit with an altitude of 415 km and an inclination of 65°. It is three-axis stabilized, with one of its axes aligned along the velocity vector and another directed toward the local zenith. Besides the spectrometric detector DS, Konus-A includes an array of four DN detectors, with anisotropic sensitivity that provides a limited source localization capability. The axes of the DN detectors are offset from the symmetry axis by 30° and separated in azimuths by 90°. The axes of the DS and the DN-array are both directed to the local zenith. In the background mode the total count rate of four DN detectors is measured with a time resolution of 2 s, with 64-channel energy spectra in the range from 10 keV to 10 MeV for each detector, with accumulation times of 64 s.

The spectrometric detectors installed on both these spacecraft are identical in function and operational modes and include omnidirectional NaI(Tl) scintillators, 130 mm in diameter and 75 mm in height, with beryllium entrance windows. Detectors DN differ from detectors S and DS in their height, which is 30 mm, and in the passive shielding of their sides. Between the NaI(Tl) crystal and the PMT, lead glass is placed in both spectrometric and DN detectors, with thicknesses of 1.9 cm and 1.6 cm respectively.

CALIBRATIONS

To investigate the properties of our detectors a detailed laboratory calibration was performed. For each detector, the energy deposition spectrum and the energy resolution in the range from 14.4 keV to 2614 keV were measured using a number of standard radioactive sources. Figure 1(a) shows energy resolution data measured for the Konus-Wind S1 detector, as an example. To determine the fluorescent response of NaI(Tl) crystals to X-rays and γ-rays the measurements with radioactive sources (^{57}Co, ^{241}Am, ^{133}Ba, ^{109}Cd, ^{152}Eu, ^{139}Ce, ^{113}Sn, ^{22}Na, ^{137}Cs, ^{54}Mn, ^{88}Y, ^{65}Zn, ^{60}Co, ^{40}K, ^{228}Th) and X-rays in energy range 6.4–71 keV (from targets: Fe, Cu, Zn, Ge, Br, Sr, Nb, Mo, Ag, Cd, In, Sn, Te, I, Cs, Ba, La, Ce, Pr, Nd, Sm, Dy, Ta, Hg) were performed. Then the fluorescent response of NaI(Tl) to monoenergetic electrons was calculated using a method similar to that described in [2]. In Figure 1(b) the experimental data (points) and calculated responses to γ-rays and electrons (solid lines) are shown.

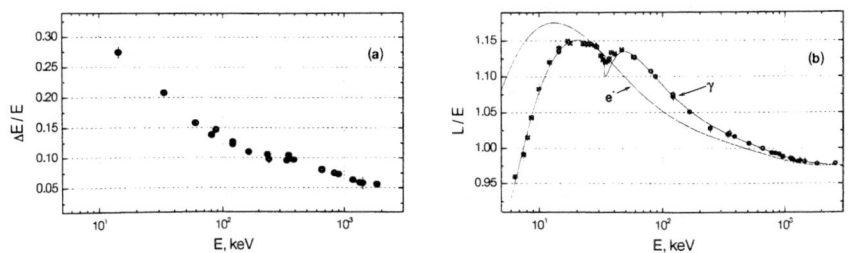

FIGURE 1. (a) - Energy resolution. (b) - Scintillation efficiency for γ and e^-.

RESPONSE FUNCTION

Detailed geometric models of both spectrometric and DN detector assemblies were developed. All detector assembly parts and mounting parts which are close to the NaI(Tl) crystals were modeled for precise dimensions and material composition. An additional spacecraft geometric model was introduced to account for scattered radiation in the response matrix calculations. In the case of the GGS Wind spacecraft for example, it was a barrel 248 cm in diameter and 186 cm in height. Upper and lower closure and equipment panels were treated separately with corresponding materials and the rest of the spacecraft volume was filled with an averaged composition.

The response function for the spectrometric and DN detectors was calculated using Monte Carlo code specially developed for this purpose. The treatment of gamma quanta propagation was done in the same way as in the well known codes EGS or GEANT. Our code is different in only two respects. The first is that we use the more recent cross-section data from the ENDF/B-VI data base. The second is in our method of fluorescent quanta generation. Fluorescent photons from different shells of an individual element in a mixture or a compound may be produced in our code. This allows us to account for characteristic features in a more detailed manner. The example in Figure 2 shows that agreement between our simulations and experimental data is quite satisfactory.

The procedure for a response matrix calculation for each detector was as follows. For angle values in the $0° - 90°$ range, with steps of five degrees, 159 monoenergetic lines equispaced in logarithmic scale were simulated. The data base so obtained provides all information necessary to interpolate a response matrix for any angle and detector energy cuts with a given amplification. Additionally, several simulations at different angles and energies were performed to check for interpolation validity. Figures 3 and 4 show channel response functions for several channels and a three-dimension view of the whole response matrix for the spectrometric detector installed on the GGS Wind spacecraft.

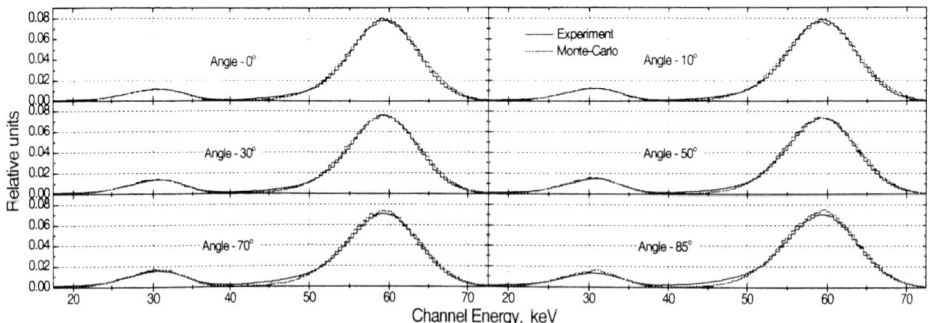

FIGURE 2. Monte Carlo simulations and experimental data for ^{241}Am (59.5 keV).

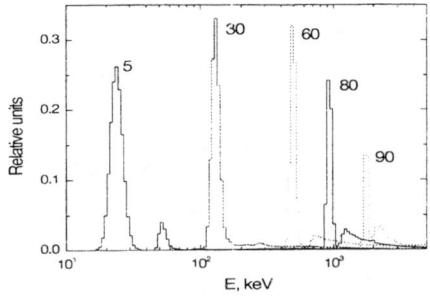 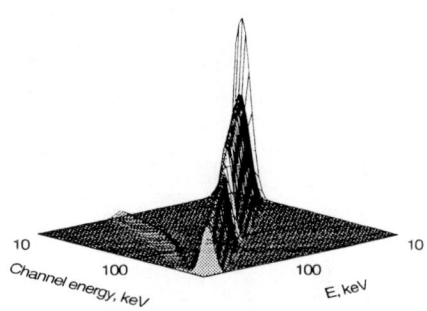

FIGURE 3. Channel response functions for spectrometric detector S1 (channel numbers indicated).

FIGURE 4. 3D view of the Konus-W response matrix.

SPECTRAL DECONVOLUTION

The count rate spectra measured by a detector are related to the intrinsic photon spectra according to an Fredholm integral equation of the first kind. So, to obtain a photon spectrum from experimentally measured count rates, we should find a solution of this equation. It is well known that this problem is a so-called ill-posed problem. This means that the solution lacks both uniqueness and stability. Unfortunately, there is no unique method for solving this problem. The most widely used approach to calculate γ-ray burst spectra is a forward-folding method based on model fitting. The resulting estimates are known to be model dependent. There are other methods for solving this problem, for example those of [3]. It is very useful to be able to compare solutions obtained by different methods, especially when some features in count rate spectra are present. So new methods for γ-ray burst spectra deconvolution are to be welcomed. In our investigations, we use both model fitting and the regularization method developed by [4]. The latter allows us to obtain model independent γ-ray burst spectra estimates. As examples Figure 5(a,b) and Figure 5(e,f) show model independent deconvolved spectra and corresponding count rate spectra for GRB961029 accumulated for 0–67 s and 0–6.9 s after the trigger respectively. Figure 5(c,d) and Figure 5(g,h) show the same spectra fitted by the Band "GRB" model [5].

CONCLUSION

The results presented here show that our studies of spectrometric detectors are complete enough and the agreement between laboratory calibration and Monte Carlo simulations is quite satisfactory. Traditional model fitting, as well as model independent spectra deconvolution methods, allow us to analyze a rich set of spectral data obtained from the Konus-Wind and Konus-A instruments.

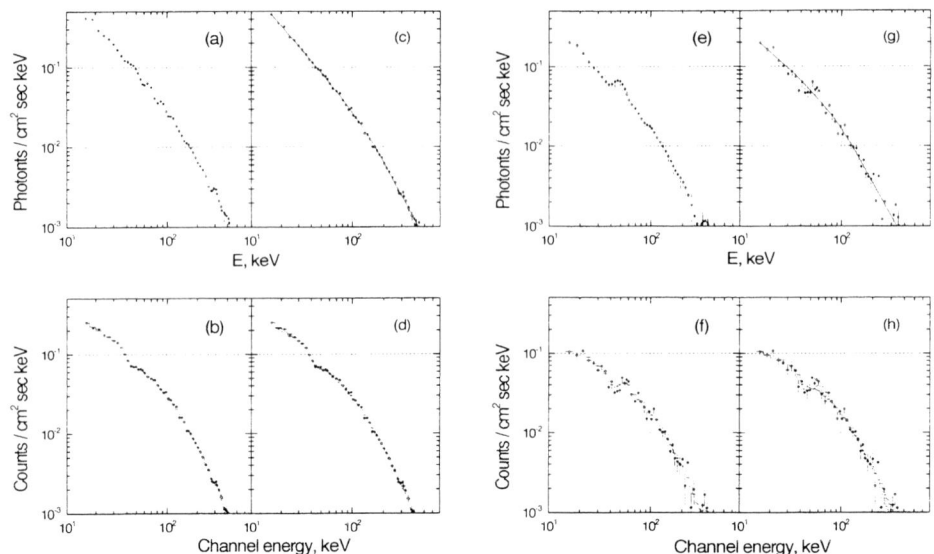

FIGURE 5. Deconvolved photon spectra for GRB961029 ((a),(b) – for 0–67 s and (e),(g) – for 0–6.9 s after the trigger; (a),(e) – model independent deconvolution and (c),(g) – fit by the Band "GRB" model) and corresponding count rate spectra (b),(d),(f),(h).

ACKNOWLEDGEMENTS

This work was supported by RSA contracts, RFBR grants N96-02-16860 and N97-02-18067 and CRDF grant RP1-236.

REFERENCES

1. Aptekar, R.L., Frederiks, D.D., Golenetskii, S.V., Ilynskii, V.N., Mazets, E.P., Panov, V.N., Sokolova, Z.Ya., Terekhov, M.M., and Sheshin, L.O., *Space Science Review* **71**, 265 (1995).
2. Collinson, A.J.L and Hill, R., *Proc. Phys. Soc.* **81**, 883 (1963).
3. Loredo, T.J. and Epstein, R.I., *Ap.J.* **336**, 898-919 (1989).
4. Tikhonov, A.N., Goncharsky, A.V., Stepanov, V.V, and Yagola, A.G., *Regularizing algorithms and apriori information* , Moscow: Nauka, 1983.
5. Band, D.L., et al., *Ap.J.* **413**, 281 (1993).

The BATSE Trigger Efficiency as a Function of Intensity and Energy Range

G. N. Pendleton*, J. Hakkila[†], C. A. Meegan[††]

*Physics Dept., University of Alabama in Huntsville, AL 35899,
[†]Physics Dept., Mankato State University, Mankato, MN 56002,
[††]Space Sciences Laboratory, NASA/MSFC, Huntsville, AL 35812

Abstract. The new intensity dependent trigger efficiency algorithm, important for accurate determination of the GRB LogN-LogP distribution at low intensities, is described. This algorithm now includes atmospheric scattering, the capacity for trigger energy range selection, and burst spectral shape sensitivity. It is used here to put tighter constraints on the low energy behavior of the LogN-LogP distribution than were previously possible, and to evaluate the consistency of the GRB populations observed with the BATSE trigger set in the 50-300 keV range, the 25-100 keV range, and the $E > 100$ keV range.

THE SKYMAP FORMALISM

The BATSE exposure sensitivity [1] available with the 3B burst catalog did not include the atmospheric scattering necessary to accurately model the instrument exposure near threshold [2]. We have developed a new formalism for calculating the efficiency as a function of burst intensity that incorporates the atmospheric scattering effects as well as allowing for the selection of a variety of burst trigger energy ranges.

Several spherical geometry coordinate systems are defined for this formalism. The sky is divided into 400 bins in J2000 right ascension and declination for the principal sky map coordinate system. In addition, a coordinate system consisting of a spherical annulus covering the surface of the earth over which CGRO passes in its orbit is defined as the terrestrial annulus. This coordinate system, fixed in the J2000 coordinate system, is necessary to uniquely define the atmospheric scattering matrices and to effectively separate regions where the instrument background is significantly different. For a particular CGRO pointing interval, data are calculated and stored for bins in these two spherical coordinate systems.

Several quantities are calculated for an individual GRO pointing for all 8 detectors separately. For a fixed CGRO pointing, the direct response matrix is specified uniquely for each principal sky map coordinate system. The direct response ma-

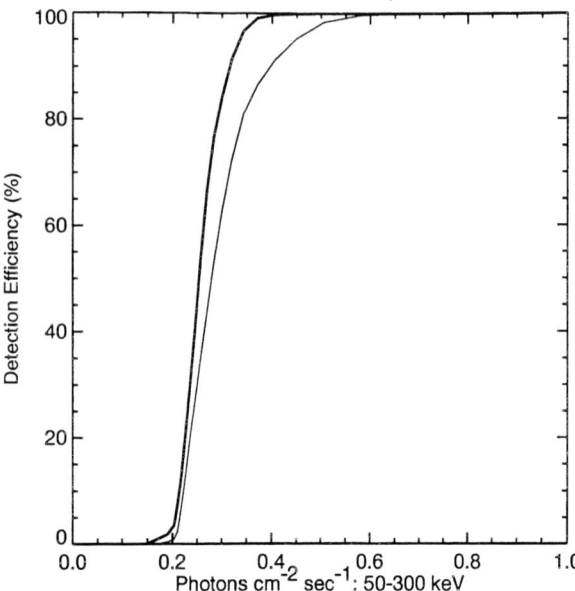

FIGURE 1. Comparison of the 3B and the new prototype efficiency calculation for the 1024 ms trigger timescale on the 50-300 keV energy range.

trix includes spacecraft scattering and represents the instrument's response at the appropriate source viewing angle. Each matrix is calculated with 16 input energy bins and 4 output counts bins with energy edges aligned with the discriminator channels used for burst triggering, so that model burst spectra can be convolved though them to calculate the expected observed count rates.

Distributions of counts are accumulated for each trigger energy range for each of the terrestrial annulus bins. For a particular spacecraft pointing, these bins separate the count rates into regions where they are systematically different. For example the high energy background correlates generally with declination due the earth's magnetic field. The low energy background correlates with the detector's earth-to-sky viewing angle. For a particular spacecraft pointing this angle is uniquely specified for each terrestrial annulus bin.

The atmospheric scattering bins have the same binning as the direct response matrices. They represent the atmospherically scattered source flux convolved through the instrument response. Each matrix is uniquely specified for each terrestrial annulus bin, for a given principal skymap bin. This is, therefore, the largest data structure in the analysis package.

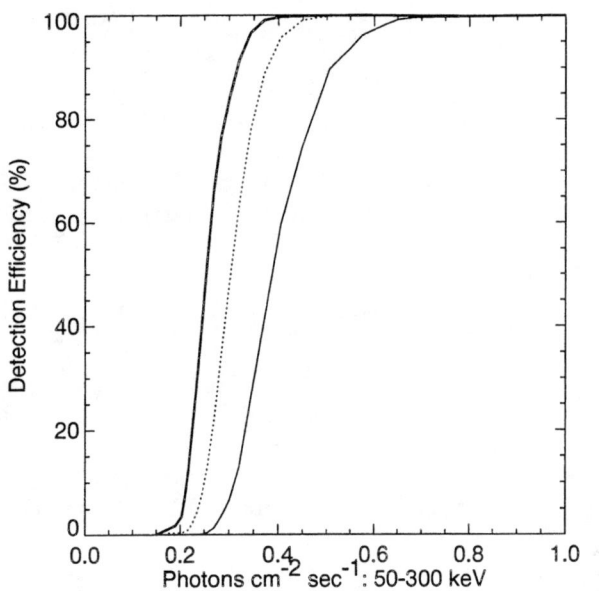

FIGURE 2. Comparison of the new prototype efficiency calculations for the 1024 ms trigger timescale in different energy ranges. Thick solid curve: 50-300 keV; dotted curve E > 100 keV; thin solid curve 25-100 KeV.

THRESHOLD CALCULATION PROCEDURE

With this formalism a burst spectrum is specified for forward folding with the response matrix structures. Then all the elements of the outer product of the principal skymap array and terrestrial annulus array contribute separately to the sensitivity calculation. For each principal skymap bin all the terrestrial annulus bins are sampled, and if a terrestrial annulus bin is above the horizon with respect to the selected principal skymap bin then an efficiency calculation is performed.

The burst spectral model is used to calculate a nominal peak flux in the peak flux energy range of interest (conventionally, 50-300 keV). For each detector the photon spectrum is folded through the sum of the direct and atmospheric scattering response matrices to calculate the counts in the selected trigger range. Then, for each bin in the counts intensity range, the calculated nominal peak flux is scaled by the ratio of the nominal peak counts to the threshold value appropriate for the value of the background count level in the counts bin, typically 5.5 × the square root of the counts value, to obtain the threshold peak flux value for that detector. Finally the second lowest threshold of the eight detectors is selected as the trigger threshold peak flux value. This procedure is performed for the product of each of the counts bins times each of the terrestrial annulus bins times each of the principal skymap bins.

The counts and response matrix databases are calculated for each spacecraft pointing. The efficiency curve for the entire mission is calculated by summing the analysis for each pointing. This formalism allows for considerable flexibility for constructing particular instrument sensitivity vs. intensity curves.

PROTOTYPE SKYMAP CALCULATION

A prototype efficiency calculation has been performed using principal skymap and terrestrial annulus coordinate systems with reduced binning and a representative counts database to determine the most significant properties of this new efficiency calculation. A typical GRB spectrum characterized by a Band function fit was used to calculate the BATSE trigger efficiency in the 50-300 keV range on the 1024 ms time scale. In Figure 1, this new efficiency curve is shown as a thick line, whereas the older efficiency curve is shown as a thinner line. The new calculation shows improved efficiency near threshold primarily due to the inclusion of atmospheric scattering in the threshold modeling.

We have used our prototype skymap efficiency algorithm to model BATSE's burst detection efficiency with the trigger set in the 25-100 keV energy range and the $E > 100$ keV range. In Figure 2, the 50-300 keV efficiency is shown (thick solid line) along with the 25-100 keV efficiency (thin solid line) and the $E > 100$ keV efficiency (dotted line). These efficiencies explain the most significant differences between the intensity distributions observed by BATSE when it triggers in these alternate energy ranges. The 25-100 keV trigger intensity distribution shows the most flattening below 0.6 photons cm^{-2} sec^{-1} as the trigger efficiency calculation would indicate. The $E > 100$ keV trigger efficiency distribution shows more flattening near threshold than the nominal 50-300 keV intensity distribution in accordance with the efficiency calculation as well.

This new efficiency is used to estimate the burst intensity distribution near threshold. To calculate this estimate correctly it is necessary to include the effects of the uncertainty in the calculated photon flux, given an observed counts flux above threshold. When the new skymap and corrections for the photon peak flux uncertainty are applied to the peak flux data on the 1024 ms timescale, the corrected differential intensity distribution shown in Figure 3 is obtained. These preliminary analyses show that the data require no significant changes in the shape of the burst intensity distribution near threshold.

However, these calculations are relatively simple and the full skymap formalism will allow for more accurate comparisons of model intensity distributions with the observations.

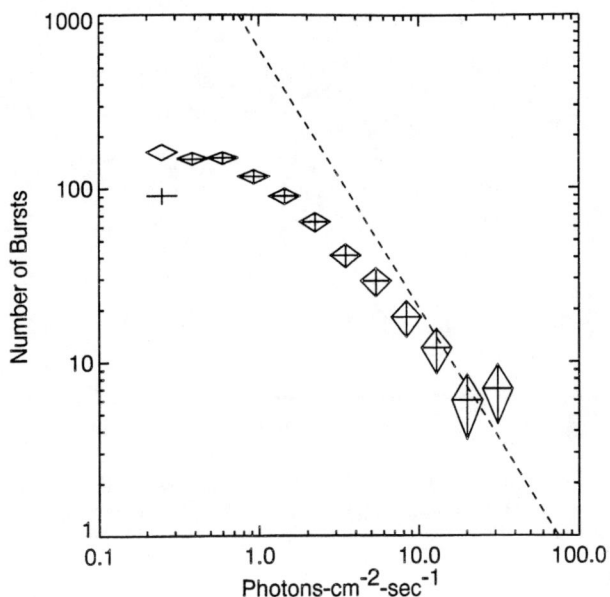

FIGURE 3. Differential peak flux distribution on the 1024 ms timescale for the 3B data. Crosses: uncorrected data; diamonds: efficiency corrected data.

REFERENCES

1. Brock, M. N., et al., in *Proceedings of 1st Huntsville Gamma-Ray Burst Workshop*, AIP Conference Proceedings No. **265**, Ed. W. Paciesas, G. Fishman, p 399 (1992).
2. Pendleton, G. N., et al., *ApJ* **464**, 606 (1996).

Wide Field X-Ray Telescopes: Detecting X-Ray Transients/Afterglows Related to GRBs

René Hudec[1], Ladislav Pina[2], Adolf Inneman[3], and Paul Gorenstein[4]

[1] Astronomical Institute of Czech Academy of Science, Observatory Ondřejov, 251 65 Ondřejov, Czech Republic,
[2] Faculty of Nuclear Engineering, Czech Technical University, Prague, Czech Republic,
[3] Department of Precision Mechanics and Optics, Faculty of Mechanical Engineering, Czech Technical University, Prague, Czech Republic,
[4] Harvard-Smithsonian Center for Astrophysics, Cambridge, MA, USA

Abstract. The recent discovery of X-ray afterglows of GRBs opens the possibility of analyses of GRBs by their X-ray detections. However, imaging X-ray telescopes in current use mostly have limited fields of view. Alternative X-ray optics geometries achieving very large fields of view have been theoretically suggested in the 70's but not constructed and used so far. We review the geometries and basic properties of the wide-field X-ray optical systems based on one- and two-dimensional lobster-eye geometry and suggest technologies for their development and construction. First results of the development of double replicated X-ray reflecting flats for use in one-dimensional X-ray optics of lobster-eye type are presented and discussed. The optimum strategy for locating GRBs upon their X-ray counterparts is also presented and discussed.

INTRODUCTION

For recently investigated GRBs with precise localization accuracy, in almost all cases variable and/or fading X-ray counterparts/afterglows have been identified. The X-ray identification of GRBs has led to great improvements in the study and understanding of these sources and especially has allowed identifications at other wavelengths due to better localization accuracy provided in X-rays as compared to gamma-ray observations. Since most of GRBs seem to be accompanied by X-ray emissions, the future systematic monitoring of these X-ray transients/afterglows is extremely important. However, these counterparts are faint in most cases, hence powerful wide-field telescopes are needed. An obvious alternative seems to be the use of wide-field X-ray optics allowing the signal-to-noise ratio to be increased as compared to non-focussing devices.

The expected limiting sensitivity of lobster-eye telescopes is roughly $\sim 10^{-12}$ erg cm^{-2} s^{-1} for a 1 day observation in the soft X-ray range [1,2]. This is roughly consistent with the peak fluxes detected for X-ray afterglows of GRBs $\sim$ hours after GRBs [3], however there is hope that the fluxes are higher at X-ray peaks and also that the X-ray flux may vary from source to source. Furthermore, the wide-field X-ray telescopes may play an important role in monitoring of faint variable X-ray sources to provide better statistics of such objects (note the recent finding of two faint fading X-ray sources inside the gamma-ray box of GRB970616, [4]).

We report on the development of first wide-field lobster-eye type X-ray telescope prototype of Schmidt geometry. Such modules are expected to allow large lobster-eye telescopes to be constructed in the near future.

THE VERY WIDE-FIELD X-RAY OPTICS

The lobster-eye geometry X-ray optics offer an excellent opportunity to achieve very wide fields of view while the classical Wolter grazing incidence mirrors are limited to about 1 deg FOV.

Schmidt objectives

One-dimensional lobster-eye geometry was originally suggested by Schmidt [5], based upon flat reflectors. The device consists of a set of flat reflecting surfaces. The plane reflectors are arranged in a uniform radial pattern around the perimeter of a cylinder of radius R. X-rays from a given direction are focussed to a line on the surface of a cylinder of radius R/2. The azimuthal angle is determined directly from the centroid of the focused image. At glancing angle of X-rays of wavelength 1 nm and longer, this device can be used for the focusing of a sizable portion of an intercepted beam of X-ray incident in parallel. Focussing is not perfect and the image size is finite. But a one dimensional focusing device offers a wide field of view, up to maximum of 2π with the coded aperture. It appears practically possible to achieve an angular resolution of the order of one tenth of a degree or better. Two such systems in sequence, with orthogonal stacks of reflectors, form a double-focusing device. Such a device should offer a field of view of up to 1000 square degrees at moderate angular resolution. It is proposed that such devices in X-ray astronomy survey or monitor the sky.

Innovative very wide field X-ray telescopes have been suggested based on these optical elements but have not been flown in space so far. One of the proposals is the All Sky Supernova and Transient Explorer (ASTRE) [1]. This proposal also includes a cylindrical coded aperture outside of the reflectors which provide angular resolution along the cylinder axis. The coded aperture contains circumferential open slits 1 mm wide in a pseudo-random pattern. The line image is modulated along its length by the coded aperture. The image is cross-correlated with the coded aperture to determine the polar angle of one or more sources. The field of view of this system can be, in principle, up to 360 deg in the azimuthal direction and nearly 90% of the solid angle in the polar direction. To create this mirror system, a development of double-sided flats is necessary while the recent X-ray foils are one-sided in most cases. There is potential for extending the wide field imaging system to higher energy by the use of multilayer coatings in analogy to those described by Joensen [7], for flat reflectors in the Kirkpatrick-Baez geometry. These coatings exert a great deal of stress upon the substrate. The system must meet severe weight limitations and so the new development of double-sided flats reinforced by composite material to keep the minimal weight still at good mechanical stability must be initiated. This is the goal of the new development in which innovative technologies for double-sided flats are tested. The basis of the sandwich-type construction of the X-ray flats is an electroformed nickel layer which is deposited on plates of float glass. The nickel-coated plates of float glass are connected by means of carbon/fibre composite material and after hardening the set of connected plates will be cut/ground off. Subsequently, the plates of glass and the produced composite sandwiches with double-sided nickel mirror foils and inner composite reinforcement will be separated. These foils are lacquered on both sides so that the surface microroughness can reach values under 1 nm. The foils will be covered by a thin gold layer at the final stage.

The first lobster–eye X-ray telescope prototype was finished in 1996 and was tested recently. The prototype consists of two perpendicular arrays of flats (36 and 42 double–sided flats 100×80 mm each). The flats are 0.3 mm thick and gold-coated. The focal distance is 400 mm from the midplane. The FOV is about 6.5 degrees. The first results of optical tests indicate the performance close to those provided by mathematical modelling (ray–tracing); the more detailed results as well as results of X-ray tests will be published later.

Angel objectives

The idea of two dimensional lobster–eye type wide–field X-ray optics was first mentioned by Angel [8]. The full lobster–eye optical grazing incidence X-ray objective consists of numerous tiny square cells located on the sphere and is similar to the reflective eyes of macruran crustaceans such as lobsters. The field of view can be made as large as desired, and it is practical to achieve good efficiency for photon energies up to 10 keV. Spatial resolution of a few seconds of arc over the full field is possible, in principle, if very small reflecting cells can be fabricated.

This idea however has never been further developed because of difficulties in the production of numerous polished square cells of very small size (about 1×1 mm or smaller at lengths of order of tens of mm). On the other hand, the very wide field imaging of the sky in X-rays would have very important consequences for a number of applications in X-ray and gamma-ray astronomy.

This demand can be also solved by electroformed replication and the first test cells have been already successfully developed this way. The recent approach is based on electroforming and composite material technology to produce identical triangular segments with square cells while these segments will be aligned in quadrants onto a sphere. The first specimen cells of 2×2 mm were successfully produced, indicating that the electroforming can be valuable tool in this very complicated and not yet solved problem.

THE STRATEGY FOR LOCATING GRBS UPON THEIR X–RAY FADING COUNTERPARTS

The optimum strategy for locating gamma-ray bursts based upon their fading X-ray counterparts that we believe is most practical and least costly is to utilize a pointed, relatively light weight, two-dimensional lobster-eye telescope with a moderate field of view, about 12 deg × 12 deg, in conjunction with a crude positioning instrument like BATSE that detects the occurance of a burst in hard X-rays and gamma rays and provides information about the temporal structure and spectrum of the burst. The 2D telescope would be then point immediately at the crude position and stay on it for about 24 hours. If the resolution is 3' × 3' and the effective area 200 cm^2 we estimate that the sensitivity is better than 10^{-14} erg cm^{-2} s^{-1} (0.5 to 3 keV), which is easily sufficient to position the fading X-ray afterglow to better than an arcminute in many cases, as well as obtain a light curve. If half the gamma–ray bursts have X-ray counterparts, this system would detect and position a burst every three or four days. However, one would not have the the wide-field telescope to perform intermediate X-ray surveys with large photon collection during the 24 hour pointings. If the gamma-ray instrument provides better positions than BATSE then the 2D telescope could have a smaller field of view and be lighter and less costly. However, one would probably pay the price by diminishing the performance of the gamma-ray instrument in other respects such as less area, timing and bandwidth or greater cost. Hence, there would be a tradeoff.

DISCUSSION

The use of a very wide-field X-ray imaging system could be without doubt very valuable for many areas of X-ray and gamma-ray astrophysics. Results of analyses and simulations of lobster-eye X-ray telescopes have indicated that they will be able to monitor the X-ray sky at an unprecedented level of sensitivity, an order of magnitude better than any previous X-ray all-sky monitor. Limits as faint as $\sim 10^{-12}$ erg cm^{-2} s^{-1} for a 1 day observation in the soft X-ray range are expected to be achieved, allowing monitoring of all classes of X-ray sources, not only X-ray binaries, but also fainter classes such as AGNs, coronal sources, cataclysmic variables, as well as fast X-ray transients, including gamma–ray bursts and the nearby Type II supernovae. For pointed observations, limits better than 10^{-14} erg cm^{-2} sec^{-1} (0.5 to 3 keV) could be obtained, sufficient to detect X-ray afterglows to GRBs.

The production of corresponding optical elements can be reasonably achieved by methods of electroforming and composite replication as an alternative to other methods. The results obtained with the development of technology for production of large area and high quality one-sided X-ray foils are very promising and together with composite material technologies represent an important input for the development of double-sided flats needed for lobster eye geometries of X-ray optics.

ACKNOWLEDGEMENTS

The development of double-sided reflecting X-ray foils was supported by a grant within the US-Czech Science and Technology program, No. 930 37. The development of the Angel objective is supported by the grant provided by the Grant Agency of the Czech Republic No. 205/97/1223.

REFERENCES

1. Gorenstein, P., "All sky supernova and transient explorer (ASTRE)", in *Variability of Galactic and Extragalactic X-ray Sources*, A. Treves Ed. Associazione per L'Avanzamento dell'Astronomia, Milano-Bologna (1987).
2. Priedhorsky, W. C., Peele, A. G. and Nugent, G. A., *Mon. Not. R. Astron. Soc.* **279**, 733 (1996).
3. van Paradijs, J. et al., *Nature* **386**, 686 (1987).
4. Greiner, J., *IAUC* No. 6722 (1997).
5. Schmidt, W. H. K., *Nucl. Instr. and Methods* **127**, 285 (1975).
6. Byrnak, B. et al. "XSPECT – An X-ray Spectroscopy and Timing Mission Concept", submitted to Interkosmos by DSRI Lyngby (1987).
7. Joensen, K. D., Gorenstein, P., Wood, J., Christensen, F. E. and Høghøj, P., *SPIE* **2279** (1994).
8. Angel, J. R. P., *Astroph. J.* **233**, 364 (1979).

A Search for Gamma-Ray Burst Optical Emission with the Automated Patrol Telescope

Bruce Grossan[*], Saul Perlmutter[*] and Michael Ashley[†]

[*] University of California at Berkeley, Lawrence Berkeley National Laboratory
[†] University of New South Wales Dept. of Physics and Astronomy

Abstract. The Automated Patrol Telescope (APT) is a wide-field ($5° \times 5°$), modified Schmidt capable of covering large gamma-ray burst (GRB) localization regions to produce a high rate of GRB optical emission measurements. Accounting for factors such as bad weather and incomplete overlap of our field and large GRB localization regions, we estimate our search will image the actual location of 20–41 BATSE GRB sources each year. Long exposures will be made for these images, repeated for several nights, to detect delayed optical transients (OTs) with light curves similar to those already discovered. The APT can also respond within about 20 s to GRB alerts from BATSE to search for prompt emission from GRBs. We expect to image more than 2.4 GRBs yr.$^{-1}$ during γ-ray emission. More than 5.1 will be imaged yr.$^{-1}$ within $\sim$20 s of emission. The APT's 50 cm aperture is much larger than other currently operating experiments used to search for prompt emission, and the APT is the only GRB dedicated telescope in the Southern Hemisphere. Given the current rate of $\sim$25% OTs per X/γ localization, we expect to produce a sample of $\sim$10 OTs for detailed follow-up observations in 1–2 years of operation.

INTRODUCTION

This meeting featured observations of the two optical transients (OTs) associated with gamma-ray bursts (GRBs) so far. These important observations have produced some controversial results (see presentations by Caraveo, Lamb, and Fruchter), and they have not yet yielded either explanations of the physics of the bursts or an identification of the bursting object(s). An obvious next step is to find and study a *sample* of OTs, not just two examples as we have now. Observations of $\sim$10 or more OTs are needed to address the controversies and give a statistical basis to host galaxy incidence and other properties.

We have begun to use the wide-field ($5° \times 5°$) Automated Patrol Telescope (APT) to image GRB positions for the identification of GRB-associated OTs. The telescope, located at Siding Spring Observatory in Australia, is fully dedicated to this project. The wide field is essential to search the large GRB localizations from

BATSE, the majority of all GRB localizations. In one mode of operation, we will study delayed emission similar to the OTs already discovered. In a second mode of operation, we will search for prompt optical emission, actually during or shortly after gamma-ray emission.

PROJECT DESCRIPTION

Our telescope performs a regular observing program unless a GRB alert is received. The Gamma-Ray Coordinates Network (GCN) sends out alerts 1–3 s after a BATSE GRB reaches threshold; these alerts then reach our site via the Internet in 200–400 msec. The APT then interrupts its observations, slews to the GRB position, and begins integration.

1) Prompt Emission Search: Because the APT can make long slews in < 20 s, we can make optical observations during the tail end of γ-ray emission for many GRBs. Figure 1 shows that more than 47% of all bursts are longer than our 20 s slew time. Shorter bursts would still be observed less than 20 s after γ emission.

After the initial slew, a series of exposures of 10, 20, 40, ... seconds are taken to sample different time scales up to $\sim$2 hrs. Our CCD can be read in 6–10 s. We expect to reach a sensitivity of V>17.7 mag at 5σ in 10 s exposures.

2) Delayed Emission Search: For up to 4 nights after the GRB (double the time-to-peak for OT970508), 1 hr. of exposures are acquired every other hour the source position can be observed. In this way, we will sample light curves at good time resolution and build up long total exposure times, but still do some scheduled

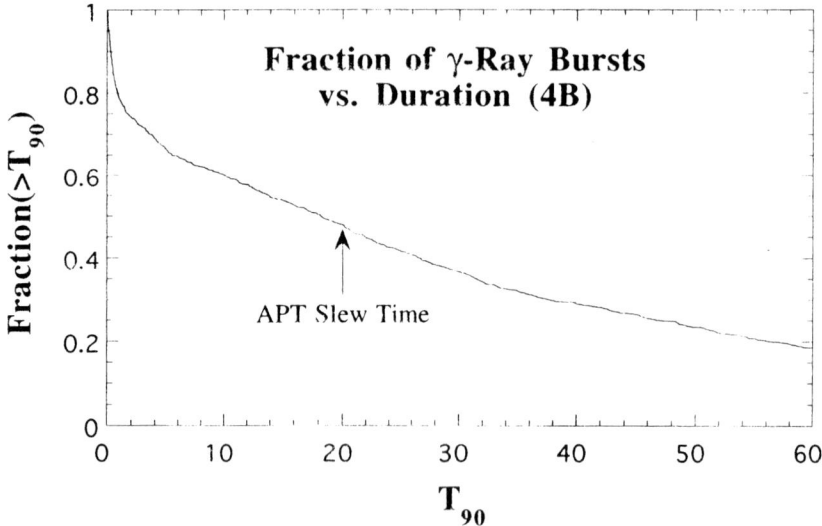

FIGURE 1. Duration (T_{90} = time to measure 90% fluence) of GRBs in the 4B BATSE catalog; many are longer than 20 s.

observing. The APT will reach a sensitivity of V>20.9 at 5σ in 1 hr. of co-added exposures. Fig. 2 shows that OT970508 would be detectable by the APT at better than 5σ in less than 1/2 hr. of exposure.

We will search for OTs by subtracting images taken just after the alert, and those taken later at the same position. Automated software will then examine the subtractions for transients, measure their position and brightness, and eliminate false candidates. To facilitate follow-up, our positions (better than $\sim 5''$) will be rapidly publicized using the GCN e-mail list.

Status

During March – April of 1997, a short pilot search was undertaken to demonstrate and test our system. While the APT was performing unrelated observations, and while no operator was present, the system responded to GRBs 970326b and 970329. The system worked as planned, producing a series of images at the alert locations for both bursts; one of them is shown in Figure 3. (The BACODINE original positions for these two "test" events were of very poor accuracy, and the GRB position was unlikely to be on our single-pointing fields.) During another run, GRB 970616 was observed for >1 hr. At this time, the APT requires an operator to open and close the telescope each night. Also, our CCD covers only a $3° \times 2°$ field at a quantum efficiency of only 28% in V. To implement our full search, we plan to upgrade the camera with a 79% quantum efficiency (at V) 2048×2048 detector to cover our $5°$ square field. We will also automate the telescope open/closing procedure and weather monitoring to enable unsupervised searching every clear night.

Event Rate

Important improvements are expected in the GCN which will provide LOCBURST quality positions (4°–11° 68% error diameter, see Kippen et al. and Briggs et al., this volume) as soon as most of the photons in a burst are received. These "updated positions" are expected to be in place by early 1998. Our "prompt emission search" event rate was calculated by taking into account time lost due to weather (35% for our site), and time lost due to the moon increasing our background by more than a mag per square arcsecond ($\sim$5% of the time not already counted), and the fraction (48%) of GRBs longer than 20 s (our estimated average time before pointing). Of the 27 updated positions per year for our site, we calculate that our system will slew to about 18 bursts per year, more than 8 longer than 20 s. Taking into account the overlap of our field with the updated position errors, **our prompt emission search will image more than 2.4 GRBs yr.$^{-1}$ during γ-ray emission, and more than 5.1 yr.$^{-1}$ within $\sim$20 s of emission.** (We use "image a GRB" here to mean that the actual position of the GRB is on the detector, resulting in either a measurement or an upper limit.)

Our "delayed emission search" event rate was calculated by taking into account weather and moon as above, but considering GRBs of all durations, and all of the sky accessible any time during the night (~64%). From 52–104 BATSE LOCBURST positions yr.$^{-1}$, **we estimate our delayed emission search will image 20–41 LOCBURST sources yr.$^{-1}$** by covering the larger error boxes with 2×2 mosaics. We will also acquire 4.7 images from RXTE ASM alerts, and 3.5 images from SAX alerts each year.

Comparison To Other Efforts

The LOTIS and ROTSE are currently operating rapid-slewing, wide-field 11 cm aperture instruments custom-built to search for prompt emission. They have faster response and a higher event rate; however, the APT's 50 cm aperture yields much greater sensitivity. LONEOS (see Wagner & Shrader, these proceedings) has a comparable aperture to the APT, but has a limited commitment to GRB observations, and lacks automated response. The APT is complementary to these projects, however, as they are located in the Northern Hemisphere, and the APT is located in the Southern Hemisphere for access to the far southern sky.

Conventional telescopes have been used to search for OTs, however, the APT program has many advantages over these searches. Most telescopes suffer from scheduling problems and can therefore miss bursts, they have a low event rate (mostly SAX alerts), and they lack the installed and maintained software required to identify OTs on the same night as their discovery. The latter capability is very important, in order that follow-up spectroscopy and other measurements may be made while the OT is near peak brightness. The rapid, dedicated APT program avoids these problems, and has an event rate more than six times that for typical telescopes (due to its BATSE follow-up capability). In addition, the APT is immune to SAX malfunctions. The most recent SAX problems include gyro failure, which means delays in producing sub-arcminute imager positions. This would make searches by typical telescopes much more time consuming.

CONCLUSION

A significant-sized sample of OT light curves and more detailed follow-up observations (e.g. spectroscopy) would be likely to yield significant, rapid progress in this field. The data would clarify controversies in the existing observations, such as host galaxy frequency and possible variable extended emission. A frequent lack of hosts would cause modifications to the neutron star – neutron star event scenario; confirmation of variable extended emission would locate some GRBs near our own galaxy. If the current trend continues, and $\sim 25\%$ of localized bursts yield OTs, our project's high event rate will produce a sample of observations of 10 OTs after only 1–2 years of operation.

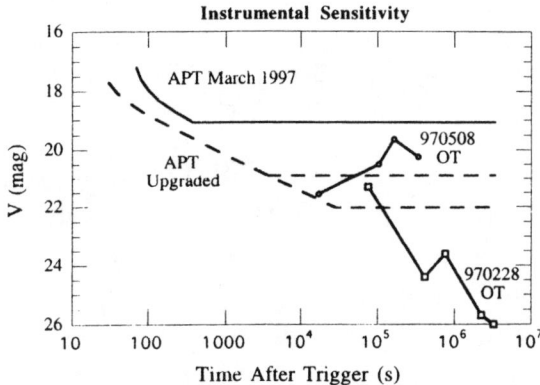

FIGURE 2. The 5σ sensitivity of the APT with increasing delay (and integration) time after a GRB trigger. The sensitivity of the APT is shown both truncated at 1 and 6 hours of co-added exposures; final verification of the sensitivity of our longest co-added exposures is now in progress.

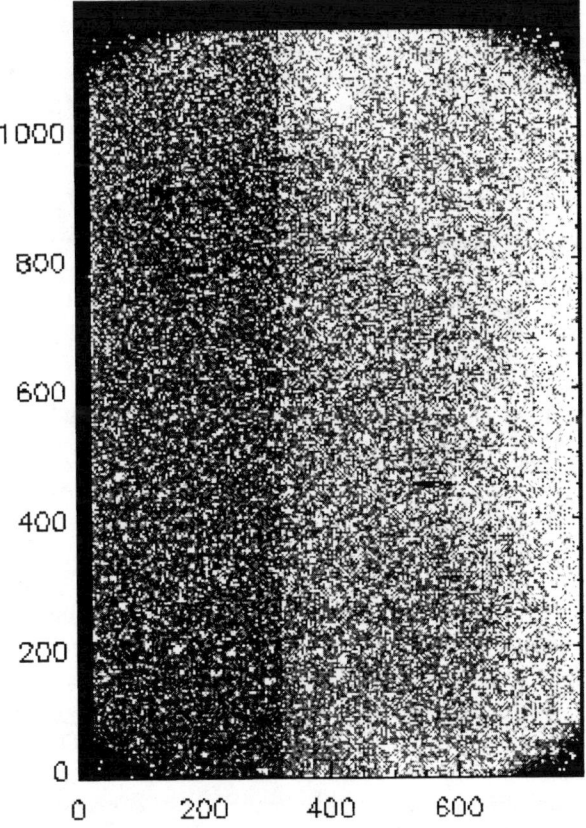

FIGURE 3. A 320 s image taken with the APT at the position of GRB970326b. The sensitivity is better than V= 19.6 at 5σ. Many sources are present in the wide field.

Search for GeV GRBs at Chacaltaya

A. Castellina[1], P.L. Ghia[1], C. Morello[1], G. Trinchero[1], P. Vallania[1], S. Vernetto[1], G. Navarra[2], O. Saavedra[2], H. Yoshii[3], T. Kaneko[4], K. Kakimoto[5], K. Nishi[6], R. Cabrera[7], D. Urgasti[7], A. Velarde[7], S.D. Barthelmy[8], P. Butterworth[8], T.L. Cline[8], N. Gehrels[8], G.J. Fishman[9], C. Kouveliotou[9] and C.A. Meegan[9]

[1] *Istituto di Cosmogeofisica del C.N.R., Torino, Italy*
[2] *Dipartimento di Fisica Generale dell'Universita' di Torino, Italy*
[3] *Department of Physics, Ehime University, Ehime 790, Japan*
[4] *Department of Physics, Okayama University, Okayama 700, Japan*
[5] *Department of Physics, Tokyo Institute of Technology, Meguro, Tokyo 152, Japan*
[6] *Institute of Physical and Chemical Research, Wako, Saitama 351-01, Japan*
[7] *Instituto de Investigaciones Fisicas, Universidad Mayor de San Andres, La Paz, Bolivia*
[8] *NASA Goddard Space Flight Center, Greenbelt, MD 20771, USA*
[9] *NASA Marshall Space Flight Center, Huntsville, AL 35812, USA*

Abstract. In this paper we present the results of a search for GeV Gamma Ray Bursts made by the INCA experiment during the first 9 months of operation. INCA, an air shower array located at Mount Chacaltaya (Bolivia) at 5200 m a.s.l., has been searching for GRBs since December 1996. Up to August, 1997, 34 GRBs detected by BATSE occurred in the field of view of the experiment. For any burst, the counting rate of the array in the 2 hours interval around the burst trigger time has been studied. No significant excess has been observed. Assuming for the bursts a power low energy spectrum extending up to 1 TeV with a slope $\alpha = -2$ and a duration of 10 s, the obtained 1 GeV – 1 TeV energy fluence upper limits range from $7.9 \cdot 10^{-5}$ erg cm^{-2} to $3.5 \cdot 10^{-3}$ erg cm^{-2} depending on the event zenith angles.

THE INCA EXPERIMENT

Gamma Ray Bursts (GRBs) in the 1 GeV – 1 TeV energy range can be observed by ground-based detectors measuring the secondary particles generated by primary photons in the atmosphere. Since at these energies the number of secondary particles reaching the detector is too small to measure the primary direction, a GRB is observable as an excess in the single-particle counting rate. The search for GRBs by measuring the single particle counting rate with Extensive Air Shower (EAS) experiments has been already exploited by the Plateau Rosa array at 3500 m a.s.l. [1] and by the EAS-TOP array at Gran Sasso at 2000 m a.s.l. [2-4].

The same technique has been recently applied to an EAS array working at Mount Chacaltaya, Bolivia, at 5200 m a.s.l. The geographic location (lat. 16.32S, long. 291.85E) allows a large part of the less-known southern celestial hemisphere to be monitored.

Moreover the altitude above the sea level (atmospheric depth 538 g cm^{-2}) noticeably increases the sensitivity to GeV photons. In fact, the effective area $A_{\text{eff}}(E,\theta)$ of an EAS array made by M modules of area A_d to detect a photon of energy E and zenith angle θ is extremely dependent on the atmospheric depth. If the number of secondary particles is small and the area of the modules is smaller than the typical size of a shower, $A_{\text{eff}}(E,\theta) \approx M\ A_d\ cos\theta\ \overline{N}_e(E,\theta)$, where $\overline{N}_e(E,\theta)$ is the mean number of charged particles (with energy larger then the detection threshold) reaching the level of observation, produced by a photon of energy E and zenith angle θ. The value of $\overline{N}_e(E,\theta)$ depends strongly on the atmospheric depth (i.e., it increases by a factor 30 going from 2000 m to 5000 m a.s.l. for a gamma primary of energy $E = 50$ GeV).

The experiment sensitivity is proportional to $A_{\text{eff}}(E,\theta)$ and inversely proportional to $\sqrt{B}$, where B is the background flux, i.e. the flux of secondary particles generated by cosmic rays of energy just above the geomagnetic cutoff. This is modulated by the atmospheric pressure, the solar activity and the 24 hours anisotropy, but the time scales of these phenomena are much larger than the typical GRB durations, hence they are not a source of noise for the the burst search. Since the background flux slightly grows with altitude (i.e. only by a factor $\sim$ 1.5-2 going from 2000 m to 5000 m a.s.l.), the sensitivity noticeably increases working at very high mountain altitude.

The INCA (INvestigating Cosmic Anomalies) experiment consists of 12 scintillator detectors of 2×2 m^2 area and 5 cm thickness, distributed over a $\sim 15 \times 15$ m^2 area. Each module is viewed by a 12 cm diameter phototube and has an energy threshold of 5 MeV. The average background rate is ~ 900 counts m^{-2} s^{-1}.

The detailed calculations of the sensitivity of INCA to GRBs as a function of different burst parameters (duration, energy spectral slope and energy cutoff) has been evaluated simulating electromagnetic cascades in the atmosphere generated by primary gamma rays [5,6]. Assuming a differential energy spectrum $dI/dE = K\ E^{-\alpha}$ ph cm^{-2} s^{-1} GeV^{-1} extending from $E = 1$ GeV up to a cutoff energy E_{max}, Figure 1 shows the minimum energy fluence in the range 1 GeV $\div$ E_{max} of a detectable Gamma Ray Burst (i.e. 4 standard deviations significance) with duration $\Delta t = 1$ s, as a function of E_{max}. The curves are given for different slopes of the spectrum and zenith angle $\theta = 0°$. The fluence for a generic duration Δt can be obtained by multiplying the given values by a factor $\sqrt{\Delta t}$. The sensitivity decreases by a factor $\sim$3-4 at $\theta = 30°$ and by a factor $\sim$10-20 at $\theta = 50°$, in the E_{max} and α range considered.

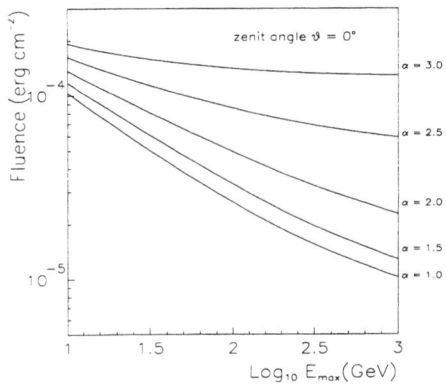 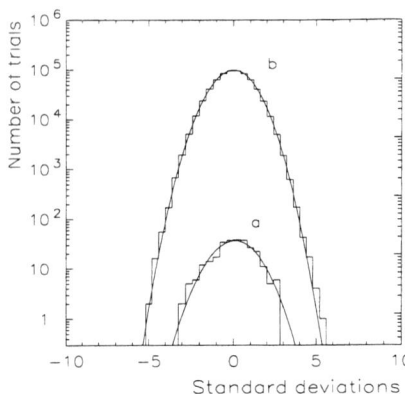

FIGURE 1. The minimum value F of the energy fluence in the range 1 GeV $< E < E_{\max}$, for a detectable burst of duration $\Delta t = 1$ s, calculated for 5 different values of the spectral slope α and for the zenith angle $\theta = 0°$.

FIGURE 2. The distribution of the differences between the number of counts and the expected background measured by INCA during 34 BATSE events, for 8 different duration *a)* in coincidence with BATSE time, *b)* in 2 hours around the BATSE time.

DATA ANALYSIS AND RESULTS

The search is performed by studying the fluctuations of the sum of the 12 detectors' single-particle counting rates, recorded every second. The Universal Time of each measurement is known with an accuracy of 100 μs. Working in single-particle mode requires good stability of the experiment and continuous and precise monitoring of the instrumental conditions. Hence, to reject possible electronic noise localized in some detectors, the counting rates of each module are recorded independently.

The INCA single-particle counting rate is studied during the occurrence of the events seen by the BATSE detector and transmitted to several experiments by the BACODINE system (BATSE COordinate DIstribution NEtwork [7]. Only about half of the BACODINE notifications are true astrophysical GRBs, the other ones being solar flares, electron precipitation events, etc. In this analysis we considered only the BACODINE events that had been later recognized by the BATSE group as real GRBs and with arrival directions corresponding to a zenith angle $\theta < 60°$ at the Chacaltaya location. After this selection 34 BACODINE events recorded from December 16, 1996 up to August 8, 1997 were taken into account.

For each BACODINE event, the INCA data recorded during 10000 s around the burst time were considered. In this time interval, the counting rates of each detector were carefully studied in order to identify possible electronic noise or anomalous behaviour. On average, 29% of the data were excluded from the analysis. Finally, the detector counts were summed and the time distribution of the total counting rate studied to single out statistically significant fluctuations.

Since BACODINE notices do not contain information about the duration of the bursts, we looked for excesses of different durations $\Delta t = 1, 2, 6, 10, 20, 50, 100, 200$ s, setting the excess start time in time coincidence with the BATSE trigger time. The counts C recorded in Δt were compared with the expected background C_B calculated using the counts measured in 30 minutes around Δt. Curve a in Figure 2 shows the distribution of the $C - C_B$ difference in units of standard deviations obtained in the 272 trials (34 events × 8 durations). We found no statistically significant excess for any burst and duration.

Looking for possible delayed or anticipated excesses with respect to the BATSE burst, the same search was performed shifting the time window Δt by steps of $l = \Delta t/2$ (except for $\Delta t=1$ s, $l = 1$ s) in a 2 hour time interval centered around the BATSE time. The distribution of the $C - C_B$ difference in units of standard deviations obtained in the 6.6×10^5 trials is plotted in Figure 2, curve b. In this case, as well, we found no significant excesses. Curves a and b are fitted by Gauss distributions respectively with r.m.s. = 1.06 and 1.17, showing the stability of the counting rates.

CONCLUSIONS

The upper limit to the energy fluence in the energy range 1 GeV – 1 TeV in coincidence with the BATSE trigger was calculated for the 25 events with a zenith angle $\theta < 50°$, assuming for the burst a power-law differential spectrum extending from 1 GeV to 1 TeV with a slope $\alpha = -2.0$ and a time duration $\Delta t = 10$ s. The obtained limits, calculated at 4 standard deviations level, range from 7.9×10^{-5} erg cm^{-2} to 3.5×10^{-3} erg cm^{-2}. This relatively large range is due to the change in the experiment sensitivity at different zenith angles.

These values, and the fluence limits given in Fig. 1, show that INCA can, in principle, observe GRBs with comparable intensity as the most powerful ones so far detected, if the energy spectrum does not cut off below $\sim$ 100 GeV – 1 TeV and the zenith angle is $< 30°$.

Fluences as large as few 10^{-5} erg cm^{-2} have been measured by EGRET in the 30 MeV – 5 GeV energy range during GRB940217 [8] and in the 1 MeV – 1 GeV energy range during GRB930131 [9]. In both cases the spectral slope was ~ -2 (that implies equal amount of power per decade of energy) and no energy cutoff was visible up to several GeV, the maximum energy that the instrument is sensitive to detect. Even if GRBs are located at cosmological distances, as recent observations suggest [10,11], the absorbtion due to the interactions with the starlight photon field should not affect seriously the flux of these very intense GRBs in the region 1 GeV – 1 TeV, since the distances of the sources (considering GRBs as standard candles) are probably less than $z \sim 0.1$.

REFERENCES

1. Morello, C., et al., *Il Nuovo Cimento* **7C**, 682 (1984).
2. Aglietta, M., et al., *Il Nuovo Cimento* **15C**, 441 (1992).
3. Aglietta, M., et al., *Astrophys.and Space Science* **231**, 351 (1995).
4. Aglietta, M., et al., *ApJ* **469**, 305 (1996)
5. Castellina, A., et al., *Il Nuovo Cimento* **20C**, no.2, 137 (1997)
6. Barthelmy, S. D., et al., *Proc. XXV ICRC, Durban* **3**, 73(1997)
7. Barthelmy, S. D., et al., *Proc. of the Second Huntsville GRB Workshop* **307**, 643 (1994).
8. Hurley, K., et al., *Nature* **372**, 652 (1994).
9. Sommer, M., et al., *ApJ* **422**, L63 (1994).
10. Costa et al., IAU Circ. 6572 & IAU Circ 6576 (1997).
11. Piro et al., IAU Circ. 6617 (1997).

SOFT GAMMA-RAY REPEATERS

Observations of SGR1806-20 with the Konus-Wind and Konus-A Experiments in 1996–97

D.D. Frederiks, R.L. Aptekar, S.V. Golenetskii, V.N. Il'inskii, E.P. Mazets, and M.M. Terekhov

Ioffe Physical-Technical Institute, St.Petersburg, 194021, Russia.
E-mail: fred@mz.ioffe.rssi.ru

Abstract. In November 1996 the Konus-Wind instrument on board the GGS-Wind spacecraft and the Konus-A instrument on board Kosmos-2326 observed 19 short events with energy spectra softter than typical gamma-ray bursts. Some of these were located at the position of the SGR 1806-20 using the Konus-Wind to Konus-A time delay and, independently, from the angular responses of the array of four DN detectors of Konus-A. Several similar events were detected later in June and September 1997. Time histories and energy spectra of the strongest events are presented. Some similarities in the time histories and spectral behaviour of the observed events are noted.

INTRODUCTION

Among known soft gamma repeaters observed in their periods of activity, SGR 1806-20 exhibits the highest burst rate. A number of short soft bursts were observed in 1996–97 by many experiments including Konus-Wind [1] and Konus-A [2]. A list of events is presented in Table 1. A majority of these events belong to SGR 1806-20. However, strong evidence was obtained that in at least two events, namely the bursts of 970629 and 970912, a new SGR manifests itself [3]. We present here the time histories, the energy spectra and their variability for observed events, noting evident intriguing similarities and characteristics.

TIME HISTORIES

From Table 1 one can see whether the listed bursts were detected in the triggered burst or in the background mode, depending on the intensity of the event, and by both or by only one of the instruments. Figure 1 presents the time histories of the events recorded in the burst mode with a resolution of 2 ms to 8 ms. In spite of earlier notions [4], the time histories reveal often complex structures with well

TABLE 1. Summary of the observations.

Date	Konus-Wind UT, s	Konus-A UT, s	Burst mode
961105a	61338	-	No
961105b	63250	-	No
961118a	17555	-	No
961118b	19510.181	19510.160	Yes
961118c	19556.133	19556.155	Yes
961118d	-	22843	No
961118e	41172	41172	No
961118f	57897.506	57897.547	Yes
961119a	11048	11048.608	Yes
961119b	-	16122	No
961119c	19852.966	-	Yes
961119d	-	21323	No
961119e	-	21537	No
961119f	22927.736	-	Yes
961119g	26644.152	26644	Yes
961119h	-	26961	No
961121	25539	-	No
961122	08258.596	08259	Yes
970629a	14424	-	No
970629b	23493.220	-	Yes
970902	39346.778	-	Yes
970912	22164.078	-	Yes

pronounced separate pulses. Burst activity often starts a comparatively long time before the main phase of an event. The detection of a burst by two instruments with a high temporal resolution permits us to localize the source, as in the case of the events of November 18, 1996 (Figure 2).

ENERGY SPECTRA

Photon energy spectra are shown for several of the most intense bursts in Figure 4. An accumulation time in the range of 64 to 256 ms, needed to obtain an acceptable statistical accuracy, is often longer than the rise and fall times in these events. As a result, real spectral changes may be suppressed. More obvious data on spectral variability follow from considering the hardness ratio behaviour. An example of this is shown in the Figure 3. In many cases, the hardness ratio varies in concordance with the current count rate.

CONCLUSION

The observed properties definitely show that there are some resemblances between SGRs and common GRBs, e.g. a complex time structure and spectral variability. However, the specific time scale for SGRs is shorter and the specific energy

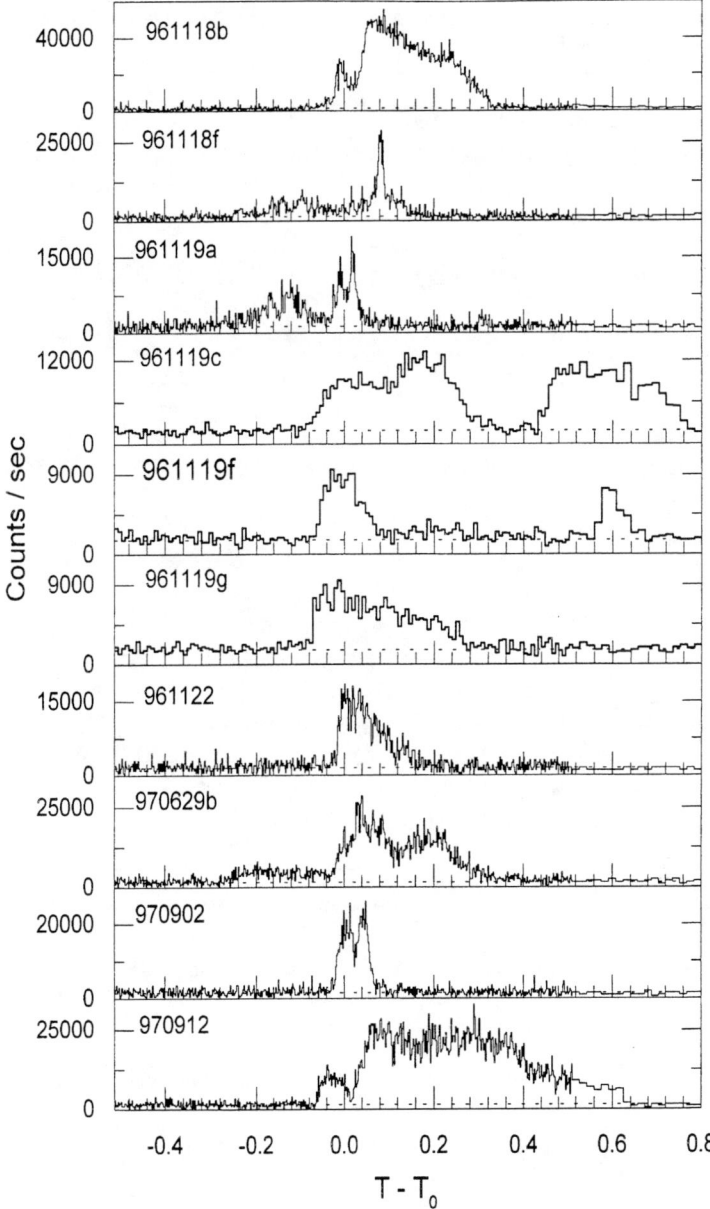

FIGURE 1. Time histories of SGR bursts in the 14 to 230 keV energy range. The 970629b and 970912 events are from the possible new SGR [3].

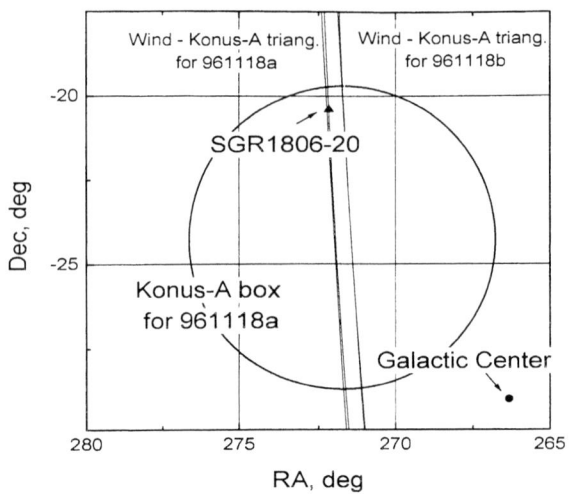

FIGURE 2. Localization of two SGR events of November 18, 1996.

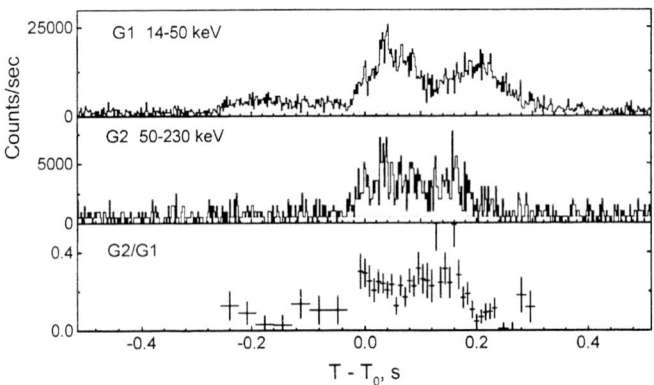

FIGURE 3. Burst of June 29b, 1997. The hardness ratio is displayed on the lower panel.

is much softer. Similarities seen in time structures and spectral behaviour of SGRs may yet prove to be signatures of the processes responsible for burst emission in the sources. A more comprehensive study of SGRs will be carried out, combining all data obtained by various experiments. This work will soon be in progress.

Acknowledgments

This work was supported partly by RSA contracts, by RFBR grants N96-02-16860 and N97-02-18067, and by CRDF grant RP1-236.

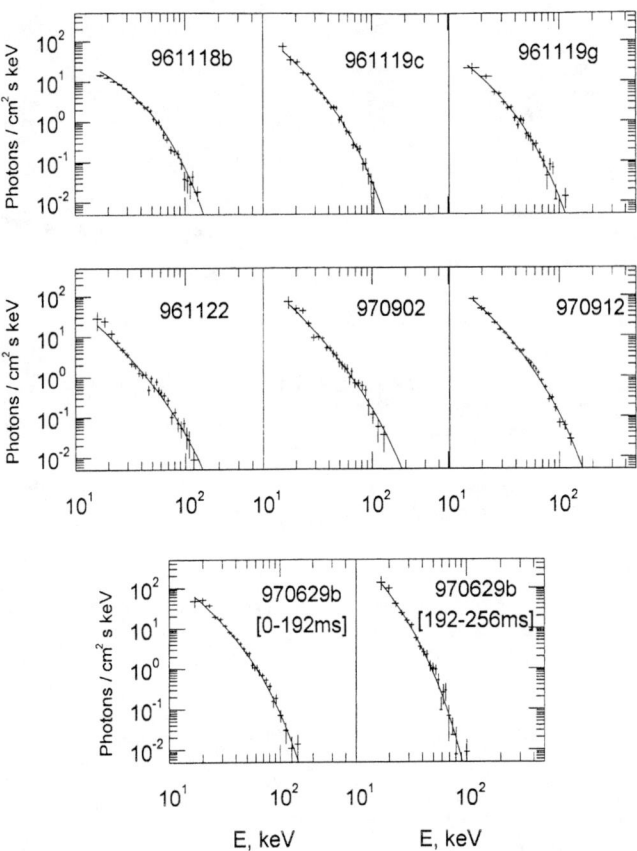

FIGURE 4. Time-averaged photon spectra of some SGR bursts are well fitted by the $E^{-\alpha}\exp(E/kT)$ law with $\alpha \sim 1$ and kT of 17–21 keV. The only exception is the spectrum of the final phase of the 970629 event with $kT \sim 8.5$ keV.

REFERENCES

1. Aptekar, R.L., et al., *Space Science Review* **71**, 265 (1995).
2. Aptekar, R.L., et al., *Astronomy Letters* **23**, 147 (1997).
3. Hurley, K., et al., *IAU Circ.* 6743 (1997).
4. Rothschild, R., in *High Velocity Neutron Stars and Gamma-Ray Bursts*, AIP Conf. Proc. 366, eds. R.E. Rothschild & R.E. Lingenfelter, 1995, p. 51.

HEXTE Observations of SGR 1806–20 During Outburst

D. Marsden[1], R. E. Rothschild[1], C. Kouveliotou[2,4], S. Dieters[2,5], J. van Paradijs[3,5]

[1] *Center for Astrophysics and Space Sciences, University of California at San Diego, La Jolla, CA 92093*
[2] *ES-84, NASA/MSFC, Huntsville, AL 35812*
[3] *University of Amsterdam, Astronomical Institute "Anton Pannekoek" Kruislaan 403, 1098 SJ, Amsterdam*
[4] *Universities Space Research Association*
[5] *University of Alabama in Huntsville*

Abstract. We discuss observations of the soft gamma repeater SGR 1806–20 during the RXTE Target of Opportunity observations made in November 1996. During the ~ 50 ksec RXTE observation, HEXTE (15 – 250 keV) detected 17 bursts from the source, with fluxes ranging from 3×10^{-9} to 2.2×10^{-7} ergs cm^{-2} s^{-1} (20 – 100 keV). We obtained spectra for the brighter HEXTE by fitting thermal bremsstrahlung and power law functions over the energy range 17 – 200 keV. The best-fit temperatures and photon indices range from 30 – 55 keV and 2.2 – 2.7, respectively. The weighted average temperature of the HEXTE bursts was 41.8 ± 1.7 keV, which is consistent with previous SGR 1806–20 burst spectra. The persistent emission from SGR 1806–20 was not detected with HEXTE.

INTRODUCTION

Soft gamma repeaters exhibit long periods of quiescence, often spanning years, punctuated by periods of intense bursting activity during which many brief (durations < 1 s) and intense (luminosities $L \sim 1 - 10^3$ L$_{\text{Edd}}$) bursts are emitted by the source [2]. Believed to be neutron stars, the mechanism(s) for both the steady and bursting X-ray emission is still not well understood [1].

SGR 1806-20 is the most prolific SGR, and it has been studied in the X-ray [3], optical [4], infrared [5], and radio [6] bands. The source became active again during the Fall of 1996, emitting many powerful bursts that were first detected with BATSE [7]. A target of opportunity observation by the *Rossi X-ray Timing Explorer* (RXTE) was initiated on November 5, 1996. The data analyzed here were taken during that 50 ksec observation, which spanned the time interval starting at 10:53:20 UT (11/5/96) and ending at 10:52:00 UT (11/6/96).

INSTRUMENTATION

The HEXTE instrument [8] aboard RXTE consists of two clusters of collimated NaI/CsI phoswich detectors with a total net area of ~ 1600 cm^2 and an effective energy range of $\sim 15 - 250$ keV. The SGR observations discussed here were taken with the HEXTE in the 16 second rocking mode, in which one cluster is always on the source, with the other pointed off-source for background accumulation. The clusters then beam-switch every 16 seconds, in such a way that one cluster is always locked on-source.

RESULTS

The on-source HEXTE data were binned into a time series for 3 energy bands ($15 - 50$ keV, $50 - 100$ keV, and $100 - 200$ keV). The $15 - 50$ keV time series for each continuous data stretch was searched for bursts using a Bayesian burst search algorithm developed at UCSD. The algorithm calculates the probability of a given number of bursts in each data stretch by incorporating the information on the expected background flux in each time bin. The burst search yielded 17 bursts, all of which correspond to bursts seen by the PCA. The burst times and durations are shown in Table 1, and the lightcurve of two of the brighter bursts (5 & 6) is shown in Figure 1. The durations of the bursts seen by HEXTE ranges from less than 0.1 to 0.6 seconds, and the weakest burst detected by HEXTE corresponds to a PCA count rate of 800 counts s^{-1}.

TABLE 1. SGR 1806–20 bright HEXTE bursts.

Burst Time[a]	Duration[b]	Photon Index[c]	kT[d]	Flux[e]
0.7106674	0.5	2.29±0.05	51.46±3.30	21.65±0.47
0.7220713	0.5	2.52±0.07	37.62±2.76	12.53±0.68
0.7222172	0.3	2.54±0.12	40.88±4.93	8.9±0.6
0.7306593	0.2	2.16±0.19	54.19±14.47	5.1±0.5
0.8404614	0.4	2.65±0.15	31.06±4.50	8.7±1.0
1.1905632	0.3	2.32±0.14	45.67±7.85	10.86±0.11

[a] Terrestrial dynamical time (MJD modulo 50392)
[b] Seconds
[c] Power-law fit
[d] Thermal bremsstrahlung fit (keV)
[e] $20 - 100$ keV flux (10^{-8} ergs cm^{-2})

All of the HEXTE bursts were fit to thermal bremsstrahlung and power-law functions using XSPEC, but only the 6 brightest bursts yielded well-constrained spectral fits. In all the burst spectral fits, background was taken from *on-source* data taken immediately preceding and following the burst. The counts spectra were fit over the energy range $17 - 175$ keV, and the resulting best-fit parameters are shown in Table 1. The weakest SGR 1806–20 burst seen by HEXTE has a $20 - 100$ keV flux of 3×10^{-9} ergs cm^{-2} s^{-1}. Co-adding the 11 dim bursts and fitting

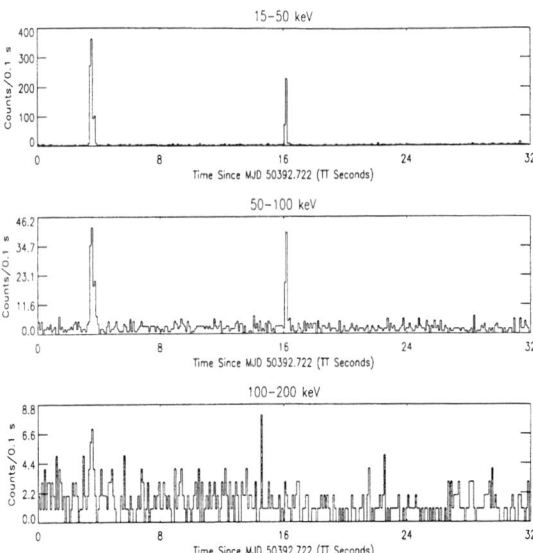

FIGURE 1. The lightcurves of SGR 1806-20 Bursts 5 & 6 as seen by HEXTE. The time resolution is 0.1 s bins^{-1}.

them with power-law and bremsstrahlung functions yields the best-fit parameters $\alpha = 2.05 \pm 0.14$ and $kT = 57 \pm 12$ keV for the mean spectrum of the weak HEXTE bursts. The HEXTE counts spectrum of a bright burst (Burst 2) is shown in Figure 2, with the best-fit thermal bremsstrahlung spectrum overplotted. The χ^2 values were typically large for the stronger bursts, suggesting that more complicated spectral models may be needed to adequately fit the burst spectra.

The weighted mean effective temperature of the six bright bursts and the fit to the mean spectrum of the weak HEXTE bursts is $kT = 41.8 \pm 1.7$ keV. This is consistent with previous measurements of SGR 1806-20 burst temperatures [9]. A chi-squared test for a constant temperature yields $\chi_\nu^2 = 3.2$ for $\nu = 6$, or a 0.5% chance that the bursts all had the same temperature, suggesting that there may be some intrinsic variability in the burst spectra.

DISCUSSION

The results of the HEXTE observations of SGR 1806-20 bursts are in general agreement with the durations [10] and bremsstrahlung temperatures [9] obtained by previous observers. The X-ray luminosities (20 – 100 keV) of the bursts, in units of the Eddington luminosity, span the range $L_X \sim 1 - 50$, assuming a distance to the source of 14.5 kpc [11], isotropic emission, and an Eddington luminosity of 10^{38} ergs s^{-1}.

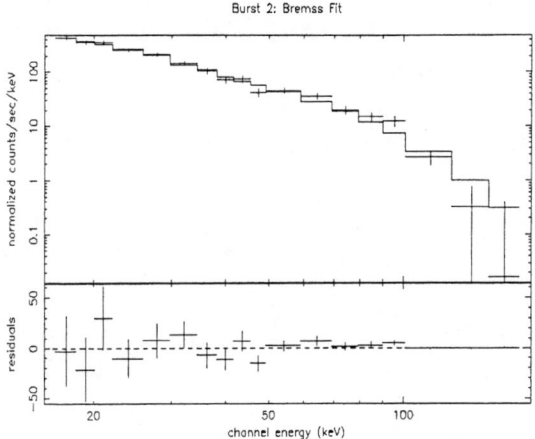

FIGURE 2. The HEXTE count spectrum of a bright SGR 1806-20 burst (Burst 2). The solid line is the best-fit thermal bremsstrahlung model, and the residuals are shown in the bottom panel.

In the future, we plan on using the data from the Proportional Counter Array (PCA) aboard RXTE, in conjunction with the HEXTE data, to fit the SGR 1806-20 burst spectra over the entire 2 − 250 keV energy range of the two instruments. This will result in a more accurate determination of the continuum spectral shape of the SGR bursts.

ACKNOWLEDGEMENTS

We thank NASA for support under grants NAS5-30720 (D.M. and R.E.R.), NAG5-2560 (S.D. and C.K.), and NAG5-4878 (JvP)

REFERENCES

1. Thompson, C., & Duncan, R.C., *MNRAS* **275**, 255 (1995).
2. Norris, J.P., et al., *ApJ* **366**, 240 (1991).
3. Sonobe, T., et al., *ApJ* **436**, L23 (1994).
4. van Kerkwijk, M.H., et al., *ApJ* **444**, L33 (1995).
5. Kulkarni, S.R., et al., *ApJ* **440**, L61 (1995).
6. Kulkarni, S.R., et al., *Nature* **368**, 129 (1994).
7. Kouveliotou, C., et al., *IAUC* 6501 (1996).

8. Gruber, D.E., et al., *Astronomy & Astrophysics* **120**, 641 (1996).
9. Atteia, J.L., et al., *ApJ* **320**, L105 (1987).
10. Kouveliotou, C., in *Towards the Source of GRBs*, 29th ESLAB Symposium, 1995, pp. 49–56.
11. Corbel, S., et al., *ApJ*, in press (1997).

Mid-Infrared Spectra of SGR 1806-20 and SGR 1900+14

William A. Mahoney[1], Stephane Corbel[2], Ph. Durouchoux[2], James C. Higdon[3], Michael E. Ressler[1], and Pierre Wallyn[2]

[1] *Jet Propulsion Laboratory 169-327, 4800 Oak Grove Drive, Pasadena, CA 91109*
[2] *DAPNIA, Service d'Astrophysique, CE Saclay, 91191 Gif sur Yvette Cedex, France*
[3] *Joint Science Department, The Claremont Colleges, Claremont, CA 91711*

Abstract. Two of the three known soft gamma repeaters, SGR 1806-20 and SGR 1900+14, have a fairly small angular separation and are visible from Mt. Palomar. During the nights of 21-23 June 1997 we observed both in the mid-infrared using SpectroCam-10 on the 5-meter Hale telescope. We obtained excellent images of the counterpart to SGR 1900+14 in 6 bands using narrow filters ($\Delta\lambda \sim 1$ μm) from approximately 8 to 13 μm. The intensities in these 6 bands, summed over the two star-like components of the suspected counterpart, were all ~2-3 Jy and together yielded the first mid-infrared spectrum of this object. A spectrum of the SGR 1806-20 counterpart was also obtained, but it displayed a significantly different shape and showed a source intensity of only ~0.1 Jy. When combined with near-infrared and *IRAS* observations, the spectra imply that the counterparts of both SGRs consist of multiple components at different temperatures.

INTRODUCTION

Soft gamma repeaters (SGRs) form a unique class of objects distinguished from classical gamma-ray bursts by (a) simple light curves, (b) short durations (typically ~ 0.1 s), (c) soft spectra (typically consistent with 30 - 40 keV thermal bremsstrahlung), and (d) clear repetitions of bursting activity [1]. Two of the three SGRs known, SGR 1806-20 and SGR 1900+14, are easily observable from Mt. Palomar. SGR 1900+14 has been tentatively identified with a peculiar double infrared source [2] also seen by *IRAS*, while SGR 1806-20 has been associated with the supernova remnant G10.0-0.3 and with a highly luminous blue variable star [3]. Both SGRs are located near the Galactic plane at estimated distances of ~14 kpc [4,5] and the suspected counterparts of both show heavily reddened spectra that peak in the infrared. The possible association of SGRs with supernovae remnants suggests a neutron star connection, however, the mechanism responsible for the bursting activity of SGRs, as well as classical gamma-ray bursts, remains a mystery. To further illuminate the nature of these objects, we observed both in the mid-infrared during the nights of 21–23 June 1997 using SpectroCam-10 on the 5-meter Hale telescope. We obtained spectra by imaging with 6 narrow filters ($\Delta\lambda \sim 1$ μm) covering approximately 8 to 13 μm. The counterparts of both SGRs were clearly resolved as point-like sources in the mid-infrared region, however, their intensities and spectral shapes are quite different.

OBSERVATIONS AND DATA ANALYSIS

SpectroCam-10 is a mid-infrared camera and spectrograph built by Cornell University for the 5-meter Hale Telescope [6]. Its detector is a 128×128 Si:As Back Illuminated Blocked Impurity Band (BIBIB) array optimized for observations from 8 to 13 μm, but with useful sensitivity near 4 and 18 μm. In the camera mode the central 64×64 pixels are used for imaging a field-of-view of 16″. We obtained images using filters having ~ 1 μm bandpass and centered at 7.9, 8.8, 9.5, 10.3, 11.7, and 12.5 μm as well as $3-5$ μm and 17.9 μm filters. In addition to imaging, we made low-resolution ($R \equiv \lambda/\Delta\lambda = 100$) spectra of SGR 1900+14 using a 1″ slit aligned to include both components. As we have not yet completed analysis of the spectrograph data, this paper reports only on the results of the imaging observations.

During each of the three nights the sky was clear and the seeing fair ($\sim 1″$ at 10 μm). Background subtraction was facilitated by chopping the source at a frequency of 5.6 Hz with a displacement of 8″, about half the field-of-view. To maximize the useful observing time, the telescope was pointed so that the source was within the field-of-view for both chop angles. Data analysis was carried out with a combination of IDL routines written specifically for analysis of SpectroCam-10 data and IRAF, with all the photometry carried out using the IRAF PHOT program and an aperture radius of $\sim 2″$.

SGR 1900+14

Through a series of gamma-ray, X-ray, and infrared observations [7,8], SGR 1900+14 has been associated with a near-infrared source consisting of two variable M5 supergiants [5] separated by 3.37″ [8]. Our SpectroCam-10 images also showed two clearly resolved components consistent with point sources in all bands through 12.5 μm.

The infrared spectrum summed over the two stars is shown in Figure 1. The near-infrared fluxes are from our previous observations [8] with the Cassegrain Infrared Camera on the Hale telescope. The broad-band emission centered at 12, 25, and 60 μm was derived by van Paradijs et al. [11] using HIRAS, a maximum entropy program specifically developed to enable high-resolution imaging with *IRAS* data. For both SGR 1806–20 and SGR 1900+14, van Paradijs et al. found the emission was point-like at 12 and 25 μm while spatially extended at 60 μm. Also shown in Figure 1 is the flux measured by Smith et al. [7] at 10 μm using the Thermal

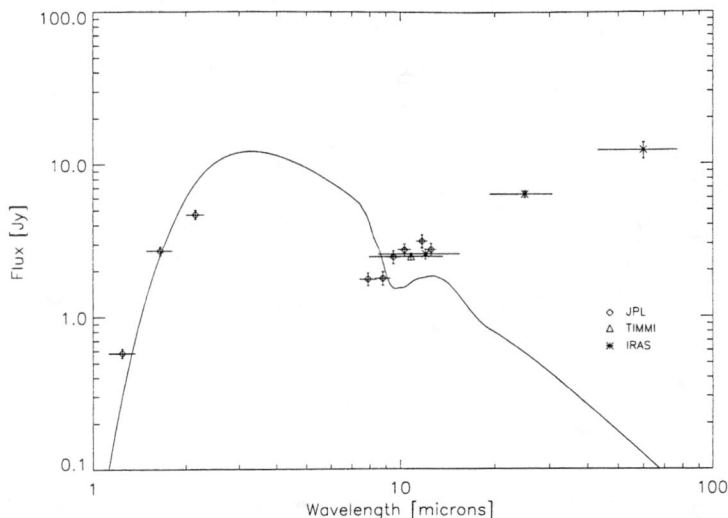

FIGURE 1. The infrared spectrum of the counterpart to SGR 1900+14 summed over both star-like components. The near-infrared fluxes (1.25 to 2.2 μm) were measured by Mahoney et al. [8] while the *IRAS* points were obtained by van Paradijs et al. [11]. Also shown is the flux near 10 μm obtained by Smith et al. [7]. The solid curve corresponds to a 3700 K blackbody convolved with an interstellar extinction of $A_V \sim 19$.

Imaging Multi-Mode Instrument (TIMMI) on the 3.6-meter telescope at ESO. All three measurements are consistent near 10 μm. The solid curve corresponds to a 3700 K blackbody spectrum attenuated by interstellar extinction ($A_V \sim 19$) using the model of Draine and Lee [9]. Not only is the overall spectrum inconsistent with a single blackbody spectrum, but so is the 8 to 13 μm region alone. The intensity increases with increasing wavelength and there is no evidence for the expected Si absorption feature near 10 μm. While also inconsistent with a single blackbody spectrum, the *IRAS* data alone could be explained by dust having a range of temperatures near 100 K.

Although not shown in Figure 1, the two individual components of SGR 1900+14 have similar spectral shapes. However, the relative intensity of Component B (the southwest of the two stars; see Mahoney et al. [8]) decreases continuously relative to that of Component A as the wavelength increases.

SGR 1806–20

Through a fortuitous simultaneous observation of a burst by both *ASCA* and BATSE [10] and subsequent infrared observations [3], SGR 1806–20 has been identified with a rare luminous blue variable star (Star A in Kulkarni et al. [3]). Our observations with SpectroCam-10 showed surprisingly low emission given the 1 Jy flux near 12 μm deduced from the analysis of *IRAS* data [11]. This is not the result

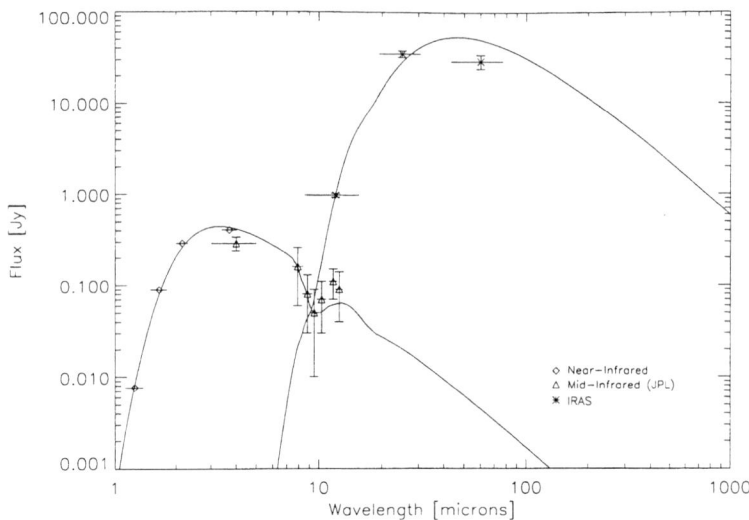

FIGURE 2. The infrared spectrum of SGR 1806–20. The near-infrared fluxes are from Kulkarni et al. [3] while the *IRAS* intensities were determined by van Paradijs et al. [11]. The solid curves correspond to blackbodies at 4300 and 120 K attenuated by an interstellar extinction of $A_V \sim 22$.

of an error in pointing which was verified accurate to within $\sim 2''$ by measuring the relative position of a nearby star with a known position using the telescope offset guider.

The highest signal-to-noise image was obtained in the 3–5 μm band where a point source was clearly evident. Consistency with the intensity measured by Kulkarni et al. [3] further confirms the correct source identification. We then used this position in the field-of-view for the photometry at the other wavelengths. Finally, similar observations were carried out on two consecutive nights (22 and 23 June) with the deduced fluxes consistent. The resulting spectrum is shown in Figure 2, together with the near-infrared [3] and the 12 to 60 μm *IRAS* intensities [11]. The solid curves correspond to blackbodies at 4300 K and 120 K convolved with interstellar extinction ($A_V \sim 22$). In contrast to SGR 1900+14, the spectrum of SGR 1806–20 is reasonably consistent with a single blackbody spectrum through 13 μm and shows evidence for the Si absorption feature near 10 μm. The *IRAS* data suggest emission from dust at a temperature of roughly 120 K.

SUMMARY

Through imaging observations with SpectroCam-10 made during the nights of 21–23 June 1997, we have obtained the first mid-infrared (8 – 13 μm) spectra of the suspected counterparts to both SGR 1806–20 and SGR 1900+14. Our preliminary analysis shows that both are consistent with point-like objects with no evidence for an extended component. While both counterparts have now been observed from approximately 1 to 60 μm, their spectral shapes are quite different and both are inconsistent with a single blackbody spectrum attenuated by interstellar extinction.

As in the near-infrared band, the two point-like components of SGR 1900+14 are clearly resolved in our 8 – 13 μm images. However, the intensity of Component A monotonically increases relative to that of Component B from a ratio of 1.5 at 1.25 μm [8] to 2.5 at 12.5 μm. Furthermore, the mid-infrared flux increases with increasing wavelength and shows no evidence for the Si absorption feature expected from interstellar extinction. We conclude that the total emission must come from multiple components having different temperatures.

By contrast, the emission from SGR 1806–20 can be fitted reasonably well by a single blackbody spectrum through about 13 μm and the spectrum does show evidence for the Si absorption feature near 10 μm at the expected depth. As with SGR 1900+14, the emission above $\sim$12 μm implies the presence of dust at a temperature of roughly 100 K.

ACKNOWLEDGMENTS

We thank Thomas Hayward for his assistance in preparing for the observations and in providing data analysis software, and Drs. Jay Goguen and Padma Yanamandra-Fisher for assisting with the observations. The research described in this paper was carried out by the Jet Propulsion Laboratory, California Institute of Technology, under contract to the National Aeronautics and Space Administration.

REFERENCES

1. Kouveliotou, C., *Astrophys. & Space Sci.* **231**, 49 (1995).
2. Hurley, K., et al., *ApJ* **463**, L13 (1996).
3. Kulkarni, S., et al., *ApJ* **440**, L61 (1995).
4. Corbel, S., et al., *ApJ* **478**, 624 (1997).
5. Vrba, F., et al., *ApJ* **468**, 225 (1996).
6. Hayward, T., et al., *Proc. SPIE* **1946**, 334 (1993).
7. Smith, I., et al., *A&A* **319**, 923 (1997).
8. Mahoney, W., et al., in *Gamma-Ray Bursts: 3rd Huntsville Symposium*, eds. C. Kouveliotou, M.S. Briggs, G.J. Fishman, New York: AIP, 1996, p. 946.
9. Draine, B.T., & Lee, H.M., *ApJ* **285**, 89 (1984).
10. Murakami, T., et al., *Nature* **368**, 127 (1994).
11. van Paradijs, J., et al., *A&A* **314**, 146 (1996).

Infrared Observations of Soft Gamma-Ray Repeaters

I. A. Smith

Department of Space Physics and Astronomy, Rice University, MS-108, 6100 South Main, Houston, TX 77005-1892

Abstract. Quiescent stellar counterparts have been suggested for the Soft Gamma-Ray Repeaters SGR 1806–20 and SGR 1900+14, while none have been found for SGR 0525–66. This paper gives a brief overview of some recent and ongoing infrared observing programs. For a more detailed review article, see Smith [2].

INFRARED SPECTRA OF SGR 1806–20

Van Kerkwijk et al. [7] showed infrared spectra of the possible stellar counterpart to SGR 1806–20. Their K-band spectrum covered 2.02 – 2.22 μm with a spectral resolution ~ 700. We obtained a higher resolution ($R \sim 2500$) K-band spectrum UT 1996 June 25 using the Cryogenic Spectrometer (CRSP) on the Kitt Peak National Observatory (KPNO) 4-m telescope [3].

Our spectrum shows the same main features as that of Van Kerkwijk et al. The most prominent emission line is that of Brγ ($\lambda_{lab} = 2.1655$ μm). Van Kerkwijk et al. suggested that this has a blueshifted absorption component (i.e. a P Cygni profile) which would indicate the presence of an outflow. This appears to be confirmed in the higher resolution data. The absorption minimum at 2.1625 μm gives an estimate of the terminal velocity of ~ 400 km s^{-1}.

INFRARED SPECTRA OF SGR 1900+14

Previous infrared spectra of the possible counterpart to SGR 1900+14 have been shown in Vrba et al. [9]. A pair of heavily reddened stars are found at this location, separated by 3.37".

We obtained higher resolution CRSP spectra of SGR 1900+14 in the same UT 1996 June 25 run [3]. As was seen before, the higher resolution spectra show that the two stars have remarkably similar spectra. Although a large number of atomic and molecular lines are present, the significant difference remains in the first overtone CO absorption band starting at 2.3 μm: the absorption is deeper in the fainter star. However, for the second overtone CO bands at 1.6 μm, the ratio of the two stars is virtually constant.

MID-INFRARED IMAGING OF SGR 0525−66

Because the other two SGR counterparts are brightest in the mid-infrared, this is the most promising place to look for one for SGR 0525−66. Our IRAS study found a source that is extended at 12, 25, and 60 μm [8]. The IRAS colors are typical of other supernova remnants, and this infrared emission likely originates from heated dust in N49.

To determine if there is a point source in addition to this extended emission, we used TIMMI UT 1995 December 9 on the 3.6-m at ESO. Images were made in the N-band ($\lambda_{\text{eff}} = 9.862$ μm) to cover most of N49 [4]. No point sources were found inside the gamma-ray error box in the TIMMI observations. Scaling the 12 μm IRAS flux densities of SGR 1806−20 or SGR 1900+14 to the distance of the LMC would give $\sim$ 100 mJy, which would have been easily detected in our TIMMI run. Similarly, no stellar counterpart has been found at any other wavelength [1].

ONGOING OBSERVING PROGRAMS

The following infrared and submillimeter observing programs are well under way, and final results should be available shortly.

Infrared Space Observatory

Using the Infrared Space Observatory this past year, we have obtained complete spectra of SGR 1806−20 and SGR 1900+14 from 2.4 to 190 μm, photometry of these sources at 25 μm, large-scale mapping of SGR 1806−20 at 60 and 200 μm and SGR 1900+14 at 60, 135, and 200 μm, and images of SGR 1806−20 at several mid-IR wavelengths. These and remaining observations will be a major improvement over the previous IRAS results [8].

Kitt Peak Phoenix

We have been making very high resolution ($R \sim 50,000$) infrared observations of small regions of the counterpart spectra using the new Phoenix spectrometer at KPNO. Velocities can be measured to ~ 1 km s^{-1}.

SCUBA on the JCMT

SCUBA is the new sub-millimeter instrument for the James Clerk Maxwell Telescope on Mauna Kea, Hawaii. For details on the instrument, see our paper in this proceedings on SCUBA observations of the gamma-ray bursters [6]. SCUBA has two arrays of bolometers, allowing complete maps of regions $\sim 1'$ in radius to be made simultaneously at two wavelengths. Observations of SGR 1806–20 and SGR 1900+14 are being made at 450 and 850 μm. In addition to improving on the point source sensitivities by a factor ~ 3 over what we found using the previous UKT14 photometer [5], these will provide orders of magnitude improvement in the sensitivity to extended emissions, for example if there are shells of dust.

ACKNOWLEDGMENTS

This work was supported by NASA grants NAG 5-3353 and 5-3824 at Rice University.

REFERENCES

1. Dickel, J.R., et al., *ApJ* **448**, 623 (1995).
2. Smith, I.A., in *The Gamma-Ray Universe Revealed by CGRO*, 1997, in press.
3. Smith, I.A., et al., in *The Transparent Universe*, ESA SP-382, 1997, p. 191.
4. Smith, I.A., et al., *Adv. Space Res.*, in press (1997).
5. Smith, I.A., Schultz, A.S.B., Hurley, K., Van Paradijs, J., & Waters, L.B.F.M., *A&A* **319**, 923 (1997).
6. Smith, I.A., et al., these proceedings.
7. Van Kerkwijk, M.H., Kulkarni, S.R., Matthews, K., & Neugebauer, G., *ApJ* **444**, L33 (1995).
8. Van Paradijs, J., et al., *A&A* **314**, 146 (1996).
9. Vrba, F.J., et al., *ApJ* **468**, 225 (1996).

Lognormal Properties of SGR1806-20 and the Possibility of a Quiescent Population of Other SGR Sources

Brian McBreen and Kevin J. Hurley

Physics Dept., University College, Dublin 4, Ireland.

Abstract. Monte Carlo simulations of long SGR event sequences based on lognormal distributions with a range of time intervals and intensity distribution parameters have been investigated. The main conclusions are that the majority of SGRs with properties similar to SGR1806-20 have been detected but SGRs with mean waiting times much longer than SGR1806-20 remain to be discovered. A large decrease in the probability for detection of an SGR source results from a relatively small increase in the distribution parameters obtained for SGR1806-20. A new breed of experiments with very long observation times are required to search for this type of source.

INTRODUCTION

The lognormal properties of the soft repeater SGR1806-20 have been previously reported [1–3]. In particular, both the time interval between repeater events and the luminosity function of the source were fit with lognormal distributions [4]. This analysis used the data-base of 111 events detected by the International Cometary Explorer (ICE) mission [5].

While the present number of events observed from the other three sources [6–8] does not allow any detailed analysis, the intervals between successive events of SGR0526-66 [9] is also suggestive of lognormal behaviour.

The relationship between the number of active (i.e observable) sources and the true number of SGRs in the galaxy is one which is the subject of some debate [10,11]. If the time intervals between SGR events proves to be lognormal with a wide range of means and variances then there may be long quiescent periods where the source could be undetectable, leading to an underestimate of the population.

SIMULATIONS

Monte Carlo simulations of SGR event sequences with a variety of distribution parameters were generated. The simulations were performed using Octave 2.0.5

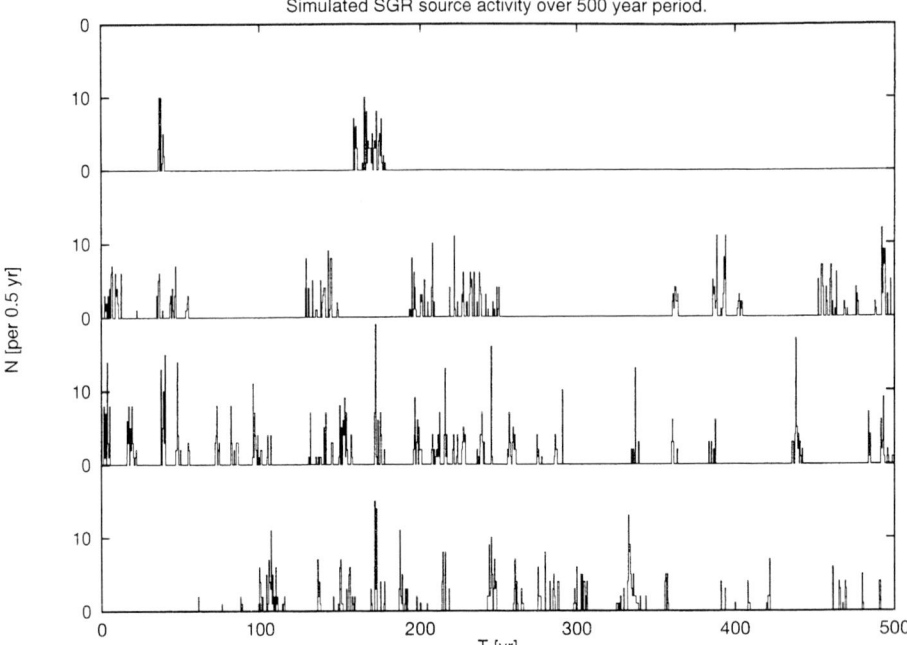

FIGURE 1. Four separate Monte Carlo simulations of duration 500 years of SGR1806-20 activity, generated from lognormal distributions for recurrence intervals and intensities with parameters as determined from SGR1806-20 [3]. The detector model is described in the text. The long gaps in activity of the source are the contributions from the tail of this highly skewed distribution.

for UNIX. Randomly generated standard normal variates, X, were transformed to lognormal variates using the relationship $Y = e^{\sigma X + \mu}$ where Y is lognormally distributed with parameters μ and σ. The effect of the detector on the observations was simulated by rejecting events below a preset intensity level to mimic a threshold and by rejecting at random some predefined fraction of events to mimic a less than 100% live time.

The parameters of the two lognormal density functions which were fit to the distribution of recurrence intervals and intensity for SGR1806-20 were respectively $\mu_W = 9.64$, $\sigma_W = 3.44$ and $\mu_I = 3.3$, $\sigma_I = 1.3$ [3]. To investigate how the probability of observation with a given time interval depended on the the parameters for the waiting times (μ_W and σ_W) and the parameters for the event intensity (μ_I and σ_I), two large simulations were performed. In the first, the waiting time parameters were varied over a small range relative to the SGR1806-20 values (while the intensity distribution was fixed), forming a grid of 14×14 cells, each with a different value for μ_W and σ_W but the same value for μ_I and σ_I. For each cell on the grid ten 500 year simulations were performed (as for the sequences produced in Figure 1) and the mean number of events per 0.5 year interval was determined. The second simulation

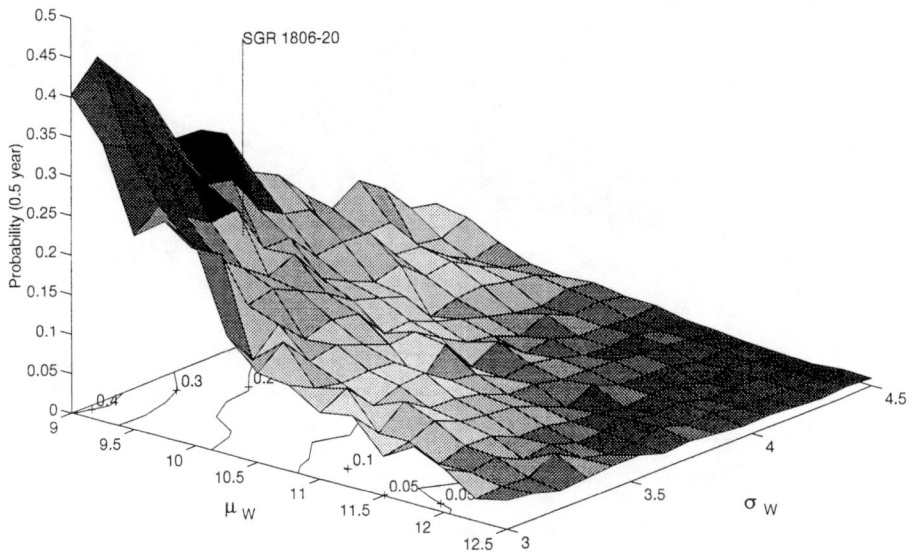

FIGURE 2. This surface plot shows how the probability of observing an SGR source varies with the parameters of the waiting time distribution (μ_W and σ_W). The probability given is per 0.5 years. The contour lines show the drop in probability with μ_W and σ_W. The indicated cell contains the value for SGR 1806-20 with $\mu_W = 9.6$ and $\sigma_W = 3.4$.

was performed with varying μ_I and σ_I (the intensity distribution parameters) but it followed the same procedure - the region of parameter space to be explored was divided into a grid and ten 500 year simulations were performed at each grid point (each grid point having a different μ_I and σ_I but the same μ_W and σ_W). The resulting distributions of probability of observation for each of the two grids are presented in Figure 2 and Figure 3.

DISCUSSION

Figure 2 presents a surface plot of the probability of detecting an SGR source within 0.5 years, given that the event intervals are distributed with a lognormal probability and the intensity distribution is fixed to that observed for SGR1806-20. Note that the parameters for SGR1806-20 (μ_W and σ_W) give an estimated probability of 30% for observation in 0.5 years and that the probability falls quickly with increasing μ_W and σ_W. For $\mu_W = 11 \approx 1.2\mu_W^{1806}, \sigma_W = 4.5 \approx 1.5\sigma_W^{1806}$, which is equivalent to a mean of 1.5×10^9s, the probability of observing the source in 0.5 years has fallen to less than 5%.

In contrast, Figure 3 shows a much flatter more uniform behaviour over a similar

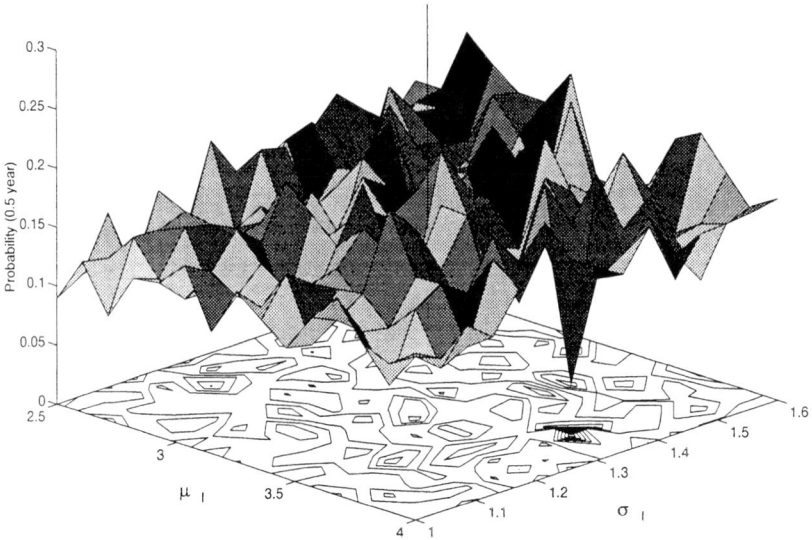

FIGURE 3. This surface plot shows how the probability of observing an SGR source varies with the parameters of the intensity distribution (μ_I and σ_I). The probability given is per 0.5 years. The indicated cell contains the value for SGR1806-20 with $\mu_I = 3.3$ and $\sigma_I = 1.3$.

variation in the intensity distribution parameters. Note that the parameters for SGR1806-20 again give us an estimated probability of 30% for observation in the 0.5 years of observation. The grainy nature of the surface is due to the much lower variation in probability across the range of intensities chosen (from 50% of SGR1806-20 to 300% for each distribution parameter).

The lognormal distribution arises in statistical processes whose completion depend on a product of probabilities, arising from a combination of independent events [12]. Lognormal statistics have previously been used in connection with gamma-ray bursts [13,14]. In the context of this investigation the physical significance of this statistical behaviour may lie in the connection between SGRs and neutron stars where a similar statistical analysis was presented for the microglitches from the Vela pulsar [1]. The time separation and the intensity of these small ($|\Delta\nu/\nu| \sim 10^{-9}$) frequency adjustments were both compatible with lognormal distributions, and there was no correlation between waiting time and intensity, just as observed with SGR1806-20 [5]. This result, combined with the identification of X-ray point sources [15,16] embedded in plerion-powered SNR [17] as counterparts to the SGR sources, suggests structural adjustments in neutron stars may be the cause of SGRs.

CONCLUSION

The activity of sources with mean recurrence times similar to or much longer than SGR1806-20 was investigated using Monte Carlo simulations. The results of the simulations indicate that there could exist a significant population of SGRs with mean waiting times considerably longer than SGR1806-20 that remain undiscovered. Structural adjustments in neutron stars may be responsible for this behaviour. Finally, a new breed of experiments with very long observation times will be required to search for this type of source.

REFERENCES

1. Cheng, B., Epstein, R. I., Guyer, R. A., and Young, A. C., *Nat.* **382**, 518 (1995).
2. Hurley, K. J., McBreen, B., Delaney, M., and Britton, A., *Ap. & Space Sci.* **231**(1-2), 81 (1995).
3. Hurley, K. J., McBreen, B., Rabbette, M., and Steel, S., *A. & A.* **288**, L49 (1994).
4. Aitchison, J., and Brown, J. A. C., *The Lognormal Distribution*, Cambridge University Press, Cambridge, 1957.
5. Laros, J. G., et al., *Ap. J.* **320**, L111 (1987).
6. Kouveliotou, C., et al., *Nat.* **362**, 728 (1993).
7. Norris, J. P., Hertz, P., and Wood, K. S., *Ap. J.* **366**, 240 (1991).
8. Hurley, K., et al., these proceedings.
9. Golenetskii, S. V., et al., *Sov. Astron. Lett.* **13**(3), 166 (1987).
10. Hurley, K. C., et al., *Ap. J.* **423**, 709 (1994).
11. Kouveliotou, C., et al., *Ap. J.* **392**, 179 (1992).
12. Montroll, E. W., and Shlesinger, M. F., *Proc. Nat. Acad. Sci.* **79**, 3380 (1982).
13. Brock, M., et al., in *Gamma-Ray Bursts: Second Workshop (Huntsville)*, AIP Conf. Proc. 307, eds. G. J. Fishman, J. J. Brainerd, and K. Hurley, 1994, p. 672.
14. McBreen, B., Hurley, K. J., Long, R., and Metcalfe, L., *MNRAS* **271**, 662 (1994).
15. Murakami, T., et al., *Nat.* **368**, 127 (1994).
16. Rothschild, R. E., Kulkarni, S. R., and Lingenfelter, R. E., *Nat.* **368**, 432 (1994).
17. Kulkarni, S. R., and Frail, D. A., *Nat.* **365**, 33 (1993).

Testing Models of the Soft Gamma Repeaters

Christopher Thompson

Physics and Astronomy, UNC CB3255, Chapel Hill, NC 27599

Abstract. Models of the Soft Gamma Repeater sources involve two key parameters: magnetic field strength and rotation period. Several relevant observational probes of the SGRs are discussed, including: their quiescent X-ray and particle emission; the X-ray afterglow from the heated surface of the neutron star following a burst; and an anomalous group of low luminosity X-ray pulsars with properties very similar to SGR 0526-66 in its quiescent state (the best studied of which is 1E2259+586) [19,5].

The strongest empirical motivation for very strong magnetic fields ($B \sim 10^{14} - 10^{15}$ G) in the SGR sources comes from: the extreme peak luminosity $L \sim 10^7 L_{\rm edd}$ [10] and large total energy $E \sim 10^{45}$ erg [13] of the 1979 March 5 superburst [6,18]. Much weaker magnetic fields do not carry enough free energy — or effectively couple a second (internal) energy source to the magnetosphere [2]. The log-normal distribution of burst intervals detected from SGR 1806-20 [3], and the enormous brightening of SGR 1806-20 during the burst detected by ASCA (with no corresponding increase in the quiescent emission to within minutes of the burst [14]) both argue against models involving accretion instabilities.

Neutron stars behave in unexpected ways when their magnetic field — rather than their rotation — is the dominant source of free energy. This is the case when the internal flux density exceeds a few $\times 10^{15}$ G at an age of $\sim 10^4$ yr. Such a "magnetar" is a self-triggering burst source, and no external mass accretion or impact is required.

The decaying magnetic field warms the interior of the neutron star, and heat conducted to the surface powers quiescent X-ray emission. Neutrino cooling of the core buffers the X-ray luminosity to the range ($10^{35} - 10^{36}$ erg s^{-1}) observed in the quiescent emission of the SGR sources 0526-66 and 1806-20, and in the anomalous X-ray pulsars [19].

Hall transport of the magnetic field through the crust excites frequent small scale fractures that can power a quasi-steady Alfvén wave flux in the magnetosphere (up to $\sim 10^{37}$ erg s^{-1} at an age of $\sim 10^4$ yr), as as well as more energetic fractures that can power optically thick bursts [19]. Below a critical amplitude, the Alfvén waves escape along open field lines before damping in the magnetosphere [20], and their energy must be converted to particles. In this way, it was suggested that the high burst rate of SGR 1806-20 has a direct physical connection to its activity as a quiescent source of relativistic particles [21], and to the non-thermal spectrum of its continuous emission [14]. In this model, short-timescale variations in the quiescent X-ray emission correspond to transient surges in the internal rate of energy dissipation.

The steady spindown of the star is interrupted by glitches a hundred to a thousand times larger in amplitude than those observed in radio pulsars of comparable age [19]. This prediction is most easily testable for 1E2259+586 [7], a relatively nearby X-ray pulsar whose 7 s spin period and quiescent X-ray luminosity of $0.5 - 1 \times 10^{35}$ erg s^{-1} [11] are close to those of SGR 0526-66. (The 5 March 1979 superburst showed a clear 8 s modulation [13], and a soft X-ray hotspot of luminosity $\simeq 7 \times 10^{35}$ erg s^{-1} detected by ROSAT overlaps its error box [16].) This X-ray pulsar also shares with SGR 0526-66 an association with a $\sim 10^4$ yr old SNR (CTB 109 vs. N49) and a soft X-ray spectrum. 1E2259+586 has been observed to spin down continuously since its discovery (albiet with irregularities in the spin-down rate [1] that at the present poor temporal resolution are consistent both with accretion torque fluctuations, and with glitches).

The anomalous X-ray pulsars are predicted to be strongly subluminous in the optical compared with LMXB's of similar X-ray luminosity. No optical source brighter than $V = 23$ is associated with 1E2259+586 [4].

A neutron star that loses energy mainly via Alfvén wave emission spins down much more rapidly than does a seismically quiet neutron star, because the magnetic field lines combed out by the Alfvén wave flux provide a longer level arm [20].

The physics of Alfvén-wave damping in neutron star magnetospheres was discussed in [20], emphasizing the strong tendency to thermalization at large compactness, and focussing on a turbulent cascade as the most effective mechanism. The simplest radiative model that accomodates the characteristic *thermal spectra* of SGR bursts [9] and their weak spectral evolution [13,12] is an optically thick pair-photon plasma trapped by the magnetic field [18]. In general, the density of pairs created during the cascade [in a volume $(\Delta R)^3$] exceeds the density associated with a LTE plasma of equal energy per volume. Thus, the plasma is guaranteed to run away to local thermodynamic equilibrium if the energy deposited by the cascade on a timescale $\Delta R/c$ corresponds to a LTE scattering depth greater than unity. The corresponding luminosity is $\sim 10^{42}$ erg s^{-1} for a cascade volume (10 km)^3 and decreases approximately in proportion to the width of the bundle of excited field lines. Energy deposition at a slower rate may be balanced continuously by radiative cooling; see [17] for discussion of relevant photon creation processes.

The trapped plasma cools via inward propagation of its outer boundary; this yields a burst luminosity that is approximately proportional to surface area [18]. If the burst trigger involves a fracture of the neutron star crust by a strong ($B > 10^{14}$ G) magnetic field, then most of the energy is injected in the form of a direct Alfvén pulse on a short ($\sim$ millisecond) timescale. Flat-topped bursts (with durations shorter than the rotation period of the source) are associated with a single large injection event. More irregular burst profiles require multiple injection events (with each injection of wave energy dissipating before the plasma can cool).

One clear prediction of this model, is that the stellar surface absorbs thermal energy during a burst, and then re-radiates it at a hyper-Eddington rate in the presence of a magnetic field stronger than $\sim 10^{13}$ G [18,15]. This afterglow emission should show strong spectral cooling, in distinction to the main burst.

The weak dependence of spectrum on burst fluence is harder to understand in a model in which elastic energy leaks gradually out of a neutron star [8], in part because the area of the open bundle of field lines varies slowly with the wave flux at the surface of the star, $\theta^2 \propto (\delta B^2)^{1/3}$.

This work was supported by NASA Grant NAG5-3100, and by the Alfred P. Sloan Foundation.

REFERENCES

1. Baykal, A., and Swank, J., *Ap. J.* **460**, 470 (1992).
2. Blaes, O., Blandford, R.D., Goldreich, P., and Madau, P., *Ap. J.* **343**, 829 (1989).
3. Cheng, B., Epstein, R.I., Guyer, R.A., and Young, C., *Nature* **382**, 518 (1996).
4. Coe, M.J., Jones, L.R., and Lehto, H., *M.N.R.A.S.* **270**, 178 (1994).
5. Corbet, R.H.D., Smale, A.P., Ozaki, M., Koyama, K., and Iwasawa, K., *Ap. J.* **433**, 786 (1995).
6. Duncan, R.C., and Thompson, C., *Ap. J.* **392**, L9 (1992).
7. Gregory, P.C., and Fahlman, G.G., *Nature* **287**, 805 (1980).
8. Fatuzzo, M., and Melia, F., *Ap. J.* **464**, 316.
9. Fenimore, E.E., Laros, J.G., and Ulmer, A., *Ap. J.* **432**, 742 (1994).
10. Fenimore, E.E., Klebesadel, R.W., and Laros, J.G., *Ap. J.* **460**, 964 (1996).
11. Iwasawa, K., Koyama, K., and Halpern, J.P., *P.A.S.J.* **44**, 9 (1992).
12. Kouveliotou, C., et al., *Ap. J.* **322**, L21 (1987).
13. Mazets, E.P., Golenetskii, S.V., Guryan, Yu. A., Ilynskii, V.N., *Ap. S.S.* **84**, 173 (1982).
14. Murakami, T., et al., *Nature* **368**, 127 (1994).
15. Paczyński, B., *Acta Astron.* **42**, 145 (1992).
16. Rothschild, R.E, Kulkarni, S.R., and Lingenfelter, R.E., *Nature* **368**, 432 (1994).
17. Thompson, C., in *Relativistic Jets in AGN*, eds. M. Ostrowski, M. Sikora, G. Madejski, and M. Begelman, Springer, in press.
18. Thompson, C., and Duncan, R.C., *M.N.R.A.S.* **275**, 255 (1995).
19. Thompson, C., and Duncan, R.C., *Ap. J.* **473**, 322 (1996).
20. Thompson, C., and Blaes, O., *Phys. Rev. D*, in press (1997).
21. Vasisht, G., Frail, D.A., and Kulkarni, S.R., *Ap. J.* **440**, L65 (1994).

X-Ray Spectroscopy of Bursts from SGR1806-20 with RXTE

Tod E. Strohmayer[1,2], and Alaa Ibrahim[3]

[1]*LHEA, NASA/GSFC, Greenbelt, MD 20771*
[2]*USRA Research Scientist*
[3]*Dept. of Physics, University of Maryland, College Park, MD*

Abstract. We report on new *Rossi X-ray Timing Explorer* (RXTE) X-ray spectral analysis of bursts from SGR1806-20, the most prolific SGR source known. Previous studies of bursts from this source noted the remarkable lack of spectral variability, both in single bursts as well as from burst to burst. Although we find that the spectrum both within and among bursts is quite uniform; we do find evidence for significant spectral changes within bursts as well as from burst to burst. We find that optically thin thermal bremsstrahlung spectra (OTTB), including photoelectric absorption, provide the best fits to most bursts, however, other models (power law, Band GRB model) can also produce statistically acceptable fits. We confirm the existence of a rolloff in the photon number spectrum below 5 keV. When modelled as photoelectric absorption and OTTB the inferred column is between $0.8 - 1.2 \times 10^{23}$ cm^{-2}. This value is larger than the $\approx 0.6 \times 10^{23}$ cm^{-2} inferred from ASCA observations of the persistent X-ray counterpart, but less than the $\approx 10.0 \times 10^{23}$ cm^{-2} indicated by ICE data.

INTRODUCTION

Soft Gamma-Ray Repeaters (SGR) are a rare class of recurrent high energy transient believed to be associated with young neutron stars [5,7]. SGR1806-20 is the most prolific object in this class. Between 1978 and 1986 more than one hundred burst events were detected from this source by X-ray detectors on the International Cometary Explorer (ICE) satellite [2,10]. Various bursts from this source have also been detected by many of the interplanetary network instruments [8]. BATSE detected a reactivation of SGR1806-20 in November, 1996 [3]. This triggered pointed observations of this source with RXTE in an effort to investigate burst spectra, search for pulsations, extend the burst size distribution to much fainter levels and investigate the spectrum of the persistent emission. We obtained a set of 4 pointed observations totaling 100 ksec of on source time. Observations on November 18, from 03:25:00 UTC to 11:30:00 UTC, revealed more than 100 SGR bursts ranging in peak flux from $\approx 2 \times 10^{-9}$ ergs cm^{-2} s^{-1} to $> 1 \times 10^{-6}$ ergs cm^{-2} s^{-1}. Here we summarize some of the X-ray spectral properties of these bursts as measured with the Proportional Counter Array (PCA).

SPECTRAL ANALYSIS

A striking feature of SGR bursts noted from previous studies is the uniformity of spectra from burst to burst and within a single event [4,2,8]. With its large area, low background, and high time resolution, PCA observations can test the uniformity of SGR burst spectra to a much greater level than previous instruments.

We have not yet completed an exhaustive investigation of burst spectra, however, we have investigated a sample of moderately bright bursts for which the PCA deadtime is not excessive. We give examples of spectral variability from burst to burst as well as evidence for significant spectral evolution in a single event. We obtained X-ray event data with 125 μs (1/8192 s) time resolution and 64 spectral channels across the 2–100 keV PCA bandpass. For each burst we estimated the background using about 20 s of pre- and post-burst data. In all cases the backgrounds were flat. For the bursts described here the background was at most a few percent of the total counts. Figure 1 shows the time history of one of the bursts detected in the PCA.

We fit spectra with optically thin thermal bremsstrahlung (OTTB) and power law models, both modified by photoelectric absorption. Both models provide statistically acceptable fits to most bursts, but the OTTB model provides a marginally better fit in almost all cases. We also fit the gamma-ray burst (GRB) model which has now been used extensively to investigate the continuum spectra of GRB with BATSE [1], but the OTTB model also provides a marginally better fit than this model.

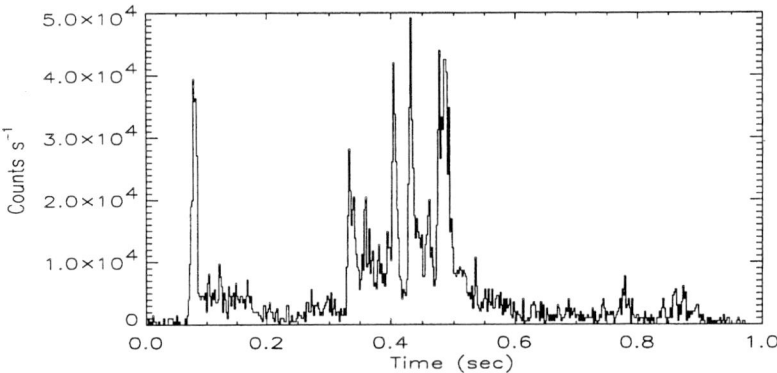

FIGURE 1. Time history of a typical burst from SGR1806-20 as seen in the PCA. The peak countrate is near 50,000 counts s^{-1}. Note that the typical PCA backgroundrate is 120 counts s^{-1}.

Figure 2 shows the best-fit OTTB model for the burst in Figure 1. The best-fit temperature, kT, and column density of Hydrogen, n_H, for this burst are 164 keV and 12×10^{22} cm^{-2}. We tested for spectral variability from burst to burst by comparing the derived confidence regions for the OTTB model parameters for different bursts. As we show below, different bursts do show statistically significant differences in the derived OTTB model parameters. To illustrate this we compare in Figure 3 the derived confidence regions for two bursts which had similar peak countrates and durations. We compare bursts with similar peak countrates in order to reduce any spectral changes that could be introduced by differential deadtime effects. Even at peak rates of 50,000 cts/s the deadtime fraction is not more than about 18%, and is reasonably well understood [9]. We show the 68, 90, and 99% confidence contours for each burst. The contours were computed by calculating the $\delta\chi^2$ appropriate for three parameters (kT, n_H, and normalization constant) [6]. The contours centered at $kT = 164$ keV and $n_H = 12 \times 10^{22}$ cm^{-2} are those for the burst shown in Figures 1 and 2. The confidence regions for these two bursts are disjoint at the 99.9% level, suggesting that these two bursts had measurably different spectra. This result is not unique to these two bursts, across the sample of bursts analyzed to date we find that values for kT and n_H can range from $\approx 20-120$ keV and $\approx 7 - 13 \times 10^{22}$ cm^{-2}, respectively.

We have also investigated spectral variations within bursts. Figure 4 shows the time history of a bright burst which shows evidence for spectral evolution. We accumulated spectra both at the rising and falling edges of this burst. The accumulation intervals are denoted by the vertical dashed lines in Figure 4. We selected the intervals to have approximately the same total number of counts as well as similar countrate profiles to again minimize any differential deadtime effects.

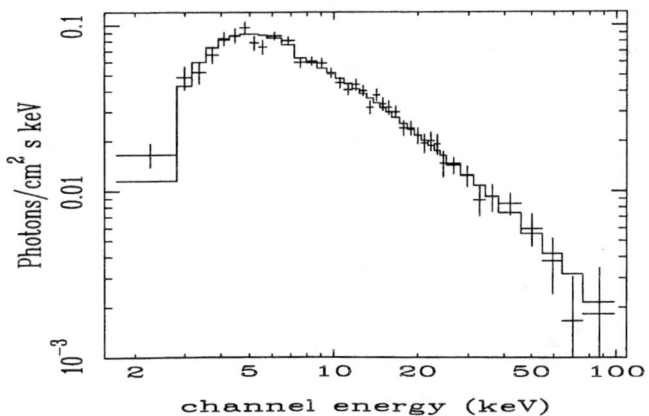

FIGURE 2. Best-fit spectral model for the burst shown in Figure 1. The model includes thermal bremsstrahlung (OTTB) modified by photoelectric absorption. The best-fit parameters and confidence region for this burst are shown in Figure 3.

Figure 5 compares the confidence regions for the OTTB model parameters for both the rising and falling portions of the burst. The rising interval is represented by the contours with the higher (harder) temperature. The regions are disjoint at the ≈ 99.5% level, suggesting that the spectrum during the rising portion of this burst was moderately harder than during the falling portion. To our knowledge this is the first substantial evidence for spectral variations during bursts from SGR1806-20.

Spectral results based on ICE data from SGR1806-20 showed strong evidence for a rolloff in the burst spectra below about 8–10 keV [2]. Our results confirm the presence of a downturn in the photon spectrum at about 4–5 keV. All models we investigated required with high significance such a rolloff in the photon number spectrum. Our modelling with a photoelectric absorption component implies an absorbing column of $8-12 \times 10^{22}$ cm^{-2}. This is higher than the $\approx 6 \times 10^{22}$ cm^{-2}

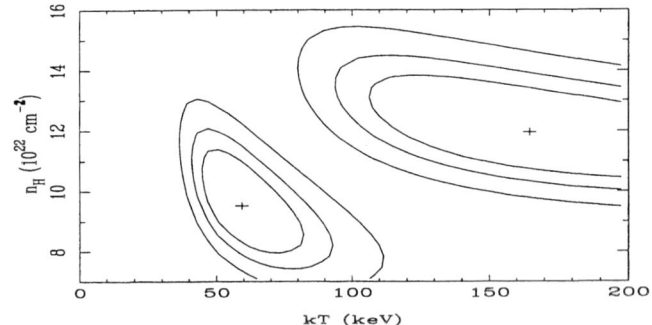

FIGURE 3. Confidence regions for kT and n_H from the OTTB model for two different bursts. The contours centered at $kT = 164$ keV and $n_H = 12$ are those for the burst shown in Figure 1.

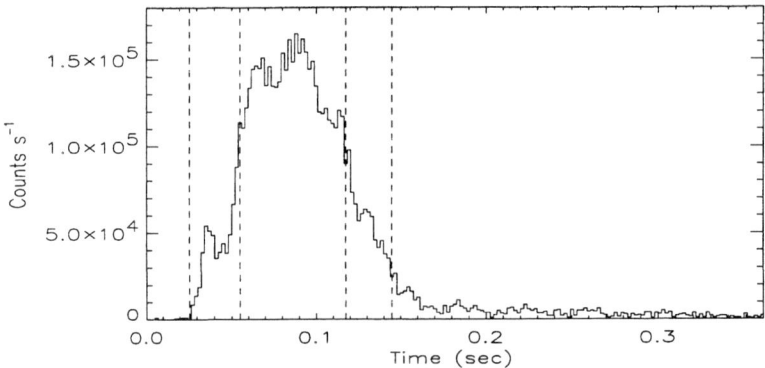

FIGURE 4. Time history of the burst for which we compared spectra during the rise and fall. The vertical dashed lines denote the intervals over which the spectra were accumulated.

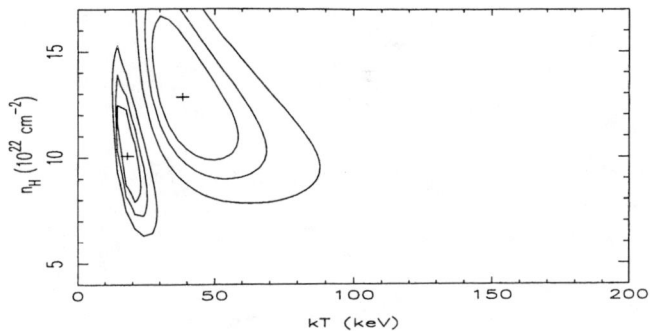

FIGURE 5. Confidence regions for the OTTB model fits to the rising and falling intervals for the burst in Figure 4. The contours denote the 68, 90, and 99% confidence regions. The rising interval contours have the higher kT.

inferred from ASCA measurements [7], but significantly less than the column of about 100.0×10^{22} cm^{-2} suggested from the analysis of ICE data [2].

REFERENCES

1. Band, D., et al., *ApJ* **413**, 281 (1993).
2. Fenimore, E.E., Laros, J.G., & Ulmer, A., *ApJ* **432**, 742 (1994).
3. Kouveliotou, C., et al., *IAUC* 6501 (1996).
4. Kouveliotou, C., et al., *Nature* **368**, 125 (1994).
5. Kulkarni, S.R., & Frail, D.A., *Nature* **365**, 33 (1993).
6. Lampton, M., Margon, B., & Bowyer, S., *ApJ* **208**, 177 (1976).
7. Murakami, T., et al., *Nature* **368**, 127 (1994).
8. Norris, J.P., et al., *ApJ* **366**, 240 (1991).
9. Strohmayer, T. E., Jahoda, K., Swank, J. H., & Stark, M. J., in *Proceedings of the Fourth Compton Symposium*, eds. C. Dermer, J. Kurfess, and M. Strickman, AIP Conference Proceedings no. 410 (1997).
10. Ulmer, A., et al., *ApJ* **418**, 395 (1993).

Symposium Participants

Akerlof, Carl, University of Michigan, akerlof@mich.physics.lsa.umich.edu

Aldridge, Susan, Universities Space Research Association/MSFC, morris@batse.msfc.nasa.gov

Anfimov, Dmitrij, IKI, Russia, dimaa@cgrsda.iki.rssi.ru

Balsano, Rick, Princeton University, rick@pulsar.princeton.edu

Band, David, CASS/University of California-San Diego, dband@ucsd.edu

Barat, C., CESR, barat@cesr.cnes.fr

Baring, Matthew, NASA/GSFC, baring@lheavx.gsfc.nasa.gov

Barthelmy, Scott, Universities Space Research Association/GSFC, scott@lheamail.gsfc.nasa.gov

Bartolini, Corrado, Università di Bologna, bartolini@astbo3.bo.astro.it

Belli, Bianca Maria, IAS NCR - Italy, bianca@saturn.ias.fra.cnr.it

Bloch, Jeffrey, Los Alamos National Laboratory, jbloch@lanl.gov

Bloom, Josh, Caltech/IoA, jsbloom@ast.cam.ac.uk

Bloom, Elliott, SLAC/Standord University, elliott@slac.stanford.edu

Boer, Michael, CESR, boer@cesr.cnes.fr

Boettcher, Markus, Rice University, Houston, mboett@spacsun.rice.edu

Bonnell, Jerry, NASA/GSFC/USRA, bonnell@grossc.gsfc.nasa.gov

Brainerd, Jim, University of Alabama in Huntsville/MSFC, jim.brainerd@msfc.nasa.gov

Briggs, Michael, University of Alabama in Huntsville/MSFC, briggs@gibson.msfc.nasa.gov

Bulik, Tomek, Copernicus Center and University of Chicago, bulik@camk.edu.pl

Bunner, Alan, NASA/HQ, abunner@hq.nasa.gov

Butterworth, Paul, NASA/GSFC, butterworth@lheavx.gsfc.nasa.gov

Caraveo, Patrizia, IFC/CNR-Milano, pat@ifctz.mi.cnr.it

Catelli, Jennifer, University of Maryland, jrc@egret.gsfc.nasa.gov

Chernenko, Anton, IKI, anton@cgrsmx.iki.rssi.ru

Cherry, Michael, Louisiana State University, cherry@phunds.phys.lsu.edu

Cline, David, University of California - Los Angeles, dcline@physics.ucla.edu

Cline, Thomas, NASA/GSFC, cline@apache.gsfc.nasa.gov

Cohen, Ehud, Hebrew University, udic@nikki.fiz.huji.ac.il

Colgate, Stirling A., Los Alamos National Laboratory, colgate@lanl.gov

Connaughton, Valerie, NRC NASA/MSFC, vc@msfc.nasa.gov

Connors, Alanna, University of New Hampshire, aconnors@comptel.sr.unh.edu

Costa, Enrico, IAS/CNR Frasacati, costa@saturn.ias.fra.cnr.it

Crider, Tony, Rice University, acrider@spacsun.rice.edu

Curry, Charles, Curry Foundation, fdtn.1000@carol.net

Daigne, Frederic, Institut d'Astrophysique de Paris, daigne@iap.fr

Deng, Ming, Yale University, ming.deng@yale.edu

Dezalay, J-P, CESR, dezalay@cesr.cnes.fr

Dieters, Stefan, University of Alabama-Huntsville/MSFC, stefan.dieters@msfc.nasa.gov

Dingus, Brenda, University of Utah, dingus@mail.physics.utah.edu

Elsner, Ronald, NASA/MSFC, elsner@avalon.msfc.nasa.gov

Fenimore, Ed, Los Alamos National Laboratory, efenimore@lanl.gov

Feroci, Marco, IAS/CNR Frascati, feroci@saturn.ias.fra.cnr.it

Finger, Mark H., Universities Space Research Assocation/MSFC, mark.finger@msfc.nasa.gov

Fishman, Jerry, NASA/MSFC, fishman@ssl.msfc.nasa.gov

Fletcher, Sandra, Los Alamos National Laboratory, sfletcher@lanl.gov

Fletcher-Holmes, David, Leicester University, dwf@star.le.ac.uk

Frail, Dale, National Radio Astronomy Observatory, dfrail@nrao.edu

Frederix, Dmitri, IOFFE Institute, fred@mz.ioffe.rssi.ru

Frontera, Filippo, Istituto TESRE, filippo@botes2.tesre.bo.cnr.it

Fruchter, Andrew, STScI, fruchter@stsci.edu

Galama, Titus, University of Amsterdam, titus@astro.uva.nl

Gehrels, Neil, NASA/GSFC, gehrels@gsfc.nasa.gov

Giblin, Tim, University of Alabama - Huntsville/MSFC, timothy.giblin@msfc.nasa.gov

Golenetski, Sergei, IOFFE Institute, golen@mz.ioffe.rssi.ru

Gorosabel, Javier, LAEFF-INTA, jgu@laeff.esa.es

Graziani, Carlo, University of Chicago, c-graziani@uchicago.edu

Greiner, Jochen, Astrophysical Institute Potsdam, jgreiner@aip.de

Groot, Paul, University of Amsterdam, paulgr@astro.uva.nl

Grossan, Bruce, LBNL/UC Berkeley, bruce@singu.lbl.gov

Gruendl, Robert, University of Illinois, gruendl@astro.uiuc.edu

Gursky, Herbert, Naval Research Labortory, herbert.gursky@nrl.navy.mil

Hakkila, Jon, Mankato State University, jhakk@msus1.msus.edu

Hanlon, Lorraine, University College Dublin, lhanlon@bermuda.ucd.ie

Harding, Alice, NASA/GSFC, harding@twinkie.gsfc.nasa.gov

Harmon, Alan, NASA/MSFC, harmon@ssl.msfc.nasa.gov

Harris, Michael J., Universities Space Research Association/GSFC, harris@tgrs2.gsfc.nasa.gov

Hartmann, Dieter, Clemson University, hartmann@grb.phys.clemson.edu

Heise, John, Space Research Utrecht, jheise@sron.ruu.nl

Henze, William, TBE/MSFC, henze@ssl.msfc.nasa.gov

Horack, John, NASA/MSFC, john.horack@msfc.nasa.gov

Howard III, William, Universities Space Research Association, whoward@usra.edu

Hudec, Rene, Astronomical Institute Ondrejov, rhudec@asu.cas.cz

Hurley, Kevin J., University College Dublin, khurley@bermuda.ucd.ir

Hurley, Kevin, University of California - Berkeley, khurley@sunspot.ssl.berkeley.edu

in't Zand, Jean, SRON, jeanz@purple.sron.ruu.nl

Joy, Marshall, NASA/MSFC, joy@ssl.msfc.nasa.gov

Kaluzienski, Louis, NASA Headquarters, lkaluzienski@gm.ossa.hq.nasa.gov

Katz, Jonathan, Washington University, katz@wuphys.wustl.edu

Katz, Charlie, Massachusetts Institute of Technology, ckatz@maggie.mit.edu

Kayser, Susan, NSF Astronomy, skayser@nsf.gov

Kehoe, Robert, University of Michigan, kehoe@nis.lanl.gov

Kippen, R. Marc, University of Alabama in Huntsville/MSFC, marc.kippen@msfc.nasa.gov

Klebesadel, Ray, Los Alamos National Lab., rklebesadel@juno.com

Klose, Sylvio, Thuringer Landessternwarte, klose@tls-tautenburg.de

Kluzniak, Wlodzimierz, University of Wisconsin, wlodek@astrog.physics.wisc.edu

Kobayashi, Shiho, Hebrew University, shiho@alf.fiz.huji.ac.il

Kommers, Jeff, Massachusetts Institute of Technology, kommers@mit.edu

Koshut, Tom, Universities Space Research Association/MSFC, tom.koshut@msfc.nasa.gov

Kouveliotou, Chryssa, Universities Space Research Assocation/MSFC, chryssa.kouveliotou@msfc.nasa.gov

Krimm, Hans, Hampden-Sydney College, hansk@pulsar.hsc.edu

Kulkarni, Shri, Caltech, srk@astro.caltec.edu

Kurczynski, Peter, University of Maryland, kraken@rosserv.gsfc.nasa.gov

Lamb, Donald, Univesity of Chicago, lamb@oddjob.uchicago.edu

Larson, Samuel, University of California - Los Angeles, sammy@astro.ucla.edu

Lee, Andrew, Stanford Linear Accelerator Center, alee@slac.stanford.edu

Lee, Brian, University of Michigan, bclee@umich.edu

Lestrade, John Patrick, Mississippi State University, lestrade@ra.msstate.edu

Lewin, Walter, Massachusetts Institute of Technology, lewin@space.mit.edu

Li, Tipei, IHEP, Beijing, litp@astrosvl.ihep.ac.cn

Li, Hui, Los Alamos National Laboratory, hli@lanl.gov

Liang, Edison, Rice University, liang@spacsun.rice.edu

Lin, Dechun, Rice University, lin@spacsun.rice.edu

Lingenfelter, Rich, University of California San Diego, rlingenfelter@ucsd.edu

Litvak, Maxim, Space Research Institute, Russia, studl608@iki.rssi.ru

Livio, Mario, Space Telescope Science Institute, mlivio@stsci.edu

Lloyd, Nicole, Stanford University, nlloyd@leland.stanford.edu

Luginbuhl, Christian, Naval Observatory Flagstaff, cbl@nofs.navy.mil

Mahoney, William, Jet Propulsion Laboratory, wam@heag4.jpl.nasa.gov

Mallozzi, Robert, University of Alabama - Huntsville, robert.mallozzi@msfc.nasa.gov

Maran, Stephen, NASA/GSFC, hsmaran@stars.gsfc.nasa.gov

Marani, Gabriela, George Mason University, gmarani@science.gmu.edu

Marsden, David, CASS/University of California - San Diego, dmarsden@mamacass.ucsd.edu

Masetti, Nicola, Università di Bologna, masetti@astbo3.bo.astro.it

Mathews, Grant, University of Notre Dame, gmathews@bootes.phys.nd.edu

Matz, Steven, Northwestern University, s-matz@nwu.edu

McBreen, Brian, University College Dublin, bmcbreen@ollamh.ucd.ie

McCollough, Michael, Universities Space Research Association/MSFC, mccollough@bowie.msfc.nasa.gov

McConnell, Mark, University of New Hampshire, mark.mcconnell@unh.edu

McKay, Timothy, University of Michigan, tamckay@umich.edu

Meegan, Charles, NASA/MSFC, charles.meegan@msfc.nasa.gov

Meier, Mike, Los Alamos National Laboratory, mmeier@lanl.gov

Meszaros, Peter, Pennsylvania State University, nnp@astro.psu.edu

Miller, James, University of Alabama - Huntsville, millerj@cspar.uah.edu

Miller, Richard S., Los Alamos National Laboratory, richard@lanl.gov

Mitrofanov, Igor, Space Research Institute, Moscow, imitrofa@iki.rssi.ru

Mochkovitch, Robert, Institute d'Astrophysics de Paris, mochko@iap.fr

Monnelly, Glen, Massachusetts Institute of Technology, monnelly@space.mit.edu

Murakami, Toshio, ISAS, murakami@astro.isas.ac.jp

Nemiroff, Robert, Michigan Tech, nemiroff@mtu.edu

Nicastro, Luciano, Instituto Te.S.R.E., nicastro@tesre.bo.cnr.it

Norris, Jay P., NASA/GSFC, norris@groax0.gsfc.nasa.gov

Paciesas, Bill, University of Alabama - Huntsville/MSFC, william.paciesas@msfc.nasa.gov

Paczynski, Bohdan, Princeton University, bp@astro.princeton.edu

Palmer, David, Universities Space Research Association/GSFC/NASA, palmer@lheamail.gsfc.nasa.gov

Panaitescu, Alin, Penn State University, apanait@astro.psu.edu

Park, Hye-Sook, Lawrence Livermore National Lab., hpark@llnl.gov

Parnell, Tom, NASA/Marshall Space Flight Center, thomas.parnell@msfc.nasa.gov

Parsons, Ann, NASA/GSFC, parsons@lheamail.gsfc.nasa.gov

Patel, Sandy, University of Alabama - Huntsville/NASA, patels@cspar.uah.edu

Pedersen, Holger, Copenhagen University Observatory, holger@astro.ku.dk

Pendleton, Geoff, University of Alabama - Huntsville, pendleton@sslmor.msfc.nasa.gov

Pengchamnan, Surasak, University of Alabama - Huntsville, surasak@bbking.msfc.nasa.gov

Peterson, Burl H., Universities Space Research Associaton/MSFC, burl.peterson@msfc.nasa.gov

Petrosian, Vahé, Stanford University, vahe@bigbang.stanford.edu

Pian, Elena, ITESRE-CNR Bologna, Italy, pian@tesre.bo.cnr.it

Piran, Tsvi, Hebrew University, tsvi@shemesh.fiz.huji.ac.il

Piro, Luigi, IAS-CNR Frascati, piro@alphal.ias.fra.cnr.it

Pizzichini, Graziella, TESRE/CNR Bologna, graziella@botesl.tesre.bo.cnr.it

Pozanenko, Alexei, Space Research Institute, Moscow, apozanen@iki.rssi.ru

Preece, Rob, University of Alabama - Huntsville, rob.preece@msfc.nasa.gov

Purcell, William, Ball Aerospace & Technologies Corp., bpurcell@ball.com

Rachen, Jorg, Penn State University, jorg@astro.psu.edu

Ramirez-Ruiz, Enrico, Los Alamas National Laboratory, enrico@nis.lanl.gov

Reichart, Daniel, University of Chicago, reichart@oddjob.uchicago.edu

Rezek, Tomas, Astronomical Institute Ondrejov, rezek@asu.cas.cz

Rhoads, James, Kitt Peak National Observatory, jrhoads@noao.edu

Richardson, Georgia, Universities Space Research Assocation /MSFC, georgia.richardson@msfc.nasa.gov

Ricker, George, Massachusetts Institute of Technology, grr@space.mit.edu

Robinson, Craig, Universities Space Research Association/MSFC, craig.robinson@msfc.nasa.gov

Ruffert, Maximilian, Max Planck Institut fer Astrophysik, mor@mpa-garching.mpg.de

Ruiz-Lapuente, Pilar, University of Barcelona, pilar@mizar.am.ub.es

Saavedra, Oscar, Torino University, saavedra@to.infn.it

Sahi, Maitrayee, Universities Space Research Association/MSFC, maitrayee.sahi@msfc.nasa.gov

Sahu, Kailash, STScI, ksahu@stsci.edu

Salmonson, Jay D., Lawrence Livermore National Laboratory, salmonsonl@llnl.gov

Sari, Re'em, Hebrew University - Israel, sari@nikki.fiz.huji.ac.il

Scargle, Jeffrey, NASA/Ames Research Center, jeffrey@sunshine.arc.nasa.gov

Schaefer, Brad, Yale University, schaefer@grb2.physics.yale.edu

Schmidt, Maarten, California Institute of Technology, mxs@deimos.caltech.edu

Seifert, Helmut, Universities Space Research Association, NASA/GSFC, helmut.seifert@gsfc.nasa.gov

Share, Gerald, NRL, share@osse.nrl.navy.mil

Shrader, Chris, NASA/GSFC, shrader@grossc.gsfc.nasa.gov

Shubert, Richard, Institute for Fundamental Sciences, rshubert@heag1.jpl.nasa.gov

Smith, Ian, Rice University, ian@spacsun.rice.edu

Smith, Don, Massachusetts Institute of Technology, dasmith@space.mit.edu

Soldan, Jan, Astronomical Instutite Ondrejov, jsoldan@asu.cas.cz

Stollberg, Mark, University of Alabama - Huntsville/MSFC, stollberg@gibson.msfc.nasa.gov

Strohmayer, Tod, Universities Space Research Association/LHEA/GSFC, stroh@pcasrvl.gsfc.nasa.gov

Sumner, Chip, Los Alamos National Laboratory, sumner@nis.lanl.gov

Sun, Xuejun, Los Alamos National Laboratory, xsun@nis.lanl.gov

Svensson, Roland, Stockholm Observatory, svensson@astro.su.se

Swartz, Doug, Universities Space Research Association/MSFC, swartz@xanth.msfc.nasa.gov

Takeshima, Toshiaki, NASA/GSFC, takeshim@ginpo.gsfc.nasa.gov

Tavani, Marco, Columbia University/IFCTR, tavani@ifctr.mi.cnr.it

Taylor, Greg, NRAO, gtaylor@nrao.edu

Teegarden, Bonnard, NASA/GSFC, bonnard@lheamail.gsfc.nasa.gov

Terekhov, Mikhail M., IOFFE Institute, mike@mz.ioffe.rssi.ru

Terrell, James, Los Alamos National Laboratory, jterrell@lanl.gov

Thompson, Christopher, University of North Carolina - Chapel Hill, thompson@physics.unc.edu

van Paradijs, Jan, University of Alabama - Huntsville, jvp@astro.uva.nl

Vanderspek, Roland, Massachusetts Institute of Technology, roland@space.mit.edu

Vietri, Mario, University of Rome, vietri@coma.mporzio.astro.it

Villasenor, Jesus Noel, Massachusetts Institute of Technology, jsvilla@space.mit.edu

Vrba, Frederick, USNO, Flagstaff, fjv@nofs.navy.mil

Wagner, R. Mark, Ohio State University, rmw@lowell.edu

Wang, Virginia, The Claremont Colleges, vwang@sunspot.ssl.berkeley.edu

Wijers, Ralph A.M.J., Institute of Astronomy, ramjw@ast.cam.ac.uk

Williams, G. Grant, Clemson University, ggwilli@hubcap.clemson.edu

Wilson, James R., Lawrence Livermore National Laboratory, wilson33@llnl.gov

Wilson, Robert, NASA/MSFC, wilson@gibson.msfc.nasa.gov

Wilson, Colleen, NASA/MSFC, colleen.wilson-hodge@msfc.nasa.gov

Winkler, Christoph, ESA/ESTEC, cwinkler@astro.estec.esa.nl

Woods, Peter, University of Alabama in Huntsville, peter.woods@Msfc.nasa.gov

Yamaguchi, Keiji, Hiroshima University, tsutomu@hiroh2.help.hirochima-u.as.jp

Yoshida, Atsunasa, RIKEN, ayoshida@postman.riken.go.jp

Young, C. Alex, University of New Hamsphire, ayoung@comptel.sr.unh.edu

Yu, Wenfei, IHEP, yuwf@astrosv1.inep.ac.cn

Zhang, S. Nan, Universities Space Research Association/MSFC, zhang@ssl.msfc.nasa.gov

Ziock, Klaus, Lawrence Livermore National Laboratory, kpziock@llnl.gov

AUTHOR INDEX

A

Ables, E., 837, 842
Alonso, M. V., 489
Altieri, B., 489
Amati, L., 404, 409, 446, 451
Andersen, M. I., 530
Anfimov, D. S., 20, 289, 364
Antonelli, L. A., 404, 409
Aptekar, R. L., 10, 284, 516, 894, 921
Ashley, M., 909
Atteia, J-L., 15, 92, 846

B

Balsano, R. J., 585
Band, D. L., 211, 299, 319, 329, 334, 605, 615, 837, 842
Barat, C., 15, 278
Baring, M. G., 732
Barthelmy, S. D., 99, 119, 129, 134, 139, 414, 435, 585, 837, 842, 914
Bartolini, C., 489, 540, 815
Beasley, A. J., 571
Belli, B. M., 82
Benítez, N., 489
Benn, C., 499
Bergeron, L. E., 504
Bernas, M., 864, 874
Beskin, G. M., 540, 815
Bignami, G. F., 494
Bijaoui, A., 196
Bionta, R. M., 837, 842
Birkle, K., 489
Bloom, E., 261
Boček, J., 851
Boër, M., 15, 50, 109, 625, 846
Bonnell, J. T., 166, 171, 181, 742, 884
Borovička, J., 851
Böttcher, M., 808
Bowell, E., 869
Bradt, H., 430
Brainerd, J. J., 3, 104, 226, 545, 727, 752
Brandt, S., 231

Briggs, M. S., 3, 20, 25, 30, 50, 59, 104, 109, 119, 144, 176, 186, 236, 256, 273, 289, 294, 299, 319, 329, 364, 369, 374
Bringer, M., 846
Broeils, A., 489, 530
Bromm, V., 379
Bulik, T., 682, 714
Butler, R. C., 397, 404, 409, 815
Butterworth, P. S., 10, 99, 134, 139, 516, 585, 837, 842, 914

C

Cabrera, R., 914
Calvet, G., 846
Cannizzo, J. K., 414, 435
Caraveo, P. A., 494
Carney, B., 548
Castander, F. J., 520
Castellina, A., 914
Castro-Tirado, A. J., 231, 489, 504, 530, 552, 557, 610, 864, 874
Catelli, J. R., 309, 349
Centurion, M., 499
Ceplecha, Z., 851
Chen, X., 149
Chernenko, A., 30, 196, 211, 294, 374
Cinti, M. N., 404, 409, 446, 451
Claver, C. F., 548
Cline, D. B., 221
Cline, T. L., 10, 50, 99, 109, 124, 139, 304, 324, 339, 516, 585, 625, 640, 837, 842, 914
Code, A. D., 548
Cohen, E., 747
Cole, A., 548
Coles, W. A., 585
Coletta, A., 397, 451
Colgate, S. A., 820
Collmar, W., 344
Connaughton, V., 3, 30, 99, 104, 119, 236, 414, 435
Connors, A., 30, 344
Conselice, C., 548
Cook, K. H., 869
Coppi, P. S., 760

Corbel, S., 931
Corbet, R. H. D., 414, 435
Corey, B. E., 581
Costa, E., 124, 397, 404, 409, 435, 446, 451, 489, 504, 525, 530
Crider, A., 63, 359
Cusumano, G., 404, 409
Czerny, B., 714

D

Daigne, F., 667, 677
Dal Fiume, D., 124, 404, 409, 446, 451, 489
Darracq, F., 15
de Freitas Pacheco, J., 846
Delaney, M., 191
Deng, M., 216, 251
Dey, A., 548
Dezalay, J-P., 15, 278
Dieters, S., 926
Dingus, B. L., 30, 309, 349, 884
Durouchoux, Ph., 931

E

Eastlund, B., 808
Eislöffel, J., 635
Evans, R. H., 25
Eysseric, J., 846

F

Fenimore, E. E., 379, 420, 456, 461, 657, 765
Ferguson, D. H., 837, 842
Feroci, M., 124, 397, 404, 409, 446, 451, 489, 504, 525, 530
Fiore, F., 404, 409, 435
Fishman, G. J., 3, 45, 50, 109, 119, 430, 466, 625, 640, 837, 842, 914
Florián, J., 851
Fockenbrock, R., 489, 530
Ford, L. A., 334
Forester, R., 576
Forrest, D. J., 889
Frail, D. A., 563, 571

Fransson, C., 530
Frederiks, D. D., 10, 284, 516, 894, 921
Frontera, F., 124, 397, 404, 409, 446, 451, 489, 504, 509, 530
Fruchter, A. S., 483, 504, 509
Fujimoto, R., 435

G

Galama, T. J., 478, 483, 499, 509, 557, 590
Gehrels, N., 99, 139, 304, 324, 585, 837, 842, 914
Ghia, P. L., 914
Giblin, T. W., 241
Giménez, A., 859
Golden, A., 815
Goleneteskii, S. V., 10, 284, 516, 894, 921
Gonzalez, R., 509
Gorenstein, P., 904
Gorosabel, J., 231, 489, 530, 552, 610, 874
Goupil, P., 15
Grav, T., 530
Graziani, C., 161
Greiner, J., 425, 435, 489
Greten, A., 201
Griffee, J. W., 54
Groot, P. J., 478, 483, 499, 509, 557, 590
Grossan, B., 909
Gruendl, R. A., 576
Guarnieri, A., 489, 540, 815
Gull, T., 504
Guziy, S., 489

H

Haglin, D. J., 77
Hakkila, J., 3, 25, 77, 144, 149, 236, 899
Halpern, J., 504
Hanlon, L., 191, 489
Harding, A. K., 732
Harmon, B. A., 466
Harris, M. J., 314
Hartmann, D. H., 236, 605, 615, 625, 630, 640, 837, 842
Heidt, J., 489

Heise, J., 397, 404, 409, 435, 446, 530
Henze, W., 3, 144
Hermsen, W., 344
Hewitt, J. N., 581
Higdon, J. C., 40, 931
Hjorth, J., 530
Hobbeheydar, J., 379
Hoekstra, H., 557
Hook, R., 509
Hora, J. L., 635
Horack, J. M., 236
Howell, S. B., 869
Howk, C., 548
Hroch, F., 859
Hudec, R., 473, 851, 855, 859, 864, 874, 904
Hurley, K. C., 15, 30, 50, 87, 104, 109, 114, 124, 166, 278, 304, 324, 339, 387, 414, 430, 435, 504, 625, 640, 837, 842
Hurley, K. J., 191, 939

I

Ibrahim, A., 947
Il'inskii, V. N., 10, 284, 516, 894, 921
Inneman, A., 904
in't Zand, J. J. M., 397, 404, 409, 446
Ishida, M., 435

J

Jacoby, G., 548
Janka, H.-T., 793
Jannuzi, B., 548
Jaunsen, A. O., 530
Jennings, M. C., 625, 630
Jernigan, J. G., 430

K

Kakimoto, K., 914
Kaneko, T., 914
Kappadath, S., 344
Katz, C. A., 581
Katz, J. I., 689
Kawai, N., 414, 435, 441
Kessler, M., 489

Kippen, J. J., 226
Kippen, R. M., 3, 30, 99, 104, 114, 119, 344, 414, 435
Klebesadel, R. W., 54
Klose, S., 635
Klotz, A., 846
Kluźniak, W., 798, 830
Kobayashi, S., 672
Koehn, B. W., 869
Kommers, J. M., 45, 109, 144
König, M., 211
Kopylov, A. I., 525
Koshut, T. M., 3, 466
Kouveliotou, C., 3, 45, 50, 99, 109, 119, 241, 414, 430, 435, 478, 483, 499, 509, 557, 590, 625, 640, 837, 842, 914, 926
Kovář, A., 851
Kristen, H., 489, 530
Kuiper, L., 344
Kulkarni, S. R., 571
Kurczynski, P., 304, 324
Kuznetsov, A., 15, 278

L

Lacy, M., 489, 530
Lamb, D. Q., 161, 520
Larson, S. B., 600, 620
Laureijs, R., 489
Lee, A., 261
Lee, J., 548
Lee, P., 54
Lee, W. H., 798
Lehnert, M., 557
Leroy, A., 846
Lestrade, J. P., 3, 278
Leventhal, M., 576
Levine, A. M., 414, 430, 435
Lewin, W. H. G., 45
Li, H., 420, 456, 820
Li, T., 246, 825
Liang, E. P., 63, 359, 576, 756, 813
Lin, D., 756
Lingenfelter, R. E., 40
Litvak, M. L., 20, 59, 176, 186, 256, 289, 364
Livio, M., 483, 509
Lloyd, N. M., 35, 67
Lo, K. Y., 576

Lucas, R. A., 504
Luginbuhl, C. B., 625, 630
Lund, N., 231

M

Macchetto, F. D., 483, 509
Mackay, C., 499
Macri, J., 889
Madden, N. W., 304, 324, 339
Mahoney, W. A., 931
Malina, R., 846
Mallozzi, R. S., 3, 25, 77, 273, 319, 369
Marani, G. F., 166, 742
Marsden, D., 926
Marshall, F. E., 99, 414, 435
Martínez-González, E., 489
Masetti, N., 489, 540, 815
Mas-Hesse, M., 859
Mathews, G. J., 788
Matteson, J. L., 299, 329
Matthey, C., 221
Matz, S. M., 354
Mazets, E. P., 10, 284, 516, 894, 921
McBreen, B., 191, 489, 939
McCollough, M. L., 3, 144, 154
McConnell, M. L., 344, 889
McLean, I. S., 600
Meegan, C. A., 3, 20, 25, 30, 45, 50, 59, 77, 104, 109, 119, 144, 154, 176, 186, 236, 256, 289, 294, 364, 374, 430, 625, 640, 837, 842, 899, 914
Meissonier, M., 846
Meredith, D. C., 201, 206
Mészáros, P., 647, 771, 776
Metcalfe, L., 489
Metcalfe, N., 557
Mignani, R., 494
Mignoli, M., 489
Miller, B., 808
Mitrofanov, I. G., 20, 30, 59, 176, 186, 256, 289, 364, 374
Mochkovitch, R., 667, 677
Molendi, S., 404, 409
Monaldi, L., 530
Moore, C. B., 581
Morello, C., 914

Muller, J. M., 397, 404, 409, 446
Murakami, T., 414, 435, 441, 461
Mutchler, M., 509

N

Nagase, F., 435
Namiki, M., 435, 441
Näslund, M., 530
Navarra, G., 914
Němček, M., 855
Nemiroff, R. J., 166, 171, 742
Nicastro, L., 397, 404, 409, 446, 451, 489, 504, 525, 530
Niel, M., 15
Nishi, K., 914
Norris, J. P., 166, 171, 181, 742, 884

O

Orlandini, M., 124, 404, 409, 446, 451
Orosz, J. A., 548
Østensen, R., 530
Otani, C., 435, 441
Ott, L., 837, 842
Otwinowski, S., 221
Owens, A., 404, 409

P

Paciesas, W. S., 3, 20, 25, 30, 59, 176, 186, 256, 273, 289, 294, 299, 319, 329, 364, 369, 374, 466
Paczyński, B., 783
Palazzi, E., 404, 409, 446, 451, 489, 504, 509, 525, 530
Palmer, D. M., 304, 324, 339
Panaitescu, A., 771
Panov, V. N., 10, 284
Park, H. S., 837, 842
Parker, E., 837, 842
Parmar, A. N., 404, 409
Páta, P., 864, 874
Pedersen, H., 489, 504, 530, 846
Pehl, R. H., 304, 324, 339
Peignot, C., 846
Pelaez, F., 344

Pendleton, G. N., 3, 20, 25, 30, 45, 59, 77, 104, 114, 119, 144, 154, 176, 186, 236, 256, 273, 289, 294, 299, 319, 329, 364, 369, 374, 414, 435, 899
Perez, A. M., 530
Perlmutter, S., 909
Petro, L., 483, 509
Petrosian, V., 35, 67
Pian, E., 483, 504, 509
Piccioni, A., 489, 540, 815
Pina, L., 904
Piran, T., 662, 672, 689, 704, 747
Piro, L., 124, 397, 409, 435, 489, 504, 530
Pisano, D. J., 548
Pizzichini, G., 446
Pollas, C., 846
Pozanenko, A. S., 59, 186
Preece, R. D., 3, 20, 25, 30, 59, 63, 176, 186, 256, 273, 289, 294, 299, 319, 329, 359, 364, 369, 374, 466

Q

Quashnock, J. M., 161
Quilligan, F., 191

R

Rachen, J. P., 776
Ramaty, R., 304, 324, 339
Ramirez, E., 657
Ray, P. S., 585
Redfern, M., 815
Reichart, D. E., 535
Remillard, R. A., 414, 430, 435
Ressler, M. E., 931
Rezek, T., 855, 859, 864, 874
Rhoads, J. E., 548, 557, 699
Richardson, G. A., 466
Rickett, B. J., 585
Robinson, C. R., 99, 119, 414, 435, 466
Rodriguez Espinosa, J. M., 530
Roiger, R. J., 77
Roscherr, B., 760
Rothschild, R. E., 926
Ruderman, M., 830
Ruffert, M., 793

Ryan, J. M., 201, 206, 344, 889

S

Saavedra, O., 914
Sahu, K. C., 483, 509
Salmonson, J. D., 788
Sanchez, P., 846
Sari, R., 662, 672, 704, 720
Sawyer, D., 548
Scargle, J. D., 171, 181, 261, 884
Schaefer, B. E., 216, 251, 266, 379, 595
Schneid, E. J., 309, 349
Schönfelder, V., 344
Seifert, H., 304, 324, 339
Seitz, T., 489
Share, G. H., 314, 354
Sharp, N., 548
Shearer, A., 815
Shibata, R., 435, 441
Shlyapnikov, A., 489
Shrader, C. R., 869
Sikora, M., 682
Smartt, S., 499
Smith, D. A., 414, 430, 435
Smith, I. A., 576, 590, 936
Smith, N., 489
Smith, P., 499, 548
Soffitta, P., 404, 409
Sokolov, V. V., 525
Sokolova, Z. J., 10, 284
Soldán, J., 855, 859, 864, 874
Sparks, W., 504
Spurný, P., 851
Starrfield, S. G., 869
Stecklum, B., 635
Stilwell, D. E., 10, 516
Strohmayer, T. E., 461, 947
Stubbs, C. W., 869
Studt, J., 489
Sumner, M. C., 657, 765
Sun, X., 420, 456
Sunyaev, R., 15, 278
Swank, J. H., 414, 435

T

Takeshima, T., 99, 414, 435
Talon, R., 15
Tanaka, Y., 435

Tavani, M., 72, 404, 409, 504, 509, 709
Taylor, G. B., 571
Teegarden, B. J., 304, 324, 339
Terekhov, M. M., 10, 284, 516, 894, 921
Terekhov, O., 15, 278
Terrell, J., 54
Thommes, E., 489
Thompson, C., 737, 944
Thorsett, S. E., 504, 509, 585
Tokanai, F., 435
Trinchero, G., 914
Tuffs, R., 635

U

Ueda, Y., 435, 441
Uno, S., 435, 441
Urzagasti, D., 914

V

Valinia, A., 414, 435
Vallania, P., 914
Vanderspek, R. K., 435, 557
van Paradijs, J., 45, 241, 435, 478, 483, 499, 509, 557, 590, 926
Varendorff, M., 344
Vedrenne, G., 15
Velarde, A., 914
Vernetto, S., 914
Vestrand, W. T., 889
von Hippel, T., 548

Vrba, F. J., 625, 630

W

Wagner, R. M., 869
Walker, K. C., 266
Wallyn, P., 931
Wang, Q. D., 456
Waters, L. B. F. M., 590
Wen, L., 430
Wijers, R. A. M. J., 499, 803
Williams, G. G., 837, 842
Williams, O. R., 344
Wilson, J. R., 788
Winkler, C., 344, 879
Wold, M., 489, 530
Wolf, C., 489, 530
Woods, P., 119
Wurtz, R., 837

Y

Yoshida, A., 414, 435, 441, 461
Yoshii, H., 914
Young, C. A., 201, 206, 344

Z

Zavattini, G., 451
Zhang, S. N., 466
Zharikov, S. V., 525